MARS

The NASA Mission Reports

Compiled from the NASA archives & Edited
by Robert Godwin

Historical Mars Missions

Mission, Country, Launch Date, Purpose, Results

[Unnamed], USSR, 10/10/60, Mars flyby, did not reach Earth orbit
[Unnamed], USSR, 10/14/60, Mars flyby, did not reach Earth orbit
[Unnamed], USSR, 10/24/62, Mars flyby, achieved Earth orbit only
Mars 1, USSR, 11/1/62, Mars flyby, radio failed at 65.9 million miles (106 million km)
[Unnamed], USSR, 11/4/62, Mars flyby, achieved Earth orbit only
Mariner 3, U.S., 11/5/64, Mars flyby, shroud failed to jettison
Mariner 4, U.S. 11/28/64, first successful Mars flyby 7/14/65, returned 21 photos
Zond 2, USSR, 11/30/64, Mars flyby, passed Mars but radio failed, returned no planetary data
Mariner 6, U.S., 2/24/69, Mars flyby 7/31/69, returned 75 photos
Mariner 7, U.S., 3/27/69, Mars flyby 8/5/69, returned 126 photos
Mariner 8, U.S., 5/8/71, Mars orbiter, failed during launch
Kosmos 419, USSR, 5/10/71, Mars lander, achieved Earth orbit only
Mars 2, USSR, 5/19/71, Mars orbiter/lander arrived 11/27/71, no useful data, lander destroyed
Mars 3, USSR, 5/28/71, Mars orbiter/lander, arrived 12/3/71, some data and few photos
Mariner 9, U.S., 5/30/71, Mars orbiter, in orbit 11/13/71 to 10/27/72, returned 7,329 photos
Mars 4, USSR, 7/21/73, failed Mars orbiter, flew past Mars 2/10/74
Mars 5, USSR, 7/25/73, Mars orbiter, arrived 2/12/74, lasted a few days
Mars 6, USSR, 8/5/73, Mars orbiter/lander, arrived 3/12/74, little data return
Mars 7, USSR, 8/9/73, Mars orbiter/lander, arrived 3/9/74, little data return
Viking 1, U.S., 8/20/75, Mars orbiter/lander, orbit 6/19/76-1980, lander 7/20/76-1982
Viking 2, U.S., 9/9/75, Mars orbiter/lander, orbit 8/7/76-1987, lander 9/3/76-1980; combined, the Vikings returned 50,000+ photos
Phobos 1, USSR, 7/7/88, Mars/Phobos orbiter/lander, lost 8/89 en route to Mars
Phobos 2, USSR, 7/12/88, Mars/Phobos orbiter/lander, lost 3/89 near Phobos
Mars Observer, U.S., 9/25/92, lost just before Mars arrival 8/21/93
Mars Global Surveyor, U.S., 11/7/96, Mars orbiter, arrived 9/12/97, currently conducting prime mission of science mapping
Mars 96, Russia, 11/16/96, orbiter and landers, launch vehicle failed
Mars Pathfinder, U.S., 12/4/96, Mars lander and rover, landed 7/4/97, last transmission 9/27/97
Nozomi (Planet-B), Japan, 7/4/98, Mars orbiter, currently in orbit around the Sun;
Mars Climate Orbiter, U.S., 12/11/98; lost on arrival at Mars 9/23/99
Mars Polar Lander/Deep Space 2, U.S., 1/3/99, lander/descent probes; lander and probes lost on arrival 12/3/99.

This book is dedicated to Dave Brock and Bob Calvert and to Michael Moorcock for Michael Kane.

Special thanks to Frederick Ordway, Mark Kahn, Jim Busby, Steve Polin, Fred Deaton, Ron Miller

All rights reserved under article two of the Berne Copyright Convention (1971).
We acknowledge the financial support of the Government of Canada through the
Book Publishing Industry Development Program for our publishing activities.
Published by Apogee Books an imprint of Collector's Guide Publishing Inc., Box 62034, Burlington, Ontario, Canada, L7R 4K2
Printed and bound in Canada
MARS - The NASA Mission Reports
by Robert Godwin
ISBN 1-896522-62-9
©2000 Apogee Books
All photos courtesy of NASA

INTRODUCTION

For generations the Red planet has cast its spell on humanity. The wandering speck of orange light in the sky has posed enigmatic questions for astronomers since the dawn of recorded history. In the classical era the planet was dubbed Mars in honor of the ancient Roman god of war. Almost every human culture studied the red planet and asked questions, but not until the space age would those questions really begin to be answered.

As recently as the 1950's many people still believed that an ancient civilisation had constructed vast canals on the surface of Mars. The legend of the Martian "canali" has long since been purged from the text books but the search for Martian indigenous life continues. Some respected researchers even contend that the Viking probes of the 1970's actually *did* discover evidence of life but the facts have been hidden from the public (although one has to wonder why NASA would do such a thing).

Mars is the only world in our solar system that seduces us with the possibility of a life-supporting environment. Although the conditions are harsh, with extremely low atmospheric pressure almost entirely composed of carbon dioxide and sub zero temperatures, Mars is perceived by many to be our best hope for an extraterrestrial foothold in space. Many of the obstacles to establishing a human presence on Mars are now purely engineering difficulties, but without the outstanding work done by the engineers at NASA and JPL we would still be striving to prove that the basic navigation necessary was possible. The flights of the unmanned robots from 1964 to the present day have laid the foundations and made a human trip to the red planet a serious topic of discussion in the new millennium.

When Mariner 3 and Mariner 4 were launched in 1964 they were done so only by pushing the Atlas-Agena D launch vehicle to its design limits. Double redundancy became the name of the game in Mars exploration. Each mission template relied on a pair of vehicles flying almost in parallel with the hopes that at least one would arrive at the destination intact. In the case of Mariner 3 and 4 only the latter accomplished its mission objectives with Mariner 3 failing to make the required trajectory due to a shroud/fairing problem just after lift off. The 22 grainy black and white pictures returned by Mariner 4 changed mankind's notions of Mars forever. No canals, no ancient cities and an abundance of impact craters reminiscent of our own moon. For many it was a disillusioning blow that almost killed the romance of the red planet forever. The science fiction writers would be forever forced to revise their Martian fables, no more Barsoom and beautiful Princesses, no elegant crystal cities or lush jungle landscapes. The masterworks of Burroughs and Bradbury would finally have to be appraised in a different way, as great literary works from a different and more romantic era.

The harsh reality of Mars would gradually be painted more clearly by the cameras of Mariners 6 and 7 (both succeeded), Mariner 9 (8 failed), and then finally by the remarkably successful Viking landers. It is now known that Mars almost certainly once had abundant liquid water (considered essential for life) but it is not known what went wrong with the Martian climate to have turned it into the desert it is today. The amazing pictures of the arid landscape taken by the two Viking landers in 1976 and subsequently by the Pathfinder mission in 1997 have served to open up a contentious debate on the past and future of Mars.

When Pathfinder sent its accompanying mobile robot, Sojourner, exploring amongst the surrounding rocks literally millions of people accessed the JPL web site to participate. For the first time in the history of humanity we were able to explore the surface of another planet (albeit vicariously) through the tentative and delicate motions of a machine the size of a microwave oven.

In the process of assembling this book it became quickly apparent that the wealth of information currently available about our seductive neighbouring planet would fill many volumes the size of an encyclopedia. Over half a million close-up pictures are now available with more being taken every day by Global Surveyor. The mass of scientific data collected about Mars is beginning to rival what we know about Earth. It was therefore decided to stick to the format employed by previous volumes in this series of books. In the following pages are the Press Kits of the United States' Mars probes followed by a sampling of Mission Reports. Some of these Press Kits were still not available on-line at the time of going to press and many of the pictures included on the accompanying CD-ROM still require quite a bit of patience to download. It is hoped that the book will serve a useful purpose as being one concise reference point for the basic information on NASA's study of the Red Planet. By following some of the links provided on the CD-ROM the reader is encouraged to delve as deeply as they wish through the gigabytes of data available online.

The format of this book follows NASA's Mars exploration chronologically, this gives an interesting insight into how our understanding of Mars took an exponential leap forward with each successful mission. Mariner 4 returned not only our first close-up look at Mars but also many new facts about interplanetary space. As the Mariner programme proceeded Mars gradually revealed its secrets. The enigmatic Nix Olympica, which had been observed by Earth based astronomers had evolved from part of an alien canal formation to a huge crater (Mariner 7) before finally revealing itself as the most colossal volcano known to man (Mariner 9). It was appropriately renamed Mons Olympus.

The dangers of navigating the void between Earth and Mars were clearly demonstrated when Mariner 7 almost succumbed to a meteor impact. The dangers of micro-meteors had been postulated as far back as the turn of the century and the team at JPL fully expected some problems. Mariner 7 was the first *Mission To Mars* to be struck and no doubt, Hollywood not withstanding, will not be the last.

By 1975 the technology was in hand to send landers to Mars. Viking was launched by the Titan/Centaur which used the Centaur's restart capability to good effect even though the machine was flying on a mere 16K of RAM in its flight computer. Four mid-course corrections were necessary to achieve orbit, as opposed to two in the case of the flyby Mariners. On arriving, Viking employed what was quaintly referred to as a "FAX camera". You can now buy such a device at your local electronic retailer for a couple of hundred dollars and it is probably 1000 times sharper with an equivalent increase in storage capacity.

Although Mars exploration costs a staggering amount of money the missions have been siphoned for every last drop of scientific blood. As an example — by tracking the location of a Mars lander as it rotates through day and night cycles the history of Mars magnetic poles was determined. This in turn leads to an understanding of the composition of Mars core and even ultimately to an understanding of Martian weather. This kind of ingenuity has resulted in a wealth of new information about our own Earth.

At the time of this writing both NASA and JPL are rewriting their plans for the exploration of Mars. This is undoubtedly due to the general perception that they have lost too many of the recent Mars missions. Mars Observer was lost in 1993 but what is not generally known is that almost three-quarters of that mission's experiments were picked up and flown successfully on Global Surveyor. Mars Climate Orbiter was lost in 1999 and although it is fully understood what happened, the fact is that the vehicle was launched on a shoe-string budget. There are a number of trajectories you can choose to get a vehicle in orbit around Mars but if you intend to use a Delta II launch vehicle (and save $300 million by not using a Titan) you are stuck with the problem of using Aero-braking as the only way to reduce your velocity at Mars. Also, the launch window of ±1 second (using a Delta for a mission like MGS) is so tight it is called "instantaneous".

Aero-braking was first tested favorably with the Magellan probe around Venus (which was a total success), it has since been used successfully with Mars Global Surveyor, but it is an inherently complex, time consuming and risky procedure. In the Mars Global Surveyor Mission Plan (featured in the pages ahead) the sheer complexity of an orbital mission comes into perspective. You can see just how many passes it takes bouncing in and out of the Martian atmosphere to realise a useful orbit. The fact that one single human error in calculation can mean the end of the mission during the Aero-braking process is clear evidence of exactly how meticulous the planning needs to be to successfully use this cheaper approach. *(The MGS Mission Plan is by far the most technical document in this book but its complexity serves a purpose to illustrate just how much effort is involved in planning a Mars mission.)*

And what about that elusive manned trip to Mars? Many believe it is only a matter of time before we launch humans to Mars. The mechanics of the journey have been well worked out as far back as the 1930's by pioneers such as Oberth and von Braun. The engineering is becoming easier every day, although the pitfalls are still abundant, as the recent losses of unmanned craft have proven. The hazards of such a trip are manifold. Mariner 7 was probably struck by a micro-meteorite and only just saved, Mariners 3 and 8 barely got off the ground, while Mars Observer, Climate Orbiter and Polar Lander all suffered a variety of mishaps ranging from careless mathematics to totally unknown failures. Getting to Mars is still not easy and the various proposals for sending people all involve surmounting the well known obstacle of fuel consumption.

The unmanned probes have all been one-way trips. None have needed to worry about how to carry enough fuel to get back. The simple mechanics of planetary travel dictate that if you want to increase your orbital velocity sufficiently to reach Mars you consume a lot of fuel. When you want to subsequently decrease your velocity to fall in towards the Sun (and Earth) you need to reverse that procedure. This means more fuel. NASA is now considering the idea of manufacturing the necessary fuel on the Martian surface utilising a process suggested by Dr Robert Zubrin. This involves hauling some of the ingredients to Mars and then using the CO_2 in the Martian atmosphere to make up the difference. It is an elegant and prosaic solution and may yet turn out to be the only way to go. However, in this age of cheaper, smaller and faster the grand schemes of the past are swiftly becoming forgotten historical footnotes and so in deference to the great visionary I have included the document "Manned Mars Landing" at the end of this book. It is a monumentally optimistic vision written in 1969 and we can only wonder what our world would be like today if von Braun had been given the green light.

Some have suggested that the problems which plagued many of the unmanned probes might have been solved had a human crew been aboard, and yet many new problems will undoubtedly manifest themselves on such a daring voyage. Some of the brightest thinkers alive are working towards the day when mankind will finally stride amongst the Martian dunes. Unlike the moon it will probably not be a short-lived visit motivated by political agendas. Due to the enormity of the gulf between Earth and Mars it is more likely that the first visitors will attempt to establish a permanent presence, leaving behind them an infrastructure to be used by their successors.

Meanwhile, the groundwork is being done every day by the ingenious controllers of our proxies, the robots. While we wait for the day when we can watch the first human step out on to the red soil of Mars we must be satisfied to live through the ersatz eyes and ears of our electronic surrogates.

Robert Godwin
(Editor)

MARS
The NASA Mission Reports
(from the archives of the National Aeronautics and Space Administration)

CONTENTS

VIKING POST LAUNCH MO   172

VIKING FACTS SUMMARY   175

MARS OBSERVER PRESS KIT   183

MARS POLAR LANDER/DEEP SPACE 2 REPORTS

PROPOSED FUTURE MISSIONS

MANNED MARS LANDING

Mariner 9, the first successful Mars Orbiter, is
launched at Kennedy Space Center.

Mariner 4 Press Kit

NATIONAL AERONAUTICS AND SPACE ADMINISTRATION
WASHINGTON, D.C. 20546

FOR RELEASE: THURSDAY AM   October 29, 1964

PROJECT: MARINER MARS 1964 SCHEDULED LAUNCH

Mariner C: No earlier than Nov. 4 Mariner D: Two days to one month later

Extreme demands will be placed on the accuracy and performance of the Atlas-Agena D launch vehicle. This will be the first NASA use of the improved Agena D second stage on an Atlas D booster.

Because of the difficulty of the mission, new engineering applications required for this mission include use of an improved propellant system for the Agena D launch vehicle second stage, a new spacecraft radio system, the first use of the star Canopus for spacecraft attitude reference, and a midcourse motor capable of firing twice.

Despite their complexity, these missions are being undertaken because Mars is of physical and geological interest and offers the best opportunity in our solar system for shedding light on the possibility of extraterrestrial life. These first, pioneering missions, however, are not designed to provide answers to the question of life on Mars.

The Mariners will be mated to the Atlas-Agenas on Launch Complexes 12 and 13 at Cape Kennedy. A third back-up flight spacecraft has been checked out and could be substituted if required. The first Mariner launched will be designated Mariner C. The second, which will be launched no earlier than two days but within four weeks of the first, will be Mariner D.

If launched successfully, they will be designated Mariner III and Mariner IV. Because of their trajectories, the two spacecraft will arrive at Mars about two to four days apart in mid-July of next year.

Mission objectives are to provide engineering experience on the operation of spacecraft during long-duration flights away from the Sun and to perform scientific measurements in interplanetary space between the orbits of Earth and Mars and in the vicinity of Mars.

Eight scientific investigations can be performed by each spacecraft.

Six are designed to measure radiation, magnetic fields and micrometeorites in interplanetary space and near Mars.

If all goes well, two additional experiments could be conducted by each Mariner in the vicinity of Mars. A single television camera could take up to 22 still photographs of Mars, and an occultation experiment to determine characteristics of the Martian atmospheric pressure is planned.

The resolution of the television pictures of Mars and the area of the planet they will cover are difficult to predict because they depend on the fly-by distance from the planet.

If planned trajectories are achieved, however, the pictures should be comparable in detail with photographs of the Moon taken by the best Earth-based telescopes.

If desired accuracies are obtained at launch and during midcourse maneuver, the spacecraft will fly by Mars inside a roughly oblong zone some 7,000 miles wide and 10,000 miles long centered 8,600 miles from the

MARINER TRAJECTORY TO MARS

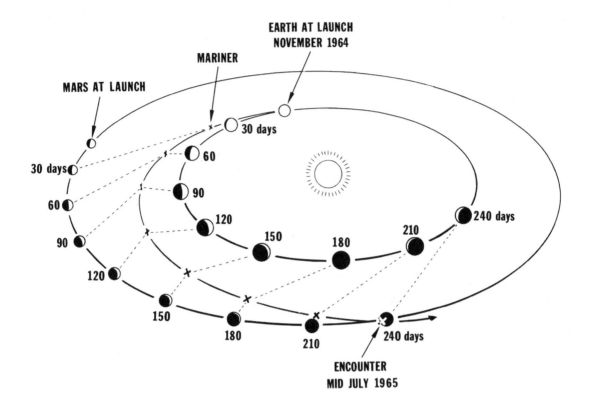

Martian surface. If on the desired trajectory, the spacecraft will pass Mars between the Martian equator and the South Pole on the trailing edge of the planet as viewed from Earth.

Useful planetary data could be obtained, however, within 54,000 miles of the planet.

Physically, the Mars Mariner spacecraft have evolved from the familiar Mariner II. Each will weigh 575 pounds including about 60 pounds of scientific instruments with a Data Automation System. Because the spacecraft will be traveling away from the Sun, each will have four solar panels instead of the two carried by Mariner II.

The Mariner program is directed by NASA's Office of Space Science and Applications. It has assigned project management to the Jet Propulsion Laboratory, Pasadena, Calif., operated by the California Institute of Technology. JPL designed, built and tested the Mariner spacecraft. NASA's Lewis Research Center, Cleveland, is responsible for the Atlas-Agena launch vehicle and Goddard Space Flight Center's Launch Operations will supervise the launch at Cape Kennedy.

Tracking and communication with the Mars Mariners will be by the NASA/JPL Deep Space Network (DSN). In addition to permanent DSN stations at Goldstone, Calif.; Woomera, Australia; and Johannesburg, South Africa, two new stations at Madrid, Spain; and Canberra, Australia, are under construction and may be activated during this mission. All data will flow from DSN stations to JPL's Space Flight Operations Facility in Pasadena which will control the mission.

Scientific investigations aboard the Mariners are provided by scientists representing eight universities and two NASA laboratories.

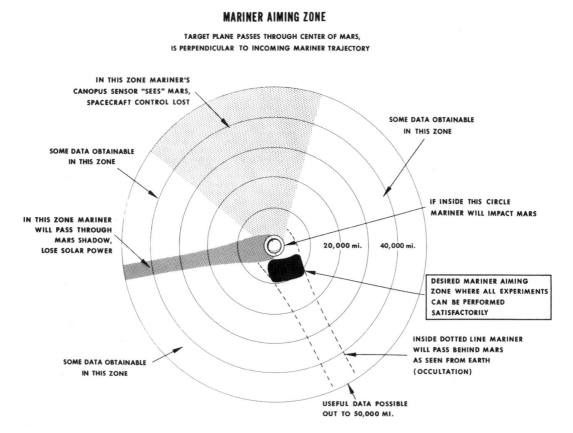

MARINER AIMING ZONE

TARGET PLANE PASSES THROUGH CENTER OF MARS,
IS PERPENDICULAR TO INCOMING MARINER TRAJECTORY

IN THIS ZONE MARINER'S CANOPUS SENSOR "SEES" MARS, SPACECRAFT CONTROL LOST

SOME DATA OBTAINABLE IN THIS ZONE

SOME DATA OBTAINABLE IN THIS ZONE

IF INSIDE THIS CIRCLE MARINER WILL IMPACT MARS

IN THIS ZONE MARINER WILL PASS THROUGH MARS SHADOW, LOSE SOLAR POWER

20,000 mi.   40,000 mi.

DESIRED MARINER AIMING ZONE WHERE ALL EXPERIMENTS CAN BE PERFORMED SATISFACTORILY

INSIDE DOTTED LINE MARINER WILL PASS BEHIND MARS AS SEEN FROM EARTH (OCCULTATION)

SOME DATA OBTAINABLE IN THIS ZONE

USEFUL DATA POSSIBLE OUT TO 50,000 MI.

MARINER MARS 1964 TECHNICAL BACKGROUND

An opportunity for a mission to Mars comes only once every 25 months. Because opportunities are rare and the mission is difficult, it was determined to launch two identical Mariner spacecraft to Mars during the 1964 opportunity.

The following spacecraft and mission description is equally applicable to Mariner C or Mariner D.

MARINER DESCRIPTION

The Mariner Mars fly-by spacecraft were designed and built by the Jet Propulsion Laboratory in Pasadena, Calif.. Industrial contractors provided subsystems and components.

Mariner's basic structure is a 30-pound eight-sided magnesium frame-work with seven electronics compartments. The midcourse rocket propulsion system takes up the eighth compartment. The compartments themselves provide structural support to the spacecraft.

Four solar panels, each 71½ inches long and 35½ inches wide, are attached to the top or sunward side of the octagon. Solar pressure vanes which will act as an auxiliary attitude control system during Mariner's flight, are located at the ends of each panel.

The interior of the octagon contains gas bottles and regulators for Mariner's dual attitude control gas system. Propellant tank for the liquid-fuel midcourse motor is supported by a cantilever arrangement inside the octagonal cavity, with the rocket nozzle protruding through one of the eight sides of the spacecraft.

Two sets of attitude control jets consisting of six jets each, which control the spacecraft on three axes, are mounted on the ends of the solar panels near the pressure vane actuators.

MARINER/MARS SPACECRAFT

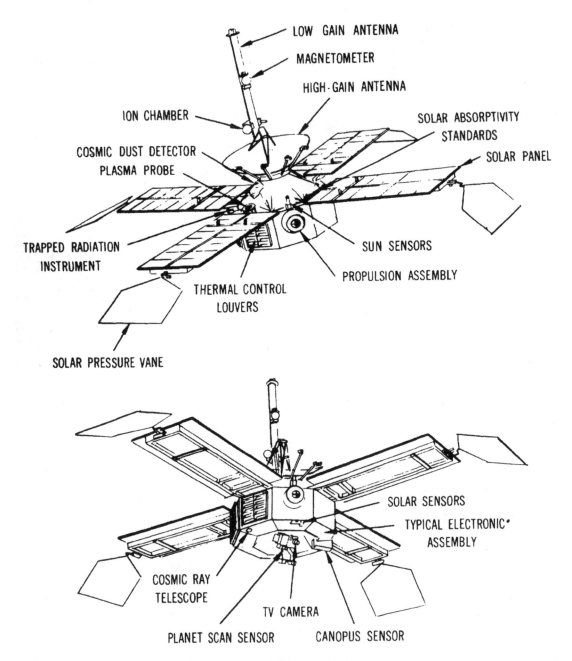

The high-gain antenna is attached to the spacecraft by an eight-legged superstructure atop the octagon. Its honeycomb dish reflector is an ellipse, 46 inches by 21 inches, and is parabolic in cross-section. The antenna, which weighs only 4½ pounds, is in a fixed position so that it will be pointed toward the Earth during the latter half of the Mariner flight including planet encounter and post-encounter phases.

The low-gain omni antenna is mounted on the end of a circular aluminum tube, 3.88 inches in diameter and extending 88 inches from the top of the octagonal structure. The tube acts as a waveguide for the low-gain antenna.

The Canopus star tracker assembly is located in the shade of the spacecraft on the lower ring structure of the octagon for a clear field of view. Sun sensors are located on both the top and bottom surfaces of the spacecraft body in order to provide spherical coverage.

The eight compartments girdling the spacecraft house the following: Bay 1, power supply and synchronizer, battery charger and squib firing assembly; Bay 2, midcourse maneuver rocket engine; Bay 3, science equipment and Data Automation System; Bay 4, data encoder (telemetry) and command subsystems; Bays 5 and 6, radio -receiver transmitters and video tape recorder; Bay 7, Central Computer and Sequencer and attitude control subsystem; Bay 8, power booster regulator and spacecraft battery.

Six of the electronics compartments are temperature controlled by light-weight louver assemblies on the outer surfaces. The octagon's interior is insulated by multi-layer fabric thermal shields at both top and bottom of the structure.

The Mariners will carry scientific instrumentation for seven interplanetary and planetary experiments. The eighth experiment, occultation, requires only the spacecraft communications system. The ion chamber and helium vapor magnetometer are mounted on the low-gain antenna support boom above the spacecraft. The trapped radiation detector and plasma probe are attached to the upper octagonal ring. The cosmic ray telescope looks through the lower ring of the octagon from one of the compartments. The cosmic dust detector is attached to the superstructure holding the high-gain antenna. The television camera and two planetary sensors are mounted on a scan platform below and to the center of the octagon.

Each Mariner weighs 575 pounds and measures 9½ feet to the top of the low-gain antenna. With solar panels extended and solar pressure vanes in up position, the spacecraft spans 22 feet, 7½ inches across. The octagonal structure is 50 inches across.

Power

Primary power source for the Mariner spacecraft is an arrangement of 28,224 photovoltaic solar cells mounted on four panels which will face the Sun during most of the flight to Mars. The cells, covering 70 square feet, will collect energy from the Sun and convert it into electrical power.

A rechargeable silver-zinc battery will provide spacecraft power during launch, midcourse maneuver and whenever the panels are turned away from the Sun. The battery will be kept in a state of full charge and will be available during planet encounter as an emergency power backup source.

Two power regulators will divide the power load and provide redundancy. In the event of a failure in one, it will be removed automatically from the line and the second will be switched in to assume the full load.

The solar panels will be folded in a near vertical position above the spacecraft during launch and will be deployed after separation from the launch vehicle. Each panel weighs 18.7 pounds, including the weight of 7056 solar cells and protective glass filters that reduce the amount of solar heat absorbed without interfering with the energy conversion. Lightweight panel structures that support the cells are made of thin-gauge aluminum approximating the thickness of kitchen foil. Panels are constructed of .0035-inch aluminum sheet formed into a corrugation that is bonded to the cell-mounting surface made of .005-inch sheet.

Nominal power from the panels is expected to be 640 watts at maximum power voltage for cruise conditions in space near Earth. This power capability decreases to about 310 watts at Mars encounter. Total power demands during the mission range from about 140 watts during post-encounter playback of television data to 255 watts during a midcourse maneuver.

The battery is a sealed unit containing 18 silver-zinc cells. Its minimum capacity ranges from 1200 watt hours at launch to about 900 watt hours at planet encounter. Load requirement on the battery may vary between zero amps and 9.5 amps with battery voltages expected to vary from 25.8 and 33.3 volts. The battery weighs 33 pounds.

The battery will be capable of delivering its required capacity and meeting all electrical requirements within an operational temperature range of 50° to 120° F. At temperatures beyond these limits, it will still function although its capability will be reduced.

To ensure maximum reliability, the power subsystem was designed to limit the need for battery power after initial Sun acquisition. Except during maneuvers, the battery will remain idle and fully charged.

Under normal flight conditions the primary power booster regulator will handle all cruise and encounter loads. A second regulator will support power loads during maneuvers. Should an out-of-tolerance voltage condition exist in the cruise regulator for 3.5 seconds or longer, the maneuver regulator will take its place on the line.

Primary form of power distributed to other spacecraft systems is 2400 cycles-per-second square wave. The gyro motors use 400 cps three-phase current and the video tape recorder motor and science scan motor are supplied with 400 cps single-phase current.

Telemetry measurements have been selected to provide the necessary information for the management of spacecraft power loads by ground command if necessary.

The battery, regulators and power distribution equipment are housed in two adjacent electronics compartments on Mariner's octagonal base.

Communications

Two-way communications with the Mariner will be accomplished with a radio link between Earth tracking stations and a dual transmitter-single receiver radio system aboard the spacecraft.

The on-board communications system also includes a telemetry subsystem, command subsystem, video tape recorder and high and low-gain antennas.

Communications will be in digital form. Radio command signals transmitted to Mariner will be decoded — translated from a binary form into electrical impulses — in the command subsystem and routed to their proper destination. Mariner is capable of accepting 29 direct commands and one stored command. The latter is a three-segment command for midcourse trajectory correction and is held in the Central Computer and Sequencer until required.

Data telemetered from the spacecraft will consist of engineering and scientific measurements prepared for transmission by the data encoder. The encoded information will indicate voltages, pressures, temperatures, louver and solar pressure vane positions and other values measured by the spacecraft telemetry sensors and scientific instruments.

The 100-channel telemetry subsystem is capable of sampling 90 engineering and science measurements and can operate in four sequences in which the data transmitted are:

· engineering only during maneuvers;
· a mix of science data and engineering during cruise;
· science data and television engineering data at planet encounter; and
· stored science data from the video tape recorder after Mars encounter with occasional insertions of engineering measurements.

All engineering data and all science data, except TV pictures, will be transmitted in real time. Science data transmitted in real-time at encounter are also recorded with the pictures on tape to be re-transmitted with pictures.

The purpose of the four-sequence operation of Mariner's telemetry system is to obtain the maximum available sampling rate on measurements required during a particular phase of the mission by not transmitting less useful information during that period.

Mariner can transmit information to Earth at two rates — 9 1/3 bits per second and 33 1/3 bits per second. The greater rate will be used for as long as the signal level from the spacecraft is high enough to allow good data recovery. As communication distance increases, a decision will be made to command the data encoder to operate at the lesser rate. The on-board Central Computer and Sequencer will back up the data rate switchover by ground command, which probably will occur sometime after the first 10 per cent of the mission is completed.

Synchronizing pulses will be spaced at regular intervals between the data signals from Mariner. Ground-based receiving equipment will generate identical pulses and match them with the pulses from the spacecraft. This will provide a reference to determine the location of the data signals, allowing receiving equipment to separate data signals from noise.

The spacecraft S-band receiver will operate continuously during the mission at 2113 megacycles. It will receive Earth commands through either the low-gain antenna or the fixed high-gain antenna.

The low-gain antenna, providing essentially uniform coverage in the forward hemisphere of the spacecraft, will provide the primary path for the Earth-to-spacecraft link. Switchover to the high-gain antenna and back to the low-gain, if desired, may be commanded from Earth.

The transmitting subsystem consists of two redundant radio frequency power amplifiers and two redundant radio frequency exciters of which any combination is possible. Only one exciter-amplifier combination will operate at any one time. Selection of the combination will be by on-board logic with ground command-backup.

Both power amplifiers will operate at a nominal output of 10 watts. Transmitter frequency is 2295 megacycles.

The operating transmitter can be connected to either antenna. Switchover will occur on command from the Central Computer and Sequencer with ground command backup. Transmission via the high-gain antenna will be required for approximately the last half of the mission. Attitude of the spacecraft will be such that the high-gain antenna is pointed at Earth during this portion of the mission. Weight of Mariner's radio subsystem, including receiver, both transmitters, cabling and both antennas, is 42.5 pounds.

Midcourse Motor

Mariner's midcourse rocket motor is a liquid monopropellant engine capable of firing twice during the Mars mission. Its function is to compensate for divergences from the planned launch injection conditions. The engine burns anhydrous hydrazine fuel and uses nitrogen tetroxide as the starting fluid.

The rocket nozzle protrudes from one of the eight sides of Mariner's octagonal base below and between two of the solar panels. The engine's direction of thrust is nearly parallel to the panels, hence perpendicular to the longitudinal or roll axis of the spacecraft.

Hydrazine is held in a rubber bladder contained inside a spherical pressure vessel. The fuel is forced into the combustion chamber by means of a pressurized cartridge. Burning is maintained by a catalyst stored in the chamber.

Firing of the engine is controlled by the Central Computer and Sequencer, which receives the time, direction and duration of firing through the ground-to-spacecraft communication link. At the command signal from the CC&S, explosively-actuated valves allow pressure-regulated nitrogen gas to enter the fuel tank.

A timer shutoff mechanism in the CC&S actuates another set of valves which stops propellant flow and fuel tank pressurization. During rocket engine firing, spacecraft attitude is maintained by autopilot-controlled jet vanes positioned in the rocket exhaust.

Re-start capability and redundancy are provided by second sets of explosive start and shutoff valves. The second midcourse maneuver may or may not be required.

The midcourse motor can burn for as little as 50 milliseconds and can alter velocity in any direction from less than 1/8 mile per hour to 188 miles per hour. Maximum burn time is 100 seconds. Thrust is continuous at 50.7 pounds.

Weight of the midcourse propulsion system, including fuel and gas pressurization system, is 47.5 pounds.

Attitude Control

Stabilization of the spacecraft during the cruise and encounter portions of the Mariner Mars mission is provided by 12 cold gas jets mounted at the outer ends of the four solar panels. They are fed by two bottles made of titanium each holding 2.5 pounds of nitrogen gas pressurized at 2470 pounds per square inch.

The jets are linked by logic circuitry to three gyroscopes, to the Canopus star sensor and to the primary and secondary Sun sensors.

There are two identical sets of six jets and one bottle in each set. Normally both sets will operate during the mission. Either system can handle the entire mission in the event the other fails.

The primary Sun sensor is mounted atop Mariner's octagonal structure on the Sunlit side and the secondary sensors are located in the shade of the spacecraft. These are light-sensitive diodes which inform the attitude control system when they see the Sun. The attitude control system responds to these signals by turning the spacecraft and pointing the solar panels toward the Sun. The nitrogen gas escapes through the appropriate jet nozzle, imparting a reaction to the spacecraft to correct its angular position.

It is planned to use the star Canopus for pointing the high-gain antenna back to Earth and to provide a celestial reference upon which to base the midcourse maneuver. The Canopus sensor will activate the gas jets to roll the spacecraft about the already fixed longitudinal or roll axis until it is "locked" in cruise position. Brightness of the Canopus sensor's target will be telemetered to the ground to verify the correct star has been acquired. An Earth detector aboard the Mariner also will provide engineers with verification of Canopus acquisition. This detector cannot see Earth if Canopus has not been acquired and is effective only near Earth.

During firing of the midcourse motor, stabilization will be affected by the use of rudder-like deflecting vanes in the rocket engine's exhaust stream. The Mariner's autopilot controls spacecraft attitude during engine firing by using the three gyros to sense motion about Mariner's three axes for positioning the jet vanes.

An auxiliary attitude control system will position sail-like vanes at the ends of the four solar panels to correct solar pressure unbalance and provide control about the pitch and yaw axes within the limit cycle of the gas jet system.

Each vane consists of seven square feet of aluminized Mylar sheet stretched over an aluminum framework. The vanes are latched in a furled position against the backs of the solar panels during launch. As the panels deploy, the vane is released and unfolds like an Oriental fan. Electromechanical and thermomechanical actuators control the vane positions. Weight of each solar vane, its deployment mechanism, latch, cable and actuator, is less than 1½ pounds.

Total weight of the attitude control system including autopilot is about 62½ pounds.

Central Computer and Sequencer

The Central Computer and Sequencer performs the timing, sequencing and computations for other subsystems aboard the Mariner spacecraft. The CC&S initiates spacecraft events in three different mission sequences — launch, midcourse and cruise/encounter.

The launch sequence includes spacecraft events from launch until the cruise mode is established, a maximum of 16 2/3 hours after liftoff. These events include deployment of solar panels and activation of the attitude control subsystem, solar pressure vanes and Canopus sensor.

The midcourse maneuver sequence controls the events necessary to perform the midcourse maneuver in trajectory. Three of these are commands radioed from Earth and stored in the CC&S prior to initiation of the maneuver. They tell the spacecraft how far and in which direction to turn on its pitch and roll axes and how long the midcourse rocket engine must fire.

The master timer sequence controls those events that occur during the cruising portion of flight and planet encounter. CC&S commands during this sequence switch the spacecraft telemetry transmission to a slower bit rate; switch the transmitter to the high-gain antenna; set the Canopus sensor at various cone angles relative to the predicted encounter time; turn on planetary science equipment prior to encounter for the 14-hour encounter sequence; and switch to the post-encounter telemetry mode for transmission of recorded video data. The CC&S weighs about 11½ pounds.

Temperature Control

If dependent solely upon direct sunlight for heat, an object in space would be approximately 125° F colder at Mars than at Earth.

For a spacecraft traveling to Mars, away from Earth and from the Sun, the primary temperature control problem, then, is maintaining temperatures within allowable limits despite the decreasing solar intensity as the mission progresses. In airless space, the temperature differential between the sunlit side and the shaded side of an object can be several hundred degrees.

Heating by direct sunlight on the Mariner spacecraft is minimized by the use of a thermal shield on its Sun side. The side away from the Sun is covered with a thermal shield to prevent rapid loss of heat to the cold of space.

The top of Mariner's basic octagon is insulated from the Sun by a shield of 30 layers of aluminized Mylar mounted to the high-gain antenna support structure. The Mylar is sandwiched between a layer of Teflon on the bottom and black Dacron on top. The entire assembly is sewn together to form a space "blanket". The bottom is enclosed by a similar shield to retain heat generated by power consumption within the spacecraft.

Temperature control of six of the electronics provided by polished metal louvers actuated by coiled bimetallic strips. The strips act as spiral-wound springs that expand and contract as they heat and cool. This mechanical action, which opens and closes the louvers is calibrated to provide an operating range from fully closed at 55°F to fully open at 85° F. A louver assembly consists of 22 horizontal louvers driven in pairs by 11 actuators. Each pair operates independently on its own local temperature determined by internal power dissipation.

Paint patterns and polished metal surfaces are used on the Mariner for passive control of temperatures outside of the protected octagon. These surfaces control both the amount of heat dissipated into space and the amount of solar heat absorbed or reflected away, allowing the establishment of temperature limits. The patterns were determined from testing a Temperature Control Model (TCM) of the spacecraft. The TCM was subjected to the variations of temperature anticipated in the Mars mission in a space simulation chamber at JPL.

The high-gain antenna dish, which is dependent upon the Sun for its surface heat, is painted green to keep it at near room temperature during planet encounter but within its upper thermal limit earlier in the mission.

SCIENTIFIC EXPERIMENTS

Data Automation System

Seven of the eight scientific experiments aboard Mariner are controlled and synchronized by the Data Automation System (DAS) and the data recorded by the instruments are converted by the DAS into a suitable digital form for transmittal to Earth.

During the mission, the DAS accumulates scientific data, reduces the data from each experiment (except the occultation experiment) to a common digital form and common rate and then feeds the data to the radio transmitter telemetry channel at proper intervals.

The telemetry channel alternately carries 280 bits of scientific data and 140 bits of engineering data except during Mars encounter when the 140 bits provide additional science data and information on performance of the scientific instruments.

Data from the interplanetary instruments are transmitted as soon as received and conditioned by the DAS. The television data, however, is recorded on a tape machine controlled by the DAS for transmission to Earth at a later time. This is required because the television pictures are recorded at a much higher rate than it is possible to transmit this data. All science data during encounter are recorded on the tape as a back-up to the same science data being transmitted in real-time.

Performance data on the television subsystem — filter position, signal level, shutter time, etc. — are continually transmitted during the encounter sequence.

The DAS is composed of four units: real-time unit, non-real time unit, buffer memory and power converter. The total weight is about 12 pounds. During cruise the power requirement is 1.35 watts; during Mars encounter, 3.06 watts.

Television

As the Mariner spacecraft passes Mars a single television camera viewing Mars through a reflecting telescope will take as many as 22 black and white, photographs of the Martian surface. The photographs will be stored on magnetic tape in digital form to be played back for transmission to Earth after Mariner passes Mars.

It is required to store the photographs on tape for later playback because the radio transmission rate from the Mars distance is 8.33 bits per second and the photographs are recorded at a rate of 10,700 bits per second.

Each photograph will consist of approximately 250,000 bits. It will take 8 1/3 hours to play back each picture. If the communication distance has not been exceeded after one playback of all pictures, each photograph will be transmitted again to provide a comparison for detection of errors in the transmission. Playback will begin about 13 to 15 hours after the last picture is taken. About 1½ hours of engineering data will be transmitted between each picture. All data from the other scientific instruments will be recorded with the pictures as a back-up to the real-time transmission of science data.

The camera head and two planet sensing devices are mounted on a movable platform. This planetary scan platform will sweep through 180 degrees until a Wide Angle Planet Sensor (50° field of view) gives a planet-in-view signal. The platform will then stop its sweeping motion and center the planet in the sensor's field of view.

Picture recording begins when Mars enters the field of view of a Narrow Angle Mars Gate sensor (1½° field of view) and the device generates a signal that is translated by the Data Automation System into a command that turns on the tape recorder.

Photographs will be taken in groups of two with a small gap between in each pair. Depending on camera distance from the planet, each pair of pictures may cover somewhat overlapping areas on the Martian surface. The number of pictures recorded will be determined by the time required at encounter to synchronize tape recorder and camera for the first picture and by lighting conditions on Mars.

The recorder will be turned off after recording each picture and then turned on again to record the next.

The tape is a continuous loop 330 feet long. Data will be recorded on two tracks.

Although the camera system will be turned on six to 10 hours prior to closest approach, pictures will not be recorded until the command from the Narrow Angle Mars Gate turns on the tape recorder. The camera itself will function as a backup to the Mars Gate. When the vidicon senses the increased illumination of Mars, a signal will be generated to order the tape recorder turned on.

It is anticipated that the camera system will sweep through a large illumination range on Mars which will include photographs near the shadow line or terminator. The camera system is equipped to increase or decrease its sensitivity to light to compensate for the changing lighting conditions. This is accomplished by a sampling circuit that detects changes in the strength of the video signal, which directly relates to the amount of light detected by the vidicon, and orders an increase or decrease in the amount of amplification in the television system's amplifier chain.

A picture is formed on the vidicon target in 1/5th of a second every 48 seconds. The scanning or readout of the 200 line picture requires 24 seconds and erasure of the image and preparation for the next requires 24 seconds.

The exposure time or shutter speed will be one-fifth of a second. If illumination levels are higher than expected at Mars the exposure time will be shortened to 2/25 of a second by an automatic switching device.

The shutter consists of a single rotating disc containing four openings for alternating filters. Two filters will be blue-green and two will be orange-red.

The filters will provide high contrast in the black and white photographs received on Earth and emphasize the difference in colors seen from Earth on Mars.

The telescope associated with the camera system is an f/8 Cassegrainian system of 12 inches equivalent focal length. The beryllium primary mirror has a diameter of 1.62 inches and an f-ratio of 2.47. The beryllium secondary mirror provides an amplification of 3.0.

The television system is divided into two parts. The camera head and a small portion of the electronic circuitry is located on the scan platform. The balance of the electronic equipment required for the television system is located in a compartment on the spacecraft bus.

A removable lens cover will protect the camera optics during the mission from micrometeorite damage.

Investigators are Professor R. B. Leighton, Prof. B. C. Murray and Prof. R. P. Sharp, of the California Institute of Technology.

Occultation Experiment

The purpose of the occultation experiment is to determine some characteristics of the atmosphere of Mars by transmitting radio signals from the Mariner spacecraft through the Martian atmosphere.

Performing this experiment does not require additional weight or power on the spacecraft. It does require, however designing a trajectory for the spacecraft that will pass behind Mars.

As the spacecraft flies behind Mars, its radio signal will be bent by the Martian atmosphere and will change in frequency and strength. This is due to refraction and diffraction in the atmosphere whose effects on the Doppler (phase shift) signal are predictable. The strength of the telemetry carrier will also fluctuate and will provide informative data.

Current estimates on the surface pressure vary over a wide range and the scale height, or how the density varies with altitude, is unknown. The surface pressure on Mars is variously estimated to be between 10 and 100 millibars (about one to 10 percent of the barometric pressure on the Earth's surface) with 25 millibars accepted by some investigators.

It is essential for landing a capsule on Mars to determine these physical characteristics. Design of an entry capsule requires knowing the rate it will be slowed by atmospheric drag and whether or not a parachute system is sufficient or if some other means of slowing the capsule is required. The scale height is also critical. This factor is of importance in the design of a capsule's heat shielding and entry shape.

The detection of the changes in the radio signal will provide a severe test of the capabilities of the Deep Space Network (DSN). The capability to accurately detect minute changes in a radio signal has been developed by the DSN in earlier missions, the Mariner II Venus mission and the Ranger VI and VII lunar flights, and by successful radar bounce experiments on the Moon, Venus, Mars and Mercury. These accuracies have been developed to a point that has made the occultation experiment feasible.

Investigators are Dr. Arvydas J. Kliore, Dan L. Cain and Gerald S. Levy of JPL, Prof. Von R. Eshelman of Stanford Electronics Laboratory, and Prof. Frank Drake of Cornell University.

Solar Plasma Probe

The solar plasma probe will measure the density, velocities, temperatures and direction of low energy (30 electron volts to 10,000 electron volts) protons that stream outward from the Sun at supersonic speeds to form what has been termed the solar wind.

Solar plasma emitted from the Sun is a boiling off of the Sun's atmosphere. It takes three forms. The solar wind, which appears to be emitted continuously in all directions; solar streams which originate in relatively small regions of the Sun; and the solar shell which is defined as bursts of solar plasma from solar flares.

The plasma itself is a thin, high velocity, high temperature gas. It is composed of the same material as the Sun, nuclei of helium and hydrogen atoms, with hydrogen nuclei (protons) being the main constituent. The atoms in the solar plasma are ionized (have lost their electrons) because of the very high temperature of the outer atmosphere (corona) of the Sun.

The study of solar plasma is important to scientists because it is the predominant and controlling factor in interplanetary apace in that it can affect magnetic fields and cosmic rays. To understand the nature of interplanetary space, it is essential to comprehend the nature of solar plasma.

Earlier solar plasma experiments in space — Mariner II, Explorer XVIII, Explorer X — have provided about eight months of total study time. The Mariner Mars mission will more than double this amount. However, the changing characteristics of the solar plasma are related to a cycle of solar activity that spans 11 years. Only a fraction of the desired information has been obtained.

The solar plasma probe is composed of three major parts; a collector cup, a high voltage power supply and circuitry to provide an output signal to the data automation system that conditions the data for transmission to Earth.

A mesh of fine tungsten wire in front of the collector is given a negative voltage that alternates rapidly between two limits. Thus, protons having a range of energies corresponding to these limits are alternately repelled and passed through to the collector, producing an intermittent current that is detected.

Low energy protons are repelled continuously. Higher energy protons pass through continuously but their continuous current is not detected.

The voltage limits on the mesh are repeatedly cycled through 32 voltage levels to allow detection of protons at 32 energy bands within a range of 30 to 10,000 electron volts.

From the measurement of current collected in each of the 32 energy ranges, the density, velocity, and temperature of the plasma can be deduced.

Improvements in this probe over those flown on Mariner II and on Explorer X are the greater energy range that can be detected and detection of the direction of motion of a proton in the solar plasma. The direction of motion is obtained by dividing the collector into three pie-shaped sectors. If the solar plasma enters the probe and impinges on the current collecting plates at an angle, the sectors will not receive equal currents. One of the sectors will receive the most current which will determine the direction of the plasma's motion.

The plasma probe is mounted on top of the spacecraft bus at an angle 10 degrees off the spacecraft-Sun line. It weighs 6.4 lbs. and utilizes three watts of power.

The investigators for the plasma probe are Prof. Herbert L. Bridge, Dr. Alan Lazarus of the Massachusetts Institute of Technology and Dr. Conway W. Snyder, JPL.

Ionization Chamber Experiment

This experiment includes an ionization chamber and Geiger-Mueller tube to measure radiation, principally galactic cosmic rays, in an energy range above 10 million electron volts for protons, ½ million electron volts for electrons and 40 million electron volts for alpha particles.

The ionization chamber will yield a measurement related to the average energy and amount of radiation at these levels and the Geiger-Mueller tube will count individual particles. Results from the two instruments will be correlated to yield data on the density of cosmic rays and their energy levels in interplanetary space between Earth and Mars and in the vicinity of Mars.

The ionization chamber is a five-inch stainless steel sphere with a wall thickness of 1/100th of an inch. The metal serves as a shielding that will only allow radiation above a given level to penetrate.

The sphere is filled with argon gas. Particles that penetrate the sphere leave a trail of ions in the gas. The ions are detected and an electric pulse is produced that is proportional to the rate of ionization in the argon gas. This pulse is processed and telemetered to Earth.

The Geiger-Mueller tube is also shielded to allow penetration by particles in the same range as detected by the chamber. The tube consists of an enclosed volume of gas with two electrodes at a different electrical potential.

The tube generates a current pulse each time a charged particle passes through the tube. The tube can count particles at a maximum rate of 50,000 per second.

This experiment weighs 2.9 lbs. and is located on the spacecraft's mast.

The investigators are Prof. H. Victor Neher of the California Institute of Technology and Dr. Hugh R. Anderson, JPL.

Trapped Radiation Detector

The purpose of this experiment is to search for magnetically trapped radiation in the vicinity of Mars that, if it exists, might be similar to the Earth's Van Allen belts of trapped radiation.

The experiment consists of four detectors, three Geiger-Mueller tubes and one solid state detector, a silicon diode covered with thin nickel foil to exclude light.

The three GM tubes are shielded so that low energy particles can only enter by passing through a window at the end of each tube. Tubes A and B will detect protons greater than 500 thousand electron volts and electrons greater than 40 thousand electron volts. Tube C will detect protons greater than 900 thousand electron volts and electrons greater than 70 thousand electron volts. The solid state detector will measure protons in two ranges: from 500 thousand electron volts to 8 million electron volts and from 900 thousand electron volts to 5.5 million electron volts.

During the cruise portion of the mission the trapped radiation detector will measure cosmic rays and electrons.

The experiment is located on the top of the spacecraft bus and weighs about 2¼ pounds.

Investigators are Dr. James A. Van Allen, Dr. Louis A. Frank and Stamatios M. Krimigas, of the State University of Iowa.

Helium Vector Magnetometer

The scientific objectives of the magnetometer experiment are to determine if Mars has a magnetic field and, if so, to map its characteristics; to investigate the interaction between planetary and interplanetary magnetic fields and measure the magnitude and direction of the interplanetary magnetic field and determine its variations.

If a magnetic field around Mars is similar to Earth's, the magnetometer should detect the transition from the interplanetary field to that of planetary field by measuring disturbances caused by the interaction of the solar wind and the planetary field. This interaction causes instabilities in the magnetic fields embedded in the solar wind and these disturbances are detectable at considerable distances.

The magnetometer will detect a general magnetic field around Mars if the spacecraft trajectory lies within the hydromagnetic cavity formed by the solar wind around a planet. In this area a planetary field is relatively undisturbed by the solar wind.

The Mariner Mars mission will provide the first scientific opportunity to measure magnetic fields in interplanetary space at distances from the Sun greater than Earth's orbit.

Knowledge of the interplanetary field is important in understanding the nature of solar cosmic rays, solar flares, galactic cosmic rays, origin of the solar wind near the Sun, solar magnetic fields and the interaction of the solar wind with the Earth and Moon.

A new type of magnetometer, developed specifically for use on planetary missions, will be flown for the first time on the Mars mission. The instrument is a low field vector helium magnetometer that measures not only the magnitude of the magnetic field but its direction as well.

A helium magnetometer is based on the principle that the amount of light that can pass through helium gas, that has been excited to a higher than normal energy level, is dependent on the angle between the light axis and the direction of the surrounding magnetic field. Measuring the amount of light passed through the helium gives a measurement of the magnetic field in magnitude and direction.

The light source is a helium lamp in the magnetometer. The light, collimated and circularly polarized, passes through a cell containing the excited helium gas and then impinges on an infrared detector that measures the amount of light passed through the helium gas.

The magnetometer is located on the low gain antenna mast to minimize the effect of the spacecraft's magnetic fields. Electronics supporting the experiment are located in a compartment on the spacecraft. The magnetometer weighs 1.25 pounds. The electronics weighs six pounds. Its sensitivity is ½ gamma per axis and the dynamic range is ± 360 gamma per axis. The experiment operates on seven watts of power.

The investigators are Dr. Edward J. Smith, JPL, Paul J. Coleman, Jr., University of California at Los Angeles, Prof. Leverett Davis, Jr., California Institute of Technology and Dr. Douglas E. Jones, Brigham Young University and JPL.

Cosmic Ray Telescope (CRT)

This experiment will detect and measure cosmic rays by type, energy levels and direction of motion. The experiment will produce a detailed analysis of portions of the energy range covered by the ionization chamber experiment and in addition, will sense particles with lower energies.

The CRT will discriminate between protons in three energy groups: .80 million to 15 million electron volts; 15 million to 80 million electron volts; and 80 million to 190 million electron volts. It will discriminate between alpha particles from 2 million to 60 million electron volts; 60 million to 320 million electron volts and 320 million to infinite million electron volts.

The CRT has three coaxial gold-silicon solid state detectors with intermediate absorbers arranged in similar fashion to a series of lenses in a telescope. Dependent upon their energy and direction, particles will pass through detector 1 only or through 1 and 2 or through detectors 1, 2 and 3. Detector 1 yields a pulse proportional to the amount of energy lost as a particle penetrates. The type of particle is determined by the pulse height that is recorded and evidence of coincidental pulses in the other detectors.

The CRT is located on the underside of an electronics compartment in the spacecraft bus and weighs 2.5 lbs.

Investigators are Dr. John A. Simpson and Joseph O'Gallagher of the University of Chicago.

Cosmic Dust Detector

The objectives of this experiment are to make direct measurements of the dust particle momentum and distribution near Earth, in interplanetary space and in the vicinity of Mars.

The experiment is composed of two sensors that measure strikes by direct penetration and by microphonic techniques.

A square aluminum plate yields a microphone signal when struck by a dust particle. Bonded to both sides of the plate are layers of non-conducting material covered with a layer of evaporated aluminum, forming penetration sensors. The penetration sensors will specify which side of the microphone plate was struck. The strength of the signal derived from the microphone plate will yield data on the momentum of the particle.

The flight paths of the two Mariner spacecraft will pass close to the orbits of four meteor streams generally believed to be composed of particles once contained in the nuclei of a comet. It is expected to require several days to pass through each stream.

The streams are the Leonid, Geminid, Ursids and the Tuttle-Giacobini-Kresak. The Leonid meteor stream will be in a period of peak activity, which occurs approximately every 33 years, near the time the Mariner spacecraft will encounter the stream.

The Mariner cosmic dust detector weighs two pounds and is mounted on the upper side of the spacecraft. The plate is approximately perpendicular to the direction the spacecraft is moving.

Investigators are W. M. Alexander, O.E. Berg, C. W. McCracken and L. Secretan, of Goddard Space Flight Center, Greenbelt, Md., and J. L. Bohn and O.P. Fuchs of Temple University, Philadelphia.

LAUNCH VEHICLE

Mariner Mars 64 is the first NASA mission to use the improved Agena-D upper stage vehicle and places maximum performance demands on the Agena and its Atlas D booster. It is also the first U.S. space mission to require a second burn of the newest model Agena-D.

Under some circumstances, it might be desirable to launch the second Mariner two days after the first so a special 36-hour data return plan has been developed. Experienced Lewis engineers will be located at a downrange tracking station during the launch. They will perform a preliminary analysis of Agena tracking and telemetry data and report the results to Cape Kennedy.

Thus project officials would be able to pinpoint a problem in the first flight in time to provide a greater margin of success for the second launch.

This model of the Agena-D — the SS-OIB — is an improved performance vehicle and can carry a heavier payload than its predecessors. Payload capability is increased by about 80 pounds.

This capability is gained by the use of lighter weight parts and by an improved propellant utilization system which leaves less residual propellant at Agena mission completion.

The launch sequence of the Atlas-Agena D is described at the beginning of the next section, "The Mission."

LAUNCH VEHICLE STATISTICS

Total lift-off weight: 280,000 pounds
Total lift-off height: 104 feet

Atlas-D Booster		Agena-D Upper Stage
Weight	260,000 pounds	15,500 pounds
Height	67 feet	24 feet
Thrust	about 370,000 pounds at sea level	16,000 pounds at altitude
Propellants	liquid oxygen and RP-1, a kerosene-type fuel	unsymmetrical dimethyl hydrazine (UDMH) and inhibited red fuming nitric acid (IRFNA)
Propulsion	two booster engines, one sustainer engine and two vernier attitude and roll control engines built by Rocketdyne Division, North American Aviation, Inc.	one engine built by Bell Aerosystems Company
Speed	about 13,000 mph at apogee for Mariner flight	about 17,500 mph after first burn, about 25,600 mph at spacecraft injection
Guidance	General Electric radio command guidance equipment; Burroughs guidance computer	Honeywell, Inc., inertial guidance and Barnes horizon ground sensors
Contractor	General Dynamics/Astronautics, San Diego, California	Lockheed Missiles and Space Company, Sunnyvale, California

Countdown Milestones

- T-395 minutes   Deliver pyrotechnics to launch complex
- T-155 minutes   Start Agena UDMH tanking
- T-135 minutes   Complete UDMH tanking
- T-130 minutes   Remove gantry
- T- 90 minutes   Start IRFNA tanking
- T- 65 minutes   Complete IRFNA tanking
- T- 60 minutes   Scheduled hold to meet launch window
- T- 40 minutes   Start Atlas LOX tanking
- T- 7 minutes   Scheduled hold to meet launch window
- T- 2 minutes   Secure LOX tanking
- T- 19 seconds   Momentary hold
- T- 2 seconds   Engines full thrust
- T- 0 seconds   Release/lift-off

THE MISSION

The Mariner spacecraft will be boosted into a parking orbit at an altitude of 115 miles and a velocity of 17,500 miles an hour by the Atlas-Agena prior to its injection on a Mars trajectory.

After lift-off, the Atlas vehicle is controlled by a combination of on-board autopilot and ground-based radio commands. For the first few minutes of flight, the vehicle attitude is controlled by gimballing the booster engine thrust nozzles.

About two minutes after launch, a radio command is transmitted which shuts down the booster engine (BECO or booster engine cut off). This event is followed by staging or release of the booster engine section. Vehicle attitude control is then determined by the sustainer engine until the transmission of the SECO command about five minutes after launch. At SECO, the vernier engines which had provided roll control during sustainer phase are activated to provide total vehicle attitude control. This continues until transmission of the VECO command less than a minute after SECO transmission.

Atlas-Agena Separation

Another series of radio commands is transmitted which turns on the Agena timer, activates the Agena guidance and control system, initiates separation of the Agena from the Atlas and fires retrorockets which retard the forward motion of the Atlas.

The Agena engine ignites for the first time about six minutes after launch on command of the Agena timer. This first burn lasts for about two minutes and is terminated when the Agena velocity meter determines that the correct increase in velocity has been achieved.

Coast Period

The Agena-Mariner coasts at an altitude around 115 miles before second burn ignition. The vehicle remains in parking orbit for a period of time determined by the final injection point — a point determined by the relative position of the planet Mars at the time and date of launch.

At the proper point in the parking orbit to ensure attaining the interplanetary trajectory, the Agena engine is started for the second time. Engine shutdown is again determined by the velocity meter. The required velocity is about 25,600 miles an hour.

On command from the Agena timer the spacecraft separation system is activated. After the Mariner spacecraft leaves the Agena, the vehicle performs a 180 degree turning maneuver followed by firing of the retro-rocket.

TYPICAL MARINER 64 FLIGHT SEQUENCE

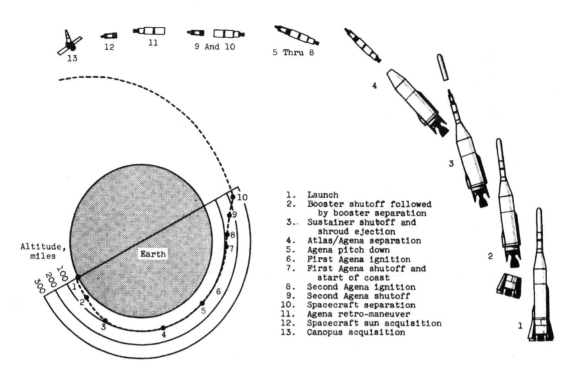

1. Launch
2. Booster shutoff followed by booster separation
3. Sustainer shutoff and shroud ejection
4. Atlas/Agena separation
5. Agena pitch down
6. First Agena ignition
7. First Agena shutoff and start of coast
8. Second Agena ignition
9. Second Agena shutoff
10. Spacecraft separation
11. Agena retro-maneuver
12. Spacecraft sun acquisition
13. Canopus acquisition

The launch vehicle has completed its part of the mission when the spacecraft is separated from the Agena. With Agena retrofire, the second stage decelerates and enters a solar orbit. Agena retro-fire assures that the vehicle will not hit Mars and that Agena's reflection will not confuse spacecraft sensors.

First Spacecraft Events

As the Mariner is separated from the Agena, a series of spacecraft events are initiated by the separation connector:

1.   Full power is applied to the spacecraft transmitter. Until this point the transmitter power has been held at low power to prevent high voltage arcing that could damage equipment in the critical area between 150,000 to 250,000 feet.

2.   The cruise scientific experiments are turned on.

3.   The CC&S is fully activated. Up to this point it has been inhibited from inadvertently giving commands.

4.   The tape recorder used to record television pictures at Mars encounter is turned off. The tape recorder runs during the launch phase to apply tension on the tape to prevent unwinding and snarling.

5.   The attitude control system is turned on and Sun acquisition is initiated. The attitude control system initiates a spacecraft roll for calibration of the magnetometer.

6.   Pyrotechnic devices are armed. A hydraulic timing device is activated which will give a back-up signal for arming the pyrotechnics and a signal for deployment of the solar panels and to unlatch the scan platform. These signals are given within a minute and 20 seconds after separation.

The Central Computer and Sequencer (CC&S) gives a back-up command, for deployment of solar panels and solar pressure vanes and for unlatching the scan platform, at liftoff plus 53 minutes. The initial commands from the hydraulic timing device occur 30 to 50 minutes from liftoff depending on the length of the parking orbit which is determined by the day and time of launch.

Sun-Canopus Acquisition

Sun acquisition will be completed in approximately 20 minutes after the first command. The time is dependent on the attitude of the spacecraft when the process begins. Any tumbling motion imparted to the spacecraft at separation is cancelled out during the acquisition process.

Sun sensors, two primary and two secondary, are mounted on the bus and will provide signals to the cold gas jets to cancel out the tumbling motion and maneuver the spacecraft until its four solar panels are pointed at the Sun. The spacecraft will now begin drawing power from the panels as they convert sunlight to electricity. The batteries, which have supplied power up until now, will only be used during the midcourse maneuver and in the event that, during the mission or at encounter, the spacecraft power demands should exceed the panels' output.

The next spacecraft event, and the next order from the CC&S, occurs at 16½ hours after launch. The Canopus Star Sensor is activated and the spacecraft begins a roll search, at a different roll rate than the magnetometer calibration roll, to allow the sensor to fix on Canopus and thus orient the spacecraft in the roll axis.

Again, the time required for this acquisition depends on the position of the sensor relative to the star at the time acquisition begins. Maximum time for completion of Canopus acquisition is 75 minutes.

It will be required during the mission to command the Canopus sensor by CC&S or Earth command to change the angle of its field of view to allow for the changing geometrical relationship of the spacecraft's roll axis and the star Canopus. This change will occur four times.

During the cruise portion of the mission, signals from the primary Sun sensor and the Canopus sensor to the attitude control system will fire the cold gas jets to maintain the spacecraft attitude — solar panels pointed at the Sun and star sensor pointed at Canopus.

At the time of Canopus acquisition, a solid state sensing device will signal that it is viewing Earth. The device cannot view Earth if the Canopus sensor is not locked onto Canopus. This will provide a check on data from the star sensor indicating it has Canopus in view. It is possible that the star sensor could lock-on to some other celestial body. The solid state sensing device is effective only near Earth.

Midcourse Maneuver

The spacecraft is now in its cruise mode which will continue until the midcourse maneuver is commanded. This probably will occur two to 10 days after launch. The midcourse maneuver will alter the spacecraft's solar orbit by changing its velocity.

Trajectory calculations at JPL's Space Flight Operations Facility will establish the actual flight path of the spacecraft for comparison with the flight path required for a Mars fly-by. Commands will be calculated for transmission to the spacecraft, to alter Mariner's flight path.

These commands, for pitch, roll and duration of motor burn, will be transmitted by a station of the Deep Space Network to the spacecraft and stored in the CC&S memory. The execute command follows and the spacecraft begins the midcourse maneuver.

MARINER/MARS MIDCOURSE MANEUVER

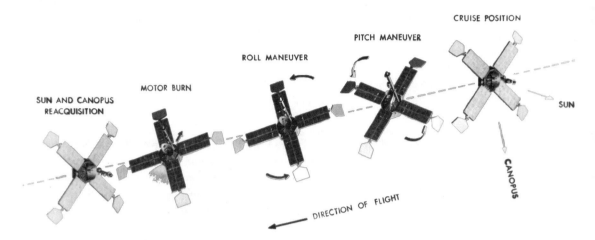

The cold gas jets will fire to roll the spacecraft in the direction and through the prescribed number of degrees as commanded by the CC&S. The pitch turn is followed by a roll turn. The spacecraft's midcourse motor is now pointing in a direction that, when fired for the proper amount of time, will alter the spacecraft's velocity and thus its flight path around the Sun.

The midcourse maneuver is a precision event. An error in the duration of the motor firing that would make a difference in the final velocity of the spacecraft of one mile per hour will make a 9000 mile difference in the flight path at the Mars fly-by.

When the maneuver is completed the spacecraft will again acquire the Sun and Canopus on command from the CC&S.

The cold gas attitude control system is not sufficient to stabilize the spacecraft during the motor firing period.

Movable jet vanes in the motor exhaust perform the stabilization function at this time. They are controlled by an autopilot that is turned on for the midcourse maneuver and used only at this time. The autopilot accepts signals from gyros to keep the motor thrust pointed in the right direction.

From execute command to reacquisition of Canopus, the maneuver will take approximately four hours. Although the telemetry will give some indication that the maneuver has been performed accurately, it will require a long period of tracking to calculate the spacecraft's new flight path.

During the midcourse maneuver the spacecraft transmits only engineering data to Earth. At completion of the maneuver the telemetry format will revert to the cruise mode of one-third engineering data and two-thirds science data.

If required, a second midcourse maneuver can be performed by the Mariner. This is the first spacecraft to have this capability.

Mariner is again in its cruise mode and with two exceptions, will continue in this mode until encounter with Mars.

The two exceptions are concerned with communications. As the distance from spacecraft to Earth increases it will be necessary to slow the transmission bit rate from 33 1/3 bits per second to 8 1/3 bits per second. This will occur, on command from the CC&S with a back-up command from Earth, at approximately nine weeks after launch.

Later in the mission, about 12 weeks after launch, the spacecraft's transmitters will be switched from the omnidirectional antenna to the high gain antenna. This is a CC&S command backed-up from Earth and will be performed at the time the omni-directional antenna nears the limits of its transmission distance capability.

The high gain antenna is rigidly fixed on the spacecraft, unlike the Rangers or Mariner II which had movable high gain antennas, because after the switchover point, Earth will be continually in the antenna's view.

The next mission event will be Mars encounter.

Encounter

The encounter sequence will begin from six to 10 hours before the spacecraft makes its closest approach to the planet. At this time the CC&S, with Earth command back-up, will turn on the television system, that portion of the Data Automation System that functions at encounter, the scan platform, the two Mars sensors and the tape recorder electronics. Although receiving power, the tape recorder is not running and will not record pictures until later.

A cover shielding the optics of the vidicon and the two planet sensors is removed at this point by the encounter command from the CC&S to a pyrotechnic device. The battery charger is also turned off at this time.

The scan platform is now sweeping through 180 degrees nearly vertical to the direction of motion of the spacecraft. The television system is functioning but pictures are not recorded at this point. Pictures will not be recorded until a signal is sent from the narrow angle Mars sensor through the DAS to the recorder that the vidicon camera is seeing the planet.

As Mariner approaches Mars the wide angle sensor will detect the planet, and send a signal that stops the 180 degree sweep of the scan platform. The scan platform will not track the planet.

Telemetry is now switched from a combination of engineering data and science data to all science, with the exception of engineering data on the TV system.

The next event is the acquisition of Mars by the narrow angle sensor. The scan platform is now locked in place and recording of pictures begins. Starting of the tape machine is synchronized with the sequencing of the television camera.

The spacecraft will sweep past Mars recording pictures and transmitting other scientific data to Earth. All science data will also be recorded on the tape. The signal being transmitted to Earth, telemetry carrier and the Doppler signal, will be transmitted through the Martian atmosphere as the Mariner crosses behind the planet to perform the occultation experiment.

Transmission from the spacecraft will not be received on Earth as Mariner passes behind Mars. When it reappears the tracking station will begin a search to reacquire the spacecraft.

About 14 hours after the CC&S or Earth command was given to begin the encounter sequence, the CC&S will command the encounter instruments turned off. Six hours later it will order the tape recorder to begin the playback of the pictures recorded.

Each picture will require approximately 8 1/3 hours to transmit to Earth. Between each picture the spacecraft will transmit 1½ hours of engineering data. The entire tape will be played back twice if the spacecraft has not exceeded the communication range after the first playback.

After the picture data has been transmitted and if the Mariner is still within communication range, the scientific instruments will be turned back and the telemetry format switched to one-third engineering data, two-thirds science data. The spacecraft is again in cruise mode and this will continue until Mariner exceeds the communication range. At that point the mission will end but the Mariner will continue to circle the Sun in a perpetual solar orbit.

Trajectory

The Mariners will be launched from Cape Kennedy at a sufficient velocity to escape Earth plus the additional velocity required to provide an encounter with Mars.

Escape velocity, 25,200 mph, would only be sufficient to place a Mariner in a solar orbit that would be near Earth's orbit. The additional velocity is carefully calculated to yield a solar orbit that will cross the path of Mars on a given date with the spacecraft properly oriented to Mars, Earth and Sun to perform its scientific experiments, communicate with Earth and receive power from the Sun.

The total required velocity is imparted the point of injection by the Agena-D second stage. The final velocity and the injection point varies from day to day throughout the launch period as the relationship between the position of Earth and Mars changes.

A typical injection velocity is 25,663 miles per hour, relative to Earth. At encounter a typical spacecraft velocity would be 11,405 mph, relative to Mars. It is required to state velocities in the relative sense because the velocity of a body in the solar system is based on the position of the observer.

To an observer on Earth the velocity of Mariner at injection would be as stated, 25,663 mph. To an observer on the Sun the velocity of the spacecraft at injection would be 87,185 mph. This is because the Earth itself is orbiting the Sun at a speed in excess of 60,000 miles per hour and this velocity plus the injection velocity is imparted to the spacecraft at injection.

The velocity of the spacecraft relative to Earth at injection will slowly diminish as Mariner heads outward from Earth and the Earth's gravitational field pulls on the spacecraft. As Mariner approaches Mars the velocity of Mariner will increase under the attraction of the planet.

TYPICAL MARINER TRAJECTORIES NEAR MARS

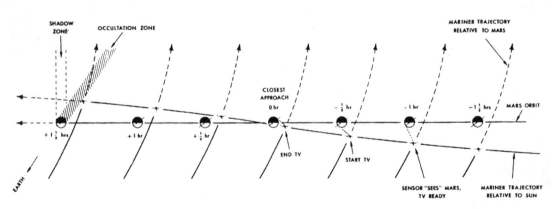

At launch Mars will be ahead of Earth. At encounter Mars will be trailing Earth by approximately 150,000,000 miles.

The spacecraft will follow a long curving path around the Sun after injection. It will pass Mars prior to encounter and, as its velocity is still decreasing, Mars will catch up with the Mariner and pass it. After this pass occurs, Mariner's direction of flight will carry it behind Mars to allow the occultation experiment. It will be behind Mars for approximately one hour.

Mariner's trajectory is planned to pass behind Mars on a line between the equator and the South Pole.

In designing trajectories for the Mars mission the trajectory engineer must satisfy numerous restrictions or constraints that influence the final trajectory. For example, the flight time must not exceed certain limits imposed by the lifetime of the spacecraft; injection velocities are prescribed by the capability of the boost vehicle, thus affecting the transit time; for power considerations the spacecraft must always be within 162 million miles of the Sun; neither Mars nor its two moons can be allowed to infringe on the field of view of the Canopus sensor nor shadow the spacecraft from the Sun; and encounter must occur during the viewing period of Goldstone, the Mojave Desert, Calif., station of the Deep Space Network.

Other factors influencing the trajectory include the effect of solar wind pressure on the flight path; the gravitational attraction of Sun, Earth, Mars, Mercury, Venus and Jupiter; and a requirement that the encounter velocity be kept as low as possible.

In selecting an aiming zone that will determine the path of the spacecraft as it passes Mars, the trajectory engineer is required to assure that the Mariner will not impact Mars, in order to prevent contamination of the planet by Earth microorganisms. The aiming zone must also satisfy the requirements of the scientific experiments aboard the spacecraft, for example, be designed to yield satisfactory television pictures.

The accuracy of the encounter with Mars will be influenced by launch accuracy and the midcourse correction accuracy.

Calculations after launch will determine if the flight path of the spacecraft is within the correction capability of the midcourse motor. Mariner has the capability of performing two midcourse corrections in the event the first does not yield the desired accuracy for encounter.

The accuracies demanded by the launch vehicle and by the midcourse motor can be illustrated by the following numbers. The injection velocity can vary only by plus or minus 30 miles per hour or the resulting trajectory will not be within the correction capability of the midcourse motor. At midcourse maneuver, an error of one mile per hour will result in moving the spacecraft at Mars by 9,000 miles.

DEEP SPACE NETWORK

The Deep Space Network (DSN) consists of four permanent space communications stations, a launch tracking station at Cape Kennedy, the Space Flight Operations Facility (SFOF) in Pasadena, Calif., and a ground communications system linking all locations.

The four permanent stations, located approximately 120 degrees apart around the Earth, are at Woomera, Australia, Johannesburg, South Africa; and two at Goldstone, Calif.

Two new stations — near Canberra, Australia, and Madrid, Spain — are under construction. The Canberra station will be operational early in 1965. The Madrid station will go on the air later in the year.

The DSN is under the technical direction of the Jet Propulsion Laboratory for the National Aeronautics and Space Administration. Its mission is to track, receive telemetry from and send commands to unmanned lunar and planetary spacecraft from the time they are injected into orbit until they complete their missions.

Dr. Eberhardt Rechtin is JPL's Assistant Director for Tracking and Data Acquisition. Dr. N.A. Renzetti is DSN Systems Manager.

The Goldstone DSN stations are operated by JPL with the assistance of the Bendix Field Engineering Corp. Walter E. Larkin is JPL's Engineer in Charge.

The Woomera and Canberra stations are operated by the Australian Department of Supply, Weapons Research Establishment. Station Manager at Woomera is William Mettyear, and JPL's DSN resident is Richard Fahnestock. Canberra station Manager and JPL DSN resident are Robert A. Leslie and Merideth S. Glenn, respectively.

The Johannesburg station is operated by the South African government through the National Institute for Telecommunications Research. Doug Hogg is station Manager and Paul Jones is DSN resident in Johannesburg.

At Madrid, JPL will operate the newest DSN station under an agreement with the Spanish government. Donald Meyer of JPL is station Manager, and Phil Tardani, also of JPL, is DSN resident in Madrid.

The 1964 Mariner mission to Mars will span a time period of about nine months. The Deep Space Net will monitor the Mariner spacecraft continuously. The permanent stations provide 360 degrees coverage around the Earth so that one or more of their 85-foot antennas can always point toward the spacecraft.

Nerve center of the Net is the Space Flight Operations Facility at JPL Headquarters in Pasadena. The overseas stations and Goldstone are linked to the SPOF by a communications network, allowing tracking and telemetry information to be sent there for analysis.

All of the Deep Space stations of the DSN are equipped with 85-foot diameter antennas, transmitting, receiving, data handling, and interstation communication equipment. Microwave frequencies (S-band) will be used in all communications with the Mariner spacecraft.

The Pioneer station at Goldstone, along with Canberra, Woomera and Johannesburg, will be primary stations for the mission. Each has a 10,000 watt transmitter. The Venus research and development station at Goldstone, with a 100,000 watt transmitter, will provide command backup capability should more power be required late in the mission. Another Goldstone station — Echo — will be used for Ranger lunar missions which may occur during Mariner transit to Mars.

The Madrid and Echo stations will probably assist in monitoring the Mariner spacecraft during planet fly-by.

Tracking data obtained early during launch will be computed both at Cape Kennedy and at the Central Computing Facility in the SFOF so that accurate predictions can be sent to the DSN stations giving the location of Mariner in the sky when it appears on the horizon.

Scientific and engineering measurements and tracking data radioed from a spacecraft are received at one of the stations, recorded on tape and simultaneously transmitted to the SFOF via high speed data lines, teletype or microwave radio. Incoming information is again recorded on magnetic tape and entered into the SFOF's computer system for processing.

At approximately halfway to Mars, the two Mariner spacecraft will both be in view of the same DSN antenna at the same time. The station viewing both spacecraft will simultaneously transmit to the SFOF real-time telemetry data from one spacecraft and record and store telemetry data from the other for later transmission or transportation to JPL.

Scientists and engineers seated at consoles in the SFOF have push button control of the displayed information they require either on TV screens in the consoles or on projection screens and automatic plotters and printers. The processed information also is stored in the computer system disc file and is available on command.

This major command center, designed for 24-hour-a-day functioning and equipped to handle two spaceflight missions concurrently while monitoring a third, is manned by some 250 personnel during critical events — launch, midcourse maneuver, planet encounter — of a Mariner mission.

In the SFOF's mission control area, stations are set up for the project manager, operations director in charge of the mission, operations manager responsible for physical operation of the SFOF, information coordinator and for representatives from the supporting technical teams.

Mission control personnel are supported by three technical teams. Space Science Analysis is responsible for evaluation of data from the scientific experiments aboard the spacecraft and for generation of commands controlling the experiments.

Flight Path Analysis is responsible for evaluation of tracking data determination of flight path and generation of commands affecting the trajectory of the spacecraft.

Spacecraft Performance Analysis evaluates the condition of the spacecraft from engineering data radioed to Earth and generates commands to the spacecraft affecting its performance.

MARINER PROJECT TEAM

The National Aeronautics and Space Administration's programs for unmanned investigation of space are directed by Dr. Homer E. Newell, Associate Administrator for Space Science and Applications. Oran W. Nicks is the Director of OSSA's Lunar and Planetary Programs and Glenn A. Reiff is the Mariner Program Manager. Andrew Edwards is NASA's Mariner program engineer and James Weldon is program scientist. Joseph B. Mahon is Agena Program Manager for OSSA's Launch Vehicle and Propulsion Programs.

NASA has assigned Mariner project management to the Jet Propulsion Laboratory, Pasadena, Calif., which is operated by the California Institute of Technology. Dr. William H. Pickering is the Director of JPL and Assistant Director Robert J. Parks heads JPL's lunar and planetary projects.

Jack N. James is Mariner Project Manager. His two assistant project managers are Wilbur A. Collier and Theodore H. Parker. In a staff capacity, Norman R. Haynes is in charge of mission analysis and planning, and John S. Reuyl, launch constraints.

Richard K. Sloan is the Mariner Project Scientist.

The project is divided into four systems:

·   Spacecraft
·   Spaceflight Operations
·   Deep Space Network
·   Launch Vehicle

The first three systems are assigned to the Jet Propulsion Laboratory. The fourth is assigned to NASA's Lewis Research Center, Cleveland, for the Atlas-Agena launch vehicle. Dr. Abe Silverstein is the Director of Lewis Research Center. Launch operations for Lewis are directed by Goddard Space Flight Center Launch Operations at Cape Kennedy.

A few of the many key personnel in each of the systems are listed.

Dan Schneiderman -	Spacecraft System Manager
John R. Casani -	Spacecraft Project Engineer
Milton T. Goldfine -	Spacecraft Operations Manager
James Maclay -	Environmental Requirements Engineer
Richard A. Welnick -	Quality Assurance Engineer
David E. Shaw -	Spacecraft Program Engineer
A. Nash Williams -	Spacecraft Launch Vehicle Integration
Herbert G. Trostle -	Space Science
James N. Bryden -	Spacecraft Telecommunications
James D. Acord -	Spacecraft Guidance and Control
James H. Wilson -	Spacecraft Engineering Mechanical
Douglas S. Hess -	Spacecraft Test Facilities
Bruce Schmitz -	Post-injection propulsion and pyrotechnics
Wade G. Earle -	Test Conductor, First flight spacecraft

Max E. Goble -	Test Conductor, Second flight spacecraft
H. Holmes Weaver -	Test Conductor, Test model spacecraft
Thomas S. Bilbo -	Spaceflight Operations Systems Manager
David W. Douglas -	Spaceflight Operation Director
Don B. Sparks -	Facility Operations Manager
Frank G. Curl -	Data Processing Project Engineer
Jay F. Helms -	Communications
Dr. Nichola A. Renzetti -	Deep Space Network System Manager
Arthur T. Burke -	Project Engineer
Clarence A. Holritz -	DSN Operations Manager
Dr. S. Himmel -	Launch Vehicle System Manager
C. Conger -	Assistant Launch Vehicle System Manager
R. Gedney -	Project Engineer
D. E. Forney -	Chief of Agena Field Engineering Branch
Robert H. Gray -	Chief of Goddard Launch Operations
Harold Zweigbaum -	Manager of Atlas-Agena Launch Operations

CONTRACTORS

The Atlas, designed and built by General Dynamics/Astronautics (GD/A), San Diego, Calif., is purchased through the Space Systems Division of the U.S. Air Force Systems Command and Rocketdyne Division of North American Aviation, Inc., of Canoga Park, Calif., builds the propulsion system. Radio command guidance is by Defense Division of General Electric Co. and ground guidance computer by the Burroughs Corp.

Some of the key General Dynamics personnel are Charles S. Ames, vice president and project director for space launch vehicles; Jim Von Der Wische, project engineer for Mariner; Tom O'Malley, GDA launch operations; Cal Fowler, launch conductor on Complex 13; and Orion Reed, launch conductor on Complex 12.

The Agena D stage and its mission modifications are purchased directly by the Lewis Center from Lockheed Missiles and Space Co. (LMSC) Sunnyvale, Calif.

Bell Aerosystems Co. Buffalo, N.Y., provides the propulsion system. Jack L. Shoenhair is manager of medium space vehicles for LMSC, Peter J. Ward is Mariner program manager, Malcom E. Avery is Mariner project engineer and Bud Zeller is Mariner operations test director.

Many contractors provided components, assemblies and personnel to the project. A list of some 61 key subcontractors to the Jet Propulsion Laboratory who provided instruments and hardware for Mariner Mars 64 follows. Their contracts amounted to $21.1 million.

SUBCONTRACTORS

Advanced Structures Division Whittaker Corp. Las Mesa, Calif.	spacecraft hi-gain antennas
Airrite Products Division of Electrada Corp. Los Angeles, Calif.	midcourse propulsion fuel tanks, nitrogen tanks
Alpha-Tronics Corp. Monrovia, Calif.	data automation system analog-to-pulse width converters
Anadite Co. Los Angeles, Calif.	surface treatment of structural elements and chassis
Anchor Plating Co. El Monte., Calif.	Gold plating

Applied Development Corp. Monterey Park, Calif.	ground telemetry decommutators, printer programmers
Astrodata Inc. Anaheim, Calif.	time code generator/translators, ground command, read-write-verify equipment, encoder simulators and spacecraft system test data system
Barnes Engineering Co. Stamford, Conn.	Canopus star tracker electronics
Bendix Corp. Scintilla Division Santa Ana, Calif.	connectors
Bergman Manufacturing Co. San Rafael, Calif.	chassis forgings
Cannon Electric Co. Los Angeles, Calif.	connectors
CBS Laboratories Division of Columbia Broadcasting System, Inc. Stamford. Conn.	image dissector tubes for Canopus star trackers
Computer Control Co, Inc. Framingham., Mass.	real time data automation system logic cards for scientific instruments, operational support equipment, DAS voltage-to-pulse width converters
Correlated Data Systems Corp. Glendale, Calif.	spacecraft external power source and solar panel simulators, voltage controlled oscillators
Data-Tronix Corp. King of Prussia, Pa.	voltage controlled oscillators
Delco Radio Division General Motors Corp. Kokomo, Ind.	telemetry format simulators
Digital Equipment Corp. Los Angeles, Calif.	data automation system operational support data system
Dunlap and Whitehead Manufacturing Co. Van Nuys, Calif.	midcourse propulsion and structural elements
Dynamics Instrumentation Co. Monterey Park,, Calif.	ground telemetry consoles, assembly of planetary scan subsystem electronics
The Electric Storage Battery Co. Raleigh, N.C.	spacecraft batteries
Electro-Optical Systems, Inc. Pasadena, Calif.	ion chamber assemblies, assembly and test of spacecraft solar panels, modification and test of spacecraft power system, spacecraft assembly cables
Electronic Memories, Inc. Los Angeles, Calif.	magnetic counter assemblies for spacecraft central computer and sequencer
Engineered Electronics Co. Santa Ana, Calif.	non-real time data automation system
Fargo Rubber Corp. Los Angeles, Calif.	midcourse propulsion fuel tank bladders
Farrand Optical New York, N.Y.	television optical systems
Franklin Electronics, Inc. Bridgeport, Pa.	ground telemetry high speed digital printers
General Dynamics Corp. General Dynamics/Electronics San Diego, Calif.	assembly of television subsystems

General Electrodynamics Corp. Garland, Texas	vidicons and television tube test set
Grindley Manufacturing Co. Los Angeles, Calif.	midcourse propulsion jet vanes, fuel manifolds, oxidizer tank shell, and supports
Hi-Shear Corp. Torrance, Calif.	squibs
Hughes Aircraft Co. Microwave Tube Division Los Angeles, Calif.	traveling wave tubes
IMC Magnetics Corp. Westbury, N.Y.	solar vane actuators
International Data Systems, Inc. Dallas, Texas	ground command modulation checkers telemetry power supplies
Kearfott Division General Precision, Inc. Los Angeles, Calif.	gyroscopes, jet vane actuators
Lawrence Industries, Inc. Burbank, Calif.	printed circuits
Lockheed Electronics Co. Division Lockheed Aircraft Corp. Los Angeles, Calif.	solar cell modules and magnetic shift register for central computer and sequencer
Lockheed Aircraft Service Co. Division Lockheed Aircraft Corp. Ontario, Calif.	spacecraft low-level positioners
Magnamill Los Angeles, Calif.	structural elements and chassis
Massachusetts Institute of Tech. Division of Sponsored Research Cambridge, Mass.	Plasma probes
Metal Bellows Corp. Chatsworth, Calif.	midcourse propulsion oxidizer bellows assembly
Milbore Co. Glendale., Calif.	midcourse propulsion engine components
Mincom Division Minnesota Mining and Manufacturing Co. Camarillo, Calif.	ground telemetry tape recorders
Motorola, Inc. Military Electronics Division Scottsdale, Arizona	spacecraft transponders, command systems and associated operational support equipment, and DSIF equivalent operational support equipment
Nortonics A Division of Northrop Corp. Palos Verdes Estates, Calif.	Development and support of attitude control electronics
Philco Corp. Palo Alto, Calif.	integrated circuit sequence generator system, spacecraft antenna feeds and spacecraft antenna subsystem tests
Proto Spec Pasadena, Calif.	chassis and subchassis
Pyronetics, Inc. Santa Fe Springs, Calif.	midcourse propulsion system explosive actuated valves
Rantec Corp. Calabasas, Calif.	S-Band circulator switches, pre-selection and band rejection filters
Raymond Engineering Laboratory, Inc.	spacecraft video storage tape recorder

Ryan Aeronautical Co. Aerospace Division San Diego, Calif.	spacecraft solar panel structure
Siemens and Halske AG Munich, West Germany	RF amplifier tubes
Space Technology Laboratories El Segundo, Calif.	spacecraft central computer and sequencer and associated operational support equipment
Sperry Utah Co. A Division of Sperry Rand Corp. Salt Lake City, Utah	magnetometer mapping fixture
State University of Iowa Iowa City, Iowa	trapped radiation detectors
Sterer Engineering and Manufacturing Co. North Hollywood, Calif.	valves and regulators for attitude control gas system
Texas Instruments, Inc. Apparatus Division Dallas, Texas	spacecraft video storage subsystem electronics, spacecraft data encoders and associated operational support equipment, helium, magnetometers, attitude control gyro, electronics assemblies, data demodulators
Textron Electronics, Inc. Heliotek Division Sylmar, Calif.	silicon photovoltaic solar cells
Thompson Ramo Wooldridge, Inc. Redondo Beach, Calif.	thermal control louvers and power converters
Univac Division of Sperry Rand Corp. St. Paul, Minn.	spacecraft data automation system buffer memory
The University of Chicago Chicago, Ill.	spacecraft cosmic ray telescopes
Wems, Inc. Hawthorne, Calif.	spacecraft television electronics modules, spacecraft attitude control electronic modules
Wyman Gordon Corp. Los Angeles, Calif.	spacecraft structural forgings

In addition to these subcontractors, there were over 1,000 individual firms contributing to Mariner. These procurements amounted to over $19 million.

On the following pages are the 22 pictures of Mars taken by Mariner 4.

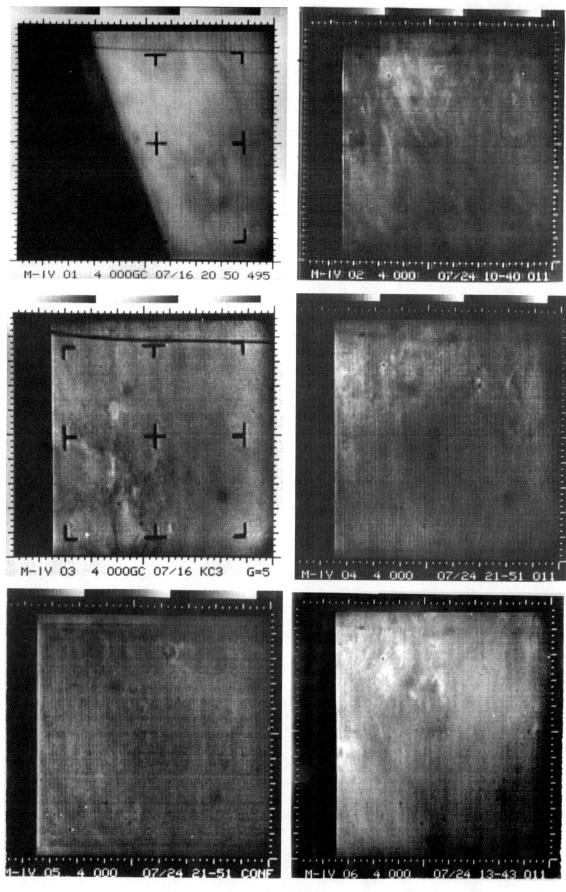

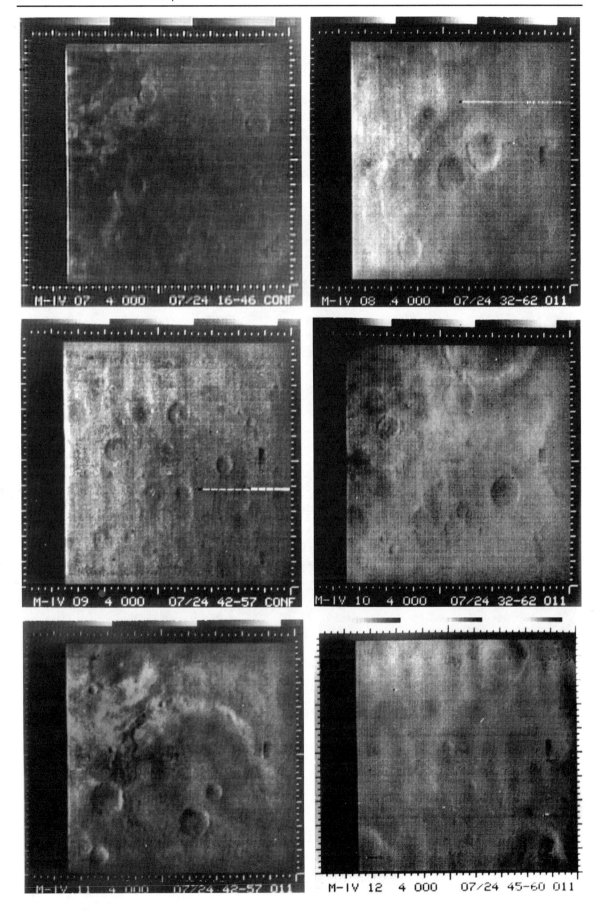

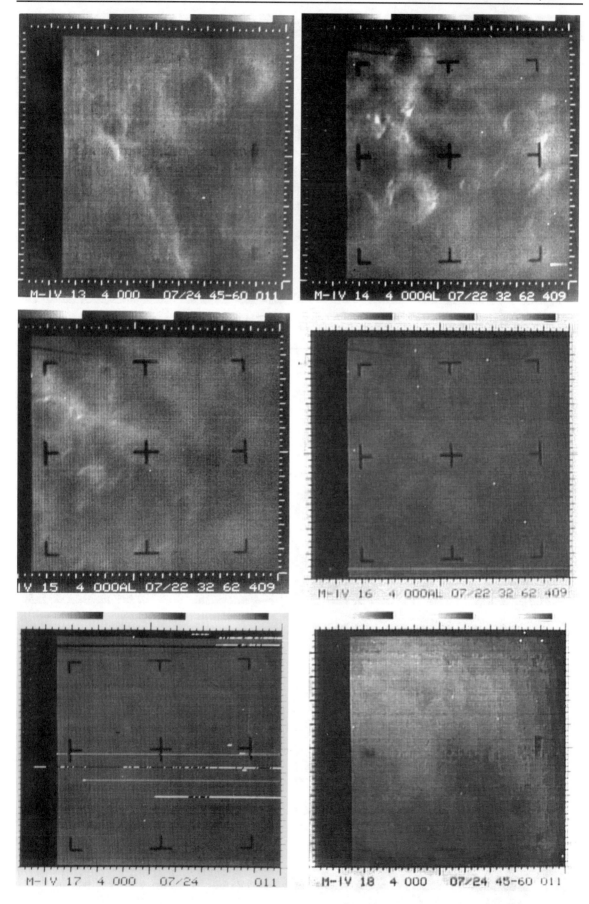

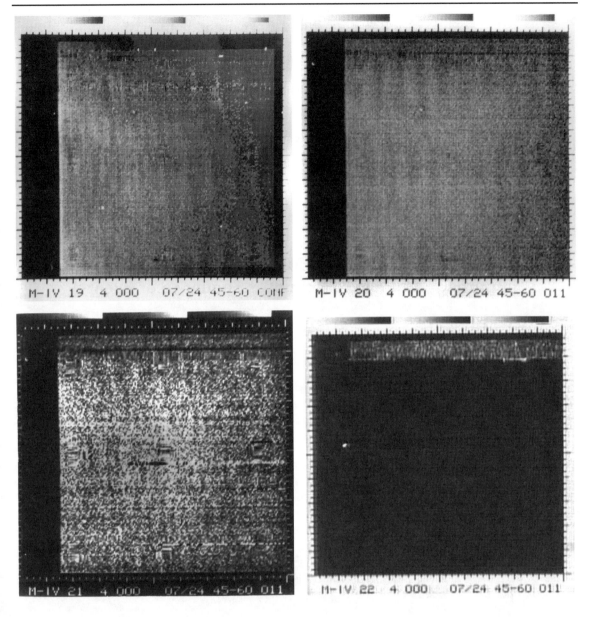

By the time Mariner 4 took pictures 21 and 22 the vehicle had flown past the Martian terminator. For further post flight details of the Mariner 4 mission refer to the accompanying CD-ROM.

MARINER '69 (6 & 7) PRESS KIT

NATIONAL AERONAUTICS AND SPACE ADMINISTRATION
WASHINGTON, D.C. 20546
TELS. WO 2-4155 - WO 3-6925

FOR RELEASE: FRIDAY P.M.
July 18, 1969
RELEASE NO: 69-26A

PROJECT: MARINER MARS '69
(Approach and Near Encounter Sequence of Events)

NOTE TO EDITORS:

The Mariner 6 and Mariner 7 spacecraft will complete their multi-million-mile flights to Mars on the nights of July 30 and August 4, respectively.

A press room will open in the von Karman Auditorium at the Jet Propulsion Laboratory, 4800 Oak Grove Drive, Pasadena, at 8:00 a.m. PDT*, Monday July 28, and will remain open on a 24-hour basis through Wednesday, August 6.

The attached pages include a summary of anticipated Mars encounter events, a day-by-day log and a computer rendering of the expected changing appearance of Mars as the Mariners photograph the planet during their approach.

A press conference outlining the expected encounter events will be held at 10 a.m. PDT Tuesday, July 29, in the JPL press room.

Please contact the JPL Public Information Office for further details if you plan to cover the Mariner encounter. Phone: Area Code 213 354-5011.

 * Because the Space Flight Operations Facility at JPL will be the command and data center for the Mariner activities and the von Karman Auditorium at JPL will be the information center, all times in this press kit supplement will be stated in PDT, California time.

MARINER 6/MARINER 7 NEARING MARS

The Mariner 6 and Mariner 7 spacecraft will fly past Mars on the nights of July 30 and August 4, 1969, respectively. Time of closest approach is now estimated at 10:18 p.m. PDT, July 30, for Mariner 6 and 10:00 p.m. PDT, August 4, for Mariner 7. Altitude at encounter will be about 2000 miles for each spacecraft.

Mariner 6, launched from Cape Kennedy on February 24, will fly a total of 241 million miles in 156 days. Communications distance from Mars at encounter will be 59.5 million miles (about 5½ light minutes).

Total Earth-to-Mars distance to be traveled by Mariner 7, launched March 27, is 197 million miles in 130 days. Communications distance at encounter will be 61.8 million miles. Both spacecraft were boosted into space by Atlas-Centaur launch vehicles.

The Mariners were developed and their missions are conducted for the National Aeronautics and Space Administration by the Propulsion Laboratory in Pasadena California.

Mariner 6 will examine the equatorial regions of Mars. Mariner 7 will cover some of the same area, but will concentrate on the southern hemisphere and a portion of the south polar cap. Together, they are expected to furnish data as different as possible from the standpoint of geography and climate.

The mission follows the 1964-65 flight to Mars by Mariner 4 is a precursor to the 1971 and 1973 Mars missions. Mariner 4 was the only other spacecraft to have photographed another planet.

In 1971 two Mariner-class vehicles will orbit Mars for three months, and in the 1973 mission, Project Viking, two spacecraft will orbit Mars and detach landing craft to descend to and operate on the surface.

Mariner 1969 mission objectives are to study the surface and atmosphere of Mars to establish the basis for future experiments in the search for extra-terrestrial life and to develop technology for future Mars missions.

The 1969 flights will not determine the presence of life on Mars but will help establish whether or not the Martian environment is suitable for life.

Television cameras aboard each spacecraft will photograph the full disc of Mars during the approach to the planet and selected surface areas at high resolution during the close Mars passage.

Thermal mapping of the areas photographed will be provided by an infrared radiometer (IRR) to correlate temperatures with surface visual appearance. A principal goal of the experiment is to determine whether the Martian polar caps are frozen carbon dioxide or frozen water.

The chemical constituents of Mars' upper atmosphere will be measured by an ultraviolet spectrometer (UVS). The experiment will identify and measure the distribution of a number of gases in the atmosphere — principally oxygen, nitrogen and perhaps hydrogen.

Composition of the lower atmosphere and possibly the surface of Mars will be determined from measurements by an infrared spectrometer (IRS). The instrument may be able to detect the presence of some organic molecules in atmospheric concentrations as small as two parts in one-million.

An occultation experiment, in which the Mariners disappear from Earth behind Mars and their radio signals pass through the Martian atmosphere, will yield information on atmospheric pressures and densities.

Radio tracking data during encounter, as well as throughout the entire flight, contributes to still another experiment — celestial mechanics — which will provide information to refine astronomical data.

The Mariner encounter can be divided into three phases — encounter or approach to Mars; near encounter or close passage by Mars; and playback of recorded near-encounter science data after the flyby.

FAR ENCOUNTER (TV LIVE FROM MARS)

As the two Mariners approach Mars, they will take a series of TV pictures while the planet revolves through several Martian days (a, Martian day is 24 hours, 37 minutes). The pictures will reveal general surface features not visible from Earth and the planet will be photographed at all longitudes. Only the north pole area will not be covered in the pictures. Some information may be obtained on the formation and motion of clouds and other Mars meteorological phenomena.

Mariner 6 will begin taking full planet pictures two days before it reaches Mars, Mariner 7 about three days before encounter.

A new high-rate telemetry system — 16,200 bits per second — on the Mariners and the use of the 210-foot antenna at the Goldstone Space Communications Station in the Mojave Desert allows the two spacecraft to record and play back an enormous amount of picture data during the approach to Mars. In the standard mission, programmed into the on-board computer prior to launch, Mariner 6 will take 50 approach pictures beginning 48 hours and 770,000 miles from Mars and ending 7 hours and 112,000 miles from Mars. Mariner 7 will take 93 approach pictures beginning 72 hours and 1,140,000 miles from Mars and ending 4 hours and 65,000 miles from Mars. Only TV camera B, the high resolution camera, will be used for taking far encounter pictures.

Each spacecraft must receive and act upon certain ground commands to initiate the standard mission sequence. These commands must be transmitted to Mariner 6 about 52 hours prior to closest approach and to Mariner 7 about 76 hours before its closest approach.

(As a backup to the standard mission in the event that certain problems occur between now and encounter, a conservative mission has been designed and programmed into each spacecraft to operate on an automatic basis or by specific command. It consists of eight approach pictures taken by each spacecraft between 22 and 11 hours before closest approach. The pictures would be stored on tape and played back at the normal science playback data rate — 270 bits per second — after the spacecraft passes Mars. The near-encounter sequence would remain the same as in the standard mission. Neither high-rate telemetry nor ground command capability is required to conduct the conservative mission.)

If both Mariner 6 and Mariner 7 conduct the standard mission, it is possible to acquire as many as 143 far encounter TV pictures. The high-rate telemetry system, the 210-foot antenna at Goldstone and a microwave link between Goldstone and the Jet Propulsion Laboratory in Pasadena, permits the real time display of the pictures as they are played back from each spacecraft.

Approximately 12 hours of real time TV may be available in five playback sessions. Every five minutes, a new picture — each containing more than half-a-million photo elements — is seen on monitors at JPL. The disc of Mars gets larger with successive pictures until the planet fills, then spills over, the edges of the frame. The five playbacks occur as follows:

     Mariner 6,   33 pictures, 7/29 6:35 p.m. - 9:27 p.m. PDT

     Mariner 6,   17 pictures, 7/30 6:00 p.m. - 7:27 p.m.

     Mariner 7,   34 pictures, 8/2 6:05 p.m. - 9:00 p.m.

     Mariner 7,   34 pictures, 8/3 7:24 p.m. - 10:19 p.m.

     Mariner 7,   25 pictures, 8/4 6:08 p.m. - 8:19 p.m.

Each of the five real-time TV playbacks occurs during the evening hours California time due to the 210-foot antenna view period.

NEAR ENCOUNTER

Mariner near encounter can be defined as a one-hour period beginning 35 minutes before closest approach to Mars and ending when the spacecraft reappears from behind the planet. Duration of near encounter including occultation is 68 minutes for Mariner 6 and 74 minutes for Mariner 7.

At about 15 minutes before closest approach, the two TV cameras — shuttering alternately every 42 seconds — the IR radiometer, IR spectrometer and UV spectrometer will begin taking planetary data, some of which is transmitted directly to Earth and all of which is recorded on board the spacecraft. During near encounter, real-time transmission of data to Earth will be at the high-rate 16,200 bits per second. It will include every seventh TV picture element for photometric measurements.

Receipt of the entire picture on Earth will occur during the post-encounter tape recorder playback. Near encounter TV totals 24 pictures — 12 high resolution and 12 medium resolution — during a period of about 17 minutes. The Mariners reach their nearest proximity to Mars during the last few minutes of the close-up TV sequence.

When the TV swath of overlapping pictures crosses the day/night terminator, picture recording ceases. The other instruments continue taking and recording dark-side data out to and beyond the limb of Mars about 10 minutes after closest approach.

Occultation — that period when Mars is between the spacecraft and Earth — begins several minutes after the end of science recording and lasts about 20 minutes for Mariner 6 and 29 minutes for Mariner 7. The occultation data, from which can be determined the density of the Martian atmosphere, is obtained at Earth tracking stations at both entry and emergence from behind the planet. Tracking data obtained throughout encounter as well as during the entire flights, contributes to the celestial mechanics experiment.

POST-ENCOUNTER PLAYBACK

Following occultation, the near encounter science data recorded on two tape recorders — one analog, one digital — aboard each Mariner is played back. The digital recorder, which stores only near encounter data, including TV, is played back at the normal science playback rate, 270 bits per second. About 19 hours after closest approach for Mariners 6 and 7, the digital playback is interrupted for two playbacks of the analog recorder totaling five hours (near encounter TV only) at the high rate, 16,200 bits per second. Mariner 6 playback is interrupted also for the Mariner 7 far encounter sequence. After both spacecraft have completed the playback several times — about August 17 — they continue to provide additional tracking and spacecraft performance information until the mission is terminated.

MARINER 6/MARINER 7 MARS ENCOUNTER LOG

PACIFIC DAYLIGHT TIME (PDT)   EVENT

Monday, July 28, 1969

	6:19 p.m.	Ground command, transmitted from Goldstone, turns on Mariner 6 science power and starts shuttering TV camera.
	8:49 p.m.	Mariner 6 scan platform is pointed at Mars so that Far Encounter Planet Sensor (FEPS) sees planet and begins tracking Mars' center of brightness to keep TV camera pointed accurately at Mars.
	10:26 p.m.	Mariner 6 takes first of 33 far encounter pictures from a distance of 771,500 miles. Entire sequence consumes 19 hours, 44 minutes, with one picture taken each 37 minutes.

Tuesday, July 29

	6:10 p.m.	Mariner 6 takes picture #33.
	6:35 p.m.	Mariner 6's first picture (M6-1) is received at JPL following high-rate transmission from the spacecraft. All 33 pictures are displayed on TV monitors at JPL as they are received about five minutes apart. Playback duration for 33 pictures is 2 hours, 52 minutes.
	9:27 p.m.	Receipt of M6-33 is completed.

Wednesday, July 30

	12:23 a.m.	Mariner 6 takes picture #34 (M6-34), the first of a series of 17 pictures. One picture is taken each 56 minutes during a duration of 15 hours, 56 minutes.
	3:19 p.m.	Mariner 6 takes picture #50 at altitude of 111,950 miles.
	6:00 p.m.	Beginning of real-time receipt and display of M6-34. Playback duration for 17 pictures is 1 hour, 27 minutes.

7:27 p.m.	Receipt of M6-50 is completed.
9:43 p.m.	Start Mariner 6 near encounter sequence with cool down of infrared Spectrometer.
10:03 p.m.	Mariner 6 begins recording data from science instruments — Infrared Spectrometer, Ultraviolet Spectrometer, Infrared Radiometer.
10:04 p.m.	High and medium resolution TV cameras each take 12 pictures with the medium resolution pictures overlapping and high resolution covering small areas within the overlaps. Twenty-four pictures are recorded in 17 minutes.
10:18 p.m.	Mariner 6 makes its nearest approach to Mars. Estimated altitude is about 2000 statute miles
10:21 p.m.	Mariner 6 takes last near encounter TV picture (M6-74). Other science instruments continue taking and recording data into the Martian night.
10:28 p.m.	End recording Mars science.
10:34 p.m.	Start playback to Earth of science data recorded on spacecraft's digital recorder during near encounter.
10:36 p.m.	Enter occultation. Ground station at Goldstone loses Mariner 6 radio signal as spacecraft disappears behind Mars.
10:56 p.m.	Exit occultation. Goldstone regains Mariner 6 radio signal as spacecraft emerges from behind Mars. Digital science playback continues.

Thursday, July 31

5:36 p.m. (apx.)	Start high-rate playback of Mariner 6 encounter pictures (M6-51 to M6-74) from spacecraft's analog tape recorder. (All near encounter pictures will be played back twice during this session. They will not be displayed on TV monitors.)
11:17 p.m. (apx.)	End high-rate playback M6-74. Resume digital playback until interrupted for Mariner 7 far encounter TV.

Friday, August 1, 1969

5:53 p.m.	Transmit ground command to Mariner 7 to turn on power for science instruments and start shuttering TV cameras.
8:23 p.m.	Mariner 7 science scan platform slews to far encounter position.
9:59 p.m.	Mariner 7 takes M7-1, first of a total of 93 far encounter pictures of mars, 34 of which are taken during this first of three sequences. Duration of 34-picture sequence is 19 hours, 48 minutes. A picture is taken each 36 minutes. M7-1 is taken from a Mars altitude of about 1,140,000 miles.

Saturday, August 2, 1969

5:47 p.m.	Mariner 7 takes M7-34
6:05 p.m.	Start playback M7-1 and continue real-time display of 34 pictures, one each five minutes. Playback sequence lasts 2 hours, 55 minutes.
9:00 p.m.	End playback M7-34.
10:59 p.m.	Mariner 7 starts second series of 34 far encounter pictures, recording one frame each 36 minutes for 19 hours, 48 minutes.

Sunday, August 3

6:47 p.m.	Mariner 7 takes picture #68.
7:24 p.m.	Start playback M7-35. Duration of playback sequence is 2 hours, 55 minutes
10:19 p.m.	End playback M7-68

Monday, August 4

1:01 a.m.	Mariner 7 takes frame M7-69. This final far encounter series numbers 25 pictures taken at 42-minute interval. The series consumes 17 hours,

48 minutes. The last picture, M7-93, is taken from a Mars distance of 65,550 miles.

5:49 p.m.	Take M7-93. At this point, Mariner 7 is 4 hours and 11 minutes from its closest approach to Mars.
6:08 p.m.	Mariner 7 plays back final series of far encounter TV pictures, M7-69 to M7-93. Receipt of all 25 pictures takes 2 hours, 11 minutes.
8:19 p.m.	End playback M7-39
9:25 p.m.	Begin Mariner 7 near encounter with IRS cooldown.
9:45 p.m.	Start recording Mars science, including 24 near encounter TV pictures (frames M7-94 to M7-117).
10:00 p.m.	Closest approach to Mars (about 2000 miles).
10:02 p.m.	Take TV frame M7-117, concluding TV recording. Continue recording other science data.
10:10 p.m.	End recording Mars science.
10:15 p.m.	Enter occultation. Goldstone loses Mariner 7 radio signal.
10:16 p.m.	Near encounter science data begins playing back, although it begins while Mariner 7 is behind Mars.
10:44 p.m.	Exit occultation. Goldstone regains Mariner 7 radio signal and receipt of data from digital tape recorder begins.

Tuesday, August 5

5:20 p.m. (apx.)	Begin high-rate playback of 24 near encounter TV pictures (M7-94 to M7-117) from Mariner 7's analog tape recorder. (All near encounter pictures will be played back twice during this session. They will not be displayed on TV monitors.)
11:22 p.m. (apx.)	End playback M7-117. Resume digital - recorder playback. Mariner 6 and Mariner 7 continue transmitting data recorded on digital tape recorders until all has been played back several times (about Aug 17).

MARS AS VIEWED BY MARINER CAMERAS IN FAR ENCOUNTER SEQUENCES

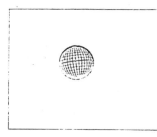 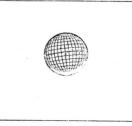 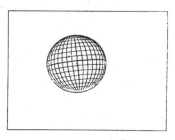

M6-1 771,500 mi.   M6-17 612,400 mi.   M6-33 453,350 mi.

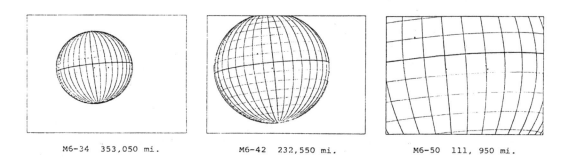

M6-34 353,050 mi.   M6-42 232,550 mi.   M6-50 111,950 mi.

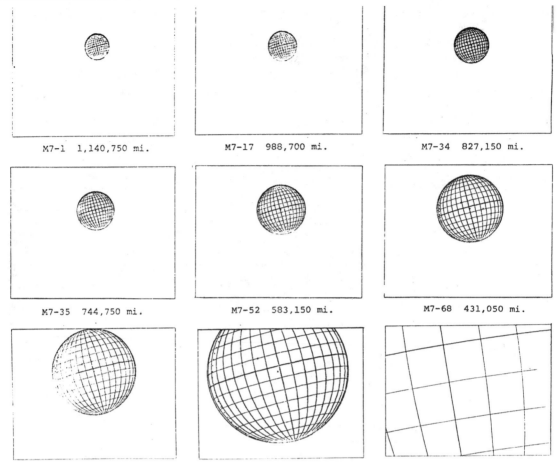

M7-1   1,140,750 mi.   M7-17   988,700 mi.   M7-34   827,150 mi.

M7-35   744,750 mi.   M7-52   583,150 mi.   M7-68   431,050 mi.

MARINER '69 RESULTS

NATIONAL AERONAUTICS AND SPACE ADMINISTRATION
WASHINGTON, D.C. 20546
TELS: (202) 963-6925 (202) 962-4155

FOR RELEASE: 9 A.M.
Sept. 11, 1969

NASA Hq- Auditorium Washington, D.C.

MARINER MARS 69 RESULTS

The Mariner 6 and Mariner 7 last month revealed Mars to be heavily cratered, bleak, cold, dry, nearly airless and generally hostile to any Earth-style life forms.

The twin spacecraft carried out the United States' first successful dual-flight scientific exploration of another planet.

Mariner 7 performed its mission objectives perfectly despite being damaged, possibly from a meteoroid impact, during the final week of its Earth-to-Mars journey.

The Mariners were developed and their missions were conducted for the National Aeronautics and Space Administration by the Jet Propulsion Laboratory in Pasadena, Calif.

Two-hundred television pictures of Mars were taken by the two Mariners, including 57 high and medium resolution views of selected Martian surface areas from an altitude of only a few thousand miles. The bulk of the photos were taken as Mariners 6 and 7 approached Mars and the planet revolved through several Martian days.

Arriving five days apart, the two instrument-laden spacecraft sampled the Martian atmosphere, for temperature, pressure and chemical constituency. Surface temperatures also were measured in an effort to correlate thermal characteristics with features observed in the TV pictures.

Mariner 6 examined at close range the equatorial regions of Mars, primarily the near-equator latitudes of the southern hemisphere. Mariner 7 covered some of the same area, but concentrated on the southern hemisphere, the south polar cap and the pole itself. Both spacecraft took a number of long range looks at the planet with telescope-equipped TV cameras.

The data, when thoroughly analyzed, should increase significantly our understanding of the atmosphere and surface of Mars, provide global and local maps and, by inference, contribute to a better understanding of the biological status of the planet.

Mission objectives of the Mariner Mars 1969 Project were to study the Martian surface and atmosphere to establish the basis for future experiments in the search for extra-terrestrial life and to develop technology for future Mars missions. The question of whether or not life is present, or did at any time exist, may be answered by these future expeditions.

Mariner 6 kept its rendezvous with Mars at 1:19 a.m. EDT on July 31, passing within 2131.4 miles of the planet's surface after a 241-million-mile flight in 156 days.

Mariner 7 made a close passage by Mars on August 5, closing to an altitude of 2130.2 miles at 1:01 a.m. EDT. Total Earth-to-Mars elapsed time and miles clocked were 130 days and 197 million miles.

Both spacecraft were launched from Cape Kennedy, Florida, by Atlas-Centaur launch vehicles — Mariner 6 on February 24, 1969, and Mariner 7 on March 27. Each Mariner required only a single mid-course-maneuver — Mariner 6 on February 28 and Mariner 7 on April 8 — to achieve its desired Earth-to-Mars trajectory. The Mariner propulsion systems provided for a second maneuver capability if required.

Following the launch phase of the flights, the spacecraft tracked by stations of the NASA-JPL Deep Space Network located on four continents. Stations assigned to the Mariner missions are in Goldstone, California; Woomera, Australia; Madrid, Spain; and Johannesburg, South Africa. Control center for the missions was the Space Flight Operations Facility at JPL in Pasadena.

The encounter phase of each flight was divided into two distinct phases — far encounter, in which full disc TV pictures of Mars were obtained, and near encounter, when the full complement of science instruments scanned Mars from close range.

Commands to activate the Mariner 6 science instruments were transmitted from the Goldstone station in the Mojave Desert on the evening of July 28. The spacecraft's scan platform was pointed at Mars so that a light sensitive sensor would track the planet's center of brightness to keep the TV camera pointed accurately at Mars.

Mariner 6 took the first of 50 approach pictures at 1:26 a.m. EDT on July 29 from a distance of 771,500 miles. Although the camera took a picture each 84 seconds, only one picture each 37 minutes was recorded on the spacecraft's analog tape recorder. (A second TV camera also was operating — alternately with the first — but its pictures were recorded only during the close passage.)

Nearly 20 hours later, a full tape load of 33 pictures had been recorded. They were played back to Earth — one each five minutes beginning at 9:35 p.m. EDT — and received at JPL via the Goldstone station.

A new high-rate telemetry system — 16,200 bits per second — on the Mariners and the use of the 210-foot-diameter antenna at Goldstone allowed the acquisition of the large number of approach pictures. Mariner 6 took 50 such pictures, in segments of 33 and 17, providing full planet pictures during two revolutions of Mars.

From August 2 to August 4, Mariner 7 took 93 far encounter pictures, recording the changing view of Mars during three planet revolutions. These were taken in increments of 34, 34 and 25.

The second tape-load of Mariner 6 pictures, taken on July 30, included a two-picture attempt to photograph one of Mars' two moons, Phobos. Pictures #40 and 41 were timed two minutes apart to "capture" Phobos as the tiny moon rose on the eastern limb a few degrees above the Martian equator. Phobos was not apparent in the raw picture data. Further picture enhancement may reveal its presence.

All 200 TV pictures received from the two Mariners were displayed on television monitors at JPL as they were received from the spacecraft. This was made possible by the high-rate telemetry system, the 210-foot antenna, a microwave link from Goldstone and specialty equipment at JPL.

Each five minutes during the playback session, a new picture containing more than half-a-million picture elements was seen on the monitors. Seven playback sessions contained a total of more than 17 hours of real-time TV display.

In addition to the 200 pictures stored on the analog recorder and played back, Mariner 6 and 7 transmitted, to Earth 1177 digital pictures as they were taken every 42 seconds by the cameras. The digital pictures contained every seventh picture element of each picture line and are used for photometric measurements.

On July 30, just seven hours before Mariner 6 was to cross the orbit of Mars, the radio signal from Mariner 7 fell silent. The Deep Space Station at Johannesburg reported loss of signal at 6:11 p.m. EDT. It was believed, and the cause is still uncertain, Mariner 7 may have been struck by a meteoroid.

One of the Deep Space Network's two Madrid stations was tracking Mariner 6. The other broke off tracking Pioneer 8 and joined the search for Mariner 7. As the Earth turned under the spacecraft and Mariner 6 neared Mars, three Goldstone stations came into view. The Pioneer Station at Goldstone picked up a very faint signal about 1:30 a.m. EDT on July 31, just a few minutes after Mariner 6 whipped by Mars. Commands were transmitted to Mariner 7 to switch antennas and 11 minutes later — round trip light time between Earth and Mariner 7 — a healthy signal was detected by stations at Goldstone and in Australia.

It was learned that Mariner 7 had lost lock with its celestial reference, the star Canopus. Hence, the high-gain antenna no longer pointed at Earth. Canopus lock was re-established and further tracking indicated Mariner 7 had been damaged, including the loss of some 20 of 90 telemetry channels.

Mariner 7 also had changed velocity. It apparently was receiving a small amount of thrust, possibly from an outgassing pressure vessel. The measured continuing acceleration of a few millimeters per second changed the spacecraft trajectory slightly, causing it to arrive 10 seconds later than predicted but very close to the predicted altitude.

The Mariners began their near encounter sequences 35 minutes before closest approach with the cryogenic cooldown of one of the science instruments — the infrared spectrometer. At encounter minus about 15 minutes, the two TV cameras — shuttering alternately every 42 seconds — the infrared spectrometer, infrared radiometer and ultraviolet spectrometer began taking planetary data, some of which was transmitted directly to Earth and all of which was recorded on board the spacecraft.

Mariner 6 took 24 near-encounter pictures — 12 high resolution and 12 medium resolution — during a

17-minute period. It reached its nearest proximity to Mars during the last few minutes of the close-up TV sequence.

Mariner 7 took 33 near-encounter pictures — 16 high resolution and 17 medium resolution.

The scan platforms on both spacecraft were slewed twice during the Mars passage, resulting in a three-segment trace across the planet. The Mariner 7 scan program had been revised on the basis of the Mariner 6 data to provide maximum coverage of the south polar cap.

When the TV swath of overlapping and nested pictures crossed the day/night terminator, picture recording ceased. The other instruments continued taking and recording dark-side data out to and beyond the limb of Mars about 10 minutes after closest approach.

All instruments aboard Mariner 7 functioned perfectly. On Mariner 6, one of the infrared spectrometer's two channels did not get cold enough to operate.

Following occultation, the near-encounter science data recorded on two tape recorders — one analog, one digital — aboard each Mariner was played back. The digital recorder, which stored all near encounter data, including TV, was played back at the normal science rate, 270 bits per second. At about 19 hours after closest approach for both Mariners, the digital playback was interrupted for a five-hour playback of the analog recorder (TV near encounter only) at the high rate, 16,200 bits per second.

By mid-August both spacecraft had played back all the science data several times. The two Mariners remain in their solar orbits and are tracked periodically by stations of the Deep Space Network.

Two experiments which required no special instrumentation were conducted by Mariner 6 and 7. One was occultation which provided atmospheric pressure measurements through the analysis of changes in the spacecraft radio signal as the spacecraft disappeared behind Mars relative to Earth.

Tracking data obtained throughout encounter, as well as during the entire flights, contributed to the celestial mechanics experiment.

The Mariner Mars 1969 Project provided the United States' second successful exploration of Mars. It followed the 1964-65 flight of Mariner 4, the only other spacecraft to have photographed another planet.

In 1971 two Mariner-class vehicles will orbit Mars for three months, and in a 1973 mission, Project Viking, two spacecraft will orbit Mars and detach landing craft to descend to and operate on the surface.

MARINER MARS 1969 SCIENCE EXPERIMENTS

The Television Experiment was designed to photograph the disc of Mars as the planet revolved through several Martian days and take high resolution pictures of the surface during the close approach. Principal science investigator is Dr. Robert B. Leighton of the California Institute of Technology. Co-investigators are Drs. Norman H. Horowitz, Bruce C. Murray and Robert P. Sharp, all of Caltech; Alan G. Herriman and Andrew T. Young of the Jet Propulsion Laboratory; Bradford A. Smith, New Mexico State University; Merton E. Davies, Rand Corporation; and Conway B. Leovy, University of Washington.

The Ultraviolet Spectrometer analyzed the composition of the upper atmosphere of Mars. Principal investigator is Dr. Charles A. Barth of the University of Colorado. Co-investigators are William G. Fastie of Johns Hopkins University; Fred C. Wilshusen, Kermit Gause, Ken K. Kelly, Ray Ruehle, Jeffrey B. Pearce and Charles W. Hord, all of the University of Colorado; and Edward F. Mackey, Packard Bell Electronics.

The Infrared Spectrometer analyzed the composition of the lower Martian atmosphere. Principal investigator is Dr. George C. Pimentel of the University of California at Berkeley. Co-investigator is Dr. Kenneth Herr, also of U.C.B.

The Infrared Radiometer made Mars surface temperature measurements. Principal investigator is Dr. Gerry Neugebauer of Caltech. Co-investigators are Dr. Guido Munch, Caltech, and Stillman C. Chase of the Santa Barbara Research Center.

The S-Band Occultation Experiment determined the pressure and density of the Martian atmosphere. Principal investigator is Dr. Arvydas J. Kliore of JPL. Co-investigators are Dr. S. I. Rasool, Goddard Institute of Space Studies; Gunner Fjeldbo, Stanford University; and Boris Seidel, JPL.

The Celestial Mechanics Experiment provided data for the continuing effort to refine astronomical values. Principal investigator is Dr. John D. Anderson of JPL. Co-investigator is Warren L. Martin, also of JPL.

PRELIMINARY SCIENCE RESULTS MARINER MARS 1969

TELEVISION

The principal results from preliminary study of the Mariner 6 pictures are: the surface of Mars appears similar to that of the moon, but there are significant differences; some features seen from Earth are characterized; the "blue haze" hypothesis is disproved; and new phenomena associated with the polar cap are discovered.

Highlights of initial study of the Mariner 7 pictures, in addition to confirming the first reported results of Mariner 6, are: additional physiographic interpretations of classically observed features are made; the surface of the south polar cap is generally visible and many large topographic forms are entirely coated with "snow;" the edge of the cap is apparently defined by local topographic configurations, but "snow-free" areas are apparent within the polar regions; and the Hellas region appears devoid of craters, thus implying the operation of more effective, more recent, and more geographically confined surface processes then heretofore evidenced.

ULTRAVIOLET SPECTROMETER

As the Mariner 6 UV spectrometer viewed the brightly illuminated limb of Mars, ultraviolet emissions were measured which showed that atomic hydrogen and atomic oxygen are present in the upper atmosphere. Additional measurements of the ultraviolet light from the limb showed the emission spectrum of expected constituents of the Martian atmosphere, carbon dioxide and carbon monoxide. The initial study of the data has shown that the very important molecule nitrogen was not defeated. If additional intensive analysis substantiates this conclusion, a very key chemical compound is missing from the Martian environment. If this is true, any life chemistry on Mars will have to be very much different than we know on Earth.

The Mariner 7 UV spectrometer measured the composition of the upper atmosphere on two passes over the limb of the planet and confirmed the results of Mariner 6 in finding no molecular nitrogen in the upper atmosphere. The new result associated with Mariner 7 comes from the pass over the polar cap. The intensity of ultraviolet light from the planet increased abruptly as the spectrometer view passed from the desert onto the polar cap showing that ultraviolet radiation at very short wavelengths penetrates to the surface of the planet. This result shows that the planet is bathed in this energetic solar radiation.

INFRARED SPECTROMETER

The Mariner 6 IR spectrometer successfully recorded data in the spectral region 2 to 6 microns. With 60 by 60 mile spatial resolution, significant thermal variations were detected, revealing temperatures up to 75°F. The data also reveal the local reflectivity of solar light and it is clear that the darker spots on Mars are warmer than the bright areas. The carbon dioxide intensity reveals topographical detail that remains to be analyzed. Neither ammonia nor nitric oxide was evident, however, carbon monoxide was detected. Perhaps the most exciting result is that the spectrum of ice was recorded. It seems unlikely that this ice could be in the spectrometer — the only possible misinterpretation — and the experimenters tentatively attribute it to a very thin ice fog. Finally, there are two or three uncertain spectral features that remain to be verified and identified.

The Mariner 7 IRS experiment produced data from both channels. These data were first believed to have contained evidence of gaseous methane and ammonia, and the suggestion that a portion of the polar cap is composed of water ice. Further analysis of the data is required before a firm conclusion can be reached.

INFRARED RADIOMETER

The heat radiated by areas approximately equal to those covered by the Mariner 6 narrow-angle TV images was measured during the Martian day and night. At the Martian noon, the temperature reaches a maximum of 60°F, while during the night it falls below -100°F. Variations in temperature have been detected which provide the absorbing properties of the soil and its ability to retain heat during the night. The preliminary analysis of the data suggests that the surface is a very good heat insulator, in fact, better than any known solid material on Earth.

Primary objective of the Mariner 7 IR radiometer experiment was to measure the temperature of the polar cap in order to ascertain whether it is made of frozen CO_2 or H_2O. The 200 measurements made while Mariner 7 pointed towards the cap show a flat temperature profile with a minimum of $-190°$ F which is in close agreement with the frost point of CO_2 at the basin pressure of Mars.

This agreement is taken as very strong circumstantial evidence in favor of the theory that the polar caps are, in fact, predominantly made of CO_2 rather than of water ice.

S-BAND OCCULTATION

Mariner 6: Near the region of Meridianii Sinus, the surface atmospheric pressure was found to be about 6.5 millibars (Earth is 1013 millibars) and the surface temperature about −9°F. At an altitude of about 82 miles, an ionosphere was observed. A smaller ledge of ionization was observed at an altitude of 63 miles.

Mariner 7: A pressure of 3.5 millibars and a temperature of 205°K, (-90°F) was obtained at a latitude of 59°S and longitude 28°E. The low value of surface pressure may indicate that this region (near Hellespontica Depressio) is substantially higher than the average (about 4 miles). An ionosphere was also observed, with a mean peak at an altitude of about 130 kilometers (81 miles).

CELESTIAL MECHANICS

By analyzing three months of data from Mariners 6 and 7, it has been possible to determine a ratio of the mass of the Earth to that of the Moon of 81.3000 with an uncertainty of 0.0015.

The mass of Mars is determined to be about one tenth the mass of the Earth; the exact figure is 0.1074469 $\pm$ 0.0000035. The only other precise determination of the mass was obtained from the Mariner 4 Doppler data. A recent re-examination of these data by George W. Null at JPL indicates a value of 0.1074464 with an uncertainty of $\pm$ 5 units in the last place; Mariners to Mars in 1964 and 1969 agree very well in their determinations of the mass.

Further analysis of the Mariner '69 data should yield information on the shape of the gravitational field and on the orbit of Mars itself. This will be important to future, more advanced missions to Mars and to an extended analysis of the data for the general relativistic effects.

MR—6

NATIONAL AERONAUTICS AND SPACE ADMINISTRATION

MARINERS

SIX AND SEVEN

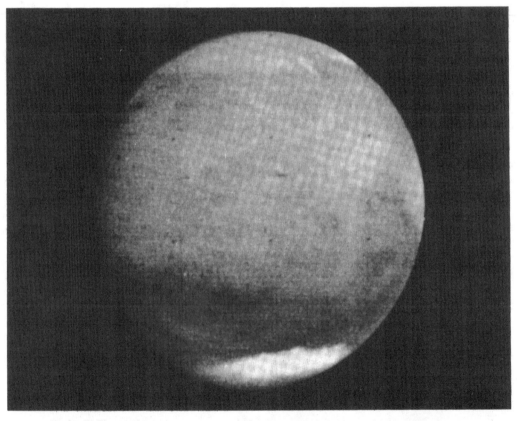

Mars from 333,700 statute miles. Large crater, upper center, is Nix Olympica, 300 miles in diameter. White area is south polar cap.

10/29/69

NATIONAL AERONAUTICS AND SPACE ADMINISTRATION

FOR RELEASE: OCTOBER 29 1969
MR-6

The planet Mars is a cold, inhospitable desert.

This preliminary summation of the voluminous data relayed back to earth by Mariners Six and Seven was offered by one of the scientists responsible for an experiment aboard the spacecraft. Additionally, the findings of the Mariners caused planetologists to revise their concept of the Red Planet for the second time.

Before the space age, Mars — with its polar caps, its seasons, its 24 hour day, and its markings — was thought to be like Earth. Mariner Four altered that view when it photographed the planet from a distance of 6,000 miles and relayed back pictures of a barren, heavily cratered surface. It was then thought that Mars might be like the Moon. But the tandem Mariner mission revealed Mars, as one experimenter put it, to be like Mars; with its own characteristic features, some of them unknown and unrecognized elsewhere in the solar system.

The latest Mariners were the first of three dual unmanned missions to Mars intended to resolve existing questions about the planet. A second pair of Mariners, scheduled for launch in 1971, will orbit the planet; and a 1973 mission, with two heavier, more complex spacecraft called Vikings, will land instrument packages on the planet's surface. The Viking mission will determine if there is life on Mars.

After flights of 241 million and 197 million miles respectively, Mariners Six and Seven made a flyby of Mars from a "miss" distance of 2,000 miles. After encounter on July 31, No. Six flew a track roughly parallel to the Martian equator while its partner craft, which arrived five days later, traveled a "NW to SE" course which carried it over a portion of the south polar cap. The tracks were so plotted in order to cover as many classically recognized features as possible - the light and dark areas, the "oases," and the polar cap. These features are distinguishable telescopically from Earth. The mission was highly successful and returned to the teams of experimenters a bonus of information and 11 more than the 189 TV pictures that were planned.

"Chaotic" Terrain

In the initial analysis of the data, the experimenters qualified their conclusions and emphasized that more intensive study of the data was required before they could describe their findings with confidence. However, certain uniquely Martian features were immediately apparent. Among these was an area of roughly 200,000 square miles of "chaotic terrain." This was a tumbled, turbulent area of short ridges and small valleys. Scientists speculated that it could have been the result of a massive slump or slide. It was as though material underlying the planet's crust might have flowed or otherwise shifted; and, with this support removed, the crust collapsed in a jumbled mass. To the experimenters, this phenomenon suggested that they were looking at evidence of a difference in processes or a difference of material, or perhaps a combination of both.

Mariners Six and Seven were not meant to supply a definitive answer to the question of whether life exists on Mars. However, thus far nothing has been discovered in the data to encourage a belief that the planet supports life, but neither do the results exclude the possibility.

Mariners Six and Seven were heavier and more advanced versions of Mariner Four. They carried six experiments, five of which were intended to develop data on the physical, chemical and thermal qualities of the planet and its environment. The sixth was designed to yield astronomical data.

The spacecraft also carried out what the mission planners called a "free experiment." This was a reading of radio signal strength as the spacecraft passed behind the planet. Called the "occultation experiment," it was "free" only in the sense that the spacecraft, themselves, needed no special on-board equipment to conduct it. As the spacecraft passed behind Mars, the radio beam from the transmitters were increasingly blocked as the Martian atmosphere interposed between the spacecraft and the giant dish antenna aimed at them from earth.

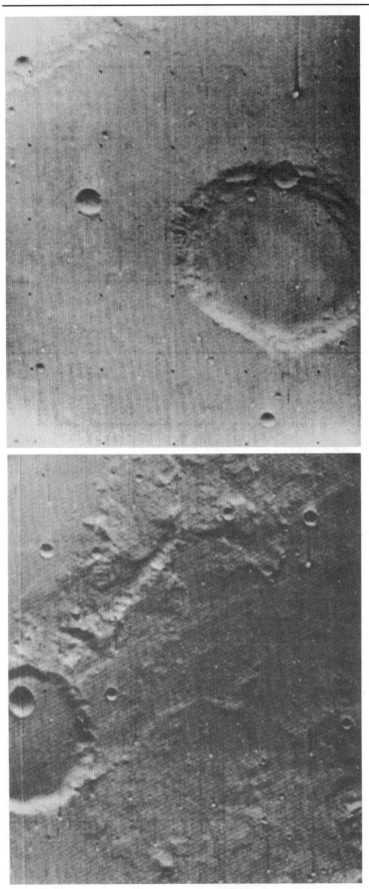

A near-encounter, high-resolution photo. Large crater, 24 miles in diameter has slump terraces on its walls.

As the mass of the planet itself interrupted the radio beam, the blockage was total for as long as the spacecraft were behind the planet. As they passed from in back of the planet, and their movement lifted the beam from the lower - and denser - layers of the Martian atmosphere, to the higher levels, signal strength progressively increased as the atmosphere's resistance lessened. Signal strength returned to normal when the beam was clear of the atmosphere. By plotting the decline and then the increase in signal strength, the experimenters got measurements of the density and pressure of the Martian environment.

Surface Photography

Each spacecraft carried two cameras — one with a wide-angle lens and the other with a high-resolution lens and a narrow view. The wide-angle cameras were equipped with red, blue and green filters. The high resolution cameras had yellow filters to reduce haze. They were timed to take pictures alternately. Enough overlap was built into the field of view so that the high-resolution pictures would be included in the area covered by the wide-angle lens.

Another near-encounter, high-resolution shot from 2150 miles. Large crater at left center is about 15 miles across. Rough terrain near top is debris which forms the rim of a much larger crater.

The TV sequences divided into two parts: "far" encounter shots beginning when the spacecraft were 770,000 miles from Mars; and "near" encounter photos which began as the Mariners approached within 6,000 miles and included their swing around the Martian far side. For Mariner Six, the "near" encounter duration was 68 minutes. For No. Seven, it was six minutes longer. Dr. Robert P. Leighton of the California Institute of Technology, principal investigator for the TV experiment, termed the pictures "spectacular."

The "far" encounter photos showed the full disc of Mars — those taken at the greatest distance showing the planet much as it would look through a large, earth-based telescope. As the spacecraft raced toward the planet at 16,000 mph, the image grew larger until it overflowed the frame of the photograph. The white polar caps stood out clearly. The "W" shaped cloud and dark areas, previously observed features, were easily seen.

Widespread Cratering

The high-resolution images were 500 times better than the best ever seen from Earth. Features with a dimension of 900' could be identified. The pictures showed that Mars had suffered a bombardment of space materials over a very long period of time. The cratering was widespread and many were very large, 30 to 50 miles in diameter, probably formed by a collision with an asteroid. At a location called Nix Olympica, long a subject of Earth observation and remarkable for its varying brightness, was a huge crater 300 miles in diameter. There was a bright spot in the center and a light-colored collar around the outside of the crater rim.

Some of the craters bore a resemblance to those on the Moon with steep sides and enough material slippage to create a terraced effect. Long winding rills similar to those photographed on the Moon were present.

The cratered areas were one of three differing types of terrain shown. The "chaotic" area was a second. A third were flat, featureless plains, such as the bright Hellas desert region, which piqued the interest of the experimenters. It was regarded as certain that these plains could not have escaped the cratering which other parts of the Martian surface had sustained. Therefore the experimenters deduced that the material composing the plain was such that craters were quickly filled in after they occurred. This smoothing process could have taken place if the plain surface were composed of dust or a loose sand-like material that would shift when exposed to the Martian wind.

The Polar Caps

The polar caps were crusted with frozen carbon dioxide possibly to a thickness of two feet. The crater slopes that faced the sun were devoid of carbon dioxide "snow," while areas in shadow were "snow" covered. Both Earth observation and "far" encounter photos depicted caps with the edges sharply delineated. This was also true of the distant views of the dark and light areas. From close-up, however, discrete boundaries were nonexistent. The edges of the caps were ragged and blurred, and the light and dark area boundaries diffused and blotchy.

The so-called Martian canals were not in evidence. The experimenters were of the opinion that the impression of canals — or more properly the Italian astronomer's "canali" which does not so translate — were possibly dark-floored impact craters in a quasi-linear alignment or a similarly positioned sequence of dark patches.

Temperature and Pressure

The spacecraft carried an array of sensors in the infrared and ultraviolet spectral range for chemical analysis of the Martian surface and environment. These also provided readings on temperatures on the surface and in the atmosphere. Atmospheric pressure on the surface of Mars ranged — according to altitude — from three and a half to six millibars. By way of comparison, Earth's atmospheric pressure at sea level is about 1,000 millibars. Pressure readings at the polar caps were at the lower end of the range, suggesting that the caps covered mountains or perhaps a high plateau. A minimal pressure reading for a feature called Hellespontica Depressio indicated that this so-called "depression" might have an altitude of 18,000 feet.

Martian temperatures covered a wide range. Data produced by the sensors did not always agree. The

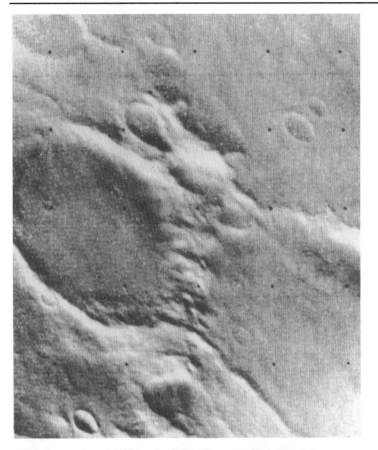

Two adjacent craters in the south polar cap make the "Giant Footprint."

discrepancies could be accounted for by the fact that they may not have been pointing at the same area. There was also a question as to whether the reading was from the planet's surface or from an atmospheric layer. The infrared radiometry experiment showed the polar cap to be -189°F — slightly above the temperature of solid carbon dioxide. It was suggested that areas of the surface not covered by the cap might be responsible for the anomaly. There was also an aerosol hood of solid carbon dioxide high in the atmosphere over the caps and at varying altitudes throughout the measured areas except over the equator.

A portion of the south polar cap. Dark area on right is the sunset shadow line — the evening terminator.

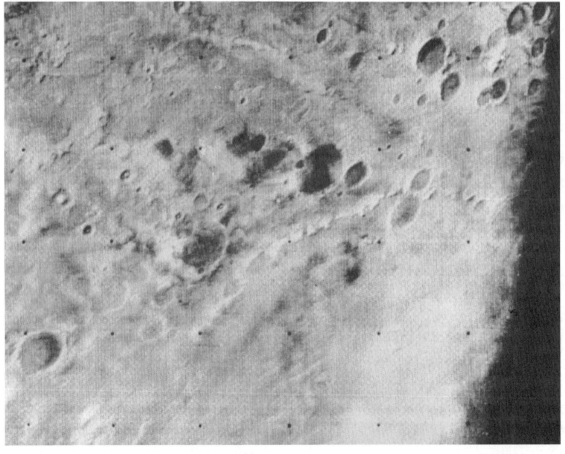

Surface temperatures on the day side — in the equatorial zone — swung from -63 to +62 F and on the night side from -153 to -63 F. in moving from the day side across the terminator to the night side, the instruments showed a cooling curve which was taken as an indication that the Martian surface might be composed of a highly insulating material such as dust. Had the material been conductive, there would have been a less marked temperature differential. The dark areas were several degrees warmer than the light, probably attributable to their absorptive color. The chaotic terrain was also several degrees warmer than adjacent areas. Experimenters offered no explanation for this beyond the comment that it was not coincidence.

The Martian Atmosphere

In determining the chemical composition of the atmosphere and the surface, the scientists took advantage of the fact that each element has a spectrum that is as individual as a fingerprint. In the controlled environment of the laboratory, the spectral signature of elements and gases was established and then the spectra were compared with the readings produced by the sensors aboard the spacecraft. The most significant finding was negative. No molecular nitrogen was found in the Martian atmosphere in any significant quantity. The experimenters, however, did not rule out the possibility that initial examination of the spectra may have overlooked the presence of traces of this gas.

The spectra showed the attenuated Martian atmosphere to be almost entirely carbon dioxide. Very thin clouds of solid carbon dioxide were found in the upper altitudes: a finding consistent with temperatures recorded that were low enough to condense the CO_2 into ice particles. At extreme altitudes there were indications of ionized carbon dioxide and definite signs of a Martian ionosphere.

The sensors reported no clear indications of any kind of water activity - no clouds, frosts or fogs. There were small amounts of water vapor in the atmosphere, but in a quantity insufficient to permit stable liquid water on the Martian surface. Extremely thin aerosol layers made up particles of dry ice, frozen CO_2 were in evidence from 10 to 30 miles above the surfaces. These were too thin to show on vertical photographs, but registered on oblique shots. Earlier reports of the presence of small quantities of methane and ammonia-gases associated with the origin of life were declared to be in error: a result of a misreading of dark bands on one of the spectra.

Intense Radiation

The ultra-violet (UV) spectrometer experiment increased pessimism over the possibility of finding life — at least as we know it — on Mars. UV returns from the surface were "bright," especially over the polar cap, indicating that a very high percentage of the ultra-violet radiation from the sun reached the surface. Whereas Earth's environment provides a shield against all but a small amount of this radiation, the Martian atmosphere offers little protection. Since such intense ultra-violet radiation destroys all but the most tightly bonded molecules, molecular nitrogen, methane, ammonia, ozone, and sulphur dioxide — gases associated with the origins of life — would have been eradicated.

Mission Profile

The Mariners were put on their Martian course by Atlas Centaur launch vehicles which lifted off at 8:29 p.m. EST February 24th and 5:22 p.m. March 27. No. Six took 156 days to make the outward trip; No. Seven, 130 days. The difference in trip time is accounted for by the orbital motion of Earth and Mars.

One mid-course correction for each spacecraft was conducted with precision, an essential in that an error of one mph in velocity of the speeding craft would have resulted in their being 5,000 miles off course on reaching the vicinity of Mars.

The outward leg went as planned save for a mishap suffered by Seven as it neared Mars. The craft was apparently hit by a micrometeorite and the blow broke its lock on Canopus, the guiding star. The tumbling that ensued shut off its communications with Earth. For five anxious hours, ground controllers wrestled with the problem and finally brought the spacecraft back to life, checked the tumbling and renewed the vital lock on

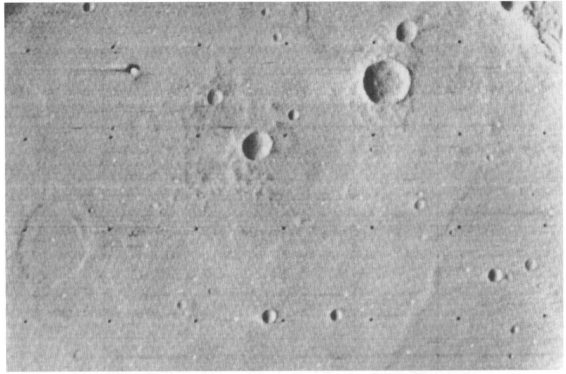

One of the bright areas of Mars, Deucalionis regio. Fresh crater in upper right center is about 3 miles in diameter.

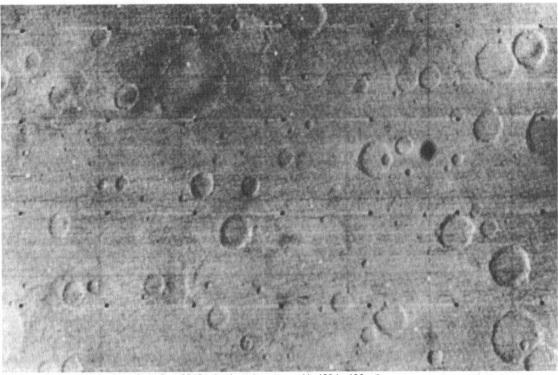

A wide angle photo of Deucalionis Regio from 2245 miles. Area shown is roughly 450 by 620 miles.

Canopus. This was achieved just in time as No. Seven's partner craft was entering its "near" encounter phase and the controllers would have had to abandon No. Seven in order to work with Six.

One measure of the improvement of these latest Mariners over their predecessor Mariner Four is found in their capacity to return data. Mariner Four had a capacity to transmit 8 1/3 bits per second. (A bit is a unit of information, digitally expressed in computer language.) In the case of the transmission of a photograph, the digits describe degrees of brightness or darkness, from dead white to totally black. Each picture has 500 lines with 900 bits to a line. Mariner Four pictures contained 340,000 bits and took more than eight hours each to transmit. Nos. Six and Seven had a high bit-rate of 16,200 per second and the pictures, which contained more than 3 million bits (which means more detail), were transmitted in minutes.

After the two Mariners completed their mission to Mars, they continued in a solar orbit. As long as the on-board power lasts, their experiments will continue to return data on the inter-planetary medium and on the sun.

The solid material on the Martian surface is of silicate origin. No indications of volcanism were found. The measurements gave the planet a radius of 2,035.8 miles at the equator and 2023.8 at 79° N latitude. Thus Mars is spherical with a slight flattening at the poles.

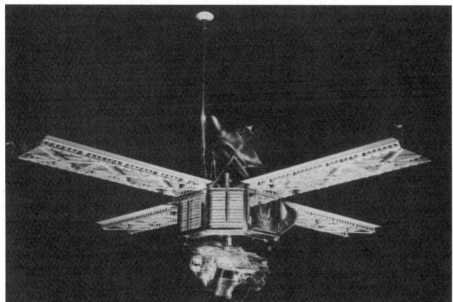

The Mariner spacecraft. The rotor-like panels are covered with solar cells which convert sunlight to electricity.

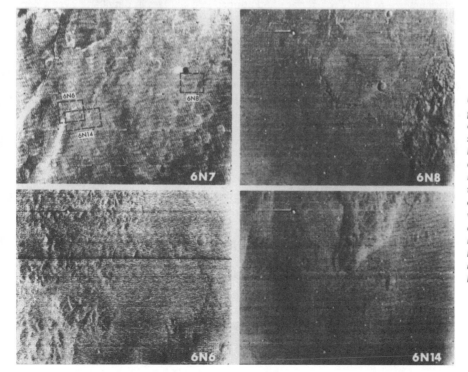

Upper left a wide angle picture of Mars taken from an altitude of 3125 miles shows an area roughly 600 by 940 statute miles. The other three photos are high resolution shots of areas that lie within the area covered in the wide angle view. Lower left is an area of "chaotic" terrain. Lower right shows a slump valley below cratered terrain, and upper right, a slump slope bordering cratered terrain.

MARINER MARS 1971 (8 & 9) PRESS KIT

NATIONAL AERONAUTICS AND SPACE ADMINISTRATION
WASHINGTON, D.C. 20546
TELS. WO 2-4155 - WO 3-6925

FOR RELEASE: FRIDAY A.M.
April 30, 1971
RELEASE NO: 71-75
Phone: 202/962-1176 (Richard T. Mittauer)

PROJECT: MARINER MARS 1971

MARINER MARS 1971 LAUNCHES

The eighth and ninth in the Mariner series of spacecraft are being prepared at Cape Kennedy, FL, for the National Aeronautics and Space Administration's first attempt to orbit another planet — Mars.

Mariner H has been mated to Atlas-Centaur 24 and is awaiting a launch window no earlier than 9:29 p.m. EDT, May 7. Mariner I will be mated to Atlas-Centaur 23 after AC-24 lifts-off, for launch no earlier than 7:57 p.m. EDT, May 17.

Arrival dates at Mars will be Nov. 14 and 24, when each spacecraft will begin a basic mission of 90 days in orbit around the planet. If one or both spacecraft survive that period, an extended mission to last up to one year is being considered.

The objectives of the Mariners are to study the surface and atmosphere of Mars in detail and over a period of time, to provide a broad picture of the history of the planet and natural processes currently shaping the Martian environment.

To accomplish these, one spacecraft will map 70% of the planet and the other will repeatedly study selected areas on Mars to observe changes on the surface and in the atmosphere. Recurring phenomena such as dust storms, clouds and seasonal changes in the appearance of the planet's surface have been observed on Mars. The orbital missions will allow scientists to study these phenomena daily at close range.

The Mariners will carry identical payloads of instruments to allow each spacecraft to conduct six scientific investigations:

- Martian topography and variable features with two television cameras, one with a wide-angle lens and one with a telephoto lens;

- surface temperature measurements with an infrared radiometer;

- composition and structure of the atmosphere with an ultraviolet spectrometer;

- studies of the planet's surface and composition and temperature of its atmosphere with an infrared interferometer spectrometer;

- atmospheric pressure and structure with an S-Band occultation experiment;

- and a more accurate description of Mars' gravity field and the orbits of its two moons, and an improved ephemeris of Mars (its position in its solar orbit at a given time).

The latter two experiments involve measurements of the Mariners' radio signals back to Earth and do not require special instruments on the spacecraft.

The scientific experiments have been teamed together to provide a maximum of correlation of the data they gather. The three instruments on the scan platform, for instance, are bore-sighted with the television cameras so that the photography can be correlated with measurements of the Martian atmospheric and surface characteristics.

The two Mariners will work as a team performing different but complementary missions. Mariner H has been assigned Mission A, a mapping mission. Mariner I has been assigned Mission B, a variable features study. If one spacecraft fails at any time up to five days before orbital insertion, the mission of the second spacecraft will be revised to perform significant parts of both missions.

Mariner H (Mission A, mapping) will orbit Mars once each 12 hours, inclined 80 degrees to the Martian equator, with a 10,000-mile (17,300-kilometer) high point in the orbit (apoapsis) and a 750-mile (1,250-kilometer) low point (periapsis).

Mariner I (Mission B, variable features) will orbit in 20 and one half hours, inclined 50 degrees, with an apoapsis of 20,500 miles (28,600 kilometers) and periapsis of 530 miles (850 kilometers). This orbit will allow the spacecraft to observe selected areas of Mars repeatedly every five days or six orbits.

The spacecraft will each weigh approximately 2,200 pounds (1,000 kilograms) at launch, with about 1,000 pounds (454 kilograms) of fuel for the 300-pound thrust retroengine. After injection into Mars orbit, the spacecraft will weigh approximately 1,200 pounds (544 kilograms).

Orbit insertion will require about a 14-minute burn of the retroengine slowing the spacecraft by about 3,250 miles-per-hour (1,450 meters-per-second). The spacecraft velocity relative to Mars prior to the burn will be about 11,000 mph (4,920 m/sec).

The launches will be direct ascent without a parking orbit. The launch aiming point will be at such a distance from Mars as to insure that neither spacecraft nor the Centaur second stage will impact Mars in the event of loss of control during the launch phases. The orbits of the two spacecraft are designed to guarantee that they will not impact Mars for at least 17 years, to avoid contamination of the planet before studies are conducted on the surface by landing spacecraft.

Following successful injection into solar orbit, two midcourse maneuvers may be performed to correct the trajectory and refine the aiming point. The retroengine will be used for midcourse maneuvers.

The accuracy required to orbit Mars is unprecedented in a flight into deep space. The aiming zone at the end of the 287-million-mile (462-million-kilometer) flight is an area about 435 miles (1,165 kilometers) square.

After insertion into Mars orbit, the spacecraft will be tracked for a sufficient period to determine the orbital corrections (trims) required to yield precise orbits. The trims will be provided by the retroengine.

The maximum data transmission rate will be 16,200 bits-per-second when the spacecraft can transmit to the sensitive 210-foot (64-meter) antenna at the Goldstone station of the Deep Space Network in the California Mojave Desert. Other stations will receive at a maximum rate of 2,025 bits-per-second.

NASA's Office of Space Science and Applications assigned project responsibility including mission operations and tracking and data acquisition to the Jet Propulsion Laboratory managed by the California Institute of Technology. The launch vehicle is the responsibility of the Lewis Research Center, Cleveland. The contractor to Lewis is General Dynamics/Convair, San Diego.

Tracking and communications is assigned to the Deep Space Net operated by JPL for NASA's Office of Tracking and Data Acquisition.

Cost of the basic 90-day Mariner Mars '71 mission is $129 million, exclusive of launch vehicles and data acquisition.

(END OF GENERAL RELEASE; BACKGROUND INFORMATION FOLLOWS)

MISSION CAPSULE

Two Mariner spacecraft to be launched to Mars in May, 1971.

Arrive Mars on November 14 and 24, 1971. Flight time approximately 190-days. Distance traveled, 287 million miles (462 million kilometers).

Mission - to orbit Mars for 90-days.

Spacecraft launch weight 2,200 lbs. (1,000 kg.). Weight in orbit, 1,200 lbs. (544 kg.).

Launch vehicle, Atlas/Centaur. Direct ascent launch.

Experiments:

1. Two television cameras, one narrow angle and one wide angle.

2. Infrared radiometer.

3. Ultraviolet spectrometer.

4. Infrared interferometer spectrometer.

5. Celestial mechanics.

6. Occultation. The latter two experiments do not require special instruments.

MARINER H   Map 70% of Mars at medium resolution and
MISSION A   provide 5% coverage at high resolution.

MARINER I   Study surface features and atmosphere as
MISSION B   they change in time.

MARINER H   10,000 miles (17,256 kilometers) apoapsis to
A ORBIT   750 miles (1,250 kilometers) periapsis. Inclination, 80 degrees. Period is 12 hours.

MARINER I   20,500 miles (28,579 kilometer) apoapsis to
B ORBIT   530 miles (850 kilometer) periapsis. Inclination, 50 degrees. Period is 20½ hours.

MARINER PLANETARY MISSIONS

	'71 MARS ORBITERS	MARINERS 6 & 7 '69 MARS (FLY-BY)	MARINER 5 '67 VENUS (FLY-BY)	MARINER 4 '65 MARS (FLY-BY)	MARINER 2 '62 VENUS (FLY-BY)
Spacecraft Weight pounds (kilograms)	Launch: 2,200 (1,000) In Orbit. 1,200 (544)	847 (384)	539 (245)	575 (261)	447 (203)
Closest Approach to Planet miles (kilograms)	Mission A: 750 (1,200) Mission B: 530 (854)	2,100 (3,380)	2,544 (4,100)	6,118 (9,850)	21,594 (34,700)
Distance from Earth at Target million miles (million kilometers)	MISSION A: Encounter: 76 (122) E+90 Days: 146 (235) MISSION B: 84 (135) 154 (248)	62 (100)	50 (80)	134 (216)	36 (58)
Communication Time (one-way)	Encounter: Approx. 7 Minutes E+90 Days: Approx. 13 Minutes	5 Minutes 22 Seconds	4 Minutes 26 Seconds	12 Minutes 1 Second	3 Minutes 14 Seconds
Time-of-Flight	Mission A: 194 Days Mission B: 192 Days	156-130 Days	127 Days	228 Days	109 Days

MARINER PLANETARY MISSIONS

	'71 MARS ORBITERS	MARINERS 6 & 7 '69 MARS (FLY-BY)	MARINER 5 '67 VENUS (FLY-BY)	MARINER 4 '65 MARS (FLY-BY)	MARINER 2 '62 VENUS (FLY-BY)
Maximum Communication Distance million miles (million kilometers)	154 (248)	250 (402)	82 (132)	216 (348)	54 (87)
Maximum Data Transmission Rates	16,200 bits per second	16,200 bps	33 1/3 bps	33 1/3 bps	33 1/3 bps

NOTE: The Mariner 1 and 3 missions were unsuccessful because of launch vehicle failures.

MARINER AIMING ZONES

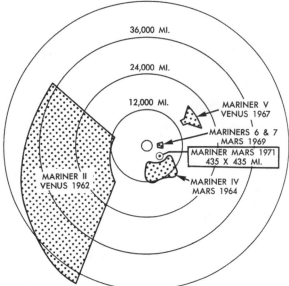

MISSIONS

There are two missions to be performed at Mars each with specific objectives and a specific orbit. Should one Mariner spacecraft fail, the other will combine as many of the objectives of both missions as possible.

Mariner H - Mapping Mission A

The objective of this mission is to map 70 per cent of Mars at a resolution of 3,280 feet (1,000 meters). High resolution photographs at a resolution of 328 feet (100 meters), nesting inside the wide-angle photographs, will cover 5 per cent of the surface.

The orbital period for the mapping mission will be fixed at 12 hours. This will allow twice-daily observations near periapsis each with a subsequent transmission of recorded data at the high bit rate of 16,200 bits per second to the 210-foot (64-meter) antenna of the Deep Space Network at Goldstone.

Following insertion into orbit, the spacecraft will be tracked for a sufficient period to allow calculation of required trimming of the orbit to secure the precise orbit desired.

In a typical orbit the spacecraft will:

1. Playback data in its tape recorder (this requires three hours) taken on the previous orbit,

2. Take TV, infrared and ultraviolet data for storage on the tape recorder (IR and UV data is also transmitted in real time at 8,000 bits per second when the large Goldstone antenna can be used);

3. Enter the occultation zone shortly after the recorder is filled;

4. Following occultation, the spacecraft will still be in view of Goldstone and the recorded data is transmitted.

On the next orbit, Goldstone is not in view. The recorder is filled again and the data held until the following orbit.

MARINER FLIGHT PATH TO MARS

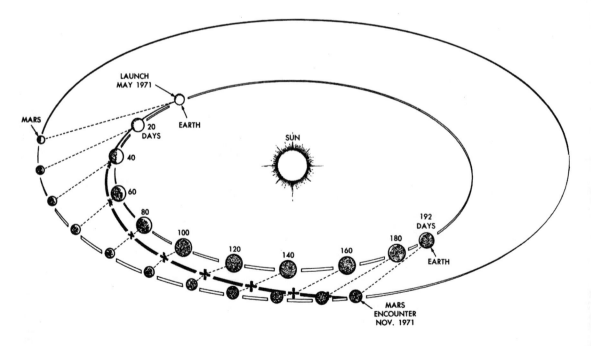

Because the orbital period is 12 hours and the rotation rate of Mars is about 24 hours 37 minutes, the planet will rotate slightly less than 180 degrees per spacecraft orbit. Mapping coverage, therefore, will alternate from one side of the planet to the other. A complete mapping circuit around Mars will be completed in about 20 days. After 90 days, the area between -60 and +40 degrees latitude will have been covered.

Areas of specific interest can be re-examined every 20 days with the full array of instruments allowing opportunity for adaptive operations, making use of previous data to plan operations.

After 90 to 100 days in orbit, depending on the actual performance of the high-rate channel, the increasing Earth-Mars distance and the spacecraft antenna orientation will cause the tape playback rate to be dropped to 8,100 bits per second, even with the use of the 210-foot (64-meter) antenna. At that time, either a full tape recorder of data will be taken on every other orbit or half the amount of data will be taken on each orbit.

Several days later data rates will have to be further reduced, as the main lobe of the spacecraft antenna beam moves away from Earth, until data rates of 4,000 bits or 2,000 bits per second stabilize for extended operations.

Mariner I - Variable Features Mission B

The objective of this mission is to study changes on the surface and in the atmosphere of Mars over a period of time.

The period of the orbit, 20.5 hours, was selected to provide repeated studies of a series of six adjacent sites girdling the planet in five days.

The relationship between the spacecraft orbital period of 20½ hours and Mar's rotational period of 24 hours 37 minutes, places the cameras over the six adjacent areas in six orbits. Repeat coverage begins with the seventh orbit at the beginning of the sixth day.

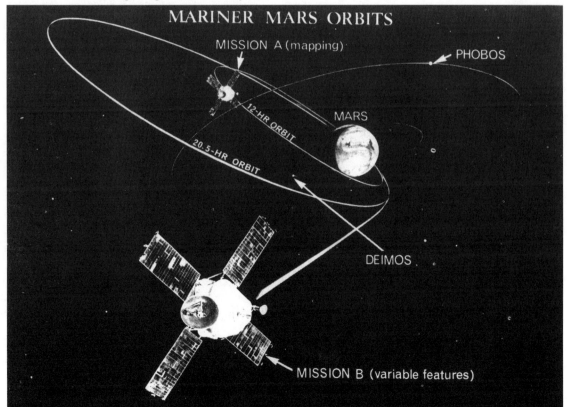

The time relationship is achieved by selecting a spacecraft velocity resulting in an orbit with a high point of 20,000 miles (28,579 kilometers) and a low point of 530 miles (850 kilometers).

A typical orbit will provide full disc coverage, high Sun angle photography for albedo studies, low and high resolution studies by the infrared and ultraviolet instruments, Earth occultation, and dark side UV and IR data.

The two moons of Mars, Phobos and Deimos, may be mapped by the television experiment and probed by the infrared and ultraviolet experiments.

Observations of starfields and Saturn also may be used to calibrate the cameras.

It is planned, when possible, to use an adaptive mode in planning science operations from orbit. In this mode, an interesting feature seen in previous photographs can be specifically re-examined in subsequent passes that cover that area. It is also possible, depending, on the amount of fuel remaining after attaining orbit, to change a spacecraft orbit to repeatedly cover an area of unusual activity or significance.

MARINER SPACECRAFT

The Mariner Mars 1971 spacecraft were designed, assembled and tested by the Jet Propulsion Laboratory, Pasadena. Industrial contractors provided the detailed design and fabrication of most of the subsystems. Component parts were provided by hundreds of manufacturers and suppliers.

Design of the spacecraft is based on continuing Mariner technology and on the 1969 Mariner 6 and 7 spacecraft in particular.

Mariner's basic structure is a 40-pound (18½ kilogram) eight-sided forged magnesium framework with eight electronics compartments. The compartments themselves provide structural support to the spacecraft.

Four solar panels, each 84½ inches (2.14 meters) long and 35½ inches (0.902 meters) wide, are attached by outrigger structures to the top or sunward side of the octagon. Each panel has a solar cell area 20.7 feet (1.92 meters) square, or a total cell surface of approximately 83 square feet (7.8 square meters) for each spacecraft.

Two sets of attitude control jets consisting of six jets each, which stabilize the spacecraft on three axes, are mounted at the tips of the four solar panels. Titanium bottles containing the nitrogen gas supply for Mariner's dual attitude control gas system and regulators for the systems are mounted on the top ring of the octagon.

Two spherical propellant tanks for the liquid-fueled rocket engine are mounted side by side atop the octagonal structure with the rocket nozzle protruding between the tanks. The titanium tanks are 30 inches (0.762 meters) in diameter.

The two-position high-gain antenna is attached to the spacecraft by a superstructure atop the octagon. Its aluminum honeycomb dish reflector is circular, 40 inches (1.02 meters) in diameter, and is parabolic in cross-section. The antenna feed is supported at the focus of the parabola by a fiberglass truss. The two-position capability allows the Earth to be centered in the antenna beam during both pre-orbit and the orbital periods of the mission.

The low-gain omnidirectional antenna is mounted at the top of a circular aluminum tube, four inches (10.2 centimeters) in diameter and extending vertically 57 inches (1.45 meters) from the top of the octagonal structure. The tube acts as a waveguide for the antenna.

A horn-shaped medium-gain antenna is mounted on a solar panel outrigger so that it is pointed toward Earth during spacecraft insertion into Mars orbit.

The Canopus star tracker assembly is located on the upper ring structure of the octagon for a clear field of

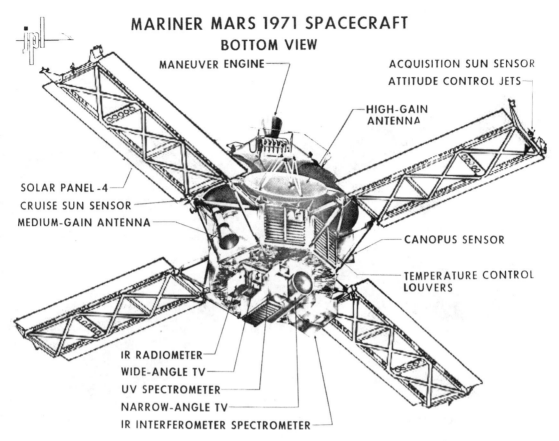

MARINER MARS 1971 SPACECRAFT
BOTTOM VIEW

MANEUVER ENGINE

ACQUISITION SUN SENSOR
ATTITUDE CONTROL JETS

HIGH-GAIN ANTENNA

SOLAR PANEL -4
CRUISE SUN SENSOR
MEDIUM-GAIN ANTENNA

CANOPUS SENSOR

TEMPERATURE CONTROL LOUVERS

IR RADIOMETER
WIDE-ANGLE TV
UV SPECTROMETER
NARROW-ANGLE TV
IR INTERFEROMETER SPECTROMETER

NOTE: PROPULSION MODULE AND SCAN PLATFORM INSULATION BLANKETS NOT SHOWN

view between two solar panels. The cruise Sun sensor and Sun gate are attached to a solar panel outrigger.

The eight electronics compartments girdling the spacecraft house the following: Bay 1, power regulators; Bay 2, power conversion equipment, scan control and IRIS electronics; Bay 3, Central Computer and Sequencer and attitude control sub-system; Bay 4, flight telemetry and command subsystems; Bay 5, tape recorder; Bay 6, radio receiver and transmitters; Bay 7, television electronics and data automation subsystem; Bay 8, spacecraft battery.

Six of the electronics compartments are temperature controlled by lightweight louver assemblies on the outer surfaces. The octagon's interior is insulated by multilayer fabric thermal blankets at both top and bottom of the structure.

To avoid puncture of the propellant tanks by micrometeoroids, the outer layer of the top thermal blanket is constructed of a tightly-woven fiberglass cloth (Armalon) designed to break up striking particles.

The Mariners will carry science instrumentation for four planetary experiments. Two additional experiments — spacecraft occultation by Mars and celestial mechanics — require only the spacecraft communications system as the source of their data.

Two television cameras, an infrared interferometer spectrometer (IRIS), ultraviolet spectrometer (UVS) and infrared radiometer (IRR) are mounted on a motor-driven two-degree-of-freedom scan platform on the bottom or shaded side of the octagon. Total rotating weight of the platform mechanism and its science instrument payload is 181 pounds (82.2 kilograms).

Each Mariner weighs 1,200 pounds (544 kilograms) unfueled and measures 7½ feet (2.29 meters) from the scan platform to the top of the low-gain antenna and rocket nozzle. With solar panels deployed, the

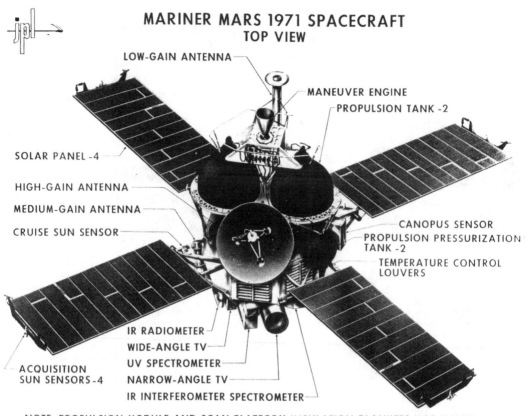

MARINER MARS 1971 SPACECRAFT
TOP VIEW

LOW-GAIN ANTENNA

MANEUVER ENGINE

PROPULSION TANK -2

SOLAR PANEL -4

HIGH-GAIN ANTENNA

MEDIUM-GAIN ANTENNA

CRUISE SUN SENSOR

CANOPUS SENSOR

PROPULSION PRESSURIZATION TANK -2

TEMPERATURE CONTROL LOUVERS

IR RADIOMETER

WIDE-ANGLE TV

UV SPECTROMETER

ACQUISITION SUN SENSORS -4

NARROW-ANGLE TV

IR INTERFEROMETER SPECTROMETER

NOTE: PROPULSION MODULE AND SCAN PLATFORM INSULATION BLANKETS NOT SHOWN

spacecraft spans 22 feet, 7½ inches (4.35 meters). The octagonal structure measures 54½ inches (1.39 meters) diagonally and 18 inches (0.457 meters) in depth.

Launch weight, including 1,000 pounds (454 kilograms) of rocket fuel and oxidizer, is more than 2,200 pounds (1,000 kilograms).

Data Automation Subsystem

The five science instruments on the spacecraft are controlled and synchronized by the Data Automation Sub-system (DAS) and data from the instruments are converted by the DAS into digital form for transmittal to Earth.

The experiments controlled by the DAS are television, infrared radiometer, infrared interferometer spectrometer and ultraviolet spectrometer. The S-band occultation experiment and the celestial mechanics experiment do not require special equipment aboard the spacecraft and are not controlled by the DAS.

The DAS controls instrument sequencing, accumulates varied science data, reduces them to a common digital form and common rate and then feeds the data to the tape recorder or to the selected science telemetry channel at proper intervals for transmission to Earth.

Attitude Control

Stabilization of the spacecraft is provided by a system of 12 cold gas jets mounted at the outer ends of the four solar panels. The jets are linked by logic circuitry to three gyroscopes (one gyro for each of the spacecraft's three axes), to the Canopus sensor and Sun sensors.

The gas system is divided into two sets of six jets, each set complete with its own gas supply, regulators, lines and valves so that a leak or valve failure will not deplete the gas and jeopardize the mission. Each system is fed by a titanium bottle containing nitrogen gas pressurized at 2,500 pounds-per-square-inch (175.7 kilograms-per-square-centimeter). Normally, both sets will operate during the mission. Either system can support the entire flight in the event of a failure in the other.

The Sun sensors are light-sensitive diodes which inform the attitude control system when they see the Sun. The attitude control system responds to these signals by turning the spacecraft and pointing the solar panels toward the Sun for stabilization on two axes and for conversion of solar energy to spacecraft power. Nitrogen gas escapes through the appropriate jet nozzle, imparting a reaction to the spacecraft to correct its angular position.

The star Canopus, one of the brightest in the galaxy, will provide a second celestial reference upon which to base maneuvers, point the high gain antenna toward Earth and the instruments toward Mars. The Canopus sensor will activate the gas jets to roll the spacecraft about the already-fixed longitudinal or roll axis until it is locked in cruise position. Canopus acquisition occurs when the light intensity in the field of view of the sensor matches the intensity anticipated for the star Canopus. Brightness of the sensor's target star will be telemetered to the ground to verify the correct star has been acquired.

Periodically during the flight, the Canopus sensor will be updated to compensate for the changing angular relationship between the spacecraft and the star. The sensor's field of view or "look angle" will be changed electronically to follow Canopus throughout the mission.

Upon receipt of commands from the Central Computer and Sequencer (CC&S), the attitude control system orients the spacecraft to align the thrust axis of the rocket engine in the direction required for the trajectory correction maneuver.

During firing of the engine, stabilization of the spacecraft will be effected by gimballing the engine so that thrust direction remains through the spacecraft's center of gravity. The Mariner's autopilot controls spacecraft attitude during engine firing by using the gyros to sense motion about the spacecraft's three axes for gimballing the engine.

Propulsion

Mariner's rocket engine provides small trajectory corrections to the spacecraft during Earth-to-Mars transit, deceleration into Mars orbit and trim maneuvers to achieve the desired periapsis and orbital period. The engine, capable of at least five starts and shutdowns, provides a continuous thrust of 300 pounds (0.36 kilograms).

The rocket nozzle protrudes from the top of the spacecraft along its longitudinal or roll axis.

The fuel, monomethyl hydrazine, and the oxidizer, nitrogen tetroxide, are contained in two 30-inch (0.762 meters) diameter pressure vessels. The propellants are forced into the engine's combustion chamber by nitrogen gas compressing bladders in the spherical tanks. Hypergolic ignition of the fuel and oxidizer causes rapid expansion of hot gases in the engine.

Firing of the engine is controlled by the CC&S, which receives the time, direction and duration of required thrust through the ground-to-spacecraft communication link. At the command signal from the CC&S, solenoid-actuated engine valves allow the propellants to enter the thrust chamber from the already pressurized tanks. For termination of thrust, an accelerometer activates the solenoid, stopping propellant flow.

During engine firing, spacecraft attitude is maintained by autopilot-controlled gimballing of the engine.

Both Mariners may perform two midcourse maneuvers — the first about six days after launch and the second as late as 10 days prior to encounter with Mars.

Insertion into orbit around Mars will be achieved with a long — 14-minute — engine burn to decelerate the spacecraft. Spacecraft velocity will be decreased by about 3,250 miles per hour (1,450 meters per second).

Each spacecraft will execute one or two trim maneuvers while in orbit around Mars. Additional trim maneuvers are possible.

Each Mariner will carry more than 1,000 pounds (454 kilograms) of propellants, about 45 per cent of the total spacecraft launch weight.

Central Computer and Sequencer

Mariner is designed to operate throughout its basic mission without the need of ground commands — with the exception of spacecraft maneuvers. This automatic capability is made possible by the on-board command function of the Central Computer and Sequencer (CC&S). Critical events, however, are backed up by the ground command capability.

The CC&S performs the timing, sequencing and computations for other subsystems aboard the spacecraft. It initiates spacecraft events in six mission sequences — launch, cruise, midcourse maneuver, orbit insertion maneuver, orbit trim maneuvers and orbital science.

Timing and sequencing are programmed into the CC&S prior to launch but can be modified anytime during the flight by command from the ground.

The CC&S consists of a special purpose programmable computer and a fixed sequencer for redundancy during maneuvers.

Under normal circumstances, the programmable computer portion and the fixed sequencer portion of the CC&S operate in tandem for the midcourse maneuver and orbit trim maneuvers. If there is disagreement on any maneuver event the maneuver is aborted and the spacecraft returns to the cruise condition. A maneuver also can be performed by either portion of the CC&S alone.

The fixed sequencer will be prime for the maneuver that puts Mariner into orbit around Mars. Selected backup functions will be provided by the programmable computer.

Communications

Two-way communications with the Mariners will be by radio link between Earth tracking stations and a dual transmitter-single receiver radio system aboard each spacecraft.

The on-board communications system also includes a telemetry subsystem, command subsystem, data storage subsystem and high-gain, low-gain and medium-gain antennas.

The spacecraft S-band receiver will operate continuously during the mission at about 2,210 megahertz. (The receivers in the two Mariners will operate at slightly different frequencies. Similarly, no two transmitters will operate at exactly the same frequency.) The receiver will be used with the high-gain antenna only or a combination of the low-gain omnidirectional antenna and the medium-gain antenna. The spacecraft receives uplink command and ranging signals from ground stations of the Deep Space Network. To provide the standard doppler tracking data, the radio signal transmitted from Earth is received at the spacecraft, changed in frequency by a known ratio and re-transmitted to Earth. In addition, a JPL-developed ranging technique using an automatic coded signal provides range measurements with an accuracy of a few yards at the Mars-Earth distance. The ranging function may be commanded on and off by ground command, or off by the Central Computer and Sequencer.

When no uplink signal is being received by Mariner, the transmitted frequency of about 2,295 megahertz originates in the spacecraft transmitter. The transmitter consists of two redundant exciters and two redundant radio frequency power amplifiers of which any combination is possible. Only one exciter-amplifier combination will operate at any one time. Selection of the combination will be by on-board failure detection logic with ground command backup.

Both amplifiers on each spacecraft employ traveling wave tubes and are capable of operating at 10 watts or 20 watts output and the signal may be transmitted through either the high-gain or low-gain antenna. Transmission via the high-gain antenna will be required during the encounter and playback phases of the mission.

The high-gain antenna, with a 40-inch-diameter (1.02 meters) parabolic reflector, provides a highly directional beam for the downlink radio signal. The high-gain antenna has two positions with respect to the spacecraft. It is deployed to the second position during orbit to enhance communications during the orbital phase. The low-gain antenna provides essentially uniform coverage in the direction of Earth. The medium-gain antenna, coupled to the low-gain is used to provide telemetry to Earth during the maneuver into Mars orbit.

All communications between the Mariners and Earth will be in digital form. Command signals transmitted to the spacecraft will be decoded — translated from a binary form into electrical impulses — in the command subsystem and routed to their proper destination.

Three types of commands are transmitted to the spacecraft: a direct command (DC) results in the closure of a switch in one of the spacecraft subsystems; a coded command (CC) provides information to the Central Computer and Sequencer for maneuvers or to update the CC&S program; a quantitative command (QC) is used to position the scan platform. A coded command to the Data Automation System allows selection of TV camera filters, shutter speeds and other science instrument options. There are 82 possible DC's which back up all critical automatic spacecraft functions, choose redundant elements, initiate maneuvers and perform other functions.

Data telemetered from the spacecraft will consist of engineering and science measurements prepared for transmission by the Telemetry Subsystem, the Data Automation Subsystem (real-time TV and science) and Data Storage Subsystem (recorded science including TV). The encoded information will indicate voltages, pressures, temperatures and other values measured by the spacecraft telemetry sensors and science instruments.

There are three data channels:
- the engineering channel which operates throughout the flight;
- the low-rate science channel employed during the orbital phase of the mission;
- and the high-rate science channel.

Mariner can transmit information to Earth at eight different rates: on the engineering channel at 8 1/3 bits per second and 33 1/3 bps; on the science channel at 50 bps; and on the high-rate science channel at 16,200 bps and 8,100 bps. Data storage playback rates are 16,200 bps, 8,100 bps, 4,050 bps, 2,025 bps and 1,012.5 bps.

Certain conditions must exist in order to utilize the high-rate channel. These include the availability of the 210-foot (64-meter) diameter antenna at the Goldstone Complex of the Deep Space Network (DSN) for receiving. The DSN's 85-foot (26-meter) antennas can receive data at rates up to 2,025 bps.

Approximately 90 engineering measurements are obtained by transducers throughout the spacecraft to make up the engineering data. The engineering samples are taken continuously and can be transmitted along with science regardless of the science channel or rate in use.

The Data Storage Subsystem (tape recorder) records digital science data from the Data Automation System during the orbital phase at 132,300 bits per second until the eight tape tracks are filled. Data consists of digitized video from the TV cameras and the other science instruments. Total storage capacity for each spacecraft is 180 million bits.

Playback is initiated by either on-board or ground command. One of the five playback data rates — from 1,012.5 bits per second to 16,200 bps — is selected depending upon the availability of the 210-foot (64-meter) Goldstone antenna and the Earth-Mars distance.

Power

The Mariner power subsystem supplies electrical power to the spacecraft, switches and controls the power and provides an accurate timing source for the spacecraft. Primary power source is an arrangement of 17,472 photovoltaic solar cells mounted on four panels which will face the Sun during most of the flight to Mars and during orbital operations. The cells, covering 83 square feet (7.8 square meters), will collect solar energy and convert it into electrical power.

A rechargeable nickel-cadmium battery provides spacecraft power during launch and whenever the panels are turned away from the Sun. The battery will be kept in a state of full charge and will be available as an emergency power backup source.

Two power regulators will provide redundancy. In the event of a failure in one, it will be removed automatically from the line and the second will be switched in to assume the full load.

The solar panels will be folded in a near vertical position above the body of the spacecraft during launch and will be deployed after separation from the launch vehicle. Each panel carries 4,368 solar cells (2 x 2 cm.) and protective glass filters that reduce the amount of solar radiation absorbed without interfering with the energy conversion. The cell modules are supported by lightweight panel structures made of thin-gauge aluminum.

Nominal power capability of the panels is expected to be 800 watts at maximum power voltage for cruise conditions in space near Earth. This power capability decreases to about 500 watts at the Mars distance if there is no degradation because of solar flares. Maximum power demand is expected to be less than 400 watts during orbital operations.

Minimum capacity of the spacecraft battery is 600 watt hours. The battery will be capable of delivering its required capacity and meeting all electrical requirements within an operational temperature range of 30 to 90 degrees Fahrenheit (10 to 32 degrees Celsius). At temperatures outside this range, it will still function although its capability will be reduced.

To ensure maximum reliability, the power subsystem was designed to limit the need for battery power after initial Sun acquisition. Except during maneuvers, the battery will remain idle and fully charged. The battery charger will provide a high-rate or a trickle charge.

Under normal flight conditions, the primary power booster-regulator will handle all spacecraft loads. A second regulator will support power loads on a stand-by basis. Should an out-of-tolerance voltage condition exist in the main regulator, the stand-by regulator will take its place on the line.

Primary form of power distributed to other spacecraft systems is 2,400 hertz square wave. The gyro spin motors use 400 hertz three-phase current, and the scan motor is supplied with 400 hertz single-phase current. The transmitter amplifier tube, battery chargers and temperature control heaters use unregulated dc power from the solar panels or the battery.

A crystal oscillator in the main power inverter controls the frequency to within 0.01 per cent, assuring other spacecraft systems of a reliable, accurate frequency on their power line. A backup crystal oscillator is located in the stand-by inverter. The spacecraft Central Computer and Sequencer uses the oscillator frequency as a timing source.

Telemetry measurements have been selected to provide the necessary information for the management of spacecraft power loads by ground command if necessary.

The battery, regulators and power distribution equipment are housed in two adjacent electronics compartments on Mariner's octagonal base.

Temperature Control

For a spacecraft traveling to Mars, away from Earth and from the Sun, the primary temperature control problem is maintaining temperatures within allowable limits despite the decreasing solar intensity as the mission progresses. In airless space, the temperature differential between the sunlit side and the shaded side of an object can be several hundred degrees.

Heating by direct sunlight on the Mariner spacecraft is minimized by the use of a thermal blanket on its Sun side. The side away from the Sun is covered with a thermal blanket to prevent rapid loss of heat to the cold of space.

The top of Mariner's basic octagon and the propulsion module are insulated from the Sun by a multilayered shield of aluminized Teflon. The outer layer of the shield, constructed of a tightly-woven fiberglass cloth, also serves as a foil against micrometeoroids. The bottom is enclosed by another multilayered blanket to retain heat generated by power consumption within the spacecraft.

Temperature control of six of the electronics compartments is provided by polished metal louvers actuated by coiled bimetallic strips. The strips act as spiral-wound springs that expand and contract as they heat and cool. This mechanical action, which opens and closes the louvers, is calibrated to provide an operating range from fully closed at 55 degrees Fahrenheit (13 degrees Celsius) to fully open at 90 degrees Fahrenheit (32 degrees Celsius). A louver assembly consists of 22 horizontal louvers driven in pairs by 11 actuators. Each pair operates independently on its own local temperature determined by internal power dissipation.

The science platform and its array of instruments at the bottom of the octagon are covered by a third thermal blanket. The platform is designed to be thermally isolated from the main equipment octagon by a plastic collar on the attaching support tube. Temperature control is achieved by electrical dissipation in heaters and in the instruments themselves.

Electric heaters are located within the science platform blanket, in the propulsion module and in two of the electronics bays to provide additional heat during certain portions of the mission.

Paint patterns and polished metal surfaces are used on the Mariner for passive control of temperatures outside of the protected octagon and covered science platform. These surfaces control both the amount of heat dissipated into space and the amount of solar heat absorbed or reflected away. The patterns were determined from testing a Temperature Control Model (TCM) of the spacecraft in a space simulation chamber at JPL and from the application of actual mission data acquired during the Mariner 4 (1964-65) and the Mariner 6 and 7 (1969) missions to Mars.

The high-gain antenna dish, which is dependent upon the Sun for its surface heat, is painted green to keep it at near room temperature at the Mars distance but within its upper thermal limit earlier in the mission.

Scan Platform

Mariner's science instruments are mounted on a scan platform which can be rotated about two axes to point the instruments toward Mars during the spacecraft's approach to the planet and while in orbit. The platform is located below the octagonal base of the spacecraft.

The scan control system allows multiple pointing directions of the instruments as the orbit phase of the mission progresses. The platform's two axes of rotation are described as the clock angle motion about the axis of the tube extending vertically from the octagon and cone angle motion about an axis which is horizontal.

The platform is motor-driven and moves 215 degrees in clock and 69 degrees in cone.

The scan pointing positions are programmed in flight and during the orbital sequence by commands from the Central Computer and Sequencer or radio commands from Earth.

As Mariner approaches Mars — several days prior to going into orbit around the planet — the narrow-angle TV camera will take a series of spaced "picture pairs" of the full disc of Mars. The pre-orbit TV sequence, enabled by a command from the CC&S begins when the planet is sufficiently bright and within the field-of-view of the camera.

During the orbital phase of the mission, scan platform pointing can be updated each orbit by ground command as scientists evaluate data from previous orbits.

SCIENTIFIC EXPERIMENTS

Mars is a constantly changing world with seasonal and daily variations that have been observed with difficulty from Earth and only briefly by flyby spacecraft.

In this orbiting mission a battery of instruments will probe the planet on a daily basis for three months and possibly longer.

The surface will be examined by photography and in the infrared wavelengths. The atmosphere will be examined in the ultraviolet and infrared and by the occultation experiment.

From the top of the thin Martian gas envelope down to the surface and the interior of craters, the atmospheric instruments will record data to identify gases, plot the mixture of constituents and variations relative to time and area.

The instruments will view an early winter atmosphere and surface in the North and early summer in the South.

The information gathered should provide a broad picture of the factors that shape physical processes at work on Mars. The questions to be answered range from daily weather patterns to the history of the formation of the planet.

The cutting edge of this mission is not only the opportunity to examine Mars in detail on a daily basis but is also the carefully planned correlation of data from the instruments to yield more than the sum total of the parts.

For example, the ultraviolet experiment and the two infrared experiments are boresighted with the two TV cameras. The ultraviolet spectrometer in probing the atmosphere, provides an elevation profile of the surface under the UV scan path. This in turn, can be correlated with photographs and with occultation pressure data as was the case in the 1969 flight in the vicinity of the feature Hellas and the adjacent area of Hellespontus.

A photograph can be correlated with temperatures on the surface and the pressure and constituents of the atmosphere above the area photographed. A physical feature on the surface can thus be related to other data obtained.

Past observations have established the presence of clouds, hazes, bright spots and flares of light on the surface. Yellow clouds, thought to be dust, can grow large enough to obscure a large portion of the face of Mars and last a month or two. White clouds range from a haze, lasting a few hours, to huge dense clouds persisting for days or weeks. Dark gray clouds have been reported and thought to possibly be volcanic in origin. Four were reported in 1950 and 1952. Flares seen on the surface have also been attributed to possible volcanic activity.

Study of these phenomena by the combined instrument package is an objective of the mission.

Another specific scientific objective is a study of the "wave of darkening". Observations from Earth have established that there is a seasonal darkening of features on Mars but if it progresses at a regular rate in a waveform is today, open to question.

The spacecraft will arrive at Mars at the peak of the darkening period in the southern hemisphere. It will be observed at its maximum intensity and can be compared with observations in the northern hemisphere.

The water content of the Martian atmosphere is known to be extremely low, similar to the dry Antarctic. But it is possible that free water was frozen in the past and remains under the surface like permafrost. Heat escaping from the interior of Mars could melt the ice and provide a water source for organisms. If so, such an area would be a prime target for the Viking lander that will seek evidence of life on Mars in 1976.

In the low pressure of the Martian atmosphere, however, water can only exist in a frozen or vaporous state with perhaps a short-lived intermediate state that could moisten the soil. In the event of underground ice melting there is, then, a possibility of a cloud forming over the area. Photography of such a cloud correlated with water vapor and temperature measurements could indicate an area acceptable for life forms.

The mapping of Mars in this mission is a basic objective. It is essential in the study of a planet to establish a three dimensional shape of the planet, the figure. A persistent discrepancy, however, exists between optical observations from Earth and data derived from the orbits of the two moons of Mars and spacecraft flyby trajectories.

Studies of the '71 data may resolve that question and will establish the Martian geoid, a standard spherical reference surface for mapping. On Earth, the geoid coincides with mean sea level in the oceans.

Discrepancies of five to 10 degrees in latitude and longitude, 180 to 380 miles (290 to 610 kilometers) at the surface, are not uncommon between various published maps of Mars. A recently completed 10-year Mars Map Project, by de Vaucouleurs, using all Earth-based visual and photographic data from 1877 to 1958 may have reduced errors to about one degree, or 31 miles (50 kilometers) in regions where well defined surface markings (i.e., albedo variations) are available. One of the major mapping applications of the wide-angle camera photography will be to precisely locate surface markings to a resolution of approximately one mile (1.6 kilometers).

Television

The television experiment will provide for Mission A, a map of 70 per cent of Mars at medium resolution and detailed studies of 5 per cent of the surface at high resolution. Mariner 4 in 1965 photographed 1 per cent of the surface, and Mariners 6 and 7 in 1969, 10 per cent.

The wide-angle cameras will resolve features on the surface about 3,280 feet (1,000 meters) in length. Narrow-angle cameras will resolve features about 328 feet (100 meters) in length.

The advantages of an orbiting mission will be exploited for Mission B in repeated photographic studies of individual areas over a period of time to detect changes on the surface and in the atmosphere.

Specific areas of study are the "wave of darkening" (seasonal surface color change), polar caps and the polar cap-edge, atmospheric and surface fluorescence, atmospheric haze, blue clearings, white and yellow clouds and cloud movements.

Although not a primary objective, the two moons of Mars, Phobos and Deimos, may be photographed to obtain information on their shape, size and surface features. Prelaunch studies indicate that the spacecraft will come within sufficient proximity of the moons to obtain surface data.

MARINER NARROW ANGLE TV CAMERA

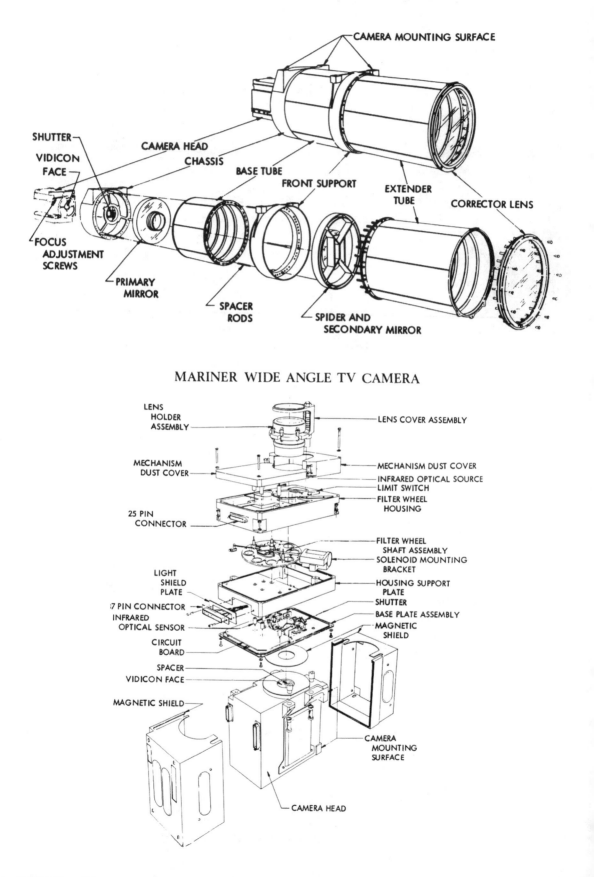

CAMERA MOUNTING SURFACE

SHUTTER
VIDICON FACE
CAMERA HEAD
CHASSIS
BASE TUBE
FRONT SUPPORT
EXTENDER TUBE
CORRECTOR LENS
FOCUS ADJUSTMENT SCREWS
PRIMARY MIRROR
SPACER RODS
SPIDER AND SECONDARY MIRROR

MARINER WIDE ANGLE TV CAMERA

LENS HOLDER ASSEMBLY
LENS COVER ASSEMBLY
MECHANISM DUST COVER
MECHANISM DUST COVER
INFRARED OPTICAL SOURCE
LIMIT SWITCH
FILTER WHEEL HOUSING
25 PIN CONNECTOR
FILTER WHEEL SHAFT ASSEMBLY
SOLENOID MOUNTING BRACKET
LIGHT SHIELD PLATE
HOUSING SUPPORT PLATE
7 PIN CONNECTOR
INFRARED OPTICAL SENSOR
SHUTTER
BASE PLATE ASSEMBLY
MAGNETIC SHIELD
CIRCUIT BOARD
SPACER
VIDICON FACE
MAGNETIC SHIELD
CAMERA MOUNTING SURFACE
CAMERA HEAD

Detection of any life forms on Mars is beyond the resolution capabilities of the cameras. However, correlation of the photographs with other data may yield information on the possibility of life forms on Mars or on suitability of Mars as a habitat for life.

Prior to insertion into orbit, the narrow cameras will provide approach photographs of Mars similar to the series taken in 1969 by Mariners 6 and 7. The sequences will begin at ranges in excess of one million miles (1,609,000 kilometers) from Mars.

As the '71 flights will arrive at Mars in a different season than the '69 flights, the approach pictures will provide project scientists with valuable data for planning the orbital operations and scientific data on changes that have occurred since 1969.

It will be early winter in Mars' northern hemisphere on arrival and early summer in the southern hemisphere. This is similar to telescope observations made in 1958 at Mars opposition.

The TV experiment on Mariner H (Mission A) will, in mapping the surface, provide a wide range of information for studies of bright and dark regions, fine structure of topographic details, and such processes as volcanism, tectonism, impact, erosion and atmospheric activity.

The objectives also include:

- A determination of the shape of the planet. (Data from trajectories of flyby spacecraft and the orbits of the two Martian moons do not agree with Earth-based observation.

- High precision geodetic coordinates of a large number of well defined topographic features for maps of the planet.

- To investigate, by photometric and photogrammetric analysis, surface slopes and relative elevation characteristics; to determine surface brightness and albedo differences; and to perform analyses related to improving the accuracy of the photometric function of Mars.

It is evident from Earth-based radar data that albedo (amount and spectrum of reflected light) and elevation differences on Mars do not have a close correlation. Detailed studies of this fact may have great significance for an interpretation of observed albedo variations.

Mariner I (Mission B) will study time-variable features on the surface and in the atmosphere of Mars in order to obtain information on the atmospheric structure and circulation, detail of diurnal and seasonal changes, and clues regarding the possibility of life on Mars. The specific areas to be studied are wave of darkening, polar caps and cap adjacent areas, nightside atmospheric and surface fluorescence, haze in the atmosphere, white clouds and patches at low latitudes, yellow clouds and storms and the Martian satellites to obtain information on their shape and surface features.

Each wide-angle camera is equipped with commandable filters allowing a choice of eight filters.

The team leader is Harold Mazursky, U. S. Geological Survey, Flagstaff, AZ. Other principal investigators are Bradford Smith, New Mexico State University; Dr. Joshua Lederberg, Stanford University; Dr. Gerard de Vaucouleurs, University of Texas; and Dr. Geoffrey A. Briggs, Bellcom.

Infrared Radiometer (IRR)

This experiment will provide temperature measurements of the surface of Mars by detection of thermal radiation in the infrared portion of the electromagnetic spectrum.

The instrument is boresighted with the television cameras to allow correlation of surface temperatures with terrain features and clouds.

MARINER INFRARED RADIOMETER
(TO MEASURE MARS SURFACE TEMPERATURES)

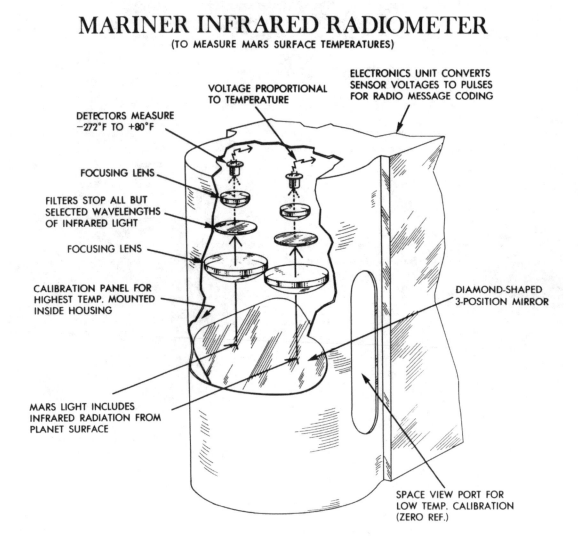

The thermophysical properties of the surface will be mapped during the length of the standard missions. Irregularities in diurnal cooling curves will be compared to photographs for similarity of areas to visible albedo variation. Existence of "hot spots", which are slightly high average temperature areas, will be studied for indications of internal heat sources.

The infrared radiometers on each Mariner '69 mission measured temperatures over broad areas of the planetary surface on a single sweeping pass. The temperatures, therefore, as a function of local time, could only be established in a mean sense.

Observations from Mariner '71 will provide cooling curves characteristic of individual areas; thus, the degree of correlation that may exist between characteristic cooling curves and major physiographic features can be found.

Essentially the same instrument flown on this mission was flown on the 1969 mission to Mars but viewed only one per cent of the surface. In the orbiting mission it is expected that as much as 20 per cent of the surface will be covered.

The wider coverage is expected to determine accurately the thermophysical properties of surface features.

The principal investigator is Dr. Gerry Neugebauer of the California Institute of Technology.

Ultraviolet Spectrometer

The objectives of this experiment are to map the surface and lower atmosphere in ultraviolet (ultraviolet cartography) and to study the temperature and structure in the upper atmosphere (ultraviolet aeronomy).

The experiment will measure: local atmospheric pressures which can be related to elevations over a major part of Mars; local concentrations of ozone; reflectability of the Martian surface in the near-ultraviolet and provide UV studies of the "wave of darkening" clouds and the blue haze-blue clearing.

Detection of ozone could yield clues as to the possibility of life forms on Mars. Biological activities can produce molecular oxygen which can lead to the formation of ozone. If the measurement of ozone is at high enough levels, then the possibility exists that life may be present. This level would have to exceed what could be explained by other non-biological processes. Final determination of the life question, however, will depend on a landed system with measurements taken at the surface.

The UV experiment will also yield data on atmospheric density, temperatures relative to altitude in the upper atmosphere and the amount of UV which strikes the surface of Mars. UV is deadly to life forms. The amount reaching the surface of Mars could have a bearing on the question of life on Mars, although shielding against UV would be relatively simple. The variation of atmospheric constituents with season will be also detectable as the mission progresses.

Principal investigator for this experiment is Dr. Charles Barth of the University of Colorado.

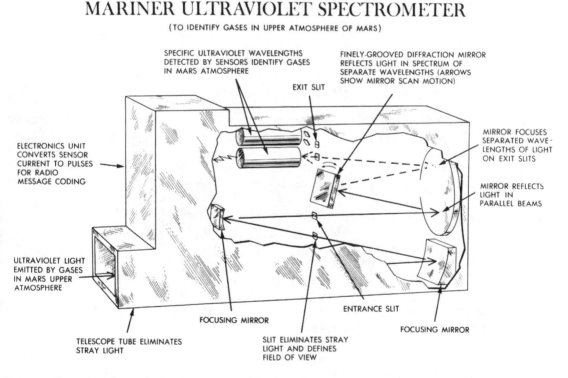

MARINER ULTRAVIOLET SPECTROMETER
(TO IDENTIFY GASES IN UPPER ATMOSPHERE OF MARS)

SPECIFIC ULTRAVIOLET WAVELENGTHS DETECTED BY SENSORS IDENTIFY GASES IN MARS ATMOSPHERE

FINELY-GROOVED DIFFRACTION MIRROR REFLECTS LIGHT IN SPECTRUM OF SEPARATE WAVELENGTHS (ARROWS SHOW MIRROR SCAN MOTION)

EXIT SLIT

MIRROR FOCUSES SEPARATED WAVE-LENGTHS OF LIGHT ON EXIT SLITS

ELECTRONICS UNIT CONVERTS SENSOR CURRENT TO PULSES FOR RADIO MESSAGE CODING

MIRROR REFLECTS LIGHT IN PARALLEL BEAMS

ULTRAVIOLET LIGHT EMITTED BY GASES IN MARS UPPER ATMOSPHERE

ENTRANCE SLIT

FOCUSING MIRROR

FOCUSING MIRROR

TELESCOPE TUBE ELIMINATES STRAY LIGHT

SLIT ELIMINATES STRAY LIGHT AND DEFINES FIELD OF VIEW

Infrared Interferometer Spectrometer (IRIS)

This experiment will measure radiation in infrared wavelengths from the surface and atmosphere of Mars to provide information on a wide range of physical characteristics of Mars.

In general, the data will provide a picture of the circulation of the atmosphere and composition of the surface.

In detail the experiment is expected to yield constituents of the atmosphere, including the important measurement of the amount of water vapor; temperatures and pressures at the surface, temperature profile

MARINER INFRARED INTERFEROMETER SPECTROMETER

(TO MEASURE GASES, PARTICLES, TEMPERATURES ON AND ABOVE MARS SURFACE)

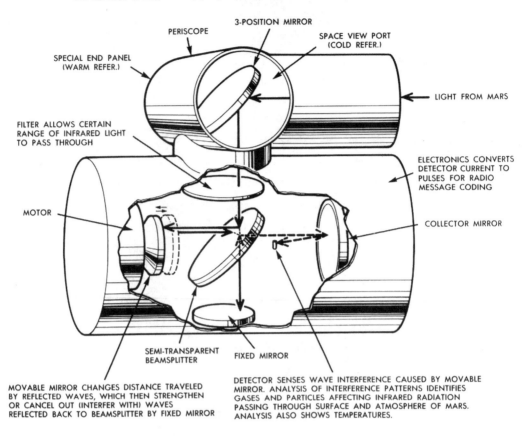

MOVABLE MIRROR CHANGES DISTANCE TRAVELED BY REFLECTED WAVES, WHICH THEN STRENGTHEN OR CANCEL OUT (INTERFER WITH) WAVES REFLECTED BACK TO BEAMSPLITTER BY FIXED MIRROR

DETECTOR SENSES WAVE INTERFERENCE CAUSED BY MOVABLE MIRROR. ANALYSIS OF INTERFERENCE PATTERNS IDENTIFIES GASES AND PARTICLES AFFECTING INFRARED RADIATION PASSING THROUGH SURFACE AND ATMOSPHERE OF MARS. ANALYSIS ALSO SHOWS TEMPERATURES.

of the atmosphere, materials on the surface, and determination of the composition of the polar caps, frozen carbon dioxide, frozen water or a mixture of both.

Although atmospheric and surface data will provide a basis for biological inferences, the question of life on Mars will only be resolved when a landed system can take measurements at the surface.

The principal investigator is Dr. Rudolph Hanel, Goddard Space Flight Center.

S-Band Occultation Experiment

This experiment has been successfully performed by flyby spacecraft at Venus in 1967 and at Mars in 1965 and 1969. The two latter missions yielded three pairs of occultation measurements. The '71 orbiting mission will provide up to 100 pairs from the two spacecraft in the 90-day missions. A pair includes measurements when the spacecraft disappears behind the planet and reappears.

The occultation experiment on the '71 mission will provide multiple measurements of the radius of Mars, atmospheric pressures, and vertical structure, electron density profile of the ionosphere and surface characteristics and radius measurements.

This experiment utilizes the radio signals transmitted from the spacecraft to Earth and does not require on-board equipment. It does require an orbital path that passes behind Mars as seen from Earth, thus occulting the spacecraft from the view of tracking stations.

As the spacecraft curves behind Mars, its radio signal will pass through the Martian atmosphere and be cut off at the surface. The signal will reappear as the spacecraft comes out from behind the planet and again the radio signal will pass through the planet's atmosphere.

The atmosphere will refract the radio waves, changing them in frequency and strength. Measurements on Earth of these changes in the radio signal yield the data on the density and pressure of the atmosphere.

Similar changes in the atmosphere of Mars are caused by electron density and are also measurable.

The cutting off of the signal at the surface and resumption of the signal as the spacecraft comes out from behind the Planet Mars, provides data for calculation of the radius and shape of Mars.

Determination of the atmospheric density of Mars is vital to the design of future landing craft, and is a critical factor in the resolution of important scientific questions on the nature of the planet.

The principal investigator is Dr. Arvydas Kliore of JPL.

Celestial Mechanics

Radio tracking of two orbiting spacecraft over a period of time is expected to provide a more accurate description of the Martian gravity field than provided by the orbits of the two moons of Mars and refinement of the astronomical unit (the distance from the Sun to Earth, a basic astronomical yard stick) to a fraction of a mile. This accuracy of the expected AU refinement is unprecedented.

This experiment derives its results from spacecraft tracking information and does not require special hardware on the spacecraft.

The objectives of the experiment includes measurements of the mass of Mars; the Earth-Moon mass ratio and the distance from Earth to Mars. Long-range objectives are to obtain an improved ephemeris of Mars (its position at given times in its solar orbit) to a high order of accuracy and to attempt to measure General Relativistic effects on orbital motions and signal propagation.

Improving the ephemeris of Mars is part of an existing NASA/JPL project to improve the ephemerides of all the inner planets. The Mariner tracking data will be combined with radar and optical telescope data to achieve the results.

The team leader is Dr. J. Lorell of JPL. The other principal investigator is Dr. Irwin I. Shapiro of the Massachusetts Institute of Technology.

MARINER MARS 1971 SCIENCE EXPERIMENTS AND INVESTIGATORS

Television

Harold Masursky-Team Leader	U.S. Geological Survey	Principal Investigator
Bradford Smith-Deputy	New Mexico State University	" "
Dr. Joshua Lederberg	Stanford University	" "
Dr. Gerard de Vaucouleurs	University of Texas	" "
Dr. Geoffrey A. Briggs	Bellcom	" "
David W. Arthur	U.S. Geological Survey	Co-Investigator
Raymond Batson	" " "	"
Warren Borgeson	" " "	"
Dr. Michael Carr	" " "	"
Dr. John F. McCauley	" " "	"
Dr. Daniel Milton .	" " "	"
Dr. Lawrence A. Soderblom	" " "	"
Dr. Robert Wildey	" " "	"
Dr. Don Wilhelms	" " "	"
Dr. Elliot Levinthal	Stanford University	"
Dr. James B. Pollack	Ames Research Center	"

Dr. Carl Sagan	Cornell University	"
Dr. Andrew T. Young	Jet Propulsion Laboratory	"
Paul L. Chandeysson	Bellcom	"
Earl Shipley	"	"
Dr. James A. Cutts	California Institute of Technology	"
Merton Davies	Rand Corporation	Co-Investigator
Dr. William Hartmann	Illinois Institute of Technology,	"
	Research Institute, Tucson	"
Dr. Robert Leighton	California Institute of Technology	"
Dr. Conway B. Leovy	University of Washington	"
Dr. Bruce C. Murray	California Institute of Technology	"
Dr. Robert P. Sharp	" " " "	"

Infrared Interferometer Spectrometer

Dr. Rudolph A. Hanel	Goddard Space Flight Center	Principal Investigator
Dr. Barney J. Conrath	" " " "	Co-Investigator
Dr. W. A. Hovis	" " " "	"
Virgil Kunde	" " " "	"
Dr. Gilbert V. Levin	" " " "	"
Dr. P. D. Lowman	" " " "	"
Dr. Cuddapah Prabhakara	" " " "	"
Benjamin Schlachman	" " " "	"

Infrared Radiometer

Dr. Gerry Neugebauer	California Institute of Technology	Principal Investigator
Stillman Chase	Santa Barbara Research Center	Co-Investigator
Dr. Hugh K. Kieffer	University of California at Los Angeles	"
Dr. Ellis D. Miner	Jet Propulsion Laboratory	"
Dr. Guido Munch	California Institute of Technology	"

Ultraviolet Spectrometer

Dr. Charles A. Barth	University of Colorado	Principal Investigator
Dr. Charles W. Hord	" " "	Co-Investigator
Jeffrey B. Pearce	" " "	"

Celestial Mechanics

Jack Lorell-Team Leader	Jet Propulsion Laboratory	Principal Investigator
Dr. John D. Anderson	" " "	Co-Investigator
Warren L. Martin	" " "	"
Dr. Irwin I. Shapiro	Massachusetts Institute of Technology	Principal Investigator
Dr. Michael E. Ash	Lincoln Lab	Co-Investigator
William B. Smith	" "	"

S-Band Occultation

Dr. Arvydas Kliore	Jet Propulsion Laboratory	Principal Investigator
Dan L. Cain	" " "	Co-Investigator
Dr. Gunner Fjeldbo	" " "	"
Dr. S. I. Rasool	Goddard Institute of Space Studies	"
Boris L. Seidel	Jet Propulsion Laboratory	"

ATLAS-CENTAUR LAUNCH VEHICLE

The launches of Mariners H and I mark the third and fourth times Atlas-Centaur launch vehicles have been used in the Mariner program. They were used to launch successfully Mariners 6 and 7 to Mars in 1969.

The AC-23 and AC-24 vehicles have been assigned to Mariner Mars '71.

Centaur, developed under the direction of the National Aeronautics and Space Administration's Lewis Research Center, was the first U.S. rocket to use the high-energy liquid hydrogen/liquid oxygen combination. Flown with an Atlas booster so far, Centaur is in the process of being integrated with the Titan III to launch Viking spacecraft to Mars in 1975.

The ability of the Atlas-Centaur to launch the much greater weight of the Mariner H and I spacecraft, approximately 2,200 pounds (998 kilograms) compared with the 850-pound (385-kilogram) weight of Mariners 6 and 7, is due primarily to the more favorable position of the planet Mars in relation to the Earth this year. Such a favorable alignment of the two planets will not occur again until the 1980's.

Some changes also have been made to the Centaur stages to increase their payload carrying capability. AC-23 and AC-24 will use a direct ascent, single burn trajectory to Mars. Because the Centaur engines will be ignited only once in space, it was possible to remove four three-pound thrust engines normally used to keep propellants settled in the tanks during coast periods. Baffles used to position propellants in a weightless environment have also been removed for this mission. Because of the single-burn mission, another small weight gain was realized by substituting a smaller helium tank. Helium is used to pressurize the propellant tanks during flight.

AC-23 and 24 consist of an Atlas SLV-3C booster combined with a Centaur second stage. The two stages are 10 feet (3.05 meters) in diameter and are connected with an interstage adapter. Both Atlas and Centaur stages rely on internal pressurization for structural integrity.

The Atlas booster develops 403,000 pounds of thrust at liftoff, using two 171,000-pound-thrust booster engines, one 60,000-pound-thrust sustainer engine and two vernier engines developing 690 pounds thrust each.

Centaur carries insulation panels and a nose fairing which are jettisoned after the vehicle leaves the Earth's atmosphere. The insulation panels, weighing about 1,200 pounds (544 kilograms), surround the second stage propellant tanks to prevent heat or air friction from causing excessive boil-off of liquid hydrogen during flight through the atmosphere. The nose fairing protects the payload.

To date Centaur has successfully launched the seven unmanned Surveyor spacecraft to the Moon, Mariners 6 and 7, and Orbiting Astronomical Observatory 2, Applications Technology Satellite 5 and the first Intelsat.

Launch Vehicle Characteristics

*Liftoff weight including spacecraft:	324,432 pounds (147,130 kilograms)
Liftoff height:	113 feet (34.4 meters)
Launch complexes:	36 A & B
Launch azimuth sector:	81 - 108 degrees

	SLV-3C Booster	Centaur Stage
**Weight:	283,577 lbs. (128,602 kg.)	37,657 lbs. (17,077 kg.)
Height:	75 feet (22.9 meters) (including interstage adapter)	48 feet (14.6 meters) (with payload fairing)
Thrust:	403,000 lbs. (sea level)	30,000 lbs. (vacuum)

* Measured at two inches (5.08 centimeters) of rise.
** Weights are based on AC-23 configuration. AC-24 varies only slightly.

Propellants:	Liquid oxygen and RP-I	Liquid hydrogen and liquid oxygen
Propulsion:	MA-5 system (two 171,000-lbs.-thrust engines, one 60,000-lb.-sustainer engine and two 690 lb.-thrust vernier engines)	Two 15,000-pound-thrust RL-10 engines. Ten small hydrogen peroxide thrusters.
Velocity:	5,766 mph (2,577 meters-per-second) at BECO 8,372 mph (3,742 m/s) at SECO	22,392 mph (10,010 m/s) at spacecraft separation
Guidance:	Pre-programmed pitch rates through BECO Switch to Centaur inertial guidance for sustainer phase.	Inertial guidance

Atlas-Centaur Flight Sequence - AC-23*

Event	Nominal Time (Seconds)	Altitude Statute Miles	- Kilometers	Surface Range Statute Miles	- Kilometers	Velocity MPH	Meters-per-second
Liftoff	0	0	0	0	0	0	
Booster Engine Cutoff	147.9	37.2	59.6	52.2	83.5	5,726.8	2,560
Booster Jettison	151.0	39.1	62.6	56.8	91.0	5,793.4	2,590
Jettison Insulation Panels	192.9	62.5	100.0	123.6	197.8	6,621.6	2,960
Jettison Nose Fairing	234.9	83.5	134.6	202.4	323.8	7,717.8	3,450
Sustainer Engine Cutoff	250.9	91.3	147.3	236.0	377.6	8,254.7	3,690
Atlas Separation	252.8	92.3	148.9	240.5	384.8	8,209.7	3,670
Centaur Engine Start	262.4	96.7	156.0	261.2	417.9	8,187.6	3,660
Centaur Engine Cutoff	716.0	109.5	176.6	1,951.1	3,121.8	24,540.3	10,970
Spacecraft Separation	811.0	116.4	187.7	2,581.1	4,129.8	24,540.3	10,970
Start Centaur Reorientation	1,271.0						
Start Centaur Retrothrust	1,366.0						

* Figures are for launch on May 17. There will be a slight variation for a launch on another day.

Atlas-Centaur Flight Sequence - AC-24*

Event	Nominal Time (Seconds)	Altitude Statute Miles	- Kilometers	Surface Range Statute Miles	- Kilometers	Velocity MPH	Meters-per-Second
Liftoff	0	0	0	0	0	0	
Booster Engine Cutoff	147.9	37.2	59.6	52.2	83.5	5,726.8	2,560
Booster Jettison	151	39.1	62.6	56.8	90.9	5,793.4	2,590
Jettison Insulation Panels	192.9	61.9	99.0	123.9	198.2	6,621.6	2,960
Jettison Nose Fairing	234.9	80.9	129.4	203.4	325.4	7,762.5	3,470
Sustainer Engine Cutoff	250.9	87.4	141.1	237.7	380.3	8,321.8	3,720
Atlas Separation	252.8	88.3	142.4	242.2	387.5	8,277.0	3,700
Centaur Engine Start	262.4	92.0	148.4	263.2	421.1	8,254.7	3,690
Centaur Engine Cutoff	716.8	106.5	171.7	1,983.4	3,173.4	24,741.7	11,060
Spacecraft Separation	811.8	155.8	251.3	2,610.6	4,171.0	24,607.4	11,000
Start Centaur Reorientation	1,271.8						
Start Centaur Retrothrust	1,366.8						

* Figures are for launch on May 7. There will be a slight variation for a launch on another day.

Flight Sequence

Flight sequences of the AC-23 and AC-24 rocket vehicles are basically the same with times varying only a second or two in some cases as noted on the accompanying charts. The following sequence description is for AC-24 on May 7.

Atlas Phase

After liftoff, AC-24 will rise vertically for about 15 seconds before beginning its pitch program. Starting at two seconds after liftoff and continuing to T+15 seconds, the vehicle will roll to the desired flight azimuth.

After 148 seconds of flight, the booster engines are shut down (BECO) and jettisoned. BECO occurs when an acceleration of 5.7 g's is sensed by accelerometers on the Centaur and the signal is issued by the Centaur guidance system. The booster package is jettisoned 3.1 seconds after BECO. The Atlas sustainer engine continues to burn for approximately another minute and 43 seconds propelling the vehicle to an altitude of about 87 miles (148 kilometers), attaining a speed of 8,300 mph (3,710 m/s).

Sustainer engine cutoff (SECO) occurs at propellant depletion. Centaur insulation panels and nose fairing are jettisoned prior to SECO.

The Atlas and Centaur stages are then separated. An explosive shaped charge slices through the interstage adapter. Retrorockets mounted on the Atlas slow the spent stage.

Centaur Phase

At four minutes, 22 seconds into the flight, the Centaur's two RL-10 engines ignite for a planned seven-minute 34-second burn. This will place Centaur and the spacecraft on an interplanetary trajectory at a speed of about 24,714 mph (11,050 m/s). After MECO, the Centaur stage and spacecraft are reoriented with the Centaur attitude control thrusters to place the spacecraft on the proper trajectory after separation.

Separation of the Mariner spacecraft is achieved by firing explosive bolts on a V-shaped metal band holding the spacecraft to the adapter. Compressed springs then push the spacecraft away from the Centaur vehicle at a rate of 2.1 feet-per-second (0.6 meters-per-second).

Retromaneuver

Seven and a half minutes after spacecraft separation, the Centaur stage attitude control thrusters are used to reorient the vehicle. The remaining liquid and gaseous propellants are then vented.

The retromaneuver insures that there is no possibility of crashing into the planet and thereby violating the Martian quarantine restraint. The spent Centaur stage will go into a solar orbit. It will pass Mars at a distance of approximately a million miles (1,600,000 kilometers).

LAUNCH OPERATIONS

The Unmanned Launch Operations (ULO) Directorate at the John F. Kennedy Space Center (KSC) is responsible for the preparation and launch of unmanned spacecraft from Florida. For the Mariner '71 missions, ULO will have an added responsibility — launching two Atlas-Centaur vehicles from Complex 36 within a 10-day period.

The spacecraft are delivered to the Cape from JPL about three months prior to launch and are placed in a clean room environment for final verification tests.

The preparations for these launches are somewhat similar to the Mariner '69 missions which were launched about a month apart. With a tighter launch frame for Mariner '71, the logistics involved in getting both vehicles ready and scheduling of key launch personnel is much more difficult. All of the hardware is fit-checked to assure that it will be adaptable to either vehicle and a pool of spares is established for both.

The key members of the launch team move from one vehicle to the other as required, and a small crew stays with the vehicle not undergoing major testing in order to monitor systems and maintain quality surveillance.

In providing launch operations, KSC handles scheduling of test milestones and review of data to assure that the launch vehicle has met all of its test requirements and is ready for launch.

Because of the relatively short launch opportunity, a special effort was made by the ULO team to develop a work schedule that permits sufficient time to check out both vehicles. Atlas-Centaur No. 23 was erected on Pad 36B in December 1970 and vehicle No. 24 was placed on Pad 36A in February, 1971. Procurement and management of the launch vehicles is directed by the NASA Lewis Research Center.

The Terminal Countdown Demonstration (TCD) was conducted about seven weeks prior to launch using an encapsulated prototype model of the Mariner spacecraft. The TCD primarily demonstrated that all of the functions leading to the actual countdown can be performed. It was an end-to-end check of all systems and included propellant loading of both launch vehicle stages to verify the tanks and facilities were ready for the countdown.

Following this, the Joint Flight Acceptance Composite Test (J-FACT) was conducted about six weeks before launch to assure that the vehicle was electrically ready for final launch preparations and that the proper connections were made with the spacecraft. The J-FACT included running the computer and programmer through post flight events and monitoring the data to assure correct response to all signals with the umbilical ejected.

After AC-24 was erected and the basic systems were checked out, the prototype model was transferred over from AC-23. The TCD on this vehicle was conducted about six weeks prior to its launch, followed by the J-FACT a week later.

In late April, the Mariner H spacecraft was mated on AC-24 and an electrical and mechanical test was conducted prior to running a second J-FACT. Mariner I was encapsulated but will not be mated to AC-23 until Mariner H is successfully launched.

At this point, it would have been possible to launch either space vehicle on either of the two missions. However, AC-24 was selected for the first mission and an electrical-mechanical test and a second J-FACT is scheduled in early May for AC-23 as an added assurance for a successful flight.

The Countdown Readiness Test is scheduled for both space vehicles about four days before launch. It verifies the ability of the launch vehicle to go through post-liftoff events and revalidates the umbilical system. The range support elements participate along with the spacecraft and launch vehicle just as during a launch.

The F-1 Day Functional Test involves final preparations in getting the entire space vehicle ready for launch, preparing ground support equipment, completing readiness procedures and installing ordnance on the launch vehicle.

The final countdown is picked up at T-450 minutes. All systems are checked against readiness procedures, establishing the integrity of the vehicle and ground support equipment interface prior to tower removal at T-120 minutes. Loading of cryogenic propellants (liquid oxygen and liquid hydrogen) begins at T-80 minutes, culminating in complete vehicle readiness at T-5 minutes. The terminal count begins monitoring all systems and topping off and venting propellant and purge systems. At T-10 seconds, the automatic release sequence is initiated and the space vehicle is clear for liftoff.

TRACKING AND DATA SYSTEM AND MISSION OPERATIONS

With facilities located around the Earth and some new items of equipment, NASA's Deep Space Network (DSN) will be able to support both Mariner '71 spacecraft almost continually through their six and one half month flight to Mars and 90-days in orbit for the basic mission, plus overlapping station coverage during critical events.

All DSN stations are equipped with new telemetry equipment capable of receiving, data synchronizing, decoding and processing the high-rate telemetry. Some telecommunications can be received at very high bit rates.

The network is tied to the Space Flight Operations Facility (SFOF), the Mariner nerve center at JPL, by NASA's Communications Network (NASCOM). New data links permit real-time transmission of nearly all data from the twin spacecraft to the SFOF.

Scientists and engineers seated at consoles in the SFOF will have push-button control of the displayed information either on television screens in the consoles or on projection screens and automatic plotters and printers. The processed information will be stored in the computer system, available on command.

Tracking and obtaining data from Mariners are part of the mission assigned to JPL. These tasks cover all phases of the flight, including telemetry from launch vehicle and spacecraft, tracking data on both launch vehicle and the Mariners, command signals and the delivery of data to SFOF.

In the launch phase of the mission, tracking will be carried out by DSN with aid of other facilities. These are radar of Air Force Eastern Test Range and downrange elements of NASA's Manned Space Flight Network (MSFN) together with the tracking ship Vanguard and an instrumented jet aircraft. Both ship and jet are operated by MSFN.

The mission is complicated by the fact that Mars, positioned 60 million miles from Earth at time of first encounter, will move beyond 100 million miles through the course of the long mission. Meanwhile, both planets continue moving in their separate orbits and turning on their axes.

The DSN consists of nine specialized tracking facilities located at four points around the Earth. Largest of these, at Goldstone, CA, has two 85-foot (26-meter) antennas and one measuring 210 feet (64-meters) in diameter. Other 85's are at Madrid (Robledo de Chavela and Cebreros), Spain; Johannesburg, South Africa; and Woomera and Canberra (Tidbinbilla) Australia. In addition a four-foot (1.2-meter) antenna at Cape Kennedy covers pre-launch and launch phases of the flight.

For all of NASA's unmanned missions in deep space, such as planetary and Sun-orbiting spacecraft, the network provides the tracking information on course and direction of the flight, velocity and range from Earth. It also receives engineering and science telemetry, including planetary television coverage, and sends commands for spacecraft operations. All communication links are in the S-band.

The 210-foot (64-meter) antenna at Goldstone, capable of receiving eight times the volume of data of the other antennas, will play a special role in Mariner '71. The 210 will transfer, on a daily basis, Mars TV pictures obtained and stored in each orbit of the planet.

DSN will support the three months of basic orbital operations by acquiring data telemetered from the spacecraft at 16,200 bits per second through the 210-foot (64-meter) antenna. (Rates with the 85's are 2,025 bits per second.) Data will be routed immediately to SFOF, distributed to computers and other specialized processing machinery, to make it ready for the experimenters.

At the same time spacecraft range and range-rate information will be relayed from both Mariners to SFOF.

A new 50,000 bps digital wideband communications line will carry data from Goldstone to SFOF. High-speed data links from all stations are capable of 4,800 bps. These new data links allow real-time transmission of almost all data from both spacecraft to JPL-SFOF.

The planning and analysis functions of Mariner '71 are carried out by special teams of engineers, including a DSN Mission Operations Team and a DSN Project Engineering Team. The Operations team is responsible not only for the operation of the Network but coordination of near-Earth phase assistance and NASCOM support. The other team has charge of resources, operations planning and configuration control.

All of NASA's networks are under the direction of the Office of Tracking and Data Acquisition. JPL manages the DSN, while the MSFN facilities and NASCOM are managed by NASA's Goddard Space Flight Center, Greenbelt, MD.

The Goldstone DSN stations are operated and maintained by JPL with the assistance of the Philco-Ford Corp.

The Woomera and Tidbinbilla stations are operated by the Australian Department of Supply.

The Johannesburg station is operated by the South African government through the National Institute for Telecommunications Research.

The two stations near Madrid are operated by the Spanish government's Instituto Nacional de Teenica Aerospacial (INTA).

Mission Operations

The Space Flight Operations Facility, designed for 24-hour-a-day functioning and equipped to handle multiple spaceflight missions concurrently, is manned by Mission Operations System (MOS) personnel of JPL and Philco-Ford Corp.

Mission operations planning and analysis functions are carried out by five teams, including the DSN Project Engineering Team:

A Navigation Team whose principal functions are spacecraft navigation and scan geometry analysis.

A Spacecraft Team whose principal functions are spacecraft systems performance evaluation and prediction through telemetry data analysis, and the development of alternative methods of utilizing the spacecraft to satisfy mission requirements.

A Science Data Handling Team to collect and coordinate science data processing requirements and to assemble and disseminate data from a data library.

A Science Recommendation Team to analyze science data and recommend science operations plans and priorities.

Real-time mission operations execution is conducted by three teams, including the DSN Mission Operations Team:

A Data Processing Team which plans, schedules, coordinates, and troubleshoots all data processing for the Project. The team is also responsible for producing Master Data Records and Experiment Data Records for which the MOS is responsible.

A Command Team continuously ensures spacecraft utilization in accordance with the mission operations plans. This team will translate mission operations plans into command sequences and will initiate their transmission to the spacecraft.

MARINER MARS '71 TEAM

Office of Space Science and Applications

Dr. John E. Naugle	Associate Administrator for OSSA
Vincent L. Johnson	Deputy Associate Administrator for OSSA
Robert S. Kraemer	Director, Planetary Programs
Earl W. Glahn	Mariner '71 Program Manager
Kenneth L. Wadlin	Mariner '71 Program Engineer
Harold F. Hipsher	Mariner '71 Program Scientist
Joseph B. Mahon	Director, Launch Vehicle Programs
T. Bland Norris	Manager, Medium Launch Vehicles
F. Robert Schmidt	Manager, Atlas-Centaur

Office of Tracking and Data Acquisition

Gerald M. Truszynski	Associate Administrator for OTDA
Arnold C. Belcher	Network Operations
Maurice E. Binkley	Network Support

Jet Propulsion Laboratory, Pasadena, CA

Dr. William H. Pickering	Laboratory Director
Adm. John E. Clark	Deputy Laboratory Director
Robert J. Parks	Assistant Laboratory Director for Flight Projects
Dan Schneiderman	Project Manager
James F. McGee	Assistant Project Manager (Near Earth)
Edwin Pounder	Assistant Project Manager (Near Planet)
Robert Forney	Spacecraft System Manager
Patrick J. Rygh	Mission Operations System Manager
Norman Haynes	Mission Analysis and Engineering Manager
Robert Steinbacher	Project Scientist
Bradford Houser	Project Control and Administration Manager
Dr. Nicholas A. Renzetti	Tracking and Data System Manager
Richard Laeser	Deep Space Network Manager
Richard T. Hayes	Chief of Mission Operations
Al Conrad	Spacecraft Systems Engineer

Lewis Research Center, Cleveland

Bruce T. Lundin	Center Director
Seymour C. Himmel	Director for Rockets and Vehicles
Edmund R. Jonash	Chief, Launch Vehicles Division
Daniel J. Shramo	Atlas-Centaur Project Manager
Edwin Muckley	Centaur Project Engineer

Kennedy Space Center, Florida

Dr. Kurt H. Debus	Center Director
John J. Neilon	Director, Unmanned Launch Operations ULO
John D. Gossett	Chief., Centaur Operations Branch, ULO
Donald C. Sheppard	Chief, Spacecraft Operations Branch, ULO

MARINER MARS 1971 SUBCONTRACTORS

Following is a list of some key subcontractors who provided instruments, hardware and services for the Mariner Mars 1971 Project:

Spacecraft Engineering Subsystem Contracts

Litton IndustriesWoodland Hills, CA	Data Automation System
Santa Barbara Research Center Goleta, CA	Infrared Radiometer
Electro-Optical Systems Pasadena, CA	Television
University of Colorado Boulder, CO	Ultraviolet Spectrometer
RCA Van Nuys, CA	Science Support Equipment
Philco - WDL Palo Alto., CA	Antenna Assembly
Martin Marietta Denver, CO	Propulsion System
Motorola Phoenix, AZ	Command System
MotorolaPhoenix, AZ	Data Storage Electronics
Lockheed Electronics Plainfield, NJ	Data Storage Tape Transport
Motorola Phoenix, AZ	Radio
Texas Instruments Dallas, TX	Telemetry and Infrared Interferometer Spectrometer
General Electric Valley Forge, PA	Attitude Control and Scan Control Electronics
Honeywell Radiation Center Lexington, MA	Canopus Sensor
Motorola Phoenix., AZ	Central Computer & Sequencer
TRW Redondo Beach CA	Batteries
Electro-Optical Systems Pasadena, CA	Power
NAA, Rocketdyne Canoga Park, CA	Engines (Flight)

MARINER 9
POST LAUNCH REPORTS

Post Launch Mission Operation Report No. S-819-71-02

MEMORANDUM   2 June 1971
TO: A/Administrator
FROM: S/Associate Administrator for Space Science and Applications
SUBJECT: Mariner 9 Post Launch #1

Mariner 9 was originally scheduled for launch on 18 May 1971. The intensive investigation of the unsuccessful launch of Mariner H on 8 May 1971 resulted in a postponement of the Mariner 9 launch. When the cause of the failure was determined and corrective action implemented, 29 May 1971 was established for the launch of Mariner 9. The launch window for this day was from 6:21 p.m. EDT to 7:25 p.m. EDT.

During the countdown on 29 May 1971 a hold was called at 4:59 p.m. EDT at T-72 minutes in the count. This hold was due to difficulty in interpreting anomalous data from the Centaur guidance package. At 6:00 p.m. EDT the difficulty was still unresolved and it was determined to postpone the launch until 30 May 1971. The launch window for this day was 6:17 p.m. EDT to 7:17 p.m. EDT.

It was determined there was no problem with the Centaur guidance package when the data obtained 29 May was fully understood. Therefore, the countdown for launch on 30 May 1971 was started. The countdown proceeded well until T-4 minutes 29 seconds (6:12 p.m. EDT) when anomalous data were received concerning the Atlas propellant utilization control. A hold was called at 6:13 p.m. EDT and the count was recycled to T-5 minutes 1 second. It was determined that the anomalous data was due to a landline telemetry problem and not any difficulty with the Atlas propellant utilization control. Therefore, the count was resumed at 6:18 p.m. EDT and continued to liftoff.

Mariner 9 was successfully launched to Mars from the Eastern Test Range at 6:23 p.m. EDT on 30 May 1971. The launch was from Launch Complex 36B utilizing an Atlas/ Centaur launch vehicle.

The solar panels were extended at 6:40 p.m. and the sun was acquired at 7:16 p.m. soon after Mariner 9 left the earth's shadow. About 4 hours after launch the Canopus sensor was energized. At that time the spacecraft rolled approximately 55° and locked on to the star Achenar. At 10:23 p.m. Canopus search was again commanded and Canopus acquisition was completed at 10:26 p.m. The spacecraft will remain in Sun-Canopus orientation throughout the remainder of the mission except during maneuvers. All spacecraft systems are operating normally. The next significant event will be the first midcourse trajectory correction maneuver which is presently planned to take place 8:30 p.m. EDT 5 June 1971, after sufficient tracking data has been obtained to compute the correction required.

As of 1 June at 8:00 a.m., the spacecraft was over 490,000 km from the earth traveling at a speed of 8 km per second.

John E. Naugle

Post-Launch Mission Operation Report No. S-819 -71-03

MEMORANDUM   7 June 1971
TO: A / Administrator
FROM: SL/ Manager, Mariner Mars '71 Planetary Programs
SUBJECT: Mariner 9 Post Launch #2

Mariner 9 continues to operate normally in all respects.

The mid-course trajectory correction maneuver that had tentatively been scheduled for June 5, 1971, was rescheduled and executed on June 4, 1971.

The Atlas Centaur launch vehicle provided such an accurate injection into the planned trans-Mars trajectory that a first trajectory correction maneuver velocity change of only one meter per second would have been required due to launch vehicle injection error. However, the initial trans-Mars trajectory was targeted away from Mars to satisfy planetary quarantine requirements.

To retarget the spacecraft trajectory to the desired aim point for injection into the planned orbit about Mars called for a 6.7 meters per second velocity change. To apply this velocity correction in the proper direction the spacecraft was re-oriented from its Sun-Canopus celestial orientation. This was done by first rolling the spacecraft 140.88°, then yawing it 44.79° prior to engine ignition. The velocity change was provided by a 5.11 second duration engine burn. After the engine burn was completed the spacecraft was re-oriented and the Sun and Canopus were re-acquired.

The entire sequence of events was programmed into the spacecraft's on-board computer 1 day before the maneuver was initiated. The sequence was started by ground command at 6:09 p.m., EDT, June 4, 1971. The actual engine ignition occurred as programmed at 8:22 p.m., EDT and the spacecraft had re-acquired the Sun and Canopus and resumed its normal cruise mode by 8:49 p.m., EDT.

Earl W. Glahn

Post Launch Mission Operation Report No. S-819-71-02

MEMORANDUM   14 June 1971
TO: A/Administrator
FROM: SL/Manager, Mariner Mars '71 Planetary Programs
SUBJECT: Mariner 9 Post Launch Report #3

Mariner 9 continues to operate normally in all respects.

At 9:00 a.m. EDT today, after 14.6 days in flight, Mariner 9 is 3,855,000 km from earth. It is travelling at a velocity of 10,966 km/hr and has traveled a total distance of 52,830,000 km since launch.

Analysis of the tracking data, received since the 4 June 1971 midcourse trajectory correction maneuver, shows Mariner 9 is on the planned trajectory.

No further activity of the on-board propulsion subsystem is planned until a possible second midcourse trajectory correction maneuver. If required this maneuver will take place about 20 days prior to arrival at Mars (about 24 October 1971).

To prevent leaks of the propellants or the pressurant gas, and to limit the absorption of pressurant gas by the propellants during this long period of inactivity, pyrotechnic valves are closed to isolate these tanks.

The valve isolating the pressurant gas supply tank was closed 8 June 1971. It is desired to minimize the pressure that could occur in the sealed propellant-filled lines caused by expansion due to thermal changes. Therefore the propellant isolation valves will not be closed until the propellants have been heated to a temperature near or above the maximum expected during the cruise phase. To do this the propellant heater has been activated. When the temperature reaches the desired level the propellant isolation valves will be closed. This should occur in the next few days.

Earl W. Glahn

Post Launch Mission Operation Report No. S-819-71-02

MEMORANDUM   21 June 1971
TO: A/Administrator
FROM: SL/Manager, Mariner Mars '71 Planetary Programs
SUBJECT: Mariner 9 Post Launch Report #4

Mariner 9 continues to operate normally.

At 9:00 a.m. EDT today, after 21.6 days in flight, Mariner 9 is 5,703,000 km from earth. It is traveling at a velocity of 11,077 km/hr and has traveled a total distance of 58,480,000 km since launch.

Since the sealing of the pressurant gas supply and the activation of the propellant heaters on 8 June 1971, the temperature of the propellants has risen to the desired level. Consequently, on 14 June 1971, the propellant isolation valves for both the fuel and oxidizer supplies were closed. The propulsion subsystem will remain in this sealed condition until just prior to the second midcourse trajectory correction, if required, on about 24 October 1971; or until insertion into Martian orbit 13 November 1971 if no second midcourse is performed.

The spacecraft will remain in a quiescent cruise mode for the next couple of months at least. Therefore, unless requested otherwise, this will be the lost regular weekly Post Launch Report. Subsequent Post Launch Reports will be issued when any significant event takes place.

Earl W. Glahn

Post Launch Mission Operation Report No. S-819-71-02

MEMORANDUM   29 June 1971
TO: A/Administrator
FROM: SL/Manager, Mariner Mars '71 Planetary Programs
SUBJECT: Mariner 9 Post Launch Report #5

Mariner 9 subsystems continue to operate within predicted performance margins.

At 9:00 a.m. EDT today, after 29.6 days in flight, Mariner 9 is 7,900,000 km from earth. It is traveling at a velocity of 11,600 km/hr relative to the earth and has traveled a total distance of 82,500,000 km since launch.

Since Post Launch Report #4 of 21 June 1971, two anomalies in spacecraft performance have been identified. Neither of these anomalies is of major concern, and although corrective action is possible, such action will not be taken at this time. Instead, spacecraft performance will continue to be closely monitored and corrective action taken, if necessary, at a later date. The two anomalies are:

The radio exciter output power has indicated a slow downward drift since launch. Some downward trend was expected, but the drift is not leveling off as rapidly as anticipated. The apparent drop in exciter output has not been sufficient to cause a measurable change in the signal strength as received at the ground antennas.

The attitude control subsystem has been consuming control gas at a more rapid rate than expected. This anomaly has been traced to a circuit design error in the cruise sun sensor circuit which supplies the error signal to the attitude control subsystem during cruise. As stated earlier, the rate of consumption is not considered critical at this time.

As of today, over 500 commands have been sent to and received by the spacecraft. Approximately 450 of these commands have been used to load the spacecraft Central Computer and Sequencer with a nominal orbital insertion and science data collection sequence. The precaution is taken so that in case of loss of command capability, the spacecraft could still carry out a "nominal" mission. This sequence of events will be updated periodically throughout cruise.

Technically, it would be possible to view the planet Jupiter during the period of July through early August without maneuvering the spacecraft. A feasibility study is now being undertaken to determine if a Jupiter science sequence should be initiated to provide inflight instrument calibrations during this time. The results of this study will be made known when available.

Earl W. Glahn

Post Launch Mission Operation Report No. S-819-71-02

MEMORANDUM   12 July 1971
TO: A/Administrator
FROM: SL/Manager, Mariner Mars '71 Planetary Programs
SUBJECT: Mariner 9 Post Launch Report #6

Mariner 9 subsystems continue to operate within acceptable performance margins.

At 9:00 a.m. EDT today, after 42.6 days in flight, Mariner is 11,600,000 km from earth. It is traveling at a velocity of 13,400 km/hr relative to the earth and has traveled a total distance of over 115,000,000 km since launch.

Since Post Launch Report #5 of 29 June 1971, one additional anomaly has appeared. Telemetry deck #410 is dropping out about 50% of the time. Obtaining measurements about 50% of the time is completely acceptable, but it is expected that the deck may drop out completely some time in the future. Deck #410 measures the following parameters:

410 -   Infrared Radiometer engineering temperature
411 -   Spacecraft equipment bay 1 temperature
412 -   Spacecraft equipment bay 3 temperature
413 -   Spacecraft equipment bay 5 temperature
414 -   Spacecraft equipment bay 7 temperature
415 -   Television camera A (wide angle camera) vidicon temperature
416 -   Ultraviolet Spectrometer detector temperature
417 -   Sun sensor temperature
418 -   Traveling wave tube amplifier #1 base temperature (we are currently using traveling wave tube amplifier #2)
419 -   +Y outboard solar panel temperature

The failure mode of deck #410 is that there appears to be an intermittent short in the field effect transistor trigger that turns on measurement #415. When this short is absent, all measurements are okay. When this short is present, only measurement #415 can be read.

Since it is expected that this deck may drop out completely some time in the future (except #415 which will probably continue to read correctly), a complete analysis is now being performed to assess the impact of losing these measurements. The initial look indicates that none of these measurements are critical.

Insofar as the two anomalies reported in Post Launch Report #5 are concerned, their status is as follows:

The radio exciter output power downward drift has decreased significantly and has remained steady for the past 2 days. It has still not dropped sufficiently to cause a detectable change in the signal strength as received at the ground antennas.

Attitude Control gas consumption continues to be higher than expected. However, worst case analysis indicates that if no corrective action is taken, the available gas will last for at least 15 months total (6 months cruise plus 9 months in orbit). Therefore, no corrective action is planned at this time. As of today, over 600 commands have been sent to and received by the spacecraft.

Earl W. Glahn

Post Launch Mission Operation Report No. S-819-71-02

MEMORANDUM   13 October 1971
TO: A/Administrator
FROM: SL/Manager, Mariner Mars '71 Planetary Programs
SUBJECT: Mariner 9 Post Launch Report #7

Mariner 9 subsystems continue to operate within acceptable performance margins.

At 9:00 a.m. EDT today (13 October 1971) after 135.6 days in flight, Mariner 9 is 80,354,500 km from earth. It is traveling at a velocity of 55,360 km/hr relative to the earth and has traveled a total distance of over 334,696,500 km since launch.

Insofar as the three anomalies reported in Post Launch Report #6 are concerned, their status is as follows:

The radio exciter output power downward drift continues. However, at the present rate of decrease it will be at least several months before it would impact the performance of the radio sufficiently to require action. If action is required a switch will be made to the redundant exciter provided for such a contingency.

The attitude control gas consumption is now normal. The correction of this anomaly is the combined result of an increase in the resistance of the sun sensor as the distance of the spacecraft from the sun increased, and the normal effects of aging of the sun sensors.

The dropouts in the 410 telemetry deck have ceased. The intermittent short in the transistor trigger has been attributed to a metal sliver. The sliver is apparently now in a benign location. In any event, if this anomaly should recur, it will not be a major problem.

Mission operations training is continuing. Weaknesses in computer software and in some procedures have been detected and are being corrected. As a whole, orbital mission operations preparedness is satisfactory at this time.

The second midcourse trajectory correction maneuver for Mariner 9 is scheduled for 26 October 1971. The engine burn associated with this maneuver is scheduled for 11:00 p.m. EDT. Mr. Wadlin of Code SL will be covering spacecraft operations during the maneuver from Room 5091 in FOB-6 beginning at 9:00 p.m. EDT. Telephone inquiries regarding status of the maneuver may be directed to Mr. Wadlin on Extension 755 3728.

The next critical event for Mariner 9 after midcourse correction will be Mars orbit insertion which will take place 13 November 1971. The exact time of the insertion will depend upon the results of the second midcourse trajectory correction. Initiation of motor burn is presently expected to be at 7:05 p.m. EST. When a more precise time is determined, another memorandum will be issued.

Earl W. Glahn

Post Launch Mission Operation Report No. S-819-71-02

MEMORANDUM   19 October 1971
TO: A/Administrator
FROM: SL/Manager, Mariner Mars '71 Planetary Programs
SUBJECT: Mariner 9 Post Launch Report #8

Mariner 9 subsystems continue to operate within acceptable performance margins.

The only significant change in status of the mission since Post Launch Report #7 is the deletion of the second midcourse trajectory correction maneuver. This maneuver had been scheduled for 26 October 1971.

Tracking data indicates that the probability of achieving an orbit within the range of satisfactory orbital parameters is sufficiently high that the second midcourse trajectory correction maneuver will not be required. The range of satisfactory orbital parameters are:

Period - 11.975 to 11.985 hours — Periapsis altitude - 1200 to 1500 km — Inclination - 60° to 70° Rotation of the line of apsides - 139° to 145°

The next critical event for Mariner 9 will be Mars orbit insertion which will take place 13 November 1971. Since the second midcourse correction has been deleted the initiation of the motor burn for this maneuver will be later than noted in Post Launch Report #7. Initiation of the motor burn is presently expected to be about 7:30 p.m. EST. When a more precise time is determined, a memorandum will be issued if the time is significantly different.

At 9:00 a.m. EDT today (20 October 1971) after 142.4 days in flight, Mariner 9 is 88,847,850 km from earth. It is traveling at a velocity of 59,220 km/hr relative to the earth and has traveled a total distance of over 349,452,100 km since launch.

E. W. Glahn

Post Launch Mission Operation Report No. S-819-71-02

MEMORANDUM   3 November 1971
TO: A/Administrator
FROM: SL/Manager, Mariner Mars '71 Planetary Programs
SUBJECT: Mariner 9 Post Launch Report #9

At 11:00 a.m. on 2 November 1971 Mariner 9 experienced an anomalous condition that caused the spacecraft to lose lock on the star Canopus which is used for celestial reference. The anomalous condition was probably caused by a bright object passing in the view of the Canopus Tracker and diverting it from Canopus.

As a result of losing lock on Canopus the spacecraft automatically went into a roll search to reacquire Canopus. This roll search lasted 10 minutes stopping when the Canopus Tracker locked on the star Sirius. This resulted in the orientation of the Mariner 9 high gain antenna being in an unfavorable orientation with respect to the earth. As a result the telemetry data from the spacecraft experienced many drop-outs and at times was unintelligible. Radio lock, however, was never lost.

Before it would be safe to command the spacecraft to its proper orientation it was necessary to determine the status of all spacecraft subsystems. To do this the downlink signal strength had to be increased to reacquire good telemetry data.

The spacecraft telemetry had been transmitting engineering data at 33-1/3 bits per second (bps) and science data at 50 bps. Lowering the bit rate and eliminating a channel of data effectively results in increased signal strength. Also the low-gain antenna can provide greater off-axis signal strength than the high-gain antenna.

The spacecraft was successively commanded to transmit via the low-gain antenna, to transmit only engineering data, and to reduce the engineering data rate to 8-1/3 bits.

After this good telemetry was received and the spacecraft was found to be in good condition, the spacecraft was then commanded to do a roll search for Canopus. At 2:51 p.m. EST confirmation was received on the ground from the spacecraft that it had acquired Canopus. The spacecraft was then commanded to transmit via the high gain antenna. This was followed by the expected large increase in downlink signal strength.

The spacecraft was then commanded to return to its pre-anomaly mode of transmitting engineering data at 33-1/3 bps and science data at 50 bps. The spacecraft confirmed its return to this mode at 3:58 p.m. EST. At this time the mission operations resumed its planned schedule, shifting into the TV photometric calibration, which involves photographing the planet Saturn. This picture taking phase of this calibration was completed successfully at 2:00 a.m. EST today (3 November). The pictures are recorded on the tape recorder and will be played back to earth today starting at 6:15 p.m.

E. W. Glahn

Post Launch Mission Operation Report No. S-819-71-02

MEMORANDUM	15 November 1971
TO:	A/Administrator
FROM:	SL/Manager, Mariner Mars '71 Planetary Programs
SUBJECT:	Mariner 9 Post Launch Report #10

Mariner 9 started obtaining Pre-Orbit Science data on 10 November 1971. These data included primarily three sequences of photographs of Mars at intervals to provide complete photography of the planet as it rotated about its axis. The first two sequences each covered a Martian day. The third covered about a third of a day. These photographs did not provide the details originally anticipated due to the "dust storm" that has prevailed on Mars since late September.

On 13 November 1971 Mariner 9 was inserted into orbit about Mars. The Mars Orbit Insertion maneuver started at 5:14 p.m. EST with turn-on of the autopilot and about ½ hour later of the gyros.

A velocity change (reduction) of 1600.5 meters per second (m/s) was required to cause the spacecraft to be captured by Mars gravity field and for it to achieve the desired orbit about Mars. To apply this velocity change in the proper direction the spacecraft was first reoriented from its Sun-Canopus celestial reference orientation. This was done by first rolling the spacecraft 42.7 degrees, then yawing it 124.9 degrees prior to engine ignition. Confirmation of engine ignition was received on Earth at 7:24 p.m. An accelerometer on board the spacecraft integrated the velocity change of the spacecraft and cut off the engine when the desired 1600.5 m/s velocity change had occurred. The engine burned for 15 minutes 14.8 seconds to produce this velocity change. After the engine burn was completed the spacecraft reoriented itself to its Sun-Canopus celestial reference. During this reorientation Mariner 9 went behind Mars and was occulted. During this period of about 37 minutes no data were received from the spacecraft. When receipt of spacecraft data was resumed the spacecraft was found to be properly oriented and all systems in normal condition.

The propulsion subsystem and the attitude control subsystem performed flawlessly and Mariner 9 achieved an orbit almost precisely than planned. The planned and actual orbital parameters were:

	PLANNED	ACTUAL
Period	12 hrs 32 min	12 hrs 34 min
Periapsis Altitude	1350 km	1398 km
Inclination	64.0 degrees	64.5 degrees

About 2 hours later the recorded data from the third Pre-Orbital Science sequence was played back to earth.

First pictures from orbit were taken during a period encompassing the time the spacecraft passed through periapsis of its first orbit. Some of these pictures were played back to earth at a slow rate through an overseas Deep Space Station at Madrid, Spain starting about 2 hours later. When the larger antenna at Goldstone, California was able to view the spacecraft the data were played back at the high rate. Some of these pictures were broadcast live by a national network. The detail of these pictures was also limited due to the "dust storm."

During the next several orbits, while the spacecraft's orbital periapsis is drifting into the desired synchronism with the zenith of the Goldstone tracking station viewing period, simplified data taking sequences will be performed. When satisfactory synchronization is achieved the orbit period of the spacecraft will be changed to 11.98 hours to maintain this synchronization. This trim maneuver is presently planned for the fourth periapsis passing. The engine burn for this trim maneuver will be at 9:44 p.m. EST 15 November 1971. The spacecraft velocity will be increased 16.5 m/s and the periapsis altitude will be reduced about 9 km.

After this trim maneuver the planned detailed science data gathering sequences will begin.

E. W. Glahn

Post Launch Mission Operation Report No. S-819-71-02

MEMORANDUM   16 November 1971
TO: A/Administrator
FROM: SL/Manager, Mariner Mars '71 Planetary Programs
SUBJECT: Mariner 9 Post Launch Report #11

On its fourth orbit about Mars, Mariner 9 successfully completed an orbit trim maneuver 15 November 1971. This maneuver changed the orbital period to be synchronous with this Mars viewing period of the large Deep Space Station antenna at Goldstone, California.

This maneuver consisted of reorienting the spacecraft so the rocket engine thrust would increase the speed of the spacecraft to reduce its orbital period from 12 hours 34 minutes to 11 hours 58 minutes 48 seconds.

To accomplish this the spacecraft was rolled 34.4 degrees, then yawed 128. 7 degrees. The engine was then burned for 6.25 seconds. This increased, the speed of the spacecraft 15.25 meters per second. The burn occurred at 9:44 p.m. EST.

Early tracking after the maneuver show the orbital parameters to be:

Period - 11 hours 57 minutes 12 seconds
Periapsis altitude - 1394 kilometers
Inclination - 64.34 degrees

Present plans are for Mariner 9 to continue the simplified mapping sequences, obtain some phase function photometric, photometry, polarimetry, and night side spectral data for the next several orbits. Orbit 9, periapsis Thursday 18 November about 9:40 a.m., is presently planned to be the first orbital science sequence following the original basic planning.

E. W. Glahn

Post Launch Mission Operation Report No. S-819-71-02

MEMORANDUM   10 January 1972
TO: A/Administrator
FROM: SL/Manager, Mariner Mars '71 Planetary Programs
SUBJECT: Mariner 9 Post Launch Report #12

Mariner 9 subsystems continue to operate within acceptable performance margins.

As of 7 January 1972, it has been 222 days since Mariner 9 was launched; and it has been in orbit about Mars for 55 days. Mariner 9 has responded to about 20,000 ground commands and has taken about 3500 pictures.

Due to the obscuration of Mars resulting from the dust storm, visual and spectral mapping of the surface of Mars has been severely limited. However, as of this date the storm appears to be abating. Now that the dust is clearing, the mission plan has been revised to meet the Mariner Mars '71 mission objectives to the greatest possible extent. Systematic mapping of Mars was started on 3 January 1972. A full 360° of longitude for a band of latitudes will be mapped every 20 days. In the first 20 days the latitudinal band will be from about 65°S to 20°S. In the second 20 days, the latitudinal band will be about 25°S to 30°N. During the second cycle, the mission results will be evaluated to determine if the third 20 day cycle should extend the mapping to more northerly latitudes or fill in areas that may have been missed during the first two cycles.

In order to accomplish the revised plan, it was necessary to trim the orbit of Mariner 9. The orbit period was increased to counter a drifting of the periapsis toward the rise of the 64 meter Goldstone antenna. The periapsis altitude was also increased to enlarge the view of the television cameras to assure side lap of the pictures. Because of the increase in periapsis altitude, the best resolution of the pictures will now be approximately 120 meters instead of approximately 100 meters.

The orbital trim maneuver took place on 30 December 1971. The significant orbital parameters are:

	Pre-trim	Target	Actual
Mean period (hr:min:sec)	11:58:13	11:59:32	11:59:26.4
Periapsis altitude (km)	1387	1650	1654

An anomaly occurred in the radio subsystem on 7 December 1971. The traveling wave tube amplifier (TWTA) power input and temperature increased markedly. The RF output power dropped somewhat and continued to drop slowly. Because of these indications, a transfer to the redundant TWTA was commanded, and radio performance has been normal since that transfer. The anomaly has been simulated at JPL and is attributed to a short from the tube collector to ground. The JPL tests also indicate that the TWTA stabilizes at a higher temperature and continues to operate at a reduced output level under these conditions. Thus, the faulty spacecraft TWTA may still be operational. However, a switch back to the faulty TWTA will not be made unless absolutely essential.

E. W. Glahn

Post Launch Mission Operation Report No. S-819-71-01/02

MEMORANDUM   15 February 1972
TO: A/Administrator
FROM: S/Associate Administrator for Space Science
SUBJECT: Mariner Mars '71 Mission, Assessment of Primary Mission

The Mariner Mars '71 primary mission has been adjudged a success based upon the results of the mission with respect to the approved prelaunch Mission objectives.

The Mariner 8 launch on 8 May 1971 was not successful due to a malfunction in the Centaur stage of the Atlas Centaur launch vehicle. Mariner 9 was successfully launched on 30 May 1971, arriving at Mars 13 November 1971. All the Mariner 9 scientific instruments operated successfully and continue to send data back to earth.

As provided in the mission objectives, when Mariner 8 was lost the mission plan for Mariner 9 was altered to best meet the primary mission objectives. A second such alteration was made due to the dust storm on Mars which initially delayed the systematic mapping of the planet.

Preliminary scientific results for the first 30 days of orbital operation are reported in detail in the following report:

Steinbacher, R. H., et al, Mariner Mars 1971 Project Final Report, Volume 11. Preliminary Science Results, Technical Report 32-1550, Jet Propulsion Laboratory, Pasadena, California, February 1, 1972.

It is anticipated that Mariner 9 will continue to return data to earth, though at a much reduced rate, during its extended mission. A final project report on the scientific data obtained will be published during the latter half of calendar year 1972.

E. Naugle

S-819-71-01/02

NASA MISSION OBJECTIVES FOR MARINER MARS '71

PROJECT OBJECTIVES

To explore, during the 1971 opportunity, the planet Mars from Martian orbit for a period of time sufficient to observe a large fraction of the planet's surface and to view selected areas during dynamic changes. Pictures of the Martian surface will be obtained as well as infrared and ultraviolet data on the planet's atmosphere and surface characteristics.

MISSION OBJECTIVES

Primary Mission

Note: Objective statements 1 and 2 will be assigned to Mariners H and I on or about May 1971. In the event that the spacecraft scheduled to arrive at the planet first fails prior to 5 days before orbital insertion of the second spacecraft, the mission objectives of the second spacecraft may be modified.

1. To map the parameters noted in the project objectives by selecting an orbit which will permit, during the primary orbital operational lifetime of 90 days, the viewing of about 70 percent of the planet's surface with the wide angle imaging camera at a resolution of about 1 km per TV line.

2. To study the dynamic characteristics and time variable features of Mars by measuring the parameters noted in the project objectives from a Martian orbit selected to permit the viewing of selected areas periodically during the primary operational lifetime of 90 days.

Extended Mission

To continue fulfilling objectives 1 and 2 of the primary mission, but at a reduced rate of data acquisition, beyond the 90 days of the primary mission to the anticipated effective operational lifetime of about 1 year for both spacecraft.

Robert S. Kraemer   John E. Naugle
Planetary Programs Director   Associate Administrator for Space Science and Applications
Date: APR 9 1971

ASSESSMENT OF MARINER MARS '71 MISSION

Based upon the results of the Mariner Mars '71 Mission with respect to the approved prelaunch mission objectives, the primary mission is adjudged a success.

Robert S. Kraemer   John E. Naugle
Planetary Programs Director   Associate Administrator for Space Science
Date: 2-11-72   Date: 11 Feb 72

Post Launch Mission Operation Report No. S-819-71-02

MEMORANDUM   10 APRIL 1972
TO: A/Administrator
FROM: SL/Manager, Mariner Mars '71 Planetary Programs
SUBJECT: Mariner 9 Post Launch Report #13

Mariner 9 subsystems continue to operate within acceptable performance margins.

As of 7 April 1972, it has been 313 days since Mariner 9 was launched; and it has been in orbit about Mars for 146 days. Mariner 9 has responded to 37,764 commands and has taken 6876 pictures.

Since completing the 90-day basic mission on 11 February 1972, Mariner 9 has been in its extended mission. Acquisition of science data continued, though at a reducing rate, during the extended mission until 14 March 1972. At that time an attempt was made to update the memory of the spacecraft central computer and sequencer. This attempt was unsuccessful. Subsequent analysis failed to reveal the reason for the failure. Therefore another attempt was made on 16 March to update the memory, again unsuccessfully. After additional analysis, a special program was loaded into the memory that checked all areas related to the problem. This special program did not identify any malfunction. Many tests were then conducted, including six simulated solar occultation exercises, all of which were successful. Therefore on 22 March 1972 normal spacecraft operation was resumed including transmission of science data. The spacecraft has been performing satisfactorily since that date. However, since we have been unable to identify the anomaly, there is a finite probability that it may recur. For that reason operational procedures have been developed that will help work around such a recurrence.

From 2 April until about 6 June the spacecraft will be undergoing solar occultation. During solar occultation no electrical power is produced by the solar panels and the solar heat input to the spacecraft is absent. This results in the on-board battery having to supply the required electrical power to support the spacecraft subsystems and provide an adequate heat balance. After solar occultation, the power produced by the solar cells is required to recharge the battery. Thus during this period the primary task is to keep the spacecraft operational. No science data will be collected during this period except selected celestial mechanics and S-band occultation information.

In June 1972 when solar occultations cease, resumption of science data acquisition will occur on a limited basis. At this time the Earth-Sun-Mars orientation is such that spacecraft maneuvers are required to reorient the high-gain antenna so that science data can be transmitted to Earth. The budgeting of spacecraft resources to keep the spacecraft active through superior conjunction of Mars in September 1972 limit this data taking to a maximum of 12 pictures per week.

The desire to keep the spacecraft active until superior conjunction of Mars in September 1972 is based upon the opportunity to carry out a relativity experiment at that time. During that period, the signals from Mariner 9 will be deflected to a varying extent by the gravity of the Sun. Analysis of this deflection will provide a check on the law of relativity to an accuracy not previously achieved,

It is anticipated that the spacecraft will survive until about mid-November 1972 when its attitude control gas will expire and the spacecraft will lose attitude control.

Earl W. Glahn

Post Launch Mission Operation Report No. S-819-71-01/02

MEMORANDUM   4 December 1972
TO: A/Administrator
FROM: S/Associate Administrator for Space Science
SUBJECT: Mariner Mars '71 Programs, Assessment of Extended Mission

The Mariner Mars '71 primary 90-day mission was adjudged a success by Post Launch Mission Operation Report No. S-819-71-01/02 dated 15 February 1972. The Mariner Mars '71 extended mission of 259 days beyond the primary mission has now been adjudged a success based upon the results of the extended mission with respect to the approved prelaunch extended mission objectives.

Mariner 9 was launched on 30 May 1971 and was inserted into its Martian orbit on 13 November 1971. On 27 October 1972 the supply of spacecraft attitude control gas was depleted and spacecraft stabilization could not be maintained. Therefore the spacecraft transmitter was turned off and the mission was officially terminated at 6:31 p.m. EDT on. 27 October 1972 when the 64-meter antenna at Goldstone lost the spacecraft signal. Although the spacecraft is no longer operational, it is expected to remain in orbit about Mars for at least 50 to 100 years.

The Mariner 9 spacecraft has mapped 100% of the planet with a visual resolution of about 1 km, and has observed large areas of the planet with the infrared and ultraviolet instruments. Many S-band occultation and celestial mechanics measurements have added to our knowledge of Mars. The results of the mission have proven Mars to be geologically and meteorologically active. These conclusions are based on observations of changing surface features; evidence of volcanic activity; dust storms both global and localized in extent; evidence of glacial activity at the polar caps; changing cloud formations; and many other phenomena.

Preliminary scientific results for the first 30 days of orbital operation were reported in detail in the Jet Propulsion Laboratory Technical Report 32-1550 dated February 1972. Preliminary scientific results for the 90-day basic mission will be published in the October 1972 issue of the Icarus magazine. The final project science report is scheduled to be published in an early 1973 issue of the Journal of Geophysical Research and as a Jet propulsion Laboratory Technical Report. In addition to these project-sponsored reports, many individual papers have already been published and we expect to see many more appear throughout the coming months and years. This will be the final Mission Operation Report on the Mariner Mars '71 Mission.

E. Naugle

NASA MISSION OBJECTIVES FOR MARINER MARS '71

PROJECT OBJECTIVES

To explore, during the 1971 opportunity, the planet Mars from Martian orbit for a period of time sufficient to observe a large fraction of the planet's surface and to view selected areas during dynamic changes. Pictures of the Martian surface will be obtained as well as infrared and ultraviolet data on the planet's atmosphere and surface characteristics.

MISSION OBJECTIVES

Primary Mission

Note: Objective statements 1 and 2 will be assigned to Mariners H and I on or about May 1971. In the event that the spacecraft scheduled to arrive at the planet first fails prior to 5 days before orbital insertion of the second spacecraft, the mission objectives of the second spacecraft may be modified.

1. To map the parameters noted in the project objectives by selecting an orbit which will permit, during the primary orbital operational lifetime of 90 days, the viewing of about 70 percent of the planet's surface with

the wide angle imaging camera at a resolution of about 1 km per TV line.

2. To study the dynamic characteristics and time variable features of Mars by measuring the parameters noted in the project objectives from a Martian orbit selected to permit the viewing of selected areas periodically during the primary operational lifetime of 90 days.

Extended Mission

To continue fulfilling objectives 1 and 2 of the primary mission, but at a reduced rate of data acquisition, beyond the 90 days of the primary mission to the anticipated effective operational lifetime of about 1 year for both spacecraft.

Robert S. Kraemer   John E. Naugle
Planetary Programs Director   Associate Administrator for Space Science and Applications

Date: APRIL 9 1971

ASSESSMENT OF MARINER MARS '71 MISSION

Based upon the results of the Mariner Mars '71 Mission with respect to the approved prelaunch mission objectives, the primary mission is adjudged a success.

Robert S. Kraemer   John E. Naugle
Planetary Programs Director   Associate Administrator for Space Science
Date: 2-11-72   Date: 11 Feb 72

ASSESSMENT OF MARINER MARS '71 EXTENDED MISSION

Based upon the results of the Mariner Mars '71 Extended Mission with respect to the approved extended mission objectives, the extended mission is adjudged a success.

Robert S . Kraemer   John E. Naugle
Planetary Programs Director   Associate Administrator for Space Science
Date: 11-12-72.   Date: 10 Dec 72

VIKING '75 REPORT

National Aeronautics and Space Administration
Ames Research Center Moffett Field, California 94035
NASA Educational Data Sheet #502 November 1971

VIKING '75 PROJECT

In 1976, man will usher in a new era in science. The National Aeronautics and Space Administration has developed a plan for investigating space and the planets in a systematic manner. Within that plan, an unmanned spacecraft called Viking will traverse some 460 million miles from Earth, destined to make a soft landing on Mars. Viking will perform scientific investigations of Mars, both from orbit and on the surface of the planet. Unique experiments on board that scientific expedition are likely to yield facts that may replace century-old hypotheses and theories. The question of life on Mars has been postulated for a hundred years. There is no conclusive way to determine its existence other than direct search by landing a vehicle on the planet. The Viking mission will do that. Besides the scientific significance, the philosophical impact of such a finding is of great importance to man's understanding of his place in the universe. This search for extraterrestrial life on Mars is only part of a larger planetary exploration goal to determine:

How did our solar system form and evolve?
How did life originate and evolve?
What are the changing processes which shape man's Earth environment?

Numerous carefully selected science experiments have been chosen specifically for the 90-day mission lifetime on Mars to help provide meaningful answers to these questions. Viking will explore the Martian surface and atmosphere, listen to the interior, take panoramic pictures and perform a series of geophysical, meteorological and physical properties experiments. Space science, thus, brings the dawn of a new period in which theories of the past are replaced with substantive data. From this will come an understanding of space and planets which will feed the mainstream of knowledge here on Earth and enrich our lives. The Viking mission is an important milestone of this period.

VIKING Spacecraft:

Each Viking spacecraft consists of an orbiter and a lander that will make a soft landing on Mars. The orbiter and lander are mated during the interplanetary trip and the Mars orbital reconnaissance phase. Then the lander separates and begins its entry into the atmosphere. Use of two orbiters and two landers provides complementary functions and allows for a flexible exploration plan. Integrated exploration using lander and orbiter combinations provides increased mission success. Should a system fail or be damaged, its twin can continue exploration and investigation.

Orbiter/lander combinations will relay information to Earth and provide scientific data of the same area from two views, orbit and surface. The four systems will provide more comprehensive information about another planet than has ever been available to man before.

MISSION Profile:

The precise launch period was selected to provide a minimum energy trajectory from Earth to Mars that will be consistent with other mission requirements. Opportunities to make such flights to Mars occur at approximate 25-month intervals.

Launch from Cape Kennedy will place the Viking spacecraft and Centaur upper stage into a 115-mile Earth "parking orbit". After a coast period of some 30 minutes, the Centaur will re-ignite to send the spacecraft on its way to Mars.

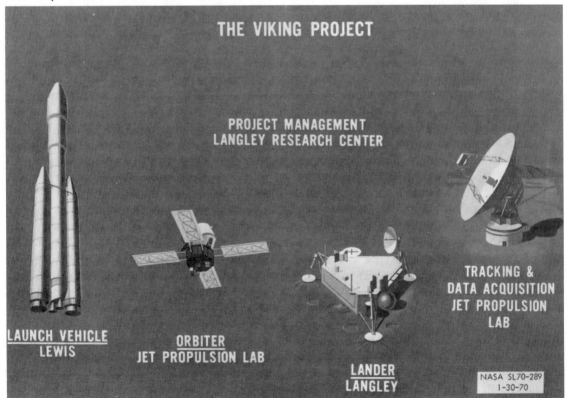

MISSION Sequence:

The Titan/Centaur will launch Viking on its way to Mars with Centaur propelling the spacecraft into Mars trajectory. The cruise to Mars will last nearly a year and will be controlled by the orbiter. Based on commands from Earth, the orbiter will perform midcourse corrections.

When Mars orbit has been achieved, data received from the spacecraft will aid in re-evaluation of preselected landing sites and may result in selection of new sites. Once a landing site has been selected, preparations for lander separation from the orbiter begin. After the lander is powered and checked out, it will separate and start descent within the aeroshell descent capsule. After landing on the Mars surface, the lander will have a direct communication link with Earth at times when the lander and Earth have direct views of one another.

Mars' atmosphere requires a heat shield and also allows use of a parachute to slow descent until the terminal propulsion system takes over. Terminal engines will shut down approximately 10 feet from the surface and the lander will gently free fall to the landing site.

ORBITER

The orbiter delivers the lander to the point of release for descent to the Mars surface, acts as a relay station, monitors the landing site and conducts investigations of the Mars atmosphere and surface. Prior to release of the lander, the orbiter searches the planet's surface to provide data necessary for detailed landing site confirmation or selection. Television cameras and thermal and water vapor mapping systems aboard the orbiter transmit data to Earth. The data are then used to construct maps of the surface.

After separation, the orbiters relay lander data to Earth and will perform scientific investigations. These investigations supplement knowledge of the Mars surface supplied by the lander, and provide data on the dynamic and physical characteristics of the planet, its surface and its atmosphere. In addition, the orbiter radio system will obtain information on the Mars atmosphere and perform certain radio science experiments.

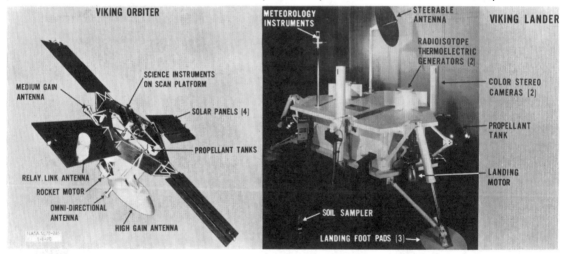

LANDER and Science:

The purpose of the Viking mission is the return to Earth of useful scientific data from Mars. The orbiter and lander will work together to assure the return of valuable scientific information. While the orbiter makes long-range studies from an orbital vantage point, the lander will make closer investigations of the planet during a 90-day period.

Scientists have carefully planned the detailed objectives to provide maximum potential from the mission. By weighing the scientific importance against technical limitations, they have defined the investigations that can be conducted by a soft landing system. Interest is obviously concentrated on the search for possible life and the study of the environment in which it would exist. To assure that Mars will not be contaminated by Earth micro-organisms, the entire lander system will be sterilized by heating after the landers are sealed in their capsules. Sterilization will assure that chances of contaminating Mars are less than one in a million.

As the lander descends through the atmosphere a mass spectrometer will measure Mars' composition. Pressure, temperature and density variations will also be noted during descent.

After landing, photographs will be taken and instruments will collect data about the surface characteristics, meteorologic activity and physical structure. Biological and organic analysis of soil samples by the lander's laboratory will then be transmitted to mission controllers on Earth, supplying answers to some of the most significant questions being asked by mission scientists.

LAUNCH, Flight & Data Tracking:

Though Viking is a highly automated system, control teams on Earth will retain the ability to command changes in the mission program while tracking and monitoring the spacecraft.

Tracking and monitoring begin at the moment of launch from Cape Kennedy. Velocity and trajectory information is received by navigation teams until the craft achieves Earth orbit. When the Centaur is in a position to inject the spacecraft into its Mars trajectory, it will fire, then separate and be deflected from the flight path to avoid encounter with Mars by some 50,000 miles.

Once the craft enters its Mars trajectory, the deep space tracking facilities will serve as the communications link. Stations located in California, Spain and Australia will maintain contact with both spacecraft during flight to Mars. At least one of the stations will have a clear communications path to the craft when in Earth view.

The 85-and 210-foot antennas will receive data from the spacecraft to assure highest quality communications throughout all phases of the mission. The Mission Control Team on Earth will monitor the flight plan at all times.

VIKING PRESS KIT

VIKING

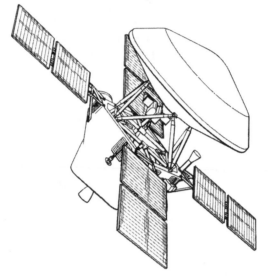

PRESS
KIT

NASA
National
Aeronautics and
Space
Administration

RELEASE NO: 75-183

NASA News
National Aeronautics and Space Administration
Washington, D.C. 20546
AC 202 755-8370

NASA RELEASE NO: 75-183

Nicholas Panagakos Headquarters, Washington, D.C. (Phone: 202/755-3680)
Maurice Parker Langley Research Center, Hampton, Va. (Phone: 804/827-3966)

VIKING MARS LAUNCH SET FOR AUGUST 11

For Release: IMMEDIATE

America's most ambitious unmanned space venture will get under way next month with the launch of two Viking spacecraft to Mars.

The year-long, 815-million-kilometer (505-million-mile) journey will culminate with the landing of an automated laboratory on the surface of the planet in the summer of 1976.

The instrument-laden craft will take pictures and conduct a detailed scientific examination of Mars, including a search for life.

The surface exploration of Mars — the planet most like Earth — should yield new knowledge on the origin and evolution of our solar system, and provide insights into the processes that have shaped our Earth.

The scientific venture begins about August 11, when Viking A is scheduled for launch from Cape Canaveral, Fla., aboard a Titan 3/Centaur rocket. Viking B is set to follow 10 days later.

The designations of Vikings A and B will be changed to Vikings 1 and 2 once each spacecraft has been successfully launched.

After a long, looping chase through space to overtake Mars, Viking 1 will arrive in Martian orbit about June 18, 1976, and may remain in orbit for as little as two weeks or as much as 50 days.

The most critical part of the mission begins when the four-ton Viking divides into an Orbiter and a Lander. The orbiter, circling the planet at distances ranging from 1500 km (930 mi.) to 32,600 km (20,200 mi.), will map the surface and take Mars' atmospheric pulse, looking for signs of life.

The Lander will survive the flashing heat of entry through the planet's atmosphere, land gently on the surface, and conduct an intricate scientific examination.

If all goes as planned, Lander 1 may touch down on Mars July 4, 1976, the 200th anniversary of the United States.

Viking 2 will arrive at Mars seven weeks after Viking 1 about August 7. Lander 2 will touch down about September 9.

Each Viking is packed with instruments that will be used to conduct 13 separate but related scientific investigations. Some instruments will do experiments, some will take measurements, others will observe, and a few will be combined into a single investigation that observes, measures and experiments.

Three investigations will be conducted from each orbiter, eight from each Lander, and two more will use equipment aboard both spacecraft.

The orbiter's three investigations will photograph the planet and map its atmospheric water vapor and thermal properties. Its instruments are two high-resolution television cameras, an infrared spectrometer and an infrared radiometer. They are mounted on a scan platform that is bore-sighted along a common axis to look at the same area of Mars. The instruments will seek suitable landing sites and provide information on Mars' atmospheric water concentration, its surface temperature, any clouds or dust storms, and terrain topography and color.

The Orbiter instruments will continue to operate during the Lander portion of each mission.

During the Lander entry to Mars, several instruments and sensors will measure the atmosphere's structure and chemical composition. In the upper atmosphere, a mass spectrometer, a retarding potential analyzer and several sensors — all mounted on the Lander's protective aeroshell — will measure atmospheric composition, temperature, pressure and density.

Continued pressure, temperature and density variations will be measured in the lower atmosphere by sensors mounted on the Lander.

The Lander's cameras will take pictures of Mars from the surface, and Lander instruments will study the planet's biology, molecular structure, inorganic chemistry, meteorology and seismology, and physical and magnetic properties.

The Lander's surface sampler, attached to a furlable boom, will be extended to dig up soil samples for incubation and analysis inside the biology instrument's three metabolism and growth experiment chambers, in a gas chromatograph-mass spectrometer (GCMS) and an X-ray fluorescence spectrometer (XRFS).

These three investigations are particularly important for understanding the biological makeup of Mars. They will also supply knowledge to chemists, planetologists and other scientists.

The meteorology instrument, located on a folding boom attached to the Lander, will periodically measure temperature, pressure, wind speed and direction during the mission.

A three-axis seismometer will measure any seismic activity that takes place during the mission, which should establish whether or not Mars is a very active planet.

The physical and magnetic properties of Mars will be studied with several small instruments and pieces of equipment located on the Lander.

The radio science investigations will make use of orbiter and Lander communications equipment to measure Mars' gravitational field, determine its axis of rotation, measure surface properties, conduct certain relativity experiments, and pinpoint the locations of both Landers on Mars. A special radio link, the X-band, will be used to study charged ion and electron particles. The Viking launch vehicle, the Titan III solid-and-liquid-fueled rocket and the liquid-fueled Centaur, is a relatively new combination, although both vehicles have been used separately for many space launches.

About 10 minutes after launch, the vehicle takes its Viking payload into an Earth parking orbit of 167 km (104 mi.). A second firing of the Centaur engine puts Viking into Trans-Mars Injection (TMI), starting the spacecraft on its long voyage to Mars.

As the cruise begins, the Orbiter adjusts its antennas toward Earth, spreads its solar panels, seeks the Sun, and acquires the bright star Canopus for guidance.

Tracked by the Deep Space Network on Earth, Viking quietly cruises toward Mars, powered by sunlight (and batteries when required). The Orbiter is the operating spacecraft during cruise, taking periodic pulse of the Lander stored in its aeroshell like a pupa inside its cocoon.

Both craft become very active once Mars orbit is achieved. The Orbiter begins its science investigations about 10 days before that time, then powers up the Lander in preparation for separation. Four landing sites have been selected for the Landers, two primary and two secondary spots, but mission controllers do not have to commit themselves to a site until they have had a chance to observe the planet for several days.

Prime target for Lander 1 is a region known as Chryse, located at the northeast end of a 4,800 km (3,000 mi.) - long rift canyon discovered by Mariner 9. The Chryse site is 19.5 degrees north and 34 degrees west.

Lander 2's primary landing site is Cydonia, in the Mare Acidalium region at the edge of the southernmost reaches of the north polar hood, a hazy veil that shrouds the region during winter and that some scientists think may carry moisture. Cydonia is 44.3 degrees north and 10 degrees west.

Lander 1's backup site is Tritonis Lacus, located at 20.5 degrees north and 252 degrees west. The Lander 2 backup is Alba, called the white region, lying 44.2 degrees north and 110 degrees west.

If any of these selected sites proves unpromising after Orbiter investigations, scientists on Earth will select other sites before committing the Landers. Once separated from its Orbiter, the Lander will be aligned for Mars entry. Its aeroshell will then bear the heat of peak acceleration. Once slowed, the Lander's parachute will pop out to gently lower the craft almost to the surface. Retro engines will finally settle the Lander to its operating base.

The Viking Project is managed by NASA's Langley Research Center, Hampton, Va. The Landers were built by Martin Marietta Aerospace of Denver, Colo., which also has integration responsibility for Viking. The Orbiters were built by NASA's Jet Propulsion Laboratory (JPL) in Pasadena, Calif.

The Titan III/Centaur launch vehicle is managed by NASA's Lewis Research Center, Cleveland, Ohio, and was built by Martin Marietta Aerospace and the General Dynamics Corp.

Viking will be controlled from NASA's Kennedy Space Center, Fla., until completion of the launch phase. Control then shifts to JPL in Pasadena for the remainder of the missions.

Nerve center of the JPL operations is the Viking Mission Control and Computing Center (VMCCC), where the 700-member Viking Flight Team of engineers, scientists and technicians will maintain constant control of the four Viking spacecraft.

(END OF GENERAL RELEASE. BACKGROUND INFORMATION FOLLOWS.)

THE SCIENTIFIC GOALS OF THE VIKING MISSION

Mars has excited man's imagination more than any other celestial body except the Sun and the Moon. Its unusual reddish color, which the ancients associated with fire and blood, gave rise to its being named for the Roman God of War.

The invention of the astronomical telescope by Galileo in 1608 opened a new era in the observation of the planet. Instead of appearing merely as a tiny disc, Mars' surface features could be resolved.

Christian Huygens made the first sketch in 1659 of the dark region, Syrtis Major ("giant quicksands"). Able to observe a distinguishable feature, Huygens could show that Mars rotated on a north-south axis like Earth, producing a day that was about half an hour longer than Earth's.

In 1666, the Italian astronomer Giovanni D. Cassini observed and sketched the Martian polar caps. Observers in the early 1700's noted changes in the surface appearance in a matter of hours, probably caused by dust storms, now known to rage periodically. In 1783, William Herschel observed that Mars' axis of rotation is inclined to its orbital plane at about the same extent as Earth's, revealing that long-term changes were often associated with seasons that would result from such inclination.

In the 17th and 18th centuries, it was commonly accepted that Mars and the other planets were inhabited, but the real excitement was created by Giovanni Schiaparelli and Percival Lowell between 1877 and 1920. As a result of extensive observations, beginning with the favorable apparition of 1877, Schiaparelli constructed detailed maps with many features, including a number of dark, almost straight lines, some of them hundreds of kilometers long. He referred to them as "canali" or channels. Through mistranslation, they became "canals" and the idea of civilized societies was propagated.

Lowell's firm opinion that the canals were not natural features but the work of "intelligent creatures, alike to us in spirit but not in form" contributed to the colorful literature. To pursue his interest in the canals and Mars, he founded the Lowell Observatory near Flagstaff, Ariz., in 1894, and his writings about the canals and possible life on Mars created great public excitement near the turn of the 20th century.

Speculation about intelligent life on Mars continued through the first part of the century, with no possibility of an unequivocal resolution, but a gradual tendency developed among scientists to be very skeptical of the likelihood.

The skepticism was reinforced by the results of two Mariner flyby missions in 1965 and 1969. The limited coverage of only about 10 per cent of the Martian surface by flyby photography indicated that Mars was a lunar-like planet with a uniformly cratered surface.

In 1971-72 the Mariner 9 orbiter revealed a completely new and different face of Mars. Whereas the flyby coverage had seen only a single geologic regime in the cratered highlands of the southern hemisphere, Mariner 9 revealed gigantic volcanoes, a rift valley that extends a fifth of the way around the planet's circumference, and possible evidence of flowing liquid water sometime in the past. Also revealed were layered terrain in the polar regions, and the effects of dust moved by winds of several hundred kilometers an hour.

In short, Mariner 9's 7,000 detailed pictures revealed a dynamic, evolving Mars completely different from the lunarlike planet suggested by the flyby evidence. That eminently successful Orbiter mission showed a fascinating subject for scientific study and also provided the maps from which the Viking sites have been selected.

The scientific goal of the Viking missions is to "increase our knowledge of the planet Mars with special emphasis on the search for evidence of extra-terrestrial life." The scientific questions deal with the atmosphere, the surface, the planetary body, and the question of bio-organic evolution. This goal ultimately means understanding the history of the planet.

The physical and chemical composition of the atmosphere and its dynamics are of considerable interest, not only because they will extend our understanding of planetary atmospheric sciences, but because of the intense focus of interest in contemporary terrestrial atmospheric problems.

We want to understand how to model our own atmosphere more accurately, and we want to know how the solar wind interacts with the upper atmosphere; to do this we must know more about its chemistry, the composition of neutral gases and charged particles.

We want to reconstruct the physics of the atmosphere and determine its density profile. We want to measure the atmosphere down to the surface and follow its changes, daily and seasonally. From these data may come clues to the atmospheric processes that have been taking place and determining the planet's character. Of special interest is the question of water on Mars. Scientific literature is sparse in data and rich in speculation. It is known that there is water in the Mars atmosphere, but the total pressure of the atmosphere (about one per cent of Earth's) will not sustain any large bodies of liquid water. Nevertheless, the presence of braided channels suggests to many geologists that they are the result of previous periods of flowing water. This idea of episodic water suggests a very dynamic planet.

The geology of Mars has attracted great interest among planetologists because of the wide variety of features seen in the Mariner photos.

Vulcanologists are intrigued by the high concentration of volcanoes near the Tharsis ridge. Scientists who study erosion are fascinated with the great rift valley (Valles Marineris) that is 100 kilometers (62 miles) wide, 3,000 km (1,800 mi.) long and 6 km (4 mi.) deep. Some geologists have focused on the polar region, which appears to be stratified terrain. The pole resembles a rosette; it has been suggested that this is evidence of precession (wobbling) of the poles. One important question that Viking is not likely to answer, due to payload limitation, is the age of the planet.

One mystery that Viking may solve is the fate of nitrogen. So far there has been no report of nitrogen on Mars. Has it been lost by outgassing? Is it locked up in the surface as nitrates or in some organic form? Chemists and biologists both look upon nitrogen, among the most cosmically abundant of the elements, as vitally important because of the clues it provides to the evolution of the atmosphere and of the planet itself.

There is the final question of life on Mars. This may be one of the most important scientific questions of our time. It is also one of the most difficult to answer.

A negative answer does not prove there is no life on Mars. The landing site may have been in the wrong place, during the wrong season, or we may have conducted the wrong experiments. Many scientists still think there is a low probability of life on Mars.

How can this extensive effort to perform the search be justified? First, it must be acknowledged that there is no evidence at present, pro or con, of the existence of life on Mars. And what we seek is evidence. The remarkable thing is that we live at a time in which we can make this first test for life, and also assemble a great store of knowledge of the planet. As has been so well stated of the importance of this goal:

"The discovery of life on another planet would be one of the momentous events of human history."

(Dr. N. H. Horowitz, Professor of Biology, Cal Tech)

Finally, we regard as of utmost importance a knowledge of the organic character of the planet. Whether life has begun or not, it is critical to our concept of chemical evolution to determine the path of carbon chemistry. Mars offers the first opportunity to gain another perspective in the cosmic history of planetary chemistry.

The scientific investigations of Viking were intentionally selected to complement one another. The Orbiter science instruments are used to help select landing sites for the Lander investigations. The Lander cameras help select soil samples for the chemical and biological analyses. The meteorology data are used to determine periods of quiet for the seismology experiment. The atmospheric data are used in determining the chemistry, which in turn is used in understanding the biological result.

But Viking's greatest asset is its flexibility. The scientist-engineer teams will be interacting, hour by hour, during the several months that Viking will be returning data. Every day will bring new discoveries and fresh ideas for improving the mission to extract the maximum benefit from this effort.

G. A. Soffen, Project Scientist
G. D. Sands, Associate Project Scientist
C. Snyder, Orbiter Scientist

VIKING SCIENCE INVESTIGATIONS

Three science investigations use instruments located on the Orbiter: orbiter imaging, atmospheric water vapor mapping and thermal mapping. One investigation, located on the Lander, is conducted while the Lander is descending to the surface of Mars: entry science.

Eight other investigations are conducted from the Lander: lander imaging, biology, molecular analysis, inorganic chemical analysis, meteorology, seismology, physical properties and magnetic properties. one investigation, radio science, has no specific instrument, but uses the Viking telecommunications system to obtain data and do certain experiments.

Orbiter Imaging

The orbiter imaging investigation has four objectives:

* Add to the geologic knowledge of Mars by providing high-resolution photographic coverage of scientifically interesting areas of the Martian surface.

* Add to the knowledge of dynamic processes on Mars by observing the planet during seasons never before seen.

* Provide high-resolution imaging data of the Viking landing sites before landing so site safety and scientific desirability can be assessed.

* Monitor the region around each landing site after landing so the dynamic environment in which Lander experiments are done is better understood.

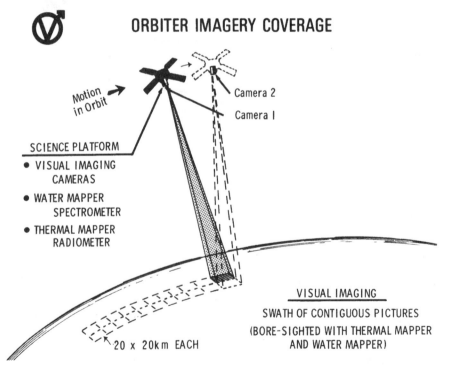

ORBITER IMAGERY COVERAGE

The Visual Imaging Subsystem (VIS) consists of two identical cameras, mounted side by side on the orbiter's scan platform.

The cameras will be used in different ways as the mission progresses. As Viking 1 approaches Mars, the planet will be photographed in three colors. The planet's atmosphere is expected to be clear, in contrast to the 1971

Mariner 9's approach to Mars, allowing the first useful approach pictures since 1969.

Between Mars orbit insertion and landing, the cameras will be used almost exclusively for examining the landing site. The intent is to characterize in detail terrain at the site and make estimates of slopes at the Lander scale; examine the region of the site for both long and short-term changes that might indicate wind action; and monitor any atmospheric activity.

For most of the period after landing, the Orbiter will be a communications relay link for the Lander and its orbit will remain synchronized with the Lander; i.e., it will pass over the Lander at the same time each day. Areas visible from the synchronous orbit will be systematically photographed during this time. Areas covered will include the large channel system upstream from the primary landing site, chaotic terrain from which many of the channels seem to originate, and areas of greatest canyon development.

The detailed observations are expected to lead to a better understanding of the origin of these features, and aid in interpreting Lander data. At the same time, activity will be monitored over all the planet that is visible from apoapsis. Any areas of unusual activity will also be examined in detail. The orbital period will be changed for short periods of time to allow the rest of the planet to be seen. During these periods, observations will be made of large volcanoes, channels and other-features.

Viking 2 will follow a similar plan, except for one major difference: shortly after Lander 2 lands, the orbital inclination of Orbiter 2 will be increased to perform polar observations. After the inclination change, a period of systematic mapping of the North Polar region is anticipated, similar to that undertaken in the canyon lands by Orbiter 1.

The succession of deposits in the polar regions, their thickness and relative ages, will be determined with these observations. This portion of the mission is particularly important because of its potential for unraveling past climatic changes and assessing the volatile inventory (primarily carbon dioxide and water) of the planet.

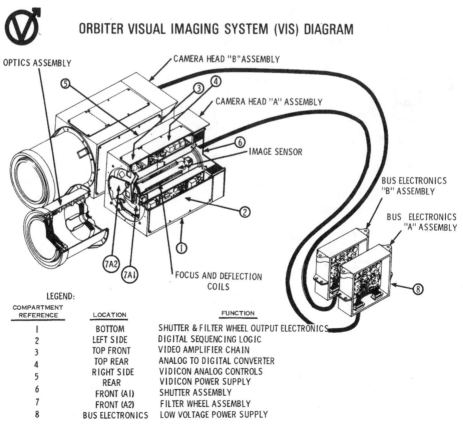

ORBITER VISUAL IMAGING SYSTEM (VIS) DIAGRAM

OPTICS ASSEMBLY
CAMERA HEAD "B" ASSEMBLY
CAMERA HEAD "A" ASSEMBLY
IMAGE SENSOR
BUS ELECTRONICS "B" ASSEMBLY
BUS ELECTRONICS "A" ASSEMBLY
FOCUS AND DEFLECTION COILS

LEGEND:

COMPARTMENT REFERENCE	LOCATION	FUNCTION
1	BOTTOM	SHUTTER & FILTER WHEEL OUTPUT ELECTRONICS
2	LEFT SIDE	DIGITAL SEQUENCING LOGIC
3	TOP FRONT	VIDEO AMPLIFIER CHAIN
4	TOP REAR	ANALOG TO DIGITAL CONVERTER
5	RIGHT SIDE	VIDICON ANALOG CONTROLS
	REAR	VIDICON POWER SUPPLY
6	FRONT (A1)	SHUTTER ASSEMBLY
7	FRONT (A2)	FILTER WHEEL ASSEMBLY
8	BUS ELECTRONICS	LOW VOLTAGE POWER SUPPLY

Each visual imaging subsystem consists of a telescope, a camera head and supporting electronics. The telescope focuses an image of the scene being viewed on the faceplate of a vidicon within the camera head. When a shutter between telescope and vidicon is activated, an imprint of the scene is left on the vidicon faceplate as a variable electrostatic charge. The faceplate is then scanned with an electronic beam and variations in charge are read in parallel onto a seven-track tape recorder.

Data are later relayed to Earth one track at a time. A picture is assembled on Earth as an array of pixels (picture elements), each pixel representing the charge at a point on the faceplate (i.e., the brightness at a point in the image). As data are read back, the pixel array is slowly assembled, complete only when all seven tracks have been read. The image is then displayed on a video screen and film copies are made.

The telescope has an all-spherical, catadioptric cassegrain lens with a 475-millimeter (18.7-inch) focal length. The sensor is a 38 mm (1.5-in.) selenium vidicon. A mechanical focal plane tape allows exposure times from 0.003 to 2.7 seconds. Between the vidicon and telescope is a filter wheel that provides color images. Each frame has 8.7 million bits (binary digits).

The cameras are mounted on the orbiter with a slight offset and the timing of each shutter is offset by one-half frame time so the cameras view slightly different fields and shutter alternately. The combined effect is to produce a swath of adjacent pictures as the motion of the spacecraft moves the fields of view across the surface of Mars. The resolution at the lowest part in the orbit (1,500 km or 810 mi.) is 37.5 meters (124 ft.) per pixel. This would allow an object the size of a football field to be resolved.

The Orbiter Imaging investigation team leader is Dr. Michael H. Carr of the U.S. Geological Survey, Menlo Park, Calif.

Water Vapor Mapping

Water vapor is a minor constituent of the Martian atmosphere. Its presence was discovered about 10 years ago from Earth-based telescopic observations. They indicated that the vapor varies seasonally, appearing and disappearing with the recession and growth of the polar cap in each hemisphere, and diurnally (daily), with its maximum close to local (Mars) noon (Mars time). Some evidence was found that water vapor is contained in the lowest layers of the atmosphere, perhaps within the first 1,000 m (3,300 ft.) above the surface.

OPTICAL CONFIGURATION DIAGRAM
(MAWD)

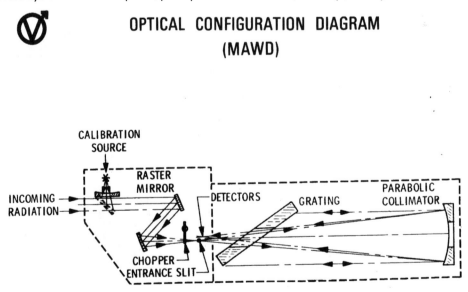

The abundance of atmospheric water vapor is usually given in units of "precipitable microns," a measure of the thickness of the ice layer or liquid that would be formed if all the vapor in the atmospheric column above the surface were condensed out.

Compared with Earth, Mars' atmosphere contains very little water. The atmosphere on Earth typically holds the equivalent of one or two centimeters (0.4 to 0.8 inches) of water, but the most water observed on Mars is only a fraction of one per cent of Earth's or 50 precipitable microns (0.002 inches).

The Mars atmosphere is a hundred times thinner than Earth's, however, and its mean or average temperature is considerably lower. The relative water concentrations, therefore, are not very different (about one part per thousand) and the relative humidity on Mars can be significant at times. It is misleading to refer to Mars as a "dry" planet.

Yet in terms of total planetary abundance, evidence suggests that there is very little water on or above the planet's surface in the form of atmospheric vapor or surface ice. Since water is cosmically one of the more abundant molecules, the question arises: Has Mars lost most of its water during its evolution, or is water present beneath the surface, a subsurface shell of ice or permafrost, or perhaps held deeper in the interior to be released by thermal and seismic activity at some future time?

Mariner spacecraft observations of Mars in 1969 and 1971 showed that while the polar hoods are predominantly frozen carbon dioxide, the visible caps left after the carbon dioxide vaporizes are water crystals,

Mariner results revealed other intriguing facts related to the history of water on Mars: The atmosphere loses hydrogen and oxygen atoms to space at a slow but steady rate, and in the relative proportions with which they make up the water molecule. Surface features exist that appear to have been formed by flowing liquid; the latter are quite different from the river-like features caused by lava flows; they appear to be wide braided channels formed from an earlier period of flooding by a more mobile liquid than volcanic lava.

Again a question: Are we now seeing the last disappearing remnants of water that was once much more plentiful on the planet, or is Mars locked in an ice age that has frozen out most of its water in the polar caps or beneath a layer of surface dust?

Martian water clearly holds many clues to the planet's history. By studying the daily and seasonal appearance and disappearance of water vapor in more detail than is possible from Earth by mapping its global distribution, and by determining the locations and mechanisms of its release into the atmosphere, scientists should understand more clearly the present water regime, and perhaps unravel some of the mystery surrounding past conditions on the planet.

In the context of Martian biology, such clarification may have great significance in establishing the existence, now or in the past, of an environment favorable to the survival and proliferation of living organisms. This presumes that Martian life is dependent on the availability of water as is life on Earth.

The Viking water vapor-mapping observations will be made with an infrared grating spectrometer mounted on the Orbiter scan platform, boresighted with the television cameras and the Infrared Thermal Mapper (IRTM). The spectrometer, called the Mars Atmospheric Water Detector (MAWD), measures solar infrared radiation reflected from the surface of the planet after it has passed through the atmosphere.

The instrument selects narrow spectral intervals coincident with characteristic water vapor absorption in the 1.4-micron wavelength region of the spectrum. Variations in the intensity of radiation received by the detectors provide a direct measure of the amount of water vapor in the atmospheric path traversed by the solar rays.

The sensitivity of the instrument enables amounts of water from a minimum of a precipitable micron to a maximum of 1,000 microns to be measured. The precise wavelengths of radiation to which the detectors respond are also selected so instrument data can be used to derive atmospheric pressure at the level where the bulk of the water vapor resides, providing an indication of its height above the surface.

At the lowest point in the orbit, the field of view of the detector is a rectangle 3 by 20 km (1.9 by 12.4 mi.) on the planet's surface. This field is swept back and forth, perpendicular to the ground track of the orbiter, by an auxiliary mirror at the entrance aperture of the instrument. In this way the water vapor over selected areas of the planet can be mapped.

A small ground-based computer, dedicated to the use of the two orbiting infrared instruments, directly reduce data to contour plots of the water vapor abundance and pressure.

During the initial orbits, and particularly through the landing site certification phase of the missions before Lander separation, MAWD observations will concentrate on an area within a few hundred kilometers around the landing sites, to help in site certification and to complement landed science measurements.

In later phases of the mission, observations will be extended to obtain global-coverage of water vapor distribution and its variation with time-of-day and seasonal progression. The search for regions of unusually high water content will be emphasized during these later stages, and areas of special interest will be studied, including volcanic ridges, the edge of the polar cap and selected topographic features.

The Water Vapor Mapping investigation team leader is Dr. Crofton B. (Barney) Farmer of the Jet Propulsion Laboratory.

Thermal Mapping

The Thermal Mapping investigation is designed to obtain temperature measurements of areas on the surface of Mars. It obtains the temperature radiometrically with an Infrared Thermal Mapper (IRTM) instrument.

Information obtained by the thermal mapper will contribute to the study of the surface and atmosphere of Mars, which is similar to and in some ways simpler to study than Earth. Mars appears to be geologically younger, and clearly is undergoing major changes. Studies of Martian geology and meteorology can have implications in tectonics (the study of crustal forces), vulcanology and understanding weathering and mineral deposition.

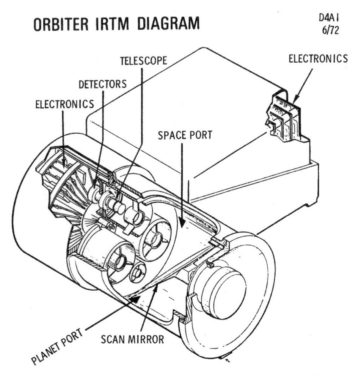

ORBITER IRTM DIAGRAM

D4A I
6/72

Just as fine beach sand cools rapidly in the evening while large rocks remain warm, daily temperature variation of the Martian surface indicates the size of individual surface particles, although the thermal mapper necessarily obtains an average value over many square kilometers. Measurements obtained just before sunrise are especially valuable (the detectors can sense the weak heat radiation from the dark part of the planet), since at that time the greatest temperature differences occur between solid and fine-grained material.

One detector is used to measure the upper atmospheric temperature. That information may be combined with surface temperatures to permit construction of meteorological models. An understanding of the important Martian wind circulation depends on such models.

Data received from the thermal mapper are intended to help establish and evaluate the site for the Viking Lander. Martian organisms would probably be affected by local water distribution and temperatures of the soil and air; these factors are either measured by the radiometer or are dependent on the soil particle characteristics determined by the thermal mapper.

The Infrared Thermal Mapper is a multi-channel radiometer mounted on the Orbiter's scan platform. It accurately measures the temperatures of the Martian surface and upper atmosphere, and also the amount of sunlight reflected by the planet. Four small telescopes, each with seven sensitive infrared detectors, are aimed parallel to the Visual Imaging optical axis. Differences of one degree Celsius (about 1.8 degrees Fahrenheit) can be measured throughout the expected temperature range of minus 130 degrees C to plus 57 degrees C (minus 202 to 135 degrees F). The instrument is 20 by 25 by 30 cm (8x10x12 in.) and has a minimum spatial resolution of 8 km (5 statute miles) on the surface.

The large number of detectors (28) is chosen to provide good coverage of the Martian surface, and to allow several infrared "colors" to be sampled. Differences in the apparent brightness of a spot on the planet in the various colors imply what kinds of rocks (granite, basalt, etc.) are present. The temperatures themselves may indicate the composition of clouds and the presence of dust in the atmosphere.

The spatial resolution available to the thermal mapper will permit reliable determination of the frost composition comprising the polar caps. The close spacing of infrared detectors and the spacecraft scanning mode improves the ability to identify possible local effects such as current volcanic activity or water condensation.

The Thermal Mapping investigation team leader is Dr. Hugh H. Kieffer of the University of California at Los Angeles.

⊘ ENTRY SCIENCES AEROSHELL INSTRUMENTATION

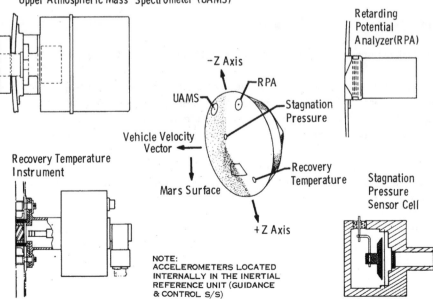

Upper Atmospheric Mass Spectrometer (UAMS)

Retarding Potential Analyzer (RPA)

−Z Axis

RPA

UAMS

Stagnation Pressure

Vehicle Velocity Vector

Recovery Temperature Instrument

Mars Surface

Recovery Temperature

Stagnation Pressure Sensor Cell

+Z Axis

NOTE: ACCELEROMETERS LOCATED INTERNALLY IN THE INERTIAL REFERENCE UNIT (GUIDANCE & CONTROL S/S)

Entry Science

The Entry Science investigation is concerned with direct measurements of the Martian atmosphere from the time the Lander and Orbiter separate until the Lander touches down on the planet's surface. Knowledge of a planet's atmosphere, both neutral and ionized components tells much about the planet's physical and chemical evolution, and it increases understanding of the history of all planets, including that of Earth.

The question of the atmospheric composition of Mars is of immediate interest to scientists. Nitrogen, believed essential to the existence of life, has never been detected on Mars by remote sensing methods, either because it is not present or because of the low sensitivity of the measurement methods. Measurements of atmospheric pressure and temperature, plus winds, are important in understanding the meteorology, just as observations made with weather balloons in Earth's atmosphere supplement surface observations.

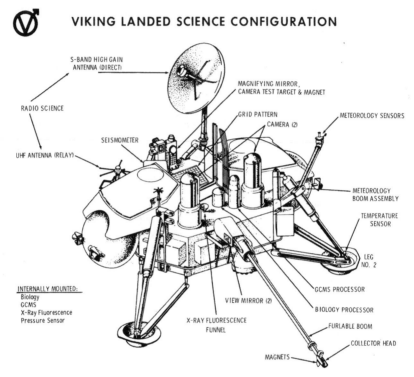

VIKING LANDED SCIENCE CONFIGURATION

Upper Atmosphere

Studies of the Martian upper atmosphere composition begin shortly after the Lander leaves the Orbiter.

The first measurements are made so high above the surface that only charged particles can be detected. These measurements are made with a Retarding Potential Analyzer (RPA) that will measure electron and ion concentrations and temperatures of these components. Measurements continue down to about 100 km (60 mi.) above the planet's surface, where the pressure becomes too high for the instrument to operate.

On Mars, which has a weak magnetic field compared with that of Earth, charged particles streaming from the Sun (called the solar wind) and interacting with the upper atmosphere may be important in determining the nature of the lower atmosphere and, therefore, the conditions for life.

At the very highest altitudes, the analyzer will study the interaction of the solar wind with the Martian atmosphere. Measurements at lower altitudes will make important contributions to knowledge of the interaction of sunlight with atmospheric gases, a matter of great significance in understanding the photochemical reactions that take place in all planetary atmospheres, including Earth's.

The analyzer will make measurements several thousand kilometers above the Martian surface, but the neutral atmosphere at high altitudes is so thin that measurements will not begin until the Lander drops to an altitude of around 300 km (180 mi.) above the surface. Measurements on the neutral constituents of the atmosphere are made with the Upper Atmosphere Mass Spectrometer (UAMS).

The mass spectrometer will sample and analyze the atmosphere as the Lander passes through. Inside the instrument, the gas to be analyzed is ionized by an electron beam, and the ions formed are sent through an appropriate combination of electric and magnetic fields to determine the amounts of the various molecular weights by which the various gases can be identified.

From remote measurements, carbon dioxide is known to be the principal atmospheric constituent on Mars. The mass spectrometer should be able to detect 0.1 per cent of nitrogen and even a smaller amount of argon. Argon's principal isotope is a radioactive decay product of potassium, an important constituent in many minerals. About one per cent of the Earth's atmosphere has come from the radioactive decay of potassium in the Earth's crust.

The mass spectrometer will also look for molecular and atomic oxygen, carbon monoxide and other common gases that may be present in the Martian atmosphere. It may tell if the isotope composition in elements such as carbon, oxygen and argon is the same as on Earth, thereby providing measurements needed to understand planetary evolution.

Lower Atmosphere

The lower atmosphere begins at about 100 km (60 mi.) altitude where the analyzer and mass spectrometer become inoperative. The bulk of the atmospheric gases reside below this altitude. On Mars the surface atmospheric pressure is only about 1.5 per cent as great as on Earth.

Measurements of Mars' surface pressure are all based on remote observations, principally alteration by the atmosphere of radio waves from Mariner spacecraft as they flew behind the planet. Viking will obtain direct pressure and temperature measurements in the lower atmosphere and on the surface.

In passing through the atmosphere from 100 km (60-mi.) to the surface, the Landers will obtain profiles of the properties of the atmosphere: pressure, density and temperature. First measurements will be by sensitive determination of the aerodynamic retardation of the Lander, from which atmospheric density can be derived.

The density profile with altitude permits the weight (pressure) of the atmosphere above any given level to be calculated. Given atmospheric composition, pressure and density will define the structure of the atmosphere from roughly 100 to 25 km (60 to 15 mi.) altitude. Below 25 km, sensors can be deployed to directly measure the pressure and temperature, although these measurements have to be done with specially designed sensors because of the low pressure in the atmosphere.

The importance of the profiles is that they are determined by solar energy absorption and vertical heat flow. Heat can be transported either radiatively or convectively (by infrared emission or absorption) or by currents and winds.

The atmosphere of Mars appears to be windy compared to Earth's lower atmosphere. This is a result of the low density of the atmosphere, which permits it to change temperature rapidly, and causes large temperature variations from day to night and seasonally. There are large contrasts in temperature of the atmosphere, precisely the condition to create winds.

One evidence for high winds is the frequently severe dust storms, such as the long-lasting one that greeted Mariner 9. These storms are a puzzle, since it takes even stronger winds than those now calculated by computer models of the atmospheric circulation (18 to 46 m per second; 40 to 100 mi. per hour) to raise dust in this tenuous atmosphere.

The vertical profiles of temperature will provide additional evidence of the thermal balance of the atmosphere and, it is hoped, of forces that drive the winds. Winds also will be measured directly in the parachute phase by tracking the motion of the Lander over the surface as it drifts, carried by local winds. These measurements will extend to an altitude of 6 km (3.7 mi.).

The Entry Science investigation team leader is Dr. Alfred O.C. Nier of the University of Minnesota.

Lander Imaging

As a person depends on sight for learning about the world, so cameras will serve as the eyes of the Lander and, indirectly, of the Viking scientists.

Pictures of the region near the Lander will be studied to select a suitable site for acquiring samples that will be analyzed by other Lander instruments. The cameras will also record that the samples have been correctly picked up and delivered. From time to time, the cameras will examine different parts of the Lander to see that components are operating correctly.

One category of the Lander imaging investigation is the study of general geology or topography. Pictures of the Martian surface visible to the cameras are of the highest scientific priority. The first pictures will be panoramic surveys, and then regions of particular interest will be imaged in high resolution, in color and in infrared.

LANDER CAMERA SYSTEM DIAGRAM

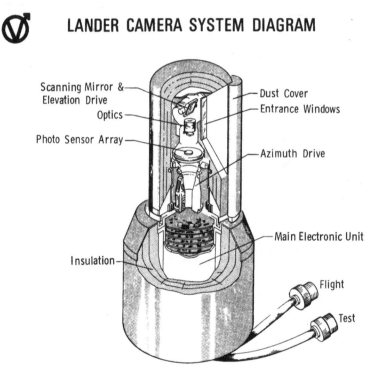

Scanning Mirror & Elevation Drive
Optics
Photo Sensor Array
Insulation

Dust Cover
Entrance Windows
Azimuth Drive
Main Electronic Unit
Flight
Test

Stereoscopic views are obtained by photographing the same object with two cameras, providing photos in which three-dimensional shapes can be distinctly resolved. Putting together this information, scientists can tell much about the character of the Martian surface and the processes that have, shaped it.

One can imagine finding shock-lithified rocks (as on the moon), igneous boulders, wind-shaped boulders (ventifacts), sand ripples, or a lag gravel deposit. Each of these possible objects could be resolved in pictures; each would bespeak a particular kind of surface modification.

The advantage of operational flexibility is important. Scientists will study the first pictures and, on the basis of what they reveal, select particular areas for more detailed examination. This method will require sending new picture commands to the Lander every few days.

Used as photometers, the cameras will yield data that permit inferences about the chemical and physical properties of Martian surface materials. Color and IR diodes will collect data in six different spectral bands. Reflectance curves constructed from these six points have diagnostic shapes for particular minerals and rocks. For example, differing degrees of iron oxidation cause varying absorption in the range from 0.9 to 1.1 micron wavelengths.

LANDER IMAGING CONCEPT

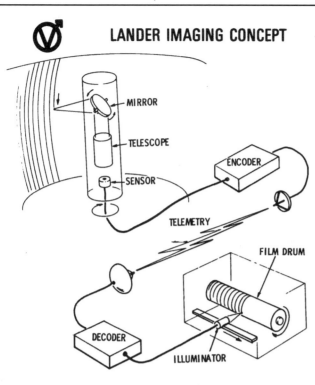

Another goal will be to spot variable features. Changes in features can be determined by taking pictures of the same region at successive times. The most probable change will be caused by the movement of sand and silt by the wind. Mariner pictures have revealed large-scale sediment movement; similar Lander observations are anticipated.

A grid target has been painted atop the Lander; one aim of the variable features investigation is to see if the target is being covered by sediment. The cameras' single-line scan will be used each day to detect any sand grains saltating (hopping) along the surface.

The most spectacular variable feature would be one of biological origin. Many scientists are skeptical about the probability of life on Mars; very few expect to see large forms that can be recognized in a picture. The possibility will not be discounted, however. If there are organic forms, they might be difficult to identify in a conventional "snapshot." Their most recognizable attribute might be motion, and this motion might be uniquely characterized by the single-line-scan mode of operation.

Another area of camera investigation is atmospheric properties. Pictures taken close to the horizon at sunset or sunrise will be used to determine the aerosol content of the atmosphere. Some pictures will also be taken of celestial objects: Venus, perhaps Jupiter, and the two Mars satellites, Phobos and Deimos. The brightness of these objects will be affected by the interference of the atmosphere, and the cameras can provide a way to measure aerosol content.

The cameras can also be used in the same way as more conventional surveying instruments. Pictures of the Sun and planets can be geometrically analyzed to determine the latitudinal and longitudinal position of the Lander on Mars.

Each Lander is equipped with two identical cameras, positioned about 1 m (39 in.) apart. They have a relatively unobstructed view across the area that is accessible to the surface sampler. The cameras are on stubby masts that extend 1.3 m (51 in.) above the surface.

The imaging instruments are called facsimile cameras. Their design is fundamentally different from that of the television cameras that have been used on most unmanned orbital and flyby spacecraft. Facsimile cameras use mechanical instead of electronic scanning.

In a television camera the entire object is simultaneously recorded as an image on the face of a vidicon tube in the focal plane. Then the image is "read" by the vidicon through the action of an electron beam as it neutralizes the electrostatic potential produced by photons when the image was recorded. In a facsimile camera, small picture elements (called pixels) that make up the total image are sequentially recorded.

In a facsimile camera an image is produced by observing the object through sequential line scans with a nodding mirror which reflects the light from a small element of the object into a diode sensor. Each time the mirror nods, one vertical line in the field of view is scanned by the diode. The entire camera then moves horizontally by a small interval and the next vertical line is scanned by the nodding mirror. Data that make up the entire picture are slowly accumulated in this way.

Because each element (spot) in the field of view is recorded on the same diode, opposed to different parts of the vidicon tube face the facsimile camera has a photometric stability that exceeds most television systems. Relatively subtle reflectance characteristics of objects in the field of view can be measured.

There are actually 12 diodes in the camera focal plane; each diode is designed to acquire data of particular spectral and spatial quality. One diode acquires a survey black-and-white picture. Three diodes have filters that transmit light in blue, green and red; together these diodes record a color picture. Three more diodes are used in essentially the same way, but have filters that transmit energy in three bands of near-infrared.

Four diodes are placed at different focal positions to get the best possible focus for high-resolution black-and-white pictures. (This results in a spatial resolution of several millimeters for the field of view closest to the camera — objects the size of an aspirin can be resolved.) The twelfth diode is designed with low sensitivity so it can image the Sun.

The survey and color pictures have a fixed elevation dimension of 60 degrees; high-resolution pictures have a fixed dimension of 20 degrees. The pictures can be positioned anywhere in a total elevation range of 60 degrees below to 40 degrees above the nominal horizon. The azimuth of the scene is adjustable; it can vary from less than one degree to almost 360 degrees to obtain a panorama.

The facsimile camera acquires data relatively slowly, line by line. Rapidly moving objects, therefore, will not be accurately recorded. They might appear as a vertical streak, recorded on only one or two lines. This apparent liability can be turned into an asset.

If the camera continues to operate while its motion is inhibited, the same vertical line is repetitively scanned. If the scene is stationary, the reflectance values between successive lines will be identical, but if an object crosses the region scanned by the single line, the reflectance values dramatically change between successive scans. The single-line-scan mode of camera operation, therefore, provides an unusual way of detecting motion.

As the mission proceeds, pictures will be acquired and transmitted three ways: the first Lander pictures will be sent directly to the Orbiter for relay to Earth. On successive days, pictures will be acquired during the day and stored on the Lander's tape recorder for later transmission to the Orbiter. Pictures can also be transmitted directly to Earth at a lower data rate.

The number of pictures that will be sent to Earth each day will vary according to the size of the pictures, amount of data to be transmitted by other instruments, and length of the transmission period. A typical daily picture budget for one Lander might be one picture directly transmitted to Earth at low data rate, two pictures transmitted real time through the Orbiter, and three pictures stored on the tape recorder and later relayed to Earth.

The Lander Imaging investigation team leader is Dr. Thomas A. (Tim) Mutch of Brown University.

Biology

Biology investigations will be performed to search for the presence of Martian organisms by looking for products of their metabolism. Three distinct investigations will incubate samples of the Martian surface under a number of different environmental conditions. Each is based on a different fundamental assumption about the possible requirements of Martian organisms; together they constitute a broad range of ideas on how to search for life on Mars. The three investigations are Pyrolytic Release (PR), Labeled Release (LR) and Gas Exchange (GEX).

Martian soil samples acquired by the surface sampler, several times during each landing, will be delivered to the Viking Biology instrument (VBI). There the samples will be automatically distributed, in measured amounts, to the three experiments for incubation and further processing.

Within the biology instrument, a complex system of heaters and thermoelectric coolers will maintain the incubation temperatures between about 8 and 17 degrees C. (46 to 63 degrees F.) in spite of external temperatures that may drop to minus 75 degrees C. (minus 103 degrees F.) or internal Lander temperatures that may rise to 35 degrees C.(95 degrees F.).

Pyrolytic Release.

The Pyrolytic Release (PR) experiment contains three incubation chambers, each of which can be used for one analysis. This experiment is designed to measure either photosynthetic or chemical fixation of carbon dioxide (CO_2) or carbon monoxide (CO). The main rationale for this is that the Martian atmosphere is known to contain CO_2, with CO as a trace component. Any Martian biota (animal or plant life) are expected to include organisms capable of assimilating one or both of these gases. It also seems reasonable that at least some organisms on Mars would take advantage of solar energy, as occurs on Earth, and that Martian soil would include photosynthetic organisms.

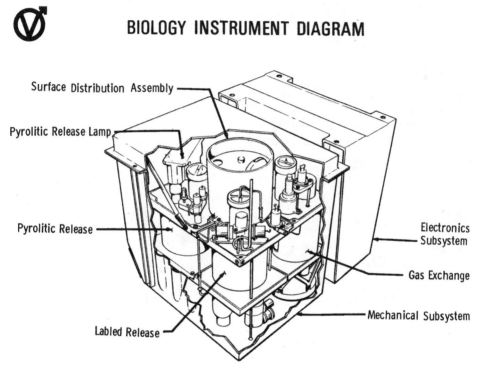

BIOLOGY INSTRUMENT DIAGRAM

Surface Distribution Assembly

Pyrolitic Release Lamp

Pyrolitic Release

Electronics Subsystem

Gas Exchange

Mechanical Subsystem

Labled Release

The experiment incubates soil in a Martian atmosphere with radioactive CO_2 added. Then, by pyrolysis (heating at high temperatures to "crack" organic compounds) and the use of an organic vapor trap (OVT), it determines whether radioactive carbon has been fixed into organic compounds. This experiment can be conducted either in the dark or light.

For an analysis, 0.25 cc of soil is delivered to a test cell, which is then moved to the incubation station and sealed. After establishing the incubation temperature, water vapor can be introduced by ground command if desired. Then the labeled CO_2/CO mixture is added from a gas reservoir and a xenon arc lamp is automatically turned on during the five-day incubation.

After incubation, the test cell is heated to 120 degrees C, (248 degrees F.) to remove residual incubation gases, which are vented to the outside. Background counts are made, after which the test cell is moved from the incubation station to another station.

Here pyrolysis is done by heating the test cell to 625 degrees C. (1,160 degrees F.), while purging the test cell with helium gas. The purged gases pass through the OVT, designed to retain organic compounds and fragments, but not CO_2 or CO. The radioactivity detector at this stage will sense a "first peak" consisting mainly of unreacted CO_2/CO. This first peak is regarded as non-biological in origin.

INTEGRATED BIOLOGY INSTRUMENT

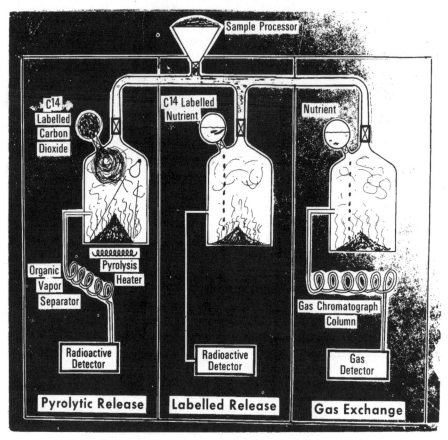

After this operation, the test cell is moved away from the pyrolysis station, the detector is heated and purged with helium, and background counts are taken once more to verify that the background radiation is down to pre-pyrolysis levels. The trapped organic compounds are then released from the OVT by heating it to 700 degrees C. (1,290 degrees F.), which simultaneously oxidizes them to CO_2. These are flushed into the detector. A second radioactive peak at this point would indicate biological activity in the original sample.

Labeled Release.

The Labeled Release (LR) experiment is designed to test metabolic activity in a soil sample moistened with a dilute aqueous solution of very simple organic compounds. The rationale for this experiment is that some Martian organisms, in contact with an atmosphere containing CO_2, should be able to break down organic compounds to CO_2. The experiment depends on the biological release of radioactive gases from a mixture of simple radioactive compounds supplied during incubation.

The test cell is provided 0.5 cc of soil sample and is moved to the incubation station and sealed. The Martian atmosphere is established in the test cell in this process. Before the radioactively-labeled nutrients (a mixture of formate, glycine, lactate, alanine, and glycolic acid; all compounds are uniformly labeled with radioactive carbon) are added, a background count is taken. Then approximately 0.15 cc of nutrients are added, and incubation proceeds in the dark for 11 days.

The atmosphere above the soil sample is continuously monitored by a separate radioactivity detector throughout the incubation, after which the test cell and detector are purged with helium. The accumulation of radioactive CO_2 (or other radioactive gases) indicates the presence of life metabolizing the nutrient. Data are collected for 12 days. These data will produce a metabolic curve as a function of time. The shape of the curve can be used to determine if growth is taking place in the test cell.

Gas Exchange.

The Gas Exchange (GEX) experiment measures the production or uptake of CO_2, nitrogen, methane, hydrogen and oxygen during the incubation of a Martian soil sample. The GEX experiment can be conducted in one of two modes: in the presence of water vapor, without added nutrients, or in the presence of a complex source of nutrients.

The first mode is based on the assumption that substrates (foodstuffs) may not be limiting in the Martian soil and that biological activity may be stimulated when only water vapor becomes available. The second mode assumes that Martian soil contains organisms metabolically similar to those found in most terrestrial soils and that these will require organic nutrients for growth.

A single test cell is used for the experiment. After receiving 1 cc of soil from the distribution assembly, the test cell is moved to its incubation station and sealed. After a helium purge, a mixture of helium, krypton and CO_2 is introduced into the incubation cell and this becomes the initial incubation atmosphere. (Krypton is used as an internal standard; helium is used to bring the test chamber pressure to approximately one-fifth of an Earth atmosphere.)

At this point either 0.5 or 2.5 cc of a rich nutrient solution can be introduced. Using the lesser quantity, the soil does not come into contact with the solution, and incubation proceeds in a "humid" mode. An additional two cc allows contact between the soil and the nutrient solution, which consists of a concentrated aqueous mixture of nineteen amino acids, vitamins, other organic compounds, and inorganic salts. Incubation initially is planned to be in the humid mode for seven days, after which additional nutrient solution will be added. For gas analyses, samples (100 microliters) of the atmosphere above the incubating soil are removed through a gas sampling tube. This occurs at the beginning of each incubation and after 1, 2, 4, 8 and 12 days.

The sample gas is placed in a stream of helium flowing through a coiled, 0.7 m (23 ft.) long, chromatograph column into a thermal conductivity detector. The system used in the GEX experiment is very sensitive and will measure changes in concentration down to about one nanomole (one-billionth of a molecule).

After a 12-day incubation cycle, a fresh soil sample can be added to the test cell to begin a new incubation cycle; the medium can be drained and replaced by fresh nutrients; and the original atmosphere is replaced with fresh incubation atmosphere. The latter procedure will be used if significant gas changes are noted in the initial incubation, on the assumption that if these changes are due to biological activity, they should be repeatable and should be enhanced. If of non-biological origin, they should not reappear.

Each incubation station also contains auxiliary heaters that can be used to heat soil samples to approximately 160 degrees C. (320 degrees F.). The heaters will be activated for three hours in case one or more of the experiments indicates a positive biological signal, after which the experiment will then be repeated on the "sterilized" soil samples. This is the control for the experiment. The detection of life would only be acknowledged if there were a significant difference between the "control" and the experiment.

An electronic system within the Viking Biology instrument, containing tens of thousands of components, will automatically sequence all events within the experiments, but will be subject to commands from Earth. The electronic subsystem will also obtain data from the experiments for transmission to Earth.

In addition to the electronic subsystem, each biology instrument contains four compartments, or modules, within a volume of just over one cubic foot. The common services module is a reservoir for the three other modules. it contains a tank of pressurized helium gas to move the incubation chambers from one place to another, to purge pneumatic lines used in the experiments, and to carry other gases as required.

The Biology investigation team leader is Dr. Harold P. Klein of NASA's Ames Research Center, Mountain View, Calif.

Molecular Analysis

The Molecular Analysis investigation will search for and identify organic (and some inorganic) compounds in the surface layer (the first few centimeters) of Mars. It will also determine the composition of the atmosphere near the surface and monitor composition changes during part of a Martian season.

Organic compounds on Earth are substances that contain carbon, hydrogen (almost always) and often oxygen, nitrogen and other elements; all are attached to one another. All but the most simple ones contain a series of carbon atoms attached to each other. On Earth carbon has the tendency to form a variety of long-chain molecules. This may happen on Mars, although it's possible that the situation may be different on another planet.

The question of whether there are organic compounds in the surface of Mars and, if so, what is their chemical structure, is of interest for several reasons. Organic substances produced by purely non-biological processes (such as thermal, photo-chemical or radiation-induced reactions) would tell something about the occurrence of these processes, and would allow speculation of precursors (carbon dioxide, carbon monoxide, ammonia, water, hydrogen sulfide, etc.) that could produce the substances detected.

The possibility also exists that the planet abounds with chemical compounds produced by living systems. Their chemical nature, distribution and structural uniqueness could be used to argue the presence of living organisms on Mars.

The wide area between these two extremes may represent manifestations of various levels of chemical evolution, or even of a planet that carried living systems that died gradually or through a catastrophic event.

Identification of the relatively small and simple organic molecules in the surface of Mars may enable comparison of the present chemistry of the planet to chemistry we assume existed on Earth a few billion years ago. Conversely, compounds may be encountered that resemble the composition of petroleum. From the distribution of individual structures, speculation can be made whether or not these hydrocarbons represent chemical fossils remaining after the decomposition of living systems of earlier times (a theory favored for the origin of petroleum on Earth).

While the major constituent of the Martian atmosphere is known to be carbon dioxide, there is a conspicuous absence of terrestrially important gases like oxygen and nitrogen, at least at the level of more than about one per cent.

Recent Soviet measurements indicate the possibility of an appreciable amount of argon. The concentration would tell something of the early history of Mars. Information about minor constituents like carbon monoxide, oxygen, nitrogen, and possibly even traces of small hydrocarbons or ammonia is important to an understanding of the chemical and possibly biological processes occurring at the surface of the planet. Periodic measurements during the day and during the entire landed phase of the mission are required for this purpose.

Finally, because of the absence of nitrogen in the atmosphere, it is of interest to search for nitrogen-containing inorganic substances such as nitrates or nitrites in the surface minerals.

A Gas Chromatograph Mass Spectrometer (GCMS) was chosen for these experiments because of its high sensitivity, high structural specificity and broad applicability to a wide range of compounds. Because mass spectra can be interpreted even in the absence of reference spectra, detection is possible of compounds not expected by terrestrial chemists.

The spectrometer will be used directly for analysis of the atmosphere before and after removal of carbon dioxide, which facilitates the identification and quantification of minor constituents.

Identification of organic substances probably present in surface material is a complex task because little is known about their overall abundance (which may be zero), and because any one of thousands of organic substances, or any combination thereof, may be present.

GCMS INSTRUMENT

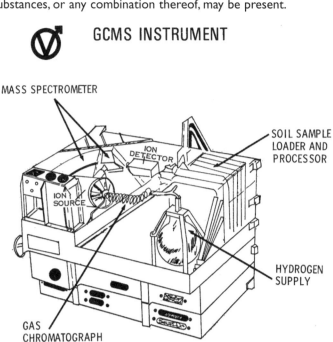

MASS SPECTROMETER

ION DETECTOR

ION SOURCE

SOIL SAMPLE LOADER AND PROCESSOR

HYDROGEN SUPPLY

GAS CHROMATOGRAPH

During the experiment, organic substances will be vaporized from the surface material by heating it to 200 degrees C. (392 degrees F.), while carbon dioxide (labeled with 13 C., a non-radioactive carbon isotope) sweeps through. The emerging material is carried into a gas chromatographic column, which is then swept by a carrier gas (hydrogen). While passing through this column (a thin tube filled with solid particles) substances entering the column are separated from each other by their different degrees of retention on this solid material.

After emerging from the column, excess carrier gas is removed by passing the stream through a palladium separator that is permeable only to hydrogen; the residual stream then moves into the mass spectrometer. This produces a complete mass spectrum (from mass 12 to 200) every 10 seconds for the entire 84 minutes of the gas chromatogram. The data are then stored and sent to Earth.

In this part of the experiment, materials that are volatile at 200 degrees C. (392 degrees F.) will be measured. The same sample is then heated to 500° C. (932° F.) to obtain less volatile materials and to pyrolyze (crack by heating) those substances that are not volatile enough to evaporate.

The results of the organic experiment will consist of three parts: interpretation of the mass spectra to identify compounds evolved from the soil sample; reconstruction of the molecular structures of those substances that were pyrolyzed and gave only mass spectra of their pyrolysis products; and correlation of the compounds detected in the surface material with hypotheses of their generation on the Martian surface.

The detection of inorganic gaseous materials such as water, carbon dioxide or nitrogen oxides, produced upon heating the soil sample, may permit conclusions on the composition of minerals that comprise the inorganic surface material. Results of the inorganic experiment are expected to help in this correlation and vice versa.

Atmospheric analyses are relatively simple and don't require much time, power or expendable supplies, but organic analyses are more involved. They consume a considerable amount of power, produce a large amount of data that must be sent to Earth, and involve materials that are limited (labeled carbon dioxide and hydrogen). For these reasons only three soil samples will be analyzed during each of the two missions. Considering the limited source (the area accessible to the surface sampler), this should be an adequate number of tests.

The Molecular Analysis investigation team leader is Dr. Klaus Biemann of the Massachusetts Institute of Technology.

Inorganic Chemistry

Scientific questions, ranging from the origin of the solar system to the metabolism of microbes, depend largely on knowledge of the elemental chemical composition of surface material. The Inorganic Chemical investigation will greatly expand present knowledge of the chemistry of Mars, and it is likely to provide a few clues to help answer some of these questions.

The conditions under which a planet condenses are thought to be reflected in its overall chemical composition. The most generally recognized relationship is that planetary bodies forming closer to the Sun should be enriched in refractory elements such as calcium, aluminum and zirconium, relative to more volatile elements such as potassium, sodium and rubidium.

To be truly diagnostic, ratios of volatile/refractory elemental pairs must represent planet-wide abundances, which will certainly be distorted by local differentiative geochemical processes (core/mantle formation, igneous and metamorphic differentiation, weathering and erosion, etc.). On the other hand, gross variations should be apparent. More detailed information on local processes (from other experiments as well as this one) will help reduce the effects of distortion.

Weathering in a watery environment (especially one highly charged with carbon dioxide, as is Mars' atmosphere) leads to fairly distinctive residual products, whose nature should be inferable from the inorganic chemistry data. This is especially so in concert with data from the Gas Chromatograph-Mass Spectrometer (GCMS) (on the presence and perhaps the identity of hydrate and carbonate minerals) and the Magnetic Properties experiment (on oxidation states of iron). We hope, therefore, to obtain data of at least a corroborative sort bearing on the question of the possible former existence of abundant liquid water on Mars.

The experiment, reduced to essentials, consists of exposing samples of Martian surface materials to x-rays from radioisotope sources, which stimulate the atoms of the sample to emit "fluorescent" x-rays. Each chemical element emits x-rays at a very few, extremely well-defined energies. This effect is analogous to the emission of visible light by certain fluorescent minerals when illuminated with "black light." By analyzing the energy of the fluorescent x-rays, the elements in the sample and their relative abundances can be ascertained.

Because of characteristics inherent in the technique, elements lighter (i.e., earlier in the Periodic Table) than magnesium are not individually measured. While several of these elements (e.g., nitrogen, carbon, oxygen) may be abundant and very important for biological processes, their precise abundance in surface materials is of relatively minor interpretative value. Gross abundances should be deducible from x-ray data combined with data from other experiments, notably the GCMS and Magnetic Properties investigations.

The sample delivered to the x-ray Fluorescence Spectrometer (XRFS) by the surface sampler may be coarse-grained material up to 1.3 cm (0.5 in.) in diameter (the opening of a screen in the funnel head) or fine-grained material that has been passed by vibratory sieving through 2 mm (0.08 in.) circular openings in the surface sampler head.

The spectrometer contains a sample analysis chamber, x-ray sources, detectors, electronics, and a dump cavity. The unit weighs two kilograms (4.5 pounds). Facing each window of the chamber are two sealed, gas-filled proportional counter (PC) detectors flanking a radioactive source.

These sources (radioactive iron and cadmium) produce x-rays of sufficient energy to excite fluorescent x-rays from the elements between magnesium and uranium in the Periodic Table. Elements before magnesium in the table can be determined only as a group, although useful estimates of their individual abundances may be indirectly achieved.

The output of the detectors is a series of electrical pulses with voltages proportional to the energy of the x-ray photons of the elements that produced them. A determination of the energy level identifies the presence of that element and the intensity (count rate) of the signal is related to its concentration.

A single-channel analyzer circuit divides the output of each detector into 128 energy levels and steps through each level, recording the accumulated count for a fixed period of time. A continuous plot of the count rate in each level produces a spectral signature of the material.

Mounted on the walls opposite the windows of the sample chamber are two calibration plaques ("A" is metallic aluminum; "B" is silver with a vertical, wedge-shaped central strip of zinc oxide). Signals from these plaques, with the chamber empty, monitor possible electronic drift, gas leakage from the detectors, radiation from the Lander's nuclear generators and/or cosmic rays, cross contamination of the windows between samples and, after sample delivery, the level of filling within the chamber. A calibration flag can be interposed on command in front of the radioactive iron source to provide additional calibration.

Spectral signatures obtained from spectrometer response to a variety of rock types are a part of an extensive library of reference spectra being accumulated, to which spectra of Martian material will be compared. Computer analysis of the spectra, to derive concentrations of the elements, is based on a semi-empirical model of the instrument response, including corrections for absorption and enhancement, PC tube response and back-scatter intensity. By matching spectra to a mathematical model of the instrument's response, it is possible to calculate element concentrations.

All spectra are normalized by reference to the back-scattered primary radiation to make comparisons uniform between spectra. The integrated intensity of the back-scatter peak also provides data on the bulk density of the sample and the amounts of elements lighter than magnesium.

Toward the end of the Lander I mission, the x-ray Fluorescence Spectrometer will repeatedly analyze a single sample, to achieve a higher order of precision than will be possible in the earlier part of the mission.

The inorganic chemistry investigation team leader is Dr. Priestley Toulmin III of the U.S. Geological Survey, Reston, Va.

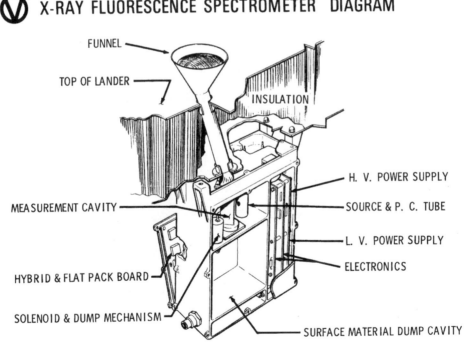

X-RAY FLUORESCENCE SPECTROMETER DIAGRAM

FUNNEL

TOP OF LANDER

INSULATION

H. V. POWER SUPPLY

MEASUREMENT CAVITY

SOURCE & P. C. TUBE

L. V. POWER SUPPLY

ELECTRONICS

HYBRID & FLAT PACK BOARD

SOLENOID & DUMP MECHANISM

SURFACE MATERIAL DUMP CAVITY

Meteorology

Meteorology science measurements on Mars will be obtained primarily from sensors mounted on a boom attached to the Lander. Measurements will include wind speed, wind direction and temperature. Atmospheric pressure will be measured by a sensor located inside the Lander and vented by a tube to the outside. Readings will be obtained during approximately 20 periods every Sol* (Mars day), each period consisting of several instantaneous measurements.

*24 hours, 39 minutes

The Meteorology investigation is designed to increase understanding of how the Martian atmosphere works. It will be man's first opportunity to directly observe the meteorology of another planet that obeys the same physical laws as does Earth's atmosphere. The opportunity to extend and refine comprehension of how an atmosphere works, driven by the Sun's radiation and subject to rotation of the planet, should give better understanding of Earth's atmosphere.

Scientific goals of the experiment are to:

* Obtain the first direct measurements of Martian meteorology with instruments placed in the atmosphere. Until now all information on wind speeds, for example, has come from theoretical calculations of the circulation of the atmosphere, or from calculations of the wind speed needed to raise dust.

* Measure and define meteorological variations during the day (Sol). The validity of existing theories that predict these diurnal (daily) variations can be compared with measurements and the theories revised as needed.

* Measure some of the turbulent characteristics of the planetary boundary layer. The boundary layer is the main brake on atmospheric circulation, and this circulation cannot be adequately understood until more is known about the turbulent dissipation of energy in the boundary layer.

*Verify whether such well-known terrestrial phenomena as weather fronts and dust devils occur on Mars by observing the behavior of the atmosphere as these things pass near the Landers.

* Support other Lander science experiments by providing information needed for other experiments. Meteorology results during the first few days, for example, should provide information on the best time of day to deploy the surface sampler boom to avoid damage from high winds.

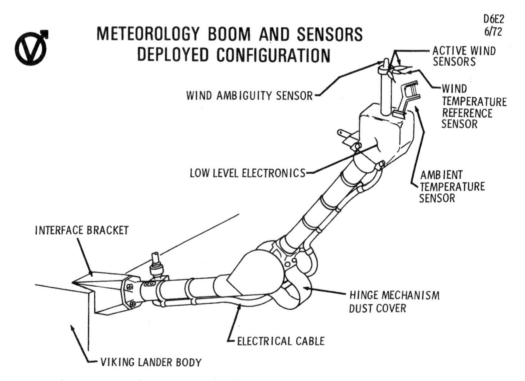

METEOROLOGY BOOM AND SENSORS
DEPLOYED CONFIGURATION

D6E2
6/72

ACTIVE WIND SENSORS

WIND AMBIGUITY SENSOR

WIND TEMPERATURE REFERENCE SENSOR

LOW LEVEL ELECTRONICS

AMBIENT TEMPERATURE SENSOR

INTERFACE BRACKET

HINGE MECHANISM DUST COVER

ELECTRICAL CABLE

VIKING LANDER BODY

The experiment's primary wind sensors are hot-film anemometers, two glass needles coated with platinum and over-coated with a protective layer of aluminum oxide. An electric current is passed through the platinum films to heat the needles, while the wind takes away the heat. Electric power needed to maintain these sensors at a fixed temperature above the surrounding air is the measure of the wind speed.

The device measures wind speed perpendicular to its length, so two devices, mounted 90 degrees apart, are necessary to find the total wind. A third identical sensor is mounted between the two, and it is used to determine air temperature and, through automatic circuitry, control the power applied to the active sensors.

The sensors give the same readings for winds from opposite directions, so an uncertainty remains as to wind direction. This problem is solved by a quadrant sensor, an electrically heated core surrounded by four thermocouples (located every 90 degrees). Heat taken away from the core by the wind affects the thermocouples enough to eliminate uncertainty about wind direction.

The quadrant sensor can also measure wind speed, so readings are combined from the hot-film anemometers and the quadrant sensor. A sophisticated computer program produces the best available determinations of both wind speed and direction.

Air temperature is measured by three fine-wire thermocouples in parallel. They are extremely thin to quickly respond to temperature fluctuations, but this makes them more subject to being broken by blowing sand. Each of the three thermocouples can operate independently, so breakage of one or two will not be catastrophic.

The pressure sensor consists of a thin metal diaphragm mounted in a case. A vacuum is maintained on one side of the diaphragm while the other side is exposed to the atmosphere. As air pressure varies, the diaphragm moves slightly in response to the fluctuating force upon it. This movement is detected by an electrical sensor and its output is converted to a pressure reading.

The Meteorology investigation team leader is Dr. Seymour L. Hess of Florida State University.

Seismology

The Seismology investigation will determine the level of seismic or tectonic (crustal forces) activity on Mars and its internal structure. Waves from naturally occurring Marsquakes spread throughout the planet and will be detected by seismometers on the surface.

Each Lander has miniature seismometers that will measure notion in three perpendicular directions. Two instruments, and the three-axis nature of each, allows a crude triangulation to be made to locate a seismic event. Regions of active tectonism can be identified and associated with surface manifestations of faulting.

The basic question: Is Mars a tectonically active planet or are the various surface features remnants of an earlier active period? The Earth is a tectonically active planet, primarily due to the motions of large crustal plates on its surface. Mars may be starting a phase of continental breakup or it may be a seismically dead planet. Either way, studying Mars will help scientists understand better the processes that cause quakes and plate motions in Earth.

If there are abundant Marsquakes, scientists can begin to unravel the internal structure of the planet. Seismic waves are used to map deep discontinuities and to determine seismic velocities as a function of depth, and would help determine if Mars has a crust and a core like Earth. This knowledge is important in understanding Earth's early evolution and the evolution of the atmosphere.

The seismology instrument consists of an approximately cubical package, about 15 cm (6 in.) on a side that weighs about 2.3 kg (5 lbs.). In the package are three miniaturized seismometers for sensing ground motion, and electronic circuitry for amplifying, conditioning and compressing data.

The seismometers are arranged in a mutually perpendicular manner to sense the components of motion in three directions. They consist of a 20-gram (0.7-ounce) mass with an attached coil, elastically pivoted from the instrument frame on a short boom, so the coil projects into a magnet mounted on the frame. Relative motion of the coil and magnet, induced by the mass's reaction to ground motion, generates a varying voltage that is applied to the input of an amplifier.

SEISMOMETER SENSOR SCHEMATIC

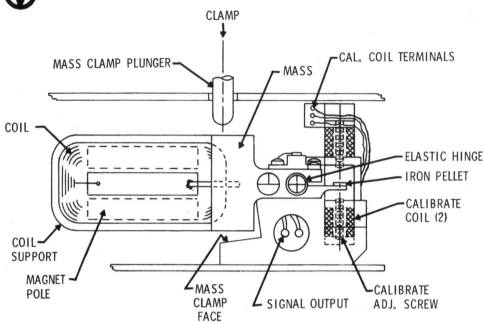

Modes of operation may be changed by command from Earth to accommodate whatever seismic environment might be found on Mars; the modes may also be automatically cycled by internal controls.

Modes include selection of various filters to determine frequency content of seismic data, or to adjust for the best possible reception of specific types of data; a low sampling rate for reading the general level of activity; a high data rate for more detailed examination of events; and a compressed, medium rate for continuous monitoring of Marsquakes. This last mode normally will be dormant, with the system operating at low rate until activated by a quake event.

Since the amount of raw data produced by the seismometer is much greater than the capacity of telemetry, data must be compressed to reduce quantity without seriously degrading quality. Normally, many samples are required for high-frequency data.

Data compression is done in two ways. First, normal ground noise (microseisms) is observed by averaging its amplitude over a 15-second period as it is passed through selectable filters. Its average amplitude and frequency content can be indicated by one sample every 15 seconds.

Second, when a Marsquake event occurs, a trigger activates a higher data rate mode that samples, not oscillations in the data, but amplitude of the overall event envelope. This varies at a much lower rate than individual oscillations and requires only one amplitude sample per second to indicate its shape.

At the same time, crossing of the zero axis by the oscillations (change in polarity of the data signal) is counted and sampled once per second. The shape of the envelope and its incremental frequency content can be transmitted to Earth with relatively few data samples and reconstructed to approximate the original event.

The Seismology investigation team leader is Dr. Don L. Anderson of the California Institute of Technology.

Physical Properties

The Physical Properties investigation group frequently has been called "the team without an instrument." While the statement is not quite true, the investigation mainly will use available engineering data. Hardware for the investigation includes two mirrors (mounted on the surface sampler boom), stroke gauges on each Lander leg, a grid on the Lander's top, ultraviolet degradable coatings, and current-measuring circuits in the surface sampler.

Besides engineering data, selected images will be taken by the Lander cameras to determine properties of the Mars surface such as grain size, bearing strength, cohesion, and eolian transportability (how easily surface material is moved by the wind). Other properties to be examined include thermal inertia (how quickly surface temperature changes) and the ultraviolet flux levels.

The bearing strength of the Mars surface will be one of the first characteristics determined. Immediately after landing, a panoramic picture will be taken that will include the Lander's number 3 footpad and its impression in the surface. This picture, data on Lander velocity and attitude at landing, and the amount of leg stroke (compression) will be used to calculate the surface bearing strength, an important fundamental parameter. The footpad impression will also give preliminary data on the cohesion of the surface material.

Early in the landed mission, the surface sampler collector head will eject its protective shroud. Following the ejection, the camera will image the spot where the shroud hits the surface, using the boom-mounted mirror (the area under the retroengine), and again photograph the footpad and its impression on the surface. This image will be analyzed like the one taken after landing to better define critical surface properties of bearing strength, cohesion and eolian transportability.

While the surface sampler is acquiring samples for the analytical instruments, the physical properties investigation will automatically be acquiring data by measuring the sampler motor currents and taking pictures of the surface markings generated by the sampler. Even the pile of excess sample dumped by the sampler after giving the instruments all they need will be of interest to the Physical Properties scientists.

When the sample for the Gas Chromatograph-Mass Spectrometer (GCMS) is comminuted (ground) the comminutor motor current will be recorded for analysis by the scientists to determine grain size, porosity and hardness.

The team has defined several unique experiments to better understand surface properties. These include digging trenches, examining material in the collector head jaw with the magnifying mirror, piling material on the grid atop the Lander, picking up and dropping a rock or clod on the surface, pressing the collector head firmly into the surface and using the collector head thermal sensor to measure surface temperatures.

Another very simple experiment for the Physical Properties investigation is the addition of ultraviolet degradable coatings on the camera reference test charts. These coatings darken in the presence of ultraviolet and the amount of darkening, to a certain limit, is proportional to the total amount of ultraviolet received.

The investigation will provide valuable information to complement the results of other studies, such as geology and mineralogy. Knowledge concerning the structure of the surface can be very helpful in understanding apparently conflicting data and grasping the significance of otherwise unexplainable findings.

The Physical Properties investigation team leader is Dr. Richard W. Shorthill of the University of Utah.

Magnetic Properties

The Magnetic Properties investigation will attempt to detect the presence of magnetic particles in the Mars surface material, and determine the identity and quantity of these particles.

Iron in magnetic minerals is usually an accessory phase in naturally occurring rocks and surface materials on Earth, on the Moon and in meteorites. The chemical form in which this magnetic iron occurs on a planetary surface may vary from elemental metal to more complex iron compounds (i.e., ferrous oxide magnetite, highly oxidized hematite, the hydrates goethite and lepidocrocite). The abundance and chemistry of the accessory iron minerals on the surface have bearing on the degree of differentiation and oxidation of the planet, the composition of its atmosphere, and the extent of interaction between the solid surface materials and the atmosphere.

PHYSICAL PROPERTIES OF MARTIAN SURFACE
USING ENGINEERING MEASUREMENTS

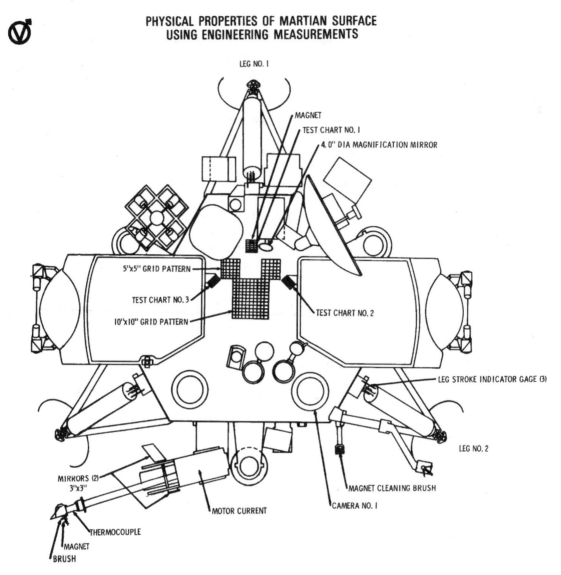

This investigation uses a set of two permanent, samarium-cobalt magnet pairs, mounted on the back of the surface sampler collector head. Each pair consists of an outer ring magnet, about the size of a quarter, with an inner core-magnet of opposite polarity. These are relatively strong magnets. (The maximum field obtained is approximately 2,500 gauss. A gauss is a unit of magnetic field intensity.)

The magnets are mounted at different depths from the outer surface of the backhoe to ensure a gradient in magnetic field strength.

In addition, a similar magnet pair is mounted on the photometric target atop the lander, where it will be automatically photographed when the camera system is calibrated. The magnets in this location should attract any magnetic particles that might be present in windblown dust.

In acquiring samples, the collector head will dig into the surface; and any magnetic particles will tend to adhere to the magnets. The collector head can be directly imaged with the camera system. A five-power magnifying mirror can also be used for maximum resolution in black-and-white or color. These images will be the scientific data return on which the conclusions will be based.

SURFACE SAMPLER COLLECTOR HEAD DIAGRAM

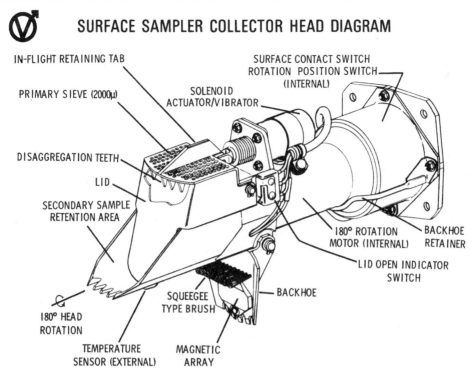

The Magnetic Properties principal investigator is Dr. Robert B. Hargraves of Princeton University.

Radio Science

The objectives of the Radio Science investigation are to conduct scientific studies of Mars using the orbiter and Lander tracking and communications systems that are required for spacecraft operations and data transmission.

Scientific uses of the systems evolved from recognition of the potential applications of the data, and developments in data analysis to extract scientific results from information contained in the radio signals.

The science investigations will provide new and improved determinations of the gravity field, figure, spin axis orientation, and surface density of Mars; pressure, temperature and electron profiles in the planet's atmosphere; and properties of the solar system.

Radio science applies the principles of celestial mechanics and electromagnetic wave propagation to relate tracking and communications systems signals to physical parameters.

The investigation has no specifically dedicated instruments except the Orbiter's X-band transmitter, which provides a dual-frequency capability on the downlink. This is unique to Viking, compared with previous Mars missions, and is especially important for the Radio Science investigations.

Radio science characteristically deals with small perturbations or changes in spacecraft orbits, deduced from tracking data analysis, and with small variations in frequency, phase or amplitude of received signals. The investigations are intimately involved with data analysis, using complicated analytical procedures and associated computer programs to determine the physical effects that produce the observed variations. Data must sometimes be collected for an extended period to produce results.

The basic tracking data consist of very precise measurements of distance (range) and line-of-sight velocity (range rate) between the spacecraft and Earth tracking stations. Range and range rate measurements are the primary data used to determine global Mars gravity field and local gravity anomalies, precise Lander locations and radii of Mars at the landing sites, spin axis (pole) orientation and motion, and the ephemerides (assigned places) of Mars and Earth. Variations in the signal and other characteristics determine Mars atmospheric and ionospheric properties during occultation experiments.

During Viking's cruise phase, properties of the interplanetary medium, particularly the total electron content and its variations, can be determined by analyses of differences in signal properties on the two downlink frequencies. From such measurements intensity, size and distribution of electron streams from the Sun and from solar storms can be studied to increase understanding of the Earth-Mars region of interplanetary space.

While the Orbiters are being gradually maneuvered to pass over the landing sites, large local gravity anomalies might be detectable in the tracking data. If such anomalies appear near the landing sites or elsewhere they will be of considerable interest with respect to the geology and internal structure of the planet.

After landing, tracking data will be used to define, precise Lander locations, including the radius of Mars at these sites. Tracking is also used to define the spin axis (pole) direction, and possibly variations in the spin axis related to the global internal density distribution of Mars.

As the Orbiters rise and set with respect to the Landers, the signal amplitude received at the Orbiter on the Lander-to-Orbiter communication link is expected to vary. An attempt will be made to analyze these variations to determine dielectric properties of the regions near the Landers; these properties can be related to surface density.

After Orbiter I has been in Mars orbit for about 80 days, it will be placed in a non-synchronous orbit to make a global survey of the planet. Tracking data taken near periapsis will be used to determine the global gravity field and local gravity anomalies.

Several times during the missions, Mars passes near the line-of-sight between Earth and a quasar (an intense extragalactic radio source). Radio signals from an Orbiter and the quasar will then be alternately recorded at two tracking stations at the same time. This is a very long baseline interferometry (VLBI) experiment that yields a precise measurement of the angular separation of the two sources.

With suitable data analysis, the results give the precise location of the spacecraft, Mars and Earth with respect to the fixed, inertial frame defined by the very distant quasar. By making such observations over a period of years, in various spacecraft missions, the precise orbits of Mars and Earth with respect to the inertial frame can be determined. One application of such information is to determine the relativistic advance of the perihelion of Mars, providing a test of the general theory of relativity.

In October 1976 Orbiter I passes behind Mars, as viewed from Earth, during a portion of its orbit. The spacecraft signals are gradually cut off or occulted, by the planet. Variations in signal properties (frequency, phase and amplitude) as the spacecraft enters or emerges from occultation are used to infer atmospheric and ionospheric properties occultations for Orbiter 2 start in January 1977.

Mars and Earth will be in conjunction on Nov. 25, 1976. As the planets approach conjunction radio signals from Viking spacecraft pass closer and closer to the Sun and are gradually more affected by the solar corona, particularly the electron content.

Signal variations, again measured with the dual frequency downlinks, will yield new information on the properties of regions close to the Sun, including the characteristics of any timely solar storms (Sun spots) or high activity events. Spacecraft signals are also affected by the intense gravitational field of the Sun, so a precise solar gravitational time-delay test of general relativity theory will be done in the conjunction time period.

Tests to resolve small differences in the Einstein formulation of general relativity, as compared with more recently proposed formulations, can have an important impact on fundamental physical laws and on studies of the Universe's evolution.

The Radio Science investigation team leader is Dr. William H. Michael, Jr. of Langley Research Center, Va.

VIKING SCIENTISTS

The Viking scientists represent an outstanding cross-section of the scientific community. They were selected from universities, research institutes, NASA centers and other government agencies.

The scientists are divided into investigation teams, each headed by a team leader or principal investigator. The teams are led by a Science Steering Group, consisting of a chairman, vice chairman, the leaders of each team and two other members.

The scientists have worked closely with Viking engineers in designing the science instruments. Considerations of weight, power, data constraints and the necessary flexibility of the investigations were developed through cooperation between the two groups.

Team leaders are listed first in each group.

Science Steering Group

Dr. Gerald A. Soffen, Chairman, Langley Research center, Hampton, Va. Dr. Richard S. Young, Vice Chairman, NASA Headquarters, Washington, D.C. A. Thomas Young, Langley Research Center Dr. Conway W. Snyder, Jet Propulsion Laboratory, Pasadena, Calif.

Orbiter Imaging

Dr. Michael H. Carr, U.S. Geological Survey, Menlo Park, Calif. Dr. William A. Baum, Lowell Observatory, Flagstaff, Ariz. Dr. Geoffrey A. Briggs, Jet Propulsion Laboratory Dr. James A. Cutts, Science Applications, Inc., Pasadena Harold Masursky, U.S. Geological Survey, Flagstaff, Ariz.

Orbiter Water Vapor Mapping

Dr. Crofton E. Farmer, Jet Propulsion Laboratory Dr. Donald W. Davies, Jet Propulsion Laboratory Daniel D. La Porte, Santa Barbara Research Center, Goleta, Calif.

Orbiter Thermal Mapping

Dr. Hugh H. Kieffer, University of California, Los Angeles Dr. Stillman Chase, Santa Barbara Research Center Dr. Ellis D. Miner, Jet Propulsion Laboratory Dr. Guido Munch, California Institute of Technology, Pasadena Dr. Gerald Neugebauer, California Institute of Technology

Entry Science

Dr. Alfred O.C. Nier, University of Minnesota, Minneapolis Dr. William B. Hanson, University of Texas, Dallas Dr. Michael B. McElroy, Harvard University, Cambridge, Mass. Alvin Seiff, Ames Research Center, Mountain View, Calif. Nelson W. Spencer, Goddard Space Flight Center, Greenbelt, Md.

Lander imaging

Dr. Thomas A. Mutch, Brown University, Providence, R.I. Dr. Alan B. Binder, Science Applications, Inc., Tucson, Ariz. Friedrich O. Huck, Langley Research Center Dr. Elliott C. Levinthal, Stanford University, Palo Alto, Calif.

*Dr. Sidney Liebes, Stanford University Dr. Elliott C. Morris, U.S. Geological Survey, Flagstaff, Ariz. Dr. James A. Pollock, Ames Research Center Dr. Carl Sagan, Cornell University, Ithaca, N.Y.
* Associate

Biology

Dr. Harold P. Klein, Ames Research Center Dr. Norman H. Horowitz, California Institute of Technology Dr. Joshua Lederberg, Stanford University Dr. Gilbert V. Levin, Biospherics, Inc., Rockville, Md. Vance I. Oyama, Ames Research Center Dr. Alexander Rich, Massachusetts Institute of Technology, Cambridge, Mass.

Molecular Analysis

Dr. Klaus Biemann, Massachusetts Institute of Technology Dr. DuWayne M. Anderson, U.S. Army Cold Regions Research Engineering Laboratory, Hanover, N.H. Dr. Alfred O.C. Nier, University of Minnesota, Minneapolis Dr. Leslie E. Orgel, Salk Institute, San Diego, Calif. Dr. John Oró, University of Houston, Tex. Dr. Tobias Owen, State University of New York, Stony Brook Dr. Priestley Toulmin III, U.S. Geological Survey, Reston, Va. Dr. Harold C. Urey, University of California at San Diego, La Jolla, Calif.

Meteorology

Dr. Seymour L. Hess, Florida State University, Tallahassee Robert M. Henry, Langley Research Center Dr. Conway Leovy, University of Washington, Seattle Dr. Jack A. Ryan, McDonnell Douglas Corp., Huntington Beach, Calif. James E. Tillman, University of Washington, Seattle

Inorganic Chemistry

Dr. Priestley Toulmin III, U.S. Geological Survey, Reston, Va. Dr. Alex K. Baird, Pomona College, Claremont, Calif. Dr. Benton C. Clark, Martin Marietta Aerospace, Denver, Colo. Dr. Klaus Keil, University of New Mexico, Albuquerque Harry J. Rose, Jr., U.S. Geological Survey, Reston, Va.

Seismology

Dr. Don L. Anderson, California Institute of Technology Dr. Robert A. Kovach, Stanford University Dr. Gary V. Latham, University of Texas, Galveston Dr. George Sutton, University of Hawaii, Honolulu Dr. M. Nafi Toksöz, Massachusetts Institute of Technology

Physical Properties

Dr. Richard W. Shorthill, University of Utah, Salt Lake City Dr. Robert E. Hutton, TRW Applied Mechanics Laboratory, Redondo Beach, Calif. Dr. Henry J. Moore II, U.S. Geological Survey, Menlo Park Dr. Ronald F. Scott, California Institute of Technology

Magnetic Properties

Dr. Robert B. Hargraves, Princeton University, Princeton, N.J.

Radio Science

Dr. William H. Michael, Jr., Langley Research Center Dan L. Cain, Jet Propulsion Laboratory Dr. John G. Davies, Jodrell Bank, MacClesfield, Cheshire, England Dr. Gunnar Fjeldbo, Jet Propulsion Laboratory Dr. Mario D. Grossi, Raytheon Co., Sudbury, Mass. Dr. Irwin I. Shapiro, Massachusetts Institute of Technology Dr. Charles T. Stelzried, Jet Propulsion Laboratory Dr. G. Leonard Tyler, Stanford University *Joseph Brenkle, Jet Propulsion Laboratory *Robert H. Tolson, Langley Research Center

Associates

MISSION DESCRIPTION

Each Viking mission is divided into five phases: launch, cruise, orbit, entry and landed operations.

(Note: Viking 1 and Viking 2 missions will be described in the singular except when there are differences between the two missions.)

Launch Phase

The Viking launch phase begins with liftoff and lasts until the Deep Space Network (DSN) antennas acquire Viking's radio signal.

Liftoff occurs two-tenths of a second after ignition of the Titan launch vehicle's solid rocket motors. The solid motors begin to shut down 109 seconds after launch, and one second after the liquid-fueled Titan first stage engine is ignited. First stage engines fire for 148 seconds and shut down 258 seconds after launch.

The Titan's second stage engine then ignites to provide thrust for 203 seconds. Ten seconds after second stage ignition, the protective shroud is jettisoned from around the Centaur and the Viking spacecraft.

The Centaur's main engine fires for two minutes (126 seconds), placing the spacecraft in an Earth parking orbit at an altitude of 167 km (104 mi.) for from 11 to 30 minutes. The parking orbit will be used to correctly position Viking for Trans-Mars Injection (TMI). TMI is achieved by reigniting the Centaur's engine for about 310 seconds.

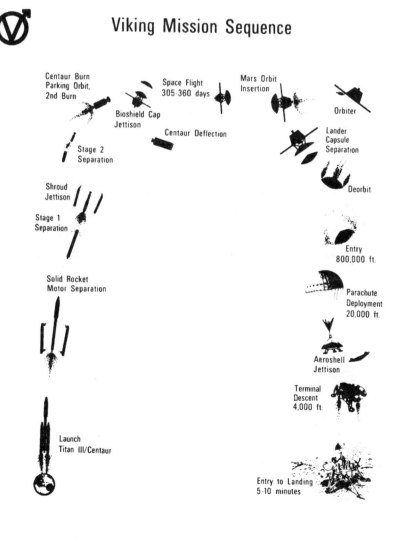

Viking Mission Sequence

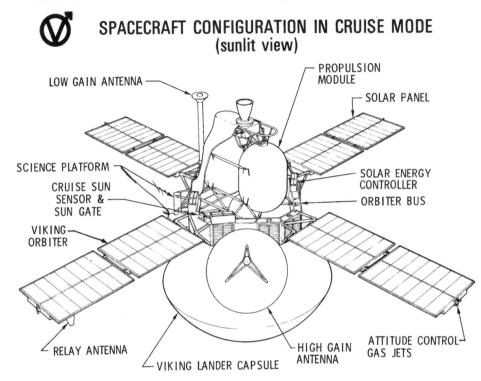

SPACECRAFT CONFIGURATION IN CRUISE MODE
(sunlit view)

LOW GAIN ANTENNA

PROPULSION MODULE

SOLAR PANEL

SCIENCE PLATFORM

CRUISE SUN SENSOR & SUN GATE

SOLAR ENERGY CONTROLLER

ORBITER BUS

VIKING ORBITER

RELAY ANTENNA

VIKING LANDER CAPSULE

HIGH GAIN ANTENNA

ATTITUDE CONTROL GAS JETS

Cruise Phase

Viking's interplanetary cruise phase from Earth to Mars will last from 305 days to 360 days, with arrival occurring during maximum Earth-to-Mars distances. The Orbiter is the operating portion of the spacecraft during this phase, but both Orbiter and Lander remain relatively inactive.

Viking will follow what is called a Type II trajectory to reach Mars, circling more than 180 degrees around the Sun as it chases the planet. Viking will travel about 815 million km (505 million mi.) in its cruise, reaching Mars during summer 1976, which is also the summer season in Mars' northern hemisphere. At that time Earth and Mars will be about 380 million km (236 million mi.) apart.

One to four midcourse maneuvers are planned during cruise to correct the launch aim bias, possible trajectory errors and to insure Viking's arrival at the proper location and at the right time for its Mars Orbit insertion (MOI) maneuver.

After the spacecraft is aligned to send and receive telemetry, it is separated from the Centaur (220 seconds after TMI). The Orbiter's four solar panels will be unfolded and the spacecraft will begin searching for the Sun. Once found, sunlight will provide power to Viking through the Orbiter's windmill-like solar panels.

At TMI plus 18 minutes, the Centaur will be deflected from a Mars flight path to one that bypasses the planet.

Soon after spacecraft separation, the DSN will acquire Viking's radio signal. DSN antennas will collect enough tracking information to determine the first mid-course maneuver, which will be done by a short firing of the Orbiter's engine.

The Lander's bioshield cap will be jettisoned shortly after the Orbiter's star sensor has acquired the star Canopus. With the acquisition of both the Sun and Canopus, the spacecraft is in a stable cruise attitude.

The DSN will track Viking, determining its position and velocity, and check the condition of both Orbiter and Lander. The combination of DSN metric data, star tracking and Sun sensing will enable Earth controllers to keep Viking on its trajectory.

Beginning about 30 days after launch, the Orbiter's scientific instruments will be periodically checked, and the Orbiter will contact the Lander about every 15 days. Instrument calibrations, battery charges, and other maintenance will be made during the cruise. A third mid-course maneuver is planned about 30 days before Mars Orbital Insertion (MOI).

About 10 days before Mars encounter, the Orbiter's scan platform will be unlatched and pointed toward Mars. Although still many thousands of kilometers away, the Orbiter will observe the planet with its cameras for calibration. Approach science observations will be made, including color photography and global infrared observations.

SPACECRAFT INTERPLANETARY ACTIVITY

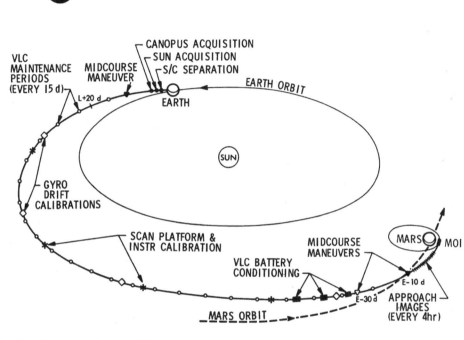

During the last three days before MOI, pictures will be taken of the Martian moon Deimos against the star background. The pictures will provide final optical navigation information to help calculate the MOI maneuver.

Orbital Phase

The initial orbital phase is the period from MOI to separation of the Lander from the orbiter. The orbital phase will continue after landed operations begin.

The first Viking is scheduled to encounter Mars June 18, 1976. Encounter may be made on or before that date if launch occurs during the first four days after Aug. 11, 1975. Launch delay beyond Aug. 18 will delay MOI beyond June 18, 1976. Delays beyond Aug. 22 will affect the second Viking and the mission profile strategy.

Viking 2 is scheduled for Mars encounter Aug. 7, 1976. This seven-week period between encounters will allow time to get Lander 1 onto the surface and complete the first cycle of landed operations before Viking 2 attains orbit.

The second Viking can be retargeted for a different landing site than its primary site up to 10 days before its MOI. Once Viking 2 is in orbit, however, its landings are restricted to a latitude band about 3 degrees wide.

If the second Viking is not launched during the first 20 days of the normal launch range (Aug. 21 through Sept. 9, 1975), its MOI will be delayed and will affect the relationship between the two missions at Mars.

As Viking nears MOI, gas jets on the Orbiter will maneuver the spacecraft to point the Orbiter's engine in the general direction of flight. The engine will fire for 40 to 50 minutes, depending on approach velocity. The spacecraft and its high-gain antenna will be aligned so communications are maintained during engine firing.

The firing will reduce Viking's velocity by as much as 4,320 km per hour (2,678 mph), and insert it into a highly elliptical orbit about Mars.

Once in orbit, site acquisition will be done by trim maneuvers to establish a synchronous orbit in which Viking I will pass daily near the site with a periapsis of 1,500 km (930 mi.), an apoapsis of 32,600 km (20,200 mi.) and a period of 24.6 hours, which is the Martian rotation period. Viking I will have a 33.4-degree inclination that will result in a 30-degree Sun elevation angle at Lander touchdown.

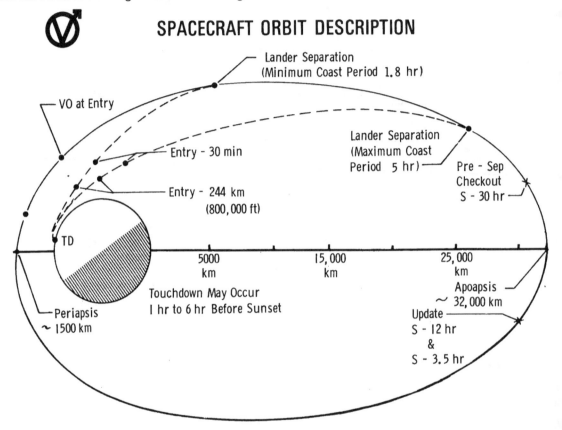

SPACECRAFT ORBIT DESCRIPTION

Viking 2 will be inserted into a super-synchronous initial orbit with a period of about 28.7 hours. This strategy will provide several opportunities for low-altitude observations of the landing sites under excellent viewing conditions. Orbiter 2's planned orbit will have a 48.9-degree inclination and a 130-degree Sun elevation angle at landing.

Viking I will make extensive observations of the Prime landing site (A-1), with particular emphasis on low-altitude coverage near periapsis, plus two or three picture pentads (groups of five) to monitor the site on most revolutions.

The A-1 site will be studied with the Orbiter's water vapor detector and thermal mapping instrument. Viking I will also observe Viking 2's prime landing site (B-1) from low-altitude with two picture swaths, and from high altitude with a pentad.

If the A-1 site is certified, Lander I will be committed. If the site is unacceptable, the backup (A-2) site will be examined. Once a site is picked, Orbiter trim maneuvers will fix periapsis near that site.

The first task of Viking 2, after its MOI, will be to certify landing site B-1. With Lander I already on the surface,

28 days have been planned for site selection. After Lander 2 touches down, an Orbiter 2 support period will be required, similar to that described for Viking 1. Alternatives for Orbiter 2 after this period will be similar to those for Viking 1.

Viking 2 will make a plane change maneuver Sept. 18, 1976, and begin a "resynchronous walk." The combined plane change/ period change maneuver will increase the orbital inclination from 49 degrees to about 75 degrees and provide a supersynchronous period of approximately 26.8 hours.

The short walk will cover about 140 degrees longitude in 3.75 revolutions. Orbiter 2 will complete the walk by resynchronizing the descending leg of the orbit approximately over the B-1 site Sept. 23, 1976. This plane change maneuver is made to allow the Orbiter instruments to view the North polar cap.

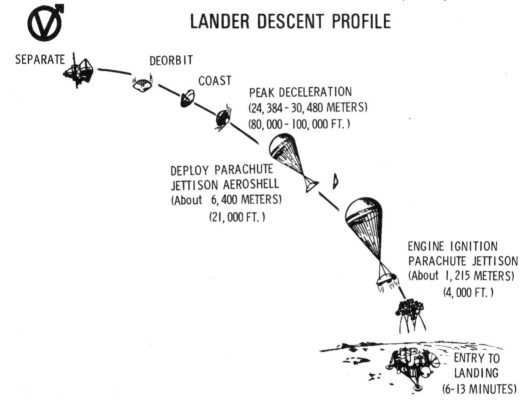

LANDER DESCENT PROFILE

SEPARATE

DEORBIT

COAST

PEAK DECELERATION
(24, 384 - 30, 480 METERS)
(80, 000 - 100, 000 FT.)

DEPLOY PARACHUTE
JETTISON AEROSHELL
(About 6, 400 METERS)
(21, 000 FT.)

ENGINE IGNITION
PARACHUTE JETTISON
(About 1, 215 METERS)
(4, 000 FT.)

ENTRY TO
LANDING
(6-13 MINUTES)

Landing Sites.

Viking landing sites must be both safe and scientifically interesting. They must have low surface elevations, winds, surface slopes and protuberances, adequate bearing strength, and radar properties.

A site must have an elevation low enough for adequate atmospheric drag on the Lander's parachute during entry. Winds less than about 234 kmph (126 mph) will not pose a landing hazard. The Lander is designed to be stable on 19-degree slopes and has a clearance above the surface of 22 cm (8.7 in.). The surface bearing strength must be enough to safely support the Lander and reduce the shock of landing.

An area with a large amount of windblown dust will be a poor landing site, and bare rock is unacceptable because of the difficulty in getting surface samples for the science instruments. Entry is also strongly dependent on the radar altimeter and terminal descent landing radars, so the surface must have properties that allow proper radar return.

Areas with the highest probability of water are of maximum scientific interest because of the biological emphasis of Viking. Although the chance of finding liquid water is extremely low, this criterion suggests areas of low elevation. Other scientifically interesting features are the geologic nature of the site and a reasonably unobstructed area that won't interfere with meteorological measurements.

Entry Phase

The entry phase is the period between separation of the Orbiter and Lander and the Lander's touchdown on Mars. The deorbit to landing sequence is completely automated, controlled by the Lander's computer. The time required for a radio signal to travel from Earth to Mars is 20 minutes, making real-time control of the descent impossible from Earth.

Once the landing site is certified and landing is approved, the Orbiter will contact the cocooned Lander 30 hours before planned separation, activate its electrical system, and initiate pre-separation checkout of all Lander systems.

Earth controllers will then put descent instructions into the Lander's Guidance and Control Sequencing Computer (GCSC). After a "go" command from Earth, a GCSC command will sever mechanical and electrical ties between the two spacecraft by energizing explosive nuts and allowing springs to separate the vehicles.

After separation the Lander will align itself for the deorbit maneuver. Several hours later the bioshield base and Lander adapter are jettisoned from the Orbiter; they will remain in Mars orbit.

A few minutes after separation, the aeroshell's small reaction control engines will fire for 24 minutes to give the Lander a 576 kmph (357 mph) velocity change (Delta V) to begin deorbit coast.

Although the Martian atmosphere is 100 times less dense than Earth's, the Lander will be initially traveling about 16,000 kmph (9,920 mph), and must be protected from the intense heat and pressure of entry.

The small aeroshell attitude control jets align the Lander for entry and provide roll control during entry to hold the Lander in the correct attitude, which produces a small amount of lift during entry. A lifting entry will be used instead of a ballistic entry because it provides several significant advantages, including increased terrain height and landed weight.

Three to five hours after separation, the Lander will enter the Mars atmosphere at an altitude of 244 km (151 mi.). The first deceleration will come from aerodynamic drag on the aeroshell. Peak deceleration occurs at between 24.4 and 30.5 km (15 and 19 mi.).

At about 5.8 km (19,000 ft.) a parachute will be deployed by a mortar in the base cover. The aeroshell will be separated by spring devices seven seconds after parachute deployment. Aerodynamic lift will cause the aeroshell to drift away from the landing site. Just after aeroshell jettison, the Lander's legs will be extended.

The parachute can be deployed within the Mach 0.5 to 1.9 range, with a velocity of 1,598 kmph (991 mph) Maximum dynamic pressures range from 239 to 383 newtons per square meter (5 to 8 pounds psf).

The parachute will take the Lander to an altitude of about 1.4 km (4,600 ft.) in one minute. The three terminal descent engines (TDE) will then be ignited and the parachute and base cover will be jettisoned.

The TDEs will fire for about 30 seconds and reduce the Lander's velocity from 207 kmph (128 mph) at parachute separation to 8.8 kmph (5.5 mph) at touchdown. From about 16.8 m (55 ft.) the Lander will be on a vertical flight path and descend to the surface at a constant velocity.

Sensors on the Lander footpads automatically shut off the TDEs when one landing leg touches the surface Shock absorbers in the legs will cushion the impact of landing.

Two entry events will be initiated from information provided by the Lander's radar altimeter: parachute deployment and TDE ignition. The terminal descent landing radar is a critical element of the guidance and control subsystem during the landing's final phase.

During the 10-minute descent through the atmosphere, the Lander will obtain data on atmospheric structure and composition. These data, plus engineering information, will be relayed to the Orbiter for immediate transmission to Earth, and storage for later transmission.

Landed Phase

Just 25 seconds after the Lander touches down, Camera 2 will take a high-resolution picture of the near field terrain and footpad 3. While this five-minute imaging is underway, the Lander begins to activate itself.

The high-gain antenna is aimed to establish contact with DSN antennas on Earth. The meteorology boom is deployed, other instruments are activated, and the Lander science platform begins its laboratory investigations.

From this time until end-of-mission, the Lander's primary functions are to be a communication relay station for scientific data on their way to Earth, directly or through the Orbiter, and to provide electrical power and a safe thermal environment for science instruments and subsystems.

The communication window that governs the Lander's direct link with Earth is open for about 12 hours a day. Lander electrical power and thermal limits will effectively restrict radio communication to about 70 minutes a day of this available window, however. This will allow a daily data volume direct to Earth of about two million bits (computer binary digits) of information early in the mission and one million bits a day later in the mission when the greatest communication distances exist. The Lander will receive all of its commands over the direct S-band link.

At least 10 million bits a day will be transmitted by ultra-high frequency (UHF) link to the Orbiter, but the transmission period can only occur when the Lander can see the Orbiter 25 degrees or more above the horizon and within 5,000 km (3,100 mi.). This window will be open anywhere from 10 to 40 minutes a day. The UHF link can send 16,000 bits per second, compared with 250 to 500 bits per second directly to Earth by the HGA.

Orbiter 1 will remain in synchronous orbit through the entire Lander 1 mission (Sol 58). It will be a communication relay station for the Lander, gather its own science data and acquire data for use in making final landing decisions for the second mission.

Flexibility will be invaluable if trouble develops with any of the spacecraft. Cross communications between orbiters and Landers, plus redundancy in many components, yield a high probability of success for Viking.

A scientific bonus also comes from joint Orbiter-Lander operations: the Orbiter provides landing site environmental information to aid landing decisions, then observe phenomena from orbit while the Lander takes surface measurements.

End of Mission.

At the end of its planned 58-day landed phase, Lander 1 will go into a reduced mode, although some of its experiments will continue. Near the end of the 120-day mission, the Landers will be powered down to a safe condition to heighten their chances of surviving the conjunction period, when the Sun is exactly between Earth and Mars and the Landers will be out of contact with Earth.

Both Landers will be placed in a safe condition before Nov. 8 to survive the conjunction communications blackout period.

VIKING ORBITER

The Viking Orbiter is a follow-on design to the Mariner class of planetary spacecraft with specific design changes for the Viking mission. Operational lifetime requirements for the Orbiter are 120 days in orbit and 90 days after the landing.

Orbiter Design

The design of the Orbiter was greatly influenced by the size of the Lander, which dictated a larger spacecraft structure than Mariner, increased propellant storage for a longer burn time for orbit insertion, and upgrading of the attitude control system with additional gas storage and larger impulse capacity.

The combined weight of the Orbiter and Lander was one factor that contributed to an 11-month transit time to Mars, instead of five months for Mariner missions. The longer flight time then dictated an increased design life for the spacecraft, larger solar panels to allow for longer degradation from solar radiation and additional attitude control gas.

Structure

The basic structure of the Orbiter is an octagon approximately 2.4 m across (8 ft.). The eight sides of the ring-like structure are 45.7 cm (18 in.) high and are alternately 1.4 by 0.6 m (55 by 22 in.).

Electronic bays are mounted to the faces of the structure and the propulsion module is attached at four points. There are 16 bays, or compartments, three on each of the long sides and one on each short side.

The Orbiter is 3.3 m (10.8 ft.) high and 9.7 m (32 ft.) across the extended solar panels. Its fueled weight is 2,325 kg (5,125 lbs.).

Combined area of the four panels is 15 square m (161 square ft.), and they provide both regulated and unregulated direct current power; unregulated power is provided to the radio transmitter and the Lander.

Two 30-ampere-hour, nickel-cadmium, rechargeable batteries provide power when the spacecraft is not facing the Sun during launch, correction maneuvers and Mars occultation.

Guidance and Control

The Orbiter is stabilized in flight by locking onto the Sun for pitch and yaw reference and onto the star Canopus for roll reference. The attitude control subsystem (ACS) keeps this attitude with nitrogen gas jets located at the solar panel tips. The jets fire to correct any drift. A cruise Sun sensor and the Canopus sensor provide error signals. Before Sun acquisition four acquisition Sun sensors are used and then turned off.

The ACS also operates in an all-inertial mode or in roll-inertial with pitch and yaw control, still using the Sun sensors. During correction maneuvers the ACS aligns the vehicle to a specified attitude in response to commands from the on-board computer. Attitude control during engine burns is provided in roll by the ACS and in pitch and yaw by an autopilot that commands engine gimballing.

If Sun lock is lost the ACS automatically realigns the spacecraft. In loss of Canopus lock, the ACS switches to roll-inertial and waits for commands from the spacecraft computer. The nitrogen gas supply for the ACS can be augmented by diverting excess helium gas from the propulsion module, if necessary.

Two on-board general-purpose computers in the computer command subsystem (CCS) decode commands and either order the desired function at once or store the commands in a 4,096-word, plated-wire memory. All Orbiter events are controlled by the CCS, including correction maneuvers, engine burns, science sequences, and high-gain antenna pointing.

Communications

The main Orbiter communications system is a two-way, S-band, high-rate radio link providing Earth command, radio tracking and science and engineering data return. This link uses either a steerable 1.5 m (59 in.) dish high-gain antenna (HGA) or an omni-directional low-gain antenna (LGA), both of them on the Orbiter. The LGA is used to send and receive near Earth, the HGA as distances increase. An X-band link is used for radio science through the HGA.

S-band transmission rates vary from 8.3 or 33.3 bits per second (bps) for engineering data to 2,000 to 16,000 bps for Lander and Orbiter science data.

Relay from the Lander is through an antenna mounted on the outer edge of a solar panel. It will be activated before separation and will receive from the Lander through separation, entry, landing, and surface operations. The bit rate during entry and landing is 4,000 bps; landed rate is 16,000 bps.

Data Storage

Data are stored aboard the orbiter on two eight-track digital tape recorders. Seven tracks are used for picture data and the eighth track for infrared data or relayed Lander data. Each recorder can store 640 million bits.

Data collected by the orbiter, including Lander data are converted into digital form by the flight data subsystem (FDS) and routed to the communications subsystem for transmission or to the tape recorders for storage. This subsystem also provides timing signals for the three Orbiter science experiments.

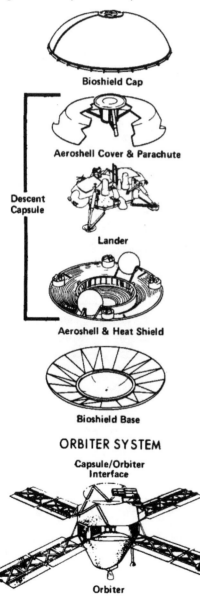

Bioshield Cap

Aeroshell Cover & Parachute

Descent
Capsule

Lander

Aeroshell & Heat Shield

Bioshield Base

ORBITER SYSTEM

Capsule/Orbiter
Interface

Orbiter

VIKING LANDER

The Lander spacecraft is composed of five basic systems: the Lander body, the bioshield cap and base, the aeroshell, the base cover and parachute system, and Lander subsystems. Operational lifetime for the Lander is 90 days after landing.

The completely outfitted Lander measures approximately 3 m (10 ft.) across and is about 2 m (7 ft.) tall. It weighs approximately 576 kg (1,270 lbs.) without fuel.

The Lander and all exterior assemblies are painted light gray to reflect solar heat and to protect equipment from abrasion. The paint is made of rubber-based silicone.

Lander Body

The body is a basic platform for science instruments and operational subsystems. It is a hexagon-shaped box with three 109-cm (43-in.) sidebeams and three 56-cm (22-in.) short sides. It looks like a triangle with blunted corners.

The box is built of aluminum and titanium alloys, and is insulated with spun fiberglass and dacron cloth to protect equipment and to lessen heat loss. The hollow container is 1.5 m (59 in.) wide and 46 cm (18 in.) deep, with cover plates top and bottom.

The Lander body is supported by three landing legs, 1.3 m (51 in.) long, attached to the short-side bottom corners of the body. The legs give the Lander a ground clearance of 22 cm (8.7 in.).

Each leg has a main strut assembly and an A-frame

assembly, to which is attached a circular footpad 30.5 cm (12 in.) in diameter. The main struts contain bonded, crushed aluminum honeycomb to reduce the shock of landing.

Bioshield Cap and Base

The two-piece bioshield is a pressurized cocoon that completely seals the Lander from any possibility of biological contamination until Viking leaves Earth's atmosphere.

The two bioshield halves generally resemble an egg, and the shield's white thermal paint heightens the resemblance. It measures 3.7 m (12 ft.) in diameter and is 1.9 m (6.4 ft.) deep. It's made of coated, woven fiberglass, 0.13 mm (0.005-in.) thin, bonded to an aluminum support structure.

The bioshield is vented to prevent over-pressurization and possible rupture of its sterile seal.

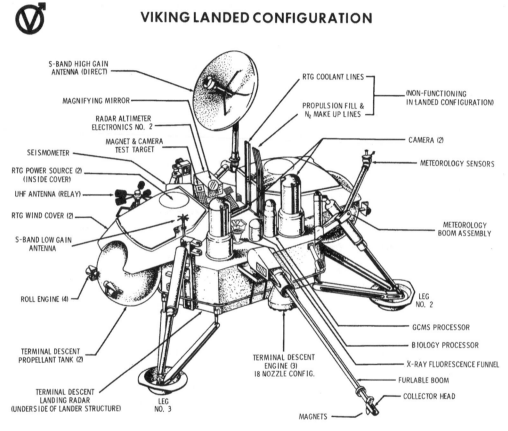

VIKING LANDED CONFIGURATION

Aeroshell

The aeroshell is an aerodynamic heat shield made of aluminum alloy in a 140-degree, flat cone shape and stiffened with concentric rings. It fits between the Lander and the bioshield base. It is 3.5 m (11.5 ft.) in diameter and its aluminum skin is 0.86 mm (0.034-in.) thin.

Bonded to its exterior is a lightweight, cork-like ablative material that burns away to protect the Lander from aerodynamic heating at entry temperatures which may reach 1,500 degrees C. (2,730 degrees F.).

The interior of the aeroshell contains twelve small reaction control engines, in four clusters of three around the aeroshell's edge, and two spherical titanium tanks that contain 85 kg (188 lbs.) of hydrazine mono-propellant.

The engines control pitch and yaw to align the Lander for entry, help slow the craft during early entry and maintain roll control.

During the long cruise phase, an umbilical connection through the aeroshell provides power from the Orbiter to the Lander; housekeeping data also flow through this connection.

The aeroshell also contains two science instruments — the Upper Atmosphere Mass Spectrometer (UAMS) and the Retarding Potential Analyzer (RPA) — plus pressure and temperature sensors.

Base Cover and Parachute System

The base cover fits between the bioshield cap and the Lander. It is made of aluminum and fiberglass; the fiberglass allows transmission of telemetry data to the Orbiter during entry. It covers the parachute and its ejection mortar, and protects the Lander's top during part of the entry phase.

The parachute is made of lightweight dacron polyester 16 m (53 ft.) in diameter. It weighs 50 kg (110 lbs.).

The parachute is packed inside a mortar 38 cm (15 in.) in diameter, mounted into the base cover. The mortar is fired to eject the parachute at about 139 km per hour (75 mph). The chute has extra-long suspension lines that trail the capsule by about 30 m (100 ft.).

Lander Subsystems

Lander subsystems are divided into six major categories: descent engines, communications equipment, power sources, landing radars, data storage, and guidance and control.

Descent Engines

Three terminal descent engines (TDE) provide attitude control and reduce the Lander's velocity after parachute separation. The 2,600-newton (600-lb.) throttleable engines are located 120 degrees apart on the Lander's sidebeams. They burn hydrazine mono-propellant.

The engines use an advanced exhaust design that won't alter the landing site environment. An unusual grouping of 18 small nozzles on each engine will spread engine exhaust over a wide angle that won't alter the surface or unduly disturb the chemical and biological experiments.

Two spherical titanium tanks, attached to opposite sides of the Lander body beneath the RTG wind covers, feed the TDEs from an 85-kg (188 lb.) hydrazine propellant supply.

Four small reaction control engines use hydrazine mono-propellant thrusters to control Lander roll attitude during terminal descent. The engines are mounted in pairs on the TDE propellant tanks and are identical to those used on the aeroshell.

Communication Equipment

The Lander is equipped to transmit information directly to Earth with an S-band communications system, or through the Orbiter with an ultra-high frequency (UHF) relay system. The Lander also receives Earth commands through the S-band system.

Two S-band receivers provide total redundancy in both command receiving and data transmission. One receiver uses the high-gain antenna (HGA), a 76-cm (30-in.) diameter parabolic reflector dish that can be pointed to Earth by computer control. The second receiver uses a fixed low-gain antenna (LGA) to receive Earth commands.

The UHF relay system transmits data to the Orbiter with a radio transmitter that uses a fixed antenna. The UHF system will operate during entry and for the first three days of landed operations. After that it will only operate during specific periods.

Landing Radars

The radar altimeter (RA) measures the Lander's altitude during the early entry phase, alerting the Lander computer to execute the proper entry commands. The RA is a solid-state pulse radar with two specially designed antennas: one is mounted beneath the Lander and one is mounted through the aeroshell. Altitude data are received from 1,370 km down to 30.5 m (740 mi. to 100 ft.).

The aeroshell antenna provides high-altitude data for entry science, vehicle control and parachute deployment. The Lander antenna is switched into operation at aeroshell separation and provides altitude data for guidance and control, and for terminal descent engine ignition.

The terminal descent landing radar (TDLR) measures the horizontal velocity of the Lander during the final landing phase. It is located directly beneath the Lander and is turned on at about 12 km (4,000 ft.). It consists of four continuous-wave Doppler radar beams that can measure velocity to an accuracy of plus or minus one meter per second.

Both radars are essential for mission success, so the terminal descent landing radar can work with any three of its four beams, and identical sets of radar altimeter electronics can be switched to either of the RA antennas.

Guidance and Control

The "brain" of the Lander is its guidance control and sequencing computer (GCSC). It commands everything the Lander does through software (computer programs) stored in advance or relayed by Earth controllers.

The computer is one of the greatest technical challenges of Viking. It consists of two general-purpose computer channels with plated-wire memories, each with an 18,000-word storage capacity. One channel will be operational while the other is in reserve.

Among other programs, the computer has instructions stored in its memory that can control the Lander's first 22 days on Mars without any contact from Earth. These instructions will be updated and modified by Earth command once communication has been established.

Power Sources

Basic power for the Lander is provided by two SNAP 19-style 35-watt radioisotope thermoelectric generators (RTGs) developed by the U. S. Energy Research and Development Administration (ERDA). They are located atop the Lander, and are connected in series to double their voltage and reduce power loss.

The SNAP 19 Viking generator is 147 cm (23 in.) across the housing fin tips, 96 cm. (15 in.) in length and weighs 15.3 kg (34 lbs.).

The first isotopic space generator was put into service in June 1961, on a Navy navigational satellite. Advances in SNAP systems were made with the development and flight of SNAP 19 aboard Nimbus III, launched in April 1969. This use of SNAP 19 represented a major milestone in the development of long-lived, highly reliable isotope power systems for space use by NASA. The SNAP 27 generator was developed to power 5 science stations left on the Moon by the Apollo 12, 14, 15, 16 and 17 astronauts. The continuing operation of these generators is providing new dimensions of data about the Moon and the universe. Four SNAP 19 nuclear generators are providing the electrical power for each of the two NASA pioneer Jupiter fly by missions (Pioneers 10 and 11) currently in space.

The generators will provide a long-lived source of electricity and heat on Mars, where sunlight is half as strong as on Earth, and is non-existent during the Martian night, when temperatures can drop as low as minus 120 degrees C. (minus 184 degrees F.).

The generators use thermoelectric elements to convert heat from decaying plutonium-238 into 70 watts of electrical power.

"Waste" or unconverted heat is conveyed by thermal switches to the Lander's interior instrument compartment, when required. Covers over the RTGs prevent excess heat dissipation into the environment.

Four nickel-cadmium, rechargeable batteries help supply Lander power requirements in peak activity periods. The batteries, mounted in pairs inside the Lander, are charged by the RTGs with power available when other Lander power requirements are less than RTG output.

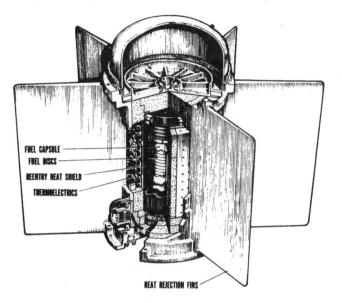

FUEL CAPSULE
FUEL DISCS
REENTRY HEAT SHIELD
THERMOELECTRICS

HEAT REJECTION FINS

SNAP 19/VIKING RADIOISOTOPE THERMOELECTRIC GENERATOR

Data Storage

This equipment collects and controls the flow of Lander scientific and engineering data. It consists of a data acquisition and processing unit (DAPU), a data storage memory and a tape recorder.

The DAPU actually collects the science and engineering information and routes it to one of three places: to Earth through the S-band HGA, to the data storage memory or to the tape recorder.

Information will be stored in the data storage memory for short periods. Several times a day the memory will transfer data to the tape recorder or back to the DAPU for further transmission. The memory has a storage capacity of 8,200 words.

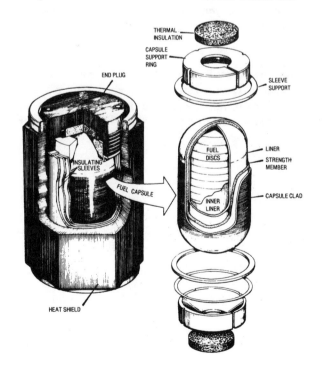

THERMAL INSULATION
CAPSULE SUPPORT RING
END PLUG
SLEEVE SUPPORT
INSULATING SLEEVES
FUEL DISCS
LINER
STRENGTH MEMBER
FUEL CAPSULE
CAPSULE CLAD
INNER LINER
HEAT SHIELD

SNAP 19/VIKING HEAT SOURCE

Data are stored on the tape recorder for long periods. The recorder can transmit at high speed back through the DAPU and the UHF link to an orbiter passing overhead. It can store as many as 40 million bits of information, and it can record at two speeds and play back at five.

LAUNCH VEHICLE

The two Viking missions are the second and third operational launches for the Titan Centaur vehicle (TC-3 and TC-4). Its first operational launch of a Helios spacecraft last December 10 was an unqualified success. In addition to launching Helios into a solar orbit, the Titan Centaur successfully performed several experiments to demonstrate its readiness to perform future missions. The first experiment demonstrated the Centaur restart capability after a one hour zero-gravity coast for application to the Mariner Jupiter Saturn mission scheduled for 1977. Subsequent experiments demonstrated its ability to directly insert spacecraft into Earth synchronous orbits.

The Titan Centaur launch vehicle combines a Titan III-E booster and Centaur D-1T third stage. It has an over all height of 48.8 m (160 ft.) and a total liftoff weight of 64,000 kg (1.4 million lbs.).

The entire Centaur third stage vehicle and its payload is covered by the Centaur Standard Shroud (CSS). The CSS measures 17.6 m (58 ft.) long and is 4.2 m (14 ft.) in diameter, and will accommodate a spacecraft nearly 8.5 m (28 ft.) in length.

Titan III-E

The Titan III-E booster consists of a two-stage liquid propellant core vehicle and two strap-on solid rocket motors, each 3 m (10 ft.) in diameter and 25.9 m (85 ft.) long. The two solids, which are made up of five segments each, provide a thrust of 10.6 million newtons (2.4 million lbs.) at liftoff.

The 3-m (10-ft.) diameter core stages are primarily constructed of aluminum alloys. They are made with aluminum skins with T-shaped aluminum stringers integrally milled. The length of the first stage is 22.2 m (72.9 ft.) and the second stage 7.1 m (23.3 ft.).

A heat shield assembly protects the Stage I engine from the high temperatures generated by the solid rocket motors. The heat shield encloses a major portion of the engine from the thrust chamber throats upward.

The two core stages burn a 50-50 blend of hydrazine and unsymmetrical dimethylhydrazine (UDMH) fuel and nitrogen tetroxide oxidizer. The first stage uses an Aerojet YLR87 OAJ-11 engine with two gimballing thrust chambers. It can burn for approximately 148 seconds and provides 2,340,000 N (520,000 lbs.) thrust. The second stage Aerojet YLR91AJ-11 engine has a single thrust chamber and provides 446,000 N (101,000 lbs.) thrust for about 208 seconds. Both the first and the second stage engines are regeneratively cooled and turbo pump fed.

Control for Stage I is achieved by gimballing the engines. The gimballed main engine provides pitch and yaw control for Stage II, with the gas generator exhaust providing roll control. Guidance commands come from the Centaur system while stability is controlled by the Titan flight control system.

The two five-segment solid strap-on motors use powdered aluminum fuel and ammonium perchlorate oxidizer in a plastic binder. Both burn for approximately 122 seconds. The two solid motors are generally referred to as Stage "0". Each carries a tank for nitrogen tetroxide mounted on the side of the motor for thrust vector control. The nitrogen tetroxide is injected through the nozzle to deflect the motor thrust for flight control.

Centaur D-1T

The Centaur vehicle is 9.1 m (30 ft.) long and 3 m (10 ft.) in diameter. The tank structure is made from pressure stabilized stainless steel 3 mm (0.014 in.) thick in the cylindrical sections. This is approximately the thickness of a dime. Pressure stabilized means that the strength of the vehicle structure depends on pressure inside the tanks and it has often been called a balloon type structure. When the vehicle is not pressurized it must be kept in a special cradle which keeps it stretched to retain its shape.

A double-walled vacuum-insulated bulkhead separates the liquid oxygen section from the liquid hydrogen tank. The forward equipment module attaches to the tank by a short conical stub adapter. The stub adapter is also used as an attach point for a truss type adapter for payloads weighing more than 1,814 kg (4,000 lbs.). Spacecraft smaller than that are supported by a payload adapter mounted on the forward end of the equipment module.

The entire cylindrical portion of the D-1T vehicle is covered with a new permanent radiation shield consisting of three separate layers of an aluminized mylar, dacron net sandwich. The forward tank bulkhead and tank access doors are insulated with a number of layers of aluminized mylar. The aft bulkhead is covered with a dacron-reinforced aluminized mylar membrane and protected further with a rigid radiation shield supported on brackets. The radiation shield is made of laminated nylon fabric with aluminized mylar on the inside and white polyvinyl fluoride on its outer surface and is necessary to limit the loss of propellants resulting from solar heating during long duration missions.

This permanent insulation system will allow the Centaur stage to coast up to 5 1/4 hours in space and restart its engines. This added capability for coast, over former Centaur vehicles, is necessary for synchronous orbit missions.

Additional hydrogen peroxide for attitude control and propellant settling as well as additional helium for tank pressurization have also been added to the D-1T vehicle to allow for extended missions. Primary propulsion for the Centaur is its two RL-10A-3-3 engines which provide 66,720 N (15,000 lbs.) thrust each.

During coast, separation and retromaneuvers, attitude control and propellant settling are provided by 12 small hydrogen peroxide thrusters rated at 26.7 N (6 lbs.) thrust.

The Centaur D-1T astrionics system consists primarily of a digital computer unit and an inertial reference unit. The 27.2 kg (60 lb.) digital computer has a 16,000-word random access memory. The inertial reference unit contains a four-gimbal, all-attitude stable platform. Three gyros stabilize this platform on which are mounted three pulse-rebalanced accelerometers.

The Centaur astrionics system handles navigation, guidance tasks, propellant and tank pressure management, telemetry formats and transmission and initiates vehicle events. The system also performs a major role in checking itself and other vehicle systems prior to launch. One of its major advantages is the increased flexibility the new astrionics system offers over the original Centaur system. In the past, hardware frequently had to be modified for each mission. Now most operational needs can be met by changing the computer software.

The Titan III vehicle previously used a radio guidance system. Modifications to mate it with Centaur were designed to retain as much Titan autopilot and programming sequence capability as possible. To keep modifications to the Titan and connections between stages as simple as possible, the Centaur guidance system feeds signals to the Titan flight computer and lets it send the proper commands to Titan systems.

Centaur Standard Shroud

The Centaur Standard Shroud provides a large payload space on the Titan Centaur. In most configurations a payload nearly 8.5 m (28 ft.) long can be accommodated. Inside clearance of the shroud is 3.8 m (12 ½ ft.). A manufacturing joint is provided which allows for future shroud growth if a need for longer payload accommodations arises.

The nose cap is made from corrosion-resistant steel attached to two conical sections of magnesium. The cylindrical sections are made of corrugated aluminum. A seal and insulation allows a clean and thermally controlled environment in the payload area.

The two halves of the Centaur Standard Shroud join along a longitudinal split line. Approximately 60 seconds after Titan Stage II ignition, the longitudinal and horizontal split lines are severed by a non-contaminating pyrotechnic system. Four compressed springs force the two halves to separate. The cone shaped bottom section of the shroud is bolted to the inter-stage adapter and is jettisoned later with the Titan stage.

That portion of the Centaur Standard Shroud which surrounds the Centaur vehicle contains 8.3 cm (3.3 in.) fiberglass insulation. This section reduces heat transfer to the Centaur liquid hydrogen and oxygen propellant on the launch pad and during ascent through the atmosphere.

TITAN CENTAUR FLIGHT SEQUENCE

Titan Phase

Liftoff occurs approximately two-tenths of a second after ignition of the solid rocket motors. At 6.5 seconds into the flight, Titan begins a programmed roll maneuver commanded by Centaur guidance. The roll maneuver and all attitude control during the solid rocket motor powered portion of the flight are accomplished by the Titan thrust vector system. Steering is accomplished by injecting liquid nitrogen tetroxide into the solid motor nozzles deflecting the rocket exhaust gases.

Stage I ignition of the Titan occurs when acceleration from the large solid motors reduces to 1.5 g. Approximately 12 seconds later, the solids are jettisoned. Titan Stage I continues thrusting until propellant depletion at approximately T plus 260 seconds.

Titan Stage II ignition occurs at Stage I propellant depletion, and separation takes place approximately one second later. During Stage I and Stage II phases of the flight, the vehicle attitude in pitch and yaw is controlled by the Titan flight control system with guidance steering corrections supplied by the Centaur guidance system.

The Centaur standard shroud is jettisoned by command from the Centaur guidance system approximately 10 seconds after Stage II ignition.

Titan Stage II boosts the vehicle until loss of acceleration due to propellant depletion, approximately 467 seconds after liftoff. The Centaur guidance system commands separation when Stage II acceleration decays to about .012 g. The Centaur interstage adapter is severed by a shaped charge and retrorockets on the Titan Stage II slow the spent stage.

Centaur Phase

Centaur first main engine start occurs approximately 10.5 seconds after Titan Centaur separation. Centaur main engine shutdown is commanded by the guidance system when the proper parking orbit is achieved.

Continuous propellant settling will be maintained during the parking orbit coast phase. During most of the coast phase the vehicle is aligned along the inertial velocity vector. Prior to second burn the vehicle is realigned to the pitch attitude required for achieving the proper trajectory to Mars.

The second Centaur burn, of approximately 310 seconds, is terminated by the Centaur guidance system.

LAUNCH WINDOW

The windows for launches to Mars are determined by the relationship between the Earth and Mars, the amount of energy required to launch a spacecraft of a given weight on each day and such factors as tracking coverage.

The energy level and velocity necessary to reach Mars is lowest when the Earth launch and Mars arrival occur almost on opposite sides of the Sun. This condition occurs during a few weeks every 25 months.

Tracking coverage is important for determining exact course and performance of the launch vehicle. A second factor is range safety. The launch azimuth, or direction of launch, must stay within certain prescribed limits to avoid overflying populated areas in case of a malfunction in the vehicle and reentry of vehicle stages and shrouds. Because of the Earth rotation, the launch azimuth for most planetary launches varies from minute to minute throughout the launch window. When the launch azimuth to reach Mars does not conform to that required for range safety, the vehicle can be launched down a safe corridor and then perform a dog-leg maneuver to the proper flight azimuth after satisfying range safety requirements.

VIKING A* LAUNCH VEHICLE CHARACTERISTICS

Liftoff weight, including spacecraft   640,827 kg (1,412,759 lbs.)
Liftoff height   48.8 m (160 ft
Launch complex   41
Launch azimuth   97 degrees (at window opening)

	Titan Booster	Centaur Stage
Weight	621,273 kg (35,556 lbs.)	16,128 kg (1,369,650 lbs.) (not including shroud & S/C)
Height	29.8 m (98 ft.)	9.6 m (31.5 ft.) (including truss payload adapter but without shroud and interstage adapter)
Propellants	Powdered aluminum and ammonium perchlorate in solid motors; Aerozine 50 and nitrogen tetroxide in Stage I and II, plus UDMH	Liquid hydrogen and liquid oxygen
Propulsion	Two solid motors provide 5.3 million N (1.2 million lbs.) thrust each. LR87AJ-11, Stage I engine, 2.3 million N (520,000 lbs.) thrust. LR91AJ-11, Stage II engine, 445,000 N (100,000 lbs.) thrust.	Two 66,720-N (15,000 lbs.) thrust RL-10 engines. 12 small hydrogen peroxide thrusters.
Velocity	4,957 kmph (3,080 mph) at Stage I ignition, 14,510 kmph (9,015 mph) at Stage II ignition, 23,278 kmph (19,439 mph) at Stage II separation.	26,596 kmph (16,525 mph) at MECO I, 41,231 kmph (25,619 mph) at MECO II, 39,576 kmph (24,596 mph) at spacecraft separation.
Guidance	Centaur inertial guidance	Inertial guidance

*Viking B launch vehicle characteristics vary slightly.

VIKING LAUNCH PREPARATIONS

The Viking spacecraft will be launched from Launch Complex 41 (LC-41) Titan III Complex, Air Force Eastern Test Range. Two Titan Centaur launches will mark the third and fourth times NASA launches have been conducted from this facility at Cape Canaveral Air Force Station. Launch will be under direction of NASA's Kennedy Space Center, Unmanned Launch Operations Directorate.

Viking launch preparations required eight months of work by a government and industrial team of about 1,200 people, representing the most extensive effort ever associated with an unmanned space project.

The Mars launch opportunity is approximately 40 days, imposing stringent scheduling requirements on the Viking team to launch both spacecraft from a single pad within a short time span. The schedule is even more demanding because of the plan to launch both Viking space vehicles 10 days apart to improve the probability of conducting both launches before the end of the "window" during September.

TYPICAL FLIGHT EVENTS FOR VIKING A*

Flight Events	Seconds	Altitude Km	Miles	Velocity kmph	mph	Range Km	Miles
Solid Motor Ignition	0	0	0	0	0	0	0
Stage I Ignition	110.6	39.3	24.4	4,957	3,080	42.2	26.2
Solid Motor Jettison	121.9	46.1	28.6	5,379	3,342	56.7	35.2
Stage I Cutoff	256.2	109	67.7	14,496	9,007	380.6	236.3
Stage II Ignition	257	109.5	68	14,510	9,015	383.8	238.3
Centaur Shroud Jettison	268	115.2	71.6	14,756	9,169	426.9	265.1
Stage II Cutoff	459.6	164.1	102	23,205	14,418	1,337.9	855.6
Stage II Jettison	465.6	164.8	102.4	23,278	14,439	1,415.6	878.9
Centaur MES-I (Main engine start)	476.1	165.9	103.1	23,230	14,436	1,481.6	920
Centaur MECO-I (Main engine cutoff)	603.3	169.7	105.4	26,596	16,525	2,336.6	1,450.9
Centaur MES-II	1,805.4	162.2	100.8	26,652	16,560	10,998	6,829.3
Centaur MECO-II	2,125.7	307.3	191	41,231	25,619	13,827.7	8,586.3
Spacecraft Separation	2,345.3	941.6	585.2	39,576	24,590	16,012	9,942.6
Start Centaur Blowdown	3,200.2	5,959.6	3,704	31,564	19,612	19,397.6	12,045
End Blowdown	3,450.2	7,663	4,762.6	29,923	18,592	18,790.2	11,667.8

*Based on launch date: 8/11/75; launch time: 4:57 p.m. EDT; launch azimuth: 97 degrees;
arrival date in Mars orbit: 6/17/76

TYPICAL FLIGHT EVENTS FOR VIKING B*

Flight Events	Seconds	Altitude Km	Miles	Velocity Kmph	mph	Range Km	Miles
Solid Motor Ignition	0	0	0	0	0	0	0
Stage I Ignition	110.7	39.4	24.5	4,970	3,088	42.3	26.3
Solid Motor Jettison	122	46.2	28.7	5,385	3,346	56.8	35.3
Stage I Cutoff	257.6	109.1	67.8	14,439	8,971	383.3	238.2
Stage II Ignition	258.4	109.6	68.1	14,452	8,980	396.5	240.2
Centaur Shroud Jettison	269	115	71.5	14,689	9,127	428.2	266.1
Stage II Cutoff	464	163.6	101.7	23,176	14,400	1,393	866
Stage II Jettison	470	164.4	102.2	23,208	14,420	1,431	889
Centaur MES-I (Main engine start)	480.5	165.5	102.9	23,203	14,417	1,497	930
Centaur MECO-I (Main engine cutoff)	609.3	169.3	105.2	26,616	16,537	2,362	1,468
Centaur MES-II	1,977.2	162.6	101	26,669	16,571	12,228	7,598
Centaur MECO-II	2,288.3	294.7	183.2	40,663	25,266	14,962	9,297
Spacecraft Separation	2,507.8	892.5	554.7	39,068	24,275	17,131	10,645

*Based on launch date: 8/22/75; launch time: 3:45 p.m. EDT; launch azimuth: 101 degrees;
arrival in Mars orbit: 8/7/76

The Titan III facility has two active pads, LC-40 and LC-41, but only Complex 41 has been modified to accommodate NASA's hydrogen-fueled Centaur.

Launch Facilities

The Titan III Complex, built on manmade islands in the Banana River, consists of solid rocket motor servicing and storage areas, a Vertical Integration Building (VIB), a Solid Motor Assembly Building (SMAB), LC-40 and LC-41, and a double-track locomotive system that transports the mated Titan core and Centaur vehicle from the VIB through the SMAB to LC-41. The rail system covers 32 km (19 mi.) to link the various facilities.

Hardware Assembly

The Titan, Centaur and Centaur shroud are erected and mated in the VIB on a mobile transporter/umbilical mast structure. Attached to the transporter are three vans, housing launch control and monitoring equipment, which remain connected to the transporter and vehicle throughout the receipt-to-launch sequence.

When integrated tests in the VIB are completed, the assembled Titan and Centaur are moved on the transporter to the SMAB. After the solid rocket motors and core stages are structurally mated, the vehicle is moved to the launch complex. A mobile service structure provides access to all mated vehicle stages, and an environmental enclosure ("white room") protects the Centaur and the spacecraft.

Because of the quick turnaround launch sequence required by Viking, the second Titan Centaur (TC-3) launch vehicle was processed first. The Titan liquid stages and Centaur upper stages were assembled atop a transporter in the VIB in January and moved to LC-41 for checkout.

The encapsulated Viking A spacecraft was mated with the launch vehicle at LC-41 in late March, and the space vehicle underwent several key tests in early April, including Flight Events Demonstration and Terminal Countdown Demonstration.

The spacecraft was then de-mated and returned to the KSC Industrial Area for further processing. The TC-3 vehicle was returned to the VIB on its transporter in late April for storage. Just before the Viking A launch on Titan Centaur-4 (TC-4), TC-3 will be moved into the SMAB for attachment of its solid booster motors. It will be moved to LC-41 immediately after launch of Viking A. Viking B will then be mated with its launch vehicle and checkout completed, heading for a launch 10 days after launch of Viking A.

TC-4 processing for the first launch began in early spring. The Titan and Centaur liquid stages were stacked in the VIB before they were moved to the SMAB in late May for attachment of solid booster motors. The vehicle was moved to LC-41 in early June; mating with the first Viking was done in late July. The space vehicle will be cycled through the normal launch preparation sequence, heading for liftoff of the Viking A August 11 at the opening of the 1975 Mars launch period.

Spacecraft Preparations

Flight elements of the Viking spacecraft began arriving at KSC in January for pre-launch operations, including assembly, checkout, fueling, sterilization and encapsulation in the payload section of the shroud. The orbiters were processed in Hangar AO at Cape Canaveral Air Force Station; the Landers were processed in Spacecraft Assembly and Encapsulation Facilities (SAEF) 1 and 2 in the KSC industrial Area.

The orbiters were later moved across the Banana River to the SAEFs for mating with the Landers. Lander processing included installation of twin 35-watt SNAP-19 radioisotope thermoelectric generators (RTG) for each spacecraft, installation and checkout of Lander science instruments and sterilization of the Landers.

The SAEF buildings are modified Apollo-related structures in the southeast sector of the KSC Industrial Area. Airlocks and clean rooms were built to provide environmental control within the structures. Each SAEF has an "oven" for spacecraft sterilization, connected to the clean room by a passageway. The sterilization chambers are built of stainless steel and insulated with polyurethane.

The Landers were sterilized in mid-June. The craft were placed in the ovens, 7 m long, 6.4 m wide and 4.3 m high (23 by 21 by 14 ft.), and sterilized at a temperature of approximately 113 degrees C. (235 degrees F.) for 40 hours. Sterilization is done in a swirling atmosphere of heated nitrogen gas, held to 97 per cent purity to control oxidation of spacecraft components.

LC-41 VEHICLE SERVICES

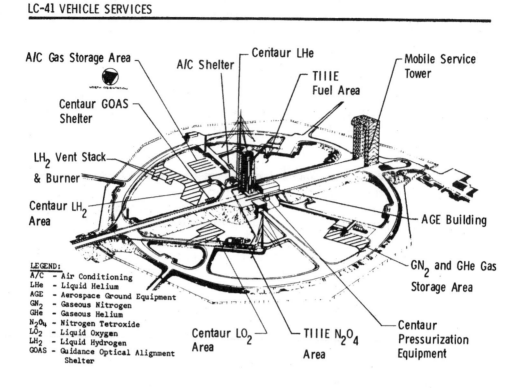

LEGEND:
A/C - Air Conditioning
LHe - Liquid Helium
AGE - Aerospace Ground Equipment
GN_2 - Gaseous Nitrogen
GHe - Gaseous Helium
N_2O_4 - Nitrogen Tetroxide
LO_2 - Liquid Oxygen
LH_2 - Liquid Hydrogen
GOAS - Guidance Optical Alignment Shelter

The RTGs were installed early in the checkout flow to permit electrical tests of Lander systems independent of ground power. The power generators remained aboard the Landers during the sterilization process; chilled water units were used to control their heat dissipation.

The Landers underwent post-sterilization testing to make sure their systems had survived the intense, long-duration heating process. Propellants for Lander deorbit and terminal descent engine subsystems were loaded before the Landers were removed from the sterilization chambers.

The Orbiters were moved to ESA-60 on Cape Canaveral Air Force Station, where their engine subsystems were assembled. Fueled propulsion modules were attached to the Orbiter buses before the orbiters were transferred to the SAEFs.

The Orbiter and Lander for the first Viking were mated and encapsulated in the payload section of the Centaur shroud in mid-July, then moved to LC-41 later in the month for final preparations.

The Orbiter and Lander for the second Viking were mated and encapsulated in late July. The second Viking will be moved to LC-41 for mating with TC-3 immediately after the first Viking is launched. Launch readiness target date is August 21.

Launch will be directed from the VIB, a 23-story structure containing nine million cubic feet. In the VIB, located 5,100 m (20,000 ft.) from LC-41, will be Launch Control Titan Centaur core vehicle assembly and initial systems checkout. The Launch Control area, consisting of three rooms, is the nerve center of the Titan Centaur Complex.

SAMSO is responsible for the design, development, procurement, acceptance, testing and delivery of the Titan III-E airborne vehicle and aerospace ground equipment to meet NASA requirements for integration with the Centaur. All Titan launch vehicles are produced for the Air Force by Martin Marietta Aerospace, Denver Division.

SAMSO, headquartered at Los Angeles Air Force Station, is the major Department of Defense development agency for this nation's present and future military space and ballistic missile programs. SAMSO's Deputy for Launch Vehicles, Col. G. J. Murphy, is responsible for the Titan III, with Maj. L. N. Johnson the program manager for Titan III-E. Its 6555th Aerospace Test Wing at Cape Canaveral AFS was responsible for activation of the ITL facility and manages its total operations. It administers the ITL's safety, scheduling, etc. The Wing supports NASA in the conduct of launch operations for Titan/Centaur.

Press Site 3, the press viewing area for Titan III launches, is located to the southeast of the VIB, slightly more than 6,100 m (20,000 ft.) from LC-41

Countdown

The Viking countdown will be conducted by a team of about 300, representing NASA's Kennedy Space Center, Langley Research Center, Lewis Research Center and Jet Propulsion Laboratory; General Dynamics/Convair, Pratt & Whitney, Martin Marietta Corp., United Technology Center, Aerojet Propulsion Co., RCA and Pan American World Airways. Representatives of the Air Force's 6555th Aerospace Test Group are consultants to NASA.

F - 6 days:	Start Readiness Count, install solid motor and Titan core ordnance and Centaur flight ordnance.
F - 5 days:	Solid rocket motor and Titan integrity inspections; start Titan and Centaur system preparations.
F - 4 days:	Load fuel for Titan first and second stages.
F - 3 days:	Load oxidizer for Titan first and second stages; install Centaur batteries.
F - 2 days:	Install and connect core vehicle ordnance.
F - 1 day:	Connect Centaur ordnance; load and pressurize solid rocket motor oxidizer tanks; conduct range safety system checks, install Titan Centaur destructors; and pressurize Titan propellant tanks.

Launch Day Count

T - 590 minutes:	Begin Titan Centaur systems launch day preparations.
T - 190 minutes:	Begin move of mobile service tower (continues to T minus 160 minutes).
T - 115 minutes:	Begin one-hour, built-in hold. At end of one-hour hold, install solid rocket motor arming plugs.
T - 100 minutes:	Begin Centaur liquid oxygen loading (topping off continues until T minus 75 seconds.)
T - 70 minutes:	Begin Centaur liquid hydrogen loading (topping off continues until T minus 90 seconds).
T - 25 minutes:	Range Safety Command test.
T - 10 minutes:	Enter 10-minute built-in hold.
T - 5 mins, 30 sec:	Flight control to launch enable.
T - 5 minutes:	Enter Terminal Count.
T - 2 minutes:	Centaur to internal power; Range Safety Command to arm.
T - 90 seconds:	Centaur hydrogen topping secured.
T - 75 seconds:	Centaur oxygen topping secured.
T - 45 seconds:	Centaur launch permit on.
T - 32 seconds:	Start automatic launch sequence.
T - 31.7 seconds:	Titan on automatic launch sequence.
T - 31 seconds:	Start Centaur Digital Computer Unit Count.
T - 21.5 seconds:	Test solid rocket motor ignition.
T - 17.5 seconds:	Open Titan pre-valves.
T - 8 seconds:	Start inertial guidance.
T - 0:	Ignite solid rocket motors.

VIKING SPACE VEHICLE CONFIGURATION

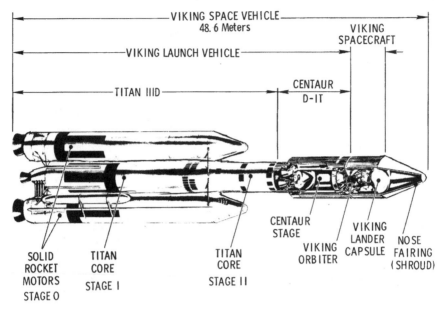

MISSION CONTROL AND COMPUTING CENTER

The focal point of all Viking flight operations is the Viking mission Control and Computing Center (VMCCC) at the Jet Propulsion Laboratory. The Viking Flight Team (VFT) is housed in the VMCCC, and through the Center will come all data from the Orbiters and Landers to be processed and presented to the Flight Team for analysis.

Housed in two buildings at JPL, the VMCCC contains all the computer systems, communication and display equipment, photo processing laboratories and mission support areas for mission controllers, spacecraft performance analysts and science investigators.

By the time the first Viking spacecraft arrives at Mars, the facilities will house more than 700 Flight Team members, plus several hundred more VMCCC people, who will operate the facilities, computers, laboratories, maintenance shops and communications networks.

The VMCCC's large and complex computer systems receive incoming Orbiter and Lander data, process them in real time, and display and organize them for further processing and analysis. Data are first received as radio signals by the Deep Space Network (DSN) stations around the world and are transmitted into the VMCCC computers, where processing begins. Software (programs) in these computers does the receiving, display and organizing of data.

Commands that cause the Orbiters and Landers to maneuver, gather science data and do other complex activities are prepared by the Flight Team. Commands are introduced into the computers through the Team's control, and are communicated to a DSN station for transmission to the appropriate spacecraft.

Three sets of computer systems are in the VMCCC. One is a complex of UNIVAC 1530, 1219 and 1616 computers that are designed to receive, process and display all Orbiter data in real time, and do preliminary image reconstruction on video data taken from Orbiter cameras.

Another set is a system of IBM 360/75 computers that receive, process and display in real time all Lander telemetry and tracking (metric) data from the tracking stations. They are the means through which commands are sent to the Orbiters and Landers. They also do early image reconstruction and display of video data from Lander cameras on the surface, and they provide computing capability for many programs that do command preparation, Lander data analysis and mission control functions.

Two large UNIVAC 1108 computers are used in non-real-time to do many detailed analyses such as navigation, science instrument data analysis and data records production.

Exposed film from the computers will be processed in the VMCCC photo processing lab; high quality prints will be quickly made available. These pictures from Mars orbit and from the surface will be analyzed by scientists housed in the mission support rooms of the VMCCC.

The VMCCC system is the responsibility of JPL's Office of Computing and Information Systems.

Image Processing Laboratory

JPL's Image Processing Laboratory (IPL) will correct all of the images (photo products) returned from the Lander and Orbiter spacecraft. Digital computer techniques are used to improve details of returned images, and to correct distortions introduced into the images by the camera systems. Large mosaics will be constructed from the Lander and Orbiter images, using the IPL products.

Special techniques developed for Viking include a program that will display Lander images for stereo viewing. The three dimensional images will be used to evaluate the terrain near the landing site before activating the surface sampler arm. IPL will also do the processing to obtain the best possible discriminability (details) of images acquired by the Orbiter during site certification before landing.

TRACKING AND DATA SYSTEM

Tracking, commanding and obtaining data from the Viking spacecraft are parts of the mission assigned to the Jet Propulsion Laboratory. The tasks cover all phases of the flight, including telemetry from launch vehicle and spacecraft, metric data on launch vehicle and Viking, command signals to the spacecraft and delivery of data to the Viking mission Control and Computing Center (VMCCC).

The Tracking and Data System (TDS) will provide elements of the world-wide NASA/JPL Deep Space Network (DSN), Air Force Eastern Test Range (AFETR), NASA Spaceflight Tracking and Data Network (STDN), and NASCOM (NASA Communications Network) to support Viking.

During the launch phase, data acquisition will be made by the Tracking and Data System through the near-Earth facilities: AFETR stations, downrange elements of STDN and instrumented jet aircraft. A communications ship in the Indian ocean may be required for launch of the second Viking.

Radar-metric data obtained immediately after liftoff and through the near-Earth phase will be delivered to and computed at AFETR's Real-Time Computer System facility in Florida, then transmitted to DSN stations, to locate the Vikings in the sky when they appear on the horizon.

Tracking and communication with Viking from the cruise phase until end-of-mission will be done by DSN. It consists of nine communications stations on three continents, a spacecraft monitoring station in Florida, the Network Operations Control Center in the VMCCC at JPL, and ground communications linking all locations.

DSN stations are strategically located around the Earth: Goldstone, Calif.; Madrid, Spain; and Canberra, Australia. Each location has a 64-m diameter (210-ft.) antenna station and two 26-m (85-ft.) antenna stations.

The three multi-station complexes are spaced at widely separated longitudes around the world so spacecraft beyond Earth orbit are never out of view. Spacecraft monitoring equipment in the STDN station at Merritt Island, Fla. covers pre-launch and launch phases of the mission. A simulated DSN station at JPL, called CTA-21, provides pre-launch compatibility support.

Each DSN station is equipped with transmitting, receiving, data handling, and inter-station communications equipment. The 64-m antenna stations in Spain and Australia have 100-kilowatt transmitters; at Goldstone the uplink signal can be radiated at up to 400 kW. Transmitter power at all six 26-m stations is 20 kW.

Nerve center of DSN is the Network Operations Control Center at JPL. All incoming data are validated here while being simultaneously transferred to computing facilities in VMCCC for real-time use by the Viking Flight Team.

The global stations are tied to the control center by NASCOM. Low-rate data from the spacecraft are transmitted over high-speed circuits at 4,800 bits per second (bps). High-rate data are carried on wideband lines at 28.5 kilobits per second (kbps) and, from Goldstone, at 50 kbps. Commands to the spacecraft are generated in the VMCCC and sent in the opposite direction to the appropriate DSN station.

Ground communications used by DSN are part of a larger network, NASCOM, which links NASA's stations around the world. For Viking NASCOM may occasionally provide a communications satellite link with the overseas stations.

For all of NASA's unmanned deep space missions, DSN provides tracking information on course and direction of flight velocity and range from Earth. It receives engineering and science telemetry and sends commands for spacecraft operations on a multi-mission basis. Concurrent with the Viking Project, the network is maintaining post-Jupiter communications with Pioneers 10 and 11, and complementing West Germany's space communications facilities on the Helios Sun-orbiting mission. DSN will support a second Helios launch, planned for December 1975, and perihelion science activities 3 months later.

Tracking and data acquisition requirements for Viking greatly exceed those of the Mariner and Pioneer projects. As many as six telemetry streams — two from both Orbiters and one or the other Lander — will be simultaneously received. Both Orbiters or an Orbiter and Lander will be tracked and commanded at any given time, although the two Landers will not be operated at the same time.

In the 16 months of the primary mission, the critical period lasts at least 5 months, beginning with Mars approach. Early in this period, two sets of antennas will be communicating with orbiter A and Lander A, separated and conducting their respective missions. A third set of antennas will be required to track Viking B, still mated and approaching at some distance. During this phase, virtually the entire capability of the DSN is occupied with Viking.

Principal communications links between the Vikings and Earth stations are in the S-band (2,100-2,300 megahertz). The Orbiters will also carry X-band (8,400 MHz) transmitters. Operating with the Orbiter S-band system, the X-band transmitter will allow the network to generate dual frequency ranging and Doppler data, and will contribute to the Radio Science investigation at Mars.

Telemetry will be immediately routed from DSN stations to the VMCCC for distribution to computers and other specialized processors for data reduction and presentation to Flight Team engineers and science investigators. Simultaneously, range and range-rate information will be generated by DSN and transmitted to the VMCCC for spacecraft navigation computations.

Commands to Viking are transmitted from the VMCCC and loaded into a command processing computer at a DSN station for transmission to the proper spacecraft. Commands may be aborted and emergency commands may be manually inserted and verified at stations after voice authorization from VMCCC.

During planetary operations the network supports celestial mechanics experiments that may use very long baseline interferometry (VLBI), using DSN and other antennas outside the network.

All of NASA's networks are under the direction of the Office of Tracking and Data Acquisition. JPL manages DSN; STDN facilities and NASCOM are managed by NASA's Goddard Space Flight Center, Greenbelt, Md.

The Goldstone DSN stations are operated and maintained by JPL with the assistance of the Aeronutronic Ford Corporation. The Canberra stations are operated by the Australian Department of Supply; the stations near Madrid are operated by the Spanish government's Instituto Nacional de Technica Aerospacial.

VIKING COMMUNICATION LINKS

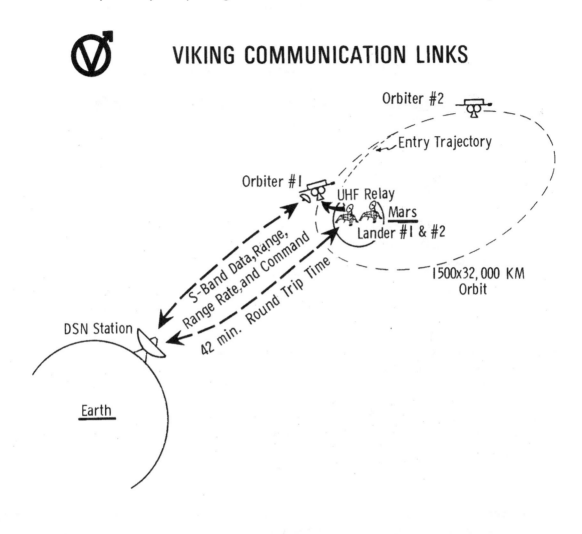

VIKING
POST LAUNCH
OPERATION REPORT

Post Launch
Mission Operation Report
No. S-815-75-01

MEMORANDUM   August 28, 1975
TO: A/Administrator
FROM: SL/Viking Program Manager
SUBJECT: Viking - 1 Post Launch Report #1

The first Viking spacecraft was successfully launched on August 20, 1975, at 5:22:00.6 p.m. EDT. When safely injected on its Trans-Mars trajectory, the spacecraft was designated Viking 1. The trajectory was within the designed 2 sigma limits. When this trajectory is translated into the Mars encounter geometry, the achieved injection point would miss the targeted point by 67,883 kilometers (see Figure 1). The targeted point was intentionally biased away from the planet to assure that the spacecraft or launch vehicle would not impact the planet due to injection errors and violate planetary quarantine restrictions.

On Wednesday, August 27, 1975, at 2:30 p.m. EDT , a trajectory maneuver of 4.684 meters per second was performed to target the spacecraft for its Mars Orbital Insertion (MOI) point. The accuracy of this maneuver will not be known for several days. However, the total uncertainty of the maneuver (i.e., orbit determination and execution errors) and orbit determination uncertainty are shown in Figure 2. A second and possibly third correction can be executed if required.

As of this report, the spacecraft is in excellent operating condition. All events were achieved as planned (see Table 1) and the spacecraft is operating in a cruise mode.

The first Viking launch was scheduled for August 11, 1975. During prelaunch tests of the Launch Vehicle's solid rocket motors thrust vector control valves, one valve was determined to be faulty. A decision was made to slip the launch to August 14 to provide time to remove the valve and perform diagnostic analyses and tests. During a scheduled orbiter battery check and recharge exercise on August 13 at 11:30 a.m., it was discovered that the orbiter had been unexplainably transferred to internal battery power and that the batteries had discharged to 9 volts, well below their safe limits. Since it was considered unsafe to attempt to recharge the batteries while on the spacecraft and since several of the subsystems were operated at voltages below their test and design limits, the project decided to remove the spacecraft from the launch vehicle and replace it with Spacecraft B. This action would require that the launch be slipped to August 24 if standard procedures were followed. After careful evaluation it was decided that the launch could take place on August 20 if a Range Safety waiver could be obtained for removal and installation of the spacecraft without detanking and depressurization of the launch vehicle. This waiver was granted and the launch occurred on August 20 at 05:22 p.m. EDT.

Walt Jakobowski

S-815-75-01

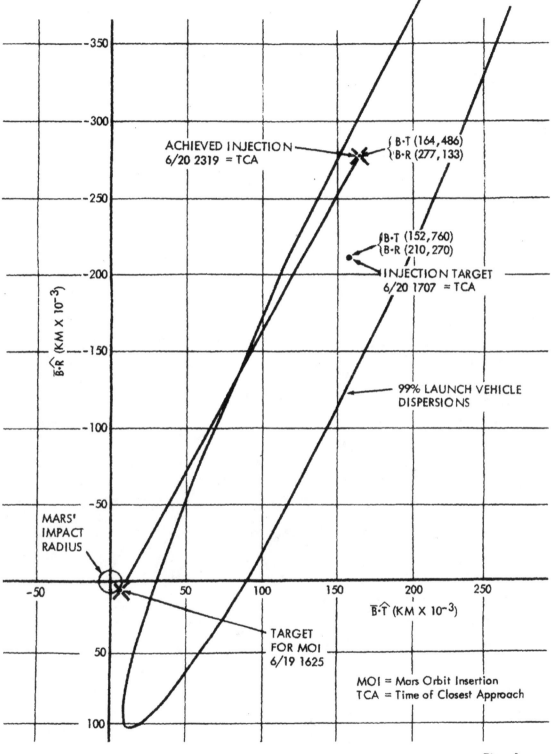

Fig. 1

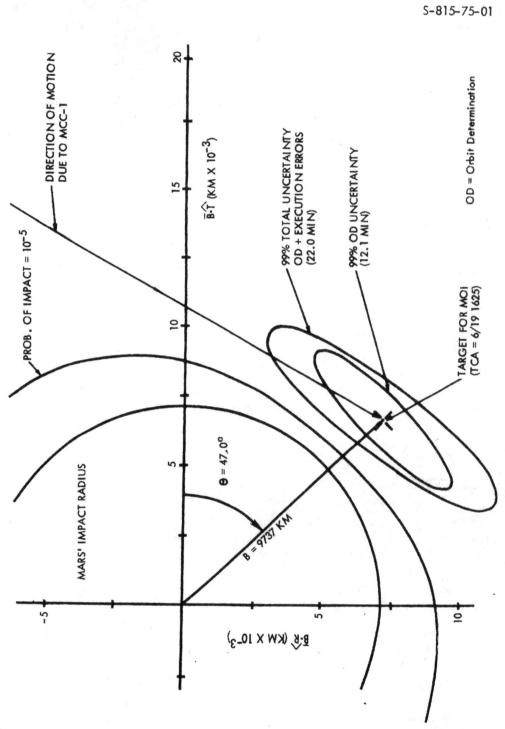

Fig. 2

TABLE I
VIKING I LAUNCH - MAJOR EVENTS

MARK NO.	EVENT	TIME (GMT)
	Liftoff	21:22:00.6
1	Heat Shield Jettison	21:23:40.2
2	Stage I Ignition	21:23:51.0
3	Stage 0/Stage I Separation	21:24:01.9
4	Stage I Shut Down	21:26:19.7
5	Stage I Jettison	21:26:20.4
6	Stage 2 Ignition	21:26:20.3
7	Centaur Standard Shroud Jettison	21:26:32.3
8	Stage 2 Shut Down	21:29:40.9
9	Stage 2 Jettison	21:29:54.65
10	Centaur Main Engine Start (MES) #1	21:30:05.9
11	Centaur Main Engine Cutoff (MECO) 01	21:32:11.6
12	MES #2	21:47:33.0
13	MECO #2	21:52:48.0
14	Centaur/Spacecroft Separation	21:56:31.0
15	Solar Panel Deployment	21:58:42.0
16	Centaur Retro Start	22:10:41.0
17	Centaur Retro End	22:14:52.0
	High Gain Antenna Unlatch	22:22:18.0
18	Bio Shield Jettison	23:29:00.0
	Sun Acquisition Command	23:40:18.0
	Canopus Acquisition	00:42:04.0

VIKING NASA Facts

National Aeronautics and Space Administration
Jet Propulsion Laboratory
California Institute of Technology Pasadena, CA 91109

Viking Mission to Mars

NASA's Viking Project was the culmination of a series of missions to explore Mars that had begun in 1964 with Mariner 4, and continued with the Manner 6 and 7 flybys in 1969, and the Mariner 9 orbital mission in 1971 and 1972. Viking found a place in history when it became the first mission to land a spacecraft safely on the surface of another planet.

Two identical spacecraft, each consisting of a lander and an orbiter, were built. Each orbiter-lander pair flew together and entered Mars orbit; the landers then separated and descended to the planet's surface.

Mission Design

Both spacecraft were launched from Cape Canaveral, Florida -Viking I on August 20, 1975, and Viking 2 on September 9, 1975. The landers were sterilized before launch to prevent contamination of Mars with organisms from Earth. The spacecraft spent nearly a year cruising to Mars. Viking I reached Mars orbit June 19, 1976; Viking 2 began orbiting Mars August 7, 1976.

After studying orbiter photos, the Viking site certification team considered the original landing site proposed for Viking I unsafe. The team examined nearby sites, and Viking I landed on Mars July 20, 1976, on the western slope of Chryse Planitia (the Plains of Gold) at 22.3 degrees north latitude, 48.0 degrees longitude.

The site certification team also decided the planned landing site for Viking 2 was unsafe after it examined high-resolution photos. Certification of a new landing site took place in time for a Mars landing September 3, 1976, at Utopia Planitia, at 47.7 degrees north latitude and 48.0 degrees longitude.

The Viking mission was planned to continue for 90 days after landing. Each orbiter and lander operated far beyond its design lifetime. Viking Orbiter 1 exceeded four years of active flight operations in Mars orbit.

The Viking project's primary mission ended November 15, 1976, 11 days before Mars' superior conjunction (its passage behind the Sun). After conjunction, in mid-December 1976, controllers reestablished telemetry and command operations, and began extended-mission operations.

The first spacecraft to cease functioning was Viking Orbiter 2, on July 25, 1978; the spacecraft had used all the gas in its attitude-control system, which kept the craft's solar panels pointed at the Sun to power the orbiter. When the spacecraft drifted off the Sun line, the controllers at JPL sent commands to shut off power to Viking Orbiter 2's transmitter.

Viking Orbiter 1 began to run short of attitude control gas in 1978, but through careful planning to conserve the remaining supply, engineers found it possible to continue acquiring science data at a reduced level for another two years. The gas supply was finally exhausted and Viking Orbiter 1's electrical power was commanded off on August 7, 1980, after 1,489 orbits of Mars.

The last data from Viking Lander 2 arrived at Earth on April 11, 1980. Lander 1 made its final transmission to Earth Nov. 11, 1982. Controllers at JPL tried unsuccessfully for another six and one-half months to regain contact with Viking Lander 1. The overall mission came to an end May 21, 1983.

Viking Orbiters

The Viking spacecraft consisted of two large orbiters, each weighing 2,325 kilograms (5,125 pounds) with fuel. Each orbiter carried a lander, weighing 576 kilograms (1,270 pounds), and their design was greatly influenced by the size of these landers.

The orbiters were a follow-on design to the Mariner class of planetary spacecraft with specific design changes for the 1976 surface mission. Operational lifetime requirements for the orbiters were 120 days in orbit and 90 days after landing.

The combined weight of the orbiter and lander was one factor that contributed to an 11-month transit time to Mars, instead of the five months for Mariner missions. The longer flight time then dictated an increased design life for the spacecraft, larger solar panels to allow for longer degradation from solar radiation and additional attitude control gas.

The basic structure of the orbiter was an octagon approximately 2.4 meters (8 feet) across. The eight sides of the ring-like structure were 45.7 centimeters (18 inches) high and were alternately 1.4 by 0.6 meters (55 by 22 inches).

Electronic bays were mounted to the faces of the structure and the propulsion module was attached at four points. There were 16 bays, or compartments, three on each of the long sides and one on each short side.

The orbiter was 3.3 meters (10.8 feet) high and 9.7 meters (32 feet) across the extended solar panels. With fuel, the orbiters weighed in excess of 2,300 kilograms (5,000 pounds).

Combined area of the four panels was 15 square meters (161 square feet), and they provided both regulated and unregulated direct current power; unregulated power was provided to the radio transmitter and the lander.

Two 30-amp-hour, nickel-cadmium, rechargeable batteries provided power when the spacecraft was not facing the Sun during launch, correction maneuvers and Mars occultation.

The orbiter was stabilized in flight by locking onto the Sun for pitch and yaw references and onto the star Canopus for roll reference. The attitude control subsystem kept this attitude with nitrogen gas jets located at the solar panel tips. The jets would fire to correct any drift. A cruise Sun sensor and the Canopus sensor provided error signals. Before Sun acquisition four acquisition Sun sensors were used and then turned off.

The attitude control subsystem also operated in an all-inertial mode or in roll-inertial with pitch and yaw control, still using the Sun sensors. During correction maneuvers, the attitude control subsystem aligned the vehicle to a specified attitude in response to commands from the on-board computer. Attitude control during engine burns was provided in roll by the attitude control subsystem and in pitch and yaw by an autopilot that commanded engine gimballing.

If Sun lock was lost the attitude control subsystem automatically realigned the spacecraft. In loss of Canopus lock, the subsystem switched to roll-inertial and waited for commands from the spacecraft computer. The nitrogen gas supply for the subsystem could be augmented by diverting excess helium gas from the propulsion module, if necessary.

Two on-board general purpose computers in the computer command subsystem decoded commands and either ordered the desired function at once or stored the commands in a 4,096-word plated-wire memory. All orbiter events were controlled by the computer command subsystem, including correction maneuvers, engine burns, science sequences and high-gain antenna pointing.

The main orbiter communications system was a two-way, S-band, high-rate radio link providing Earth command, radio tracking and science and engineering data return. This link used either a steerable 1.5-meter (59-inch) dish high-gain antenna or an omni-directional low-gain antenna, both of them on the orbiter. The low-gain antenna was used to send and receive near Earth, and the high-gain antenna was used as the orbiter journeyed farther from Earth.

S-band transmission rates varied from 8.3 or 33.3 bits per second for engineering data to 2,000 to 16,000 bits per second for lander and orbiter science data.

Relay from the lander was achieved through an antenna mounted on the outer edge of a solar panel. It was activated before separation and received from the lander through separation, entry, landing and surface operations. The bit rate during entry and landing was 4,000 bits per second; landed rate was 16,000 bits per second.

Data were stored aboard the orbiter on two eight track digital tape recorders. Seven tracks were used for picture data and the eighth track for infrared data or relayed lander data. Each recorder could store 640 million bits.

Data collected by the orbiter, including lander data, were converted into digital form by the flight data subsystem and routed to the communications subsystem for transmission or to the tape recorders for storage. This subsystem also provided timing signals for the three orbiter science experiments.

Viking Lander

The lander spacecraft was composed of five basic systems: the lander body, the bioshield cap and base, the aeroshell, the base cover and parachute system and lander subsystems.

The completely outfitted lander measured approximately 3 meters (10 feet) across and was about 2 meters (7 feet) tall. It weighed about 576 kilograms (1,270 pounds) without fuel.

The lander and all exterior assemblies were painted light gray to reflect solar heat and to protect equipment from abrasion. The paint was made of rubber based silicone.

The body was a basic platform for science instruments and operational subsystems. It was a hexagon shaped box with three 109-centimeter (43-inch) sidebeams and three 56 centimeter (22-inch) short sides. It looks like a triangle with blunted corners.

The box was built of aluminum and titanium alloys, and was insulated with spun fiberglass and dacron cloth to protect equipment and to lessen heat loss. The hollow container was 1.5 meters (59 inches) wide and 46 centimeters (18 inches) deep, with cover plates on the top and bottom.

The lander body was supported by three landing legs, 1.3 meters (51 inches) long, attached to the short-side bottom corners of the body. The legs gave the lander a ground clearance of 22 centimeters (8.7 inches).

Each leg had a main strut assembly and an A-frame assembly, to which was attached a circular footpad 30.5 centimeters (12 inches) in diameter. The main struts contained bonded, crushed aluminum honeycomb to reduce the shock of landing.

The two-piece bioshield was a pressurized cocoon that completely sealed the lander from any possibility of biological contamination until Viking left Earth's atmosphere.

The two bioshield halves generally resembled an egg, and the shield's white thermal paint heightened the resemblance. It measured 3.7 meters (12 feet) in diameter and was 1.9 meters (6.4 feet) deep. It was made of coated, woven fiberglass, 0. 13 millimeters (0.005 inches) thin, bonded to an aluminum support structure.

The bioshield was vented to prevent over-pressurization and possible rupture of its sterile seal.

The aeroshell was an aerodynamic heat shield made of aluminum alloy in a 140-degree, flat cone shape and stiffened with concentric rings. It fit between the lander and the bioshield base. It was 3.5 meters (11.5 feet) in diameter and its aluminum skin was 0.86 millimeters (0.034 inches) thin.

Bonded to its exterior was a lightweight, cork-like ablative material that burned away to protect the lander from aerodynamic heating at entry temperatures which may have reached 1,500 degrees C.

The interior of the aeroshell contained twelve small reaction control engines, in four clusters of three around the aeroshell's edge, and two spherical titanium tanks that contained 85 kilograms (18 8 pounds) of hydrazine monopropellant.

The engines controlled the pitch and yaw to align the lander for entry, help slow the craft during early entry and maintain roll control.

During the long cruise phase, an umbilical connection through the aeroshell provided power from the orbiter to the lander; housekeeping data also flowed through this connection.

The aeroshell also contained two science instruments — the upper atmosphere mass spectrometer and the retarding potential analyzer — plus pressure and temperature sensors.

The base cover fit between the bioshield cap and the lander. It was made of aluminum and fiberglass; the fiberglass allowed transmission of telemetry data to the orbiter during entry. It covered the parachute and its ejection mortar, and protected the lander's top during part of the entry phase.

The parachute was made of lightweight dacron polyester 16 meters (53 feet) in diameter. It weighed 50 kilograms (110 pounds).

The parachute was packed inside a mortar 38 centimeters (15 inches) in diameter and mounted into the base cover. The mortar was fired to eject the parachute at about 139 kilometers per hour (75 miles per hour). The chute had an extra-long suspension line that trailed the capsule by about 30 meters (100 feet).

The lander subsystems were divided into six major categories: descent engines, communications equipment, power sources, landing radars, data storage and guidance and control.

Three terminal descent engines provided attitude control and reduced the lander's velocity after parachute separation. The 2,600-newton (600-pound) throttleable engines were located 120 degrees apart on the lander's sidebeams. They burned hydrazine monopropellant.

The engines used an advanced exhaust design that wouldn't alter the landing site environment. An unusual grouping of 18 small nozzles on each engine would spread engine exhaust over a wide angle that wouldn't alter the surface or unduly disturb the chemical and biological experiments.

Two spherical titanium tanks, attached to opposite sides of the lander body beneath the RTG wind covers, fed the descent engines from an 85-kilogram (188 pound) hydrazine propellant supply.

Four small reaction control engines used hydrazine monopropellant thrusters to control lander roll attitude during terminal descent. The engines were mounted in pairs on the terminal descent engines' propellant tanks and were identical to those used on the aeroshell.

The lander was equipped to transmit information directly to Earth with an S-band communications system, or through the orbiter with an ultra-high frequency (UHF) relay system. The lander also received Earth commands through the S-band system.

Two S-band receivers provided total redundancy in both command receiving and data transmission. One receiver used the high-gain antenna, a 76-centimeter (30-inch) diameter parabolic reflector dish that could be pointed to Earth by computer control. The second receiver used a fixed low-gain antenna to receive Earth commands.

The UBF relay system transmitted data to the orbiter with a radio transmitter that used a fixed antenna. The UBF system operated during entry and during the first three days of landed operations. After that it was operated only during specific periods.

The radar altimeter measured the lander's altitude during the early entry phase, alerting the lander computer to execute the proper entry commands. The radar was a solid-state pulse radar with two specially designed antennas: one was mounted beneath the lander and one was mounted through the aeroshell. Altitude data were received from 1,370 kilometers down to 30.5 meters (740 miles to 100 feet).

The aeroshell antenna provided high-altitude data for entry science, vehicle control and parachute deployment. The lander antenna was switched into operation at aeroshell separation and provided altitude data for guidance and control, and for terminal descent engine ignition.

The terminal descent landing radar measured the horizontal velocity of the lander during the final landing phase. It was located directly beneath the lander and was turned on at about 12 kilometers (4,000 feet) above the surface. It consisted of four continuous wave Doppler radar beams that could measure velocity to an accuracy of plus or minus one meter per second.

Both radars were essential for mission success, so the terminal descent landing radar could work with any three of its four beams, and identical sets of radar altimeter electronics could be switched to either of the radar antennas.

The "brain" of the lander was its guidance control and sequencing computer. That computer commanded everything the lander did through software (computer programs) stored in advance or relayed by Earth controllers.

The computer was one of the greatest technical challenges of Viking. It consisted of two general-purpose computer channels with plated-wire memories, each with an 18,000 word storage capacity. One channel would be operational while the other was in reserve.

Among other programs, the computer had instructions stored in its memory that could control the lander's first 22 days on Mars without any contact from Earth. These instructions would be updated and modified by Earth commands once communications had been established.

Basic power for the lander was provided by two SNAP 19-style, 35-watt radioisotope thermoelectric generators (RTGs) developed by the then U.S. Energy Research and Development Administration. They were located atop the lander and were connected in series to double their voltage and reduce power loss.

The SNAP 19 Viking generator was 147 centimeters (23 inches) across the housing fin tips, 96 centimeters (15 inches) in length and weighed 15.3 kilograms (34 pounds).

The first isotopic space generator was put into service in June 1961, on a Navy navigational satellite. Advances in SNAP systems were made with the development and flight of SNAP 19 aboard Nimbus III, launched in April 1969. This use of SNAP 19 represented a major milestone in the development of long-lived, highly reliable isotope power systems for space use by NASA. The SNAP 27 generator was developed to power five science stations left on the Moon by the Apollo 12, 14, 15, 16 and 17 astronauts. The continuing operation of these generators provided new dimensions of data about the Moon and the universe. Four SNAP 19 nuclear generators later provided the electrical power for each of two NASA pioneering Jupiter flyby spacecraft known as Pioneers 10 and 11.

The generators provided a long-lived source of electricity and heat on Mars, where sunlight is half as strong as on Earth, and is non-existent during the Martian night, when temperatures can drop as low as -120 C (-184 F).

The generators used thermoelectric elements to convert heat from decaying plutonium-238 into 70 watts of electrical power.

Waste or unconverted heat was conveyed by thermal switches to the lander's interior instrument compartment, when required. Covers over the RTGs prevented excess heat dissipation into the environment.

Four nickel-cadmium, rechargeable batteries helped supply lander power requirements in peak activity periods. The batteries, mounted in pairs inside the lander, were charged by the RTGs with power available when other lander power requirements were less than RTG output.

This equipment collected and controlled the flow of lander scientific and engineering data. It consisted of a data acquisition and processing unit, a data storage memory and a tape recorder.

The data acquisition and processing unit collected the science and engineering information and routed it to one of three places: to Earth through the S-band high-gain antenna, to the data storage memory or to the tape recorder. Information was stored in the data storage memory for short periods. Several times a day the memory would transfer data to the tape recorder or back to the data acquisition and processing unit for further transmission. The memory had a storage capacity of 8,200 words.

Data were stored on the tape recorder for long periods. The recorder could transmit at high speed back through the data acquisition and processing unit, and the UHF link to an orbiter passing overhead. The recorder could store as many as 40 million bits of information and it could record at two speeds and play back at five.

Science Experiments

With a single exception — the seismic instruments — the science instruments acquired more data than expected. The seismometer on Viking Lander 1 would not work after landing, and the seismometer on Viking Lander 2 detected only one event that may have been seismic. Nevertheless, it provided data on wind velocity at the landing site to supplement information from the meteorology experiment, and showed that Mars has very low seismic background.

The three biology experiments discovered unexpected and enigmatic chemical activity in the Martian soil, but provided no clear evidence for the presence of living microorganisms in soil near the landing sites. According to mission biologists, Mars is self-sterilizing. They believe the combination of solar ultraviolet radiation that saturates the surface, the extreme dryness of the soil and the oxidizing nature of the soil chemistry prevent the formation of living organisms in the Martian soil. The question of life on Mars at some time in the distant past remains open.

The landers' gas chromatograph/mass spectrometer instruments found no sign of organic chemistry at either landing site, but they did provide a precise and definitive analysis of the composition of the Martian atmosphere and found previously undetected trace elements. The X-ray fluorescence spectrometers measured elemental composition of the Martian soil.

Viking measured physical and magnetic properties of the soil. As the landers descended toward the surface they also measured composition and physical properties of the Martian upper atmosphere.

The two landers continuously monitored weather at the landing sites. Weather in the Martian midsummer was repetitious, but in other seasons it became variable and more interesting. Cyclic variations appeared in weather patterns (probably the passage of alternating cyclones and anticyclones). Atmospheric temperatures at the southern landing site (Viking Lander 1) were as high as -14 C (7 F) at midday, and the predawn summer temperature was -77 C (107 F). In contrast, the diurnal temperatures at the northern landing site (Viking Lander 2) during midwinter dust storms varied as little as 4 degrees C (7 degrees F) on some days. The lowest predawn temperature was -120 C (- 184 F), about the frost point of carbon dioxide. A thin layer of water frost covered the ground around Viking Lander 2 each winter.

Barometric pressure varies at each landing site on a semiannual basis, because carbon dioxide, the major constituent of the atmosphere, freezes out to form an immense polar cap, alternately at each pole. The carbon dioxide forms a great cover of snow and then evaporates again with the coming of spring in each hemisphere. When the southern cap was largest, the mean daily pressure observed by Viking Lander 1 was as low as 6.8 millibars; at other times of the year it was as high as 9.0 millibars. The pressures at the Viking Lander 2 site were 7.3 and 10.8 millibars. (For comparison, the surface pressure on Earth at sea level is about 1,000 millibars.)

Martian winds generally blow more slowly than expected. Scientists had expected them to reach speeds of several hundred miles an hour from observing global dust storms, but neither lander recorded gusts over 120 kilometers (74 miles) an hour, and average velocities were considerably lower. Nevertheless, the orbiters observed more than a dozen small dust storms. During the first southern summer, two global dust storms occurred, about four Earth months apart. Both storms obscured the Sun at the landing sites for a time and hid most of the planet's surface from the orbiters' cameras. The strong winds that caused the storms blew in the southern hemisphere.

Photographs from the landers and orbiters surpassed expectations in quality and quantity. The total exceeded 4,500 from the landers and 52,000 from the orbiters. The landers provided the first close-up look at the surface, monitored variations in atmospheric opacity over several Martian years, and determined the mean size of the atmospheric aerosols. The orbiter cameras observed new and often puzzling terrain and provided clearer detail on known features, including some color and stereo observations. Viking's orbiters mapped 97 percent of the Martian surface.

The infrared thermal mappers and the atmospheric water detectors on the orbiters acquired data almost daily, observing the planet at low and high resolution. The massive quantity of data from the two instruments will require considerable time for analysis and understanding of the global meteorology of Mars. Viking also definitively determined that the residual north polar ice cap (that survives the northern summer) is water ice, rather than frozen carbon dioxide (dry ice) as once believed.

Analysis of radio signals from the landers and the orbiters — including Doppler, ranging and occultation data, and the signal strength of the lander-to-orbiter relay link — provided a variety of valuable information.

Other Significant Discoveries

Other significant discoveries of the Viking mission included:

* The Martian surface is a type of iron-rich clay that contains a highly oxidizing substance that releases oxygen when it is wetted.

* The surface contains no organic molecules that were detectable at the parts-per-billion level — less, in fact, than soil samples returned from the Moon by Apollo astronauts.

* Nitrogen, never before detected, is a significant component of the Martian atmosphere, and enrichment of the heavier isotopes of nitrogen and argon relative to the lighter isotopes implies that atmospheric density was much greater than in the distant past.

* Changes in the Martian surface occur extremely slowly, at least at the Viking landing sites. Only a few small changes took place during the mission lifetime.

* The greatest concentration of water vapor in the atmosphere is near the edge of the north polar cap in midsummer. From summer to fall, peak concentration moves toward the equator, with a 30 percent decrease in peak abundance. In southern summer, the planet is dry, probably also an effect of the dust storms.

* The density of both of Mars' satellites is low — about two grams per cubic centimeter — implying that they originated as asteroids captured by Mars' gravity. The surface of Phobos is marked with two families of parallel striations, probably fractures caused by a large impact that may nearly have broken Phobos apart.

* Measurements of the round-trip time for radio signals between Earth and the Viking spacecraft, made while Mars was beyond the Sun (near the solar conjunctions), have determined delay of the signals caused by the Sun's gravitational field. The result confirms Albert Einstein's prediction to an estimated accuracy of 0.1 percent — 20 times greater than any other test.

* Atmospheric pressure varies by 30 percent during the Martian year because carbon dioxide condenses and sublimes at the polar caps.

* The permanent north cap is water ice; the southern cap probably retains some carbon dioxide ice through the summer.

* Water vapor is relatively abundant only in the far north during the summer, but subsurface water (permafrost) covers much if not all of the planet.

* Northern and southern hemispheres are drastically different climatically, because of the global dust storms that originate in the south in summer.

The Viking Team

NASA's Langley Research Center in Hampton, VA, had management responsibility for the Viking project from its inception in 1968 until April 1, 1978, when the Jet Propulsion Laboratory, Pasadena, CA, assumed the task.

Langley and JPL managed the mission for NASA's Office of Space Science, Washington, DC. JPL designed and built the Viking orbiters, managed tracking and data acquisition through the Deep Space Network and managed Viking's mission control and computing center. The former Martin Marietta Aerospace in Denver, CO, designed and built the Viking landers. NASA's Lewis Research Center in Cleveland, OH, had responsibility for the Titan-Centaur launch vehicles.

At NASA Langley, James S. Martin Jr. was Viking project manager; Dr. Gerald A. Soffen was Viking project scientist. At NASA Headquarters, Walter Jakobowski was Viking program manager; Dr. Loyal G. Goff was Viking program scientist.

Leaders of the science teams included Dr. Michael H. Carr of the U.S. Geological Survey, Menlo Park, CA, team leader of the orbiter imaging investigation; Dr. Crofton B. "Barney" Farmer of JPL, team leader of the water vapor mapping investigation; Dr. Hugh H. Kieffer of the University of California, Los Angeles, team leader of the thermal mapping investigation; Dr. Alfred O.C. Nier of the University of Minnesota, Minneapolis, who led the entry science investigation; and Dr. Thomas A. "Tim" Mutch of Brown University, Providence, RI, team leader of the lander imaging investigation.

Investigators on the Viking biology experiments were Dr. Harold P. Klein of NASA Ames Research Center, Mountain View, CA, who led the Viking biology investigation; Dr. Klaus Biemann of the Massachusetts Institute of Technology, who led the molecular analysis investigation; and Dr. Priestley Toulmin III, team leader of the inorganic chemistry investigation.

The meteorology investigation was led by Dr. Seymour L. Hess of Florida State University; the seismology investigation was led by Dr. Don L. Anderson of the California Institute of Technology, Pasadena, CA; Dr. Richard W. Shorthill of the University of Utah led the physical properties investigation; and Dr. Robert B. Hargraves of Princeton University was principal investigator on the magnetic properties team. Radio science investigations were led by Dr. William H. Michael Jr. of NASA's Langley Research Center, Hampton, VA.

1988 DB

MARS OBSERVER PRESS KIT

MARS OBSERVER

Mars Orbit Insertion Press Kit — August 24, 1993

The National Aeronautics and Space Administration's Mars Observer spacecraft arrives at Mars at approximately 1:40 p.m. Pacific Daylight Time on Aug. 24, 1993, after an 11-month, 450-million-mile (720-million- kilometer) journey through the inner solar system. The Mars Observer mission is a global scientific mapping mission.

The spacecraft carries a complement of new instruments to the Red Planet, and it will provide a long-lived orbital platform from which these instruments will examine the entire Martian surface, atmosphere and interior. Measurements will be collected from a low-altitude, nearly polar orbit, 234 miles (378 kilometers) above the Martian surface, over the course of one complete Martian year — the equivalent of about two Earth years or 687 Earth days.

"Throughout its primary, two-year mission, Mars Observer will gather information on the geology, geophysics and climate of Mars," said JPL Project Manager Glenn E. Cunningham. "The mission will provide a global portrait of Mars as it exists today," he said. "This view will help planetary scientists to better understand the history of Mars' evolution, and will provide clues about the planet's interior and surface evolution. With this information, we will have a better understanding of the history of all of the inner planets of the solar system including our own Earth."

Mars Observer continues NASA's long exploration of the Red Planet, which began nearly 30 years ago with the Mariner IV spacecraft that produced the first pictures of the planet's cratered surface. The last U.S. spacecraft to visit Mars were the twin Viking orbiters and landers, which arrived at Mars in 1976. The lander touched down on the Martian surface and analyzed samples of the Martian surface to look for signs of life in the planet's distant past.

Mars Observer was lofted into Earth orbit at 1:05 p.m. Eastern Daylight Time (10:05 a.m. Pacific Daylight Time) on Sept. 25, 1992, aboard a Titan III launch vehicle with a Transfer Orbit Stage (TOS) booster, which later injected the spacecraft on its path to Mars. Once on course for the cruise to Mars, the spacecraft deployed four of its six solar panels to begin generating solar power, and partially extended its high-gain antenna and two booms on which science instruments are mounted. The low-gain antenna was used for initial spacecraft communications, until the spacecraft was far enough away from Earth in early January 1993 to require use of the dish-shaped, high-gain antenna. Mars Observer's seven science instruments were calibrated during the cruise to Mars. In addition, the radio science team participated in a three-week attempt to detect gravity waves using the spacecraft's own telecommunication system with those of two other interplanetary spacecraft, Galileo and Ulysses.

"All spacecraft subsystems and the instrument payload have performed well as Mars Observer headed for its destination," said Dr. Sam Dallas, Mars Observer mission manager at JPL. "Since launch, we have sent more than 500 command files to the spacecraft, containing instructions to be carried out by the spacecraft and its payload," Dallas said. "The spacecraft has executed these instructions flawlessly, except on two or three occasions when minor errors occurred. The spacecraft and payload hardware have had no failures thus far."

When the spacecraft reaches Mars on Aug. 24, onboard rocket engines will fire to slow the spacecraft's speed by more than 1,500 miles per hour (2,400 kilometers per hour) with respect to Mars and allow the craft to be captured by Mars' gravity. Mars Observer will then follow a "power in" orbit insertion strategy, using several braking and orbit trim maneuvers to insert the spacecraft into the planned mapping orbit in less than three months (see Orbit Insertion Maneuver schedule, page 186).

The first insertion maneuver, called Mars Orbit Insertion (MOI), is the largest and most critical maneuver to be performed by the spacecraft. When the spacecraft is about 1,072 miles (1,730 kilometers) above the surface of the planet, it will fire two of its four large, 490-Newton thruster rockets to slow its speed enough to enter orbit around Mars, said Suzanne Dodd, Mars Observer mission planning team chief. The burn will last 29 minutes and place the spacecraft at about 308 miles (498 kilometers) above the Martian surface at periapsis, the closest approach to the planet. "Mars Observer will be flying over the north pole and Olympus Mons, a huge Martian volcano, when it enters orbit around Mars," Dodd said. "The spacecraft will first spend 10½ days in this highly elliptical orbit around the planet, completing just one orbit around Mars every three days."

Instrument calibrations and some science measurements will take place during the elliptical orbit phase, according to Dr. Arden Albee, Mars Observer project scientist. "The spacecraft will be crossing in and out of the planet's magnetic field during this initial, three-day orbit around Mars," Albee said. "Mars Observer will be able to make unique measurements of the interactions of magnetic field lines with the solar wind that will not be observable from the lower altitude mapping orbit. "This will also be a critical period for Magnetometer and Gamma Ray Spectrometer calibrations," he said. "Noise from the spacecraft will have to be subtracted out to obtain the magnetic and gamma ray signatures of Mars, and that can only be done as the spacecraft moves closer and then farther away from this planetary boundary."

Eleven days after MOI, the first of two trim maneuvers, called Elliptical Change Maneuvers (ECM-1 and ECM-2), will be performed to slow the spacecraft into a one-day "drift orbit" around Mars. The spacecraft will spend 32 days in this 23-hour orbit, waiting for Mars to move around the sun so that it can drift into the correct lighting position for the mapping orbit (2 p.m. local Mars time) without using up valuable fuel reserves. A third ECM will adjust the orbit parameters before the spacecraft begins to descend into the low-altitude mapping orbit. The entire spacecraft payload will be turned on and calibrated during this one-day drift orbit.

From Sept. 16 to Oct. 16, all instruments except the Mars Observer Laser Altimeter (MOLA) will be operating for varying amounts of time. From this orbit, Mars Observer may also be able to make observations of Phobos, the larger of Mars' two moons. The science payload may help determine the surface mineral content of this potential asteroid as the spacecraft passes possibly as close as 248 miles (400 kilometers) from Phobos during each orbit of Mars. An instrument called the Pressure Modulator Infrared Radiometer (PMIRR) will make radiometric scans of Phobos in visible and infrared wavelengths, while the Thermal Emission Spectrometer (TES) measures the temperature of this airless body. The Magnetometer and Electron Reflectometer will measure particle disturbances in the Martian atmosphere caused by Phobos' orbit around Mars. Imaging by the Mars Observer Camera (MOC) may be attempted, depending on how closely the spacecraft passes by the moon.

As the spacecraft continues to circularize its orbit, NASA's Deep Space Network 112-foot (34-meter) antennas will be used to begin a radio science experiment, measuring small shifts in the spacecraft's velocity that will tell scientists more about the planet's gravity field. Two Transfer-to-Low-Orbit (TLO) maneuvers will then bring the spacecraft into the near-circular mapping orbit. The first, TLO-1, will be performed on Oct. 17 to bring the spacecraft into a 4.2-hour orbit. After that maneuver has been performed, the spacecraft will spend a week characterizing the complete gravity field of Mars in detail. A gravity map of the entire planet will be created from variations in the radio signals from the spacecraft due to variations in mass distribution around the planet.

TLO-2 will be performed on Oct. 28 to adjust the orbit again and bring the spacecraft into the desired 118-minute mapping orbit. A subsequent trim maneuver (Orbit Change Maneuver or OCM) will occur on Nov. 8 to correct any small errors in the orbit. On that day, 76 days after Mars orbit insertion, the spacecraft will reach its near-circular, near-polar mapping orbit, inclined 93 degrees to the planet's equator.

Starting on Nov. 12, the spacecraft will fully deploy the two remaining solar panels, the high-gain antenna and the two science booms in preparation for the start of mapping. Boom deployments will occur a month ahead of solar conjunction — the period when the spacecraft moves behind the sun as seen from Earth — to insure that any problems associated with all these deployments can be solved before spacecraft-to-Earth communications are diminished.

All of the spacecraft's instruments will be turned on Nov. 23 for a final in-orbit checkout. Mars Observer will undergo in-orbit checkout through Dec. 16, operating its payload and testing its mapping command sequences. Data from this final checkout phase will allow the spacecraft to obtain one complete global map of the surface — a process which takes 26 days — before the solar conjunction in December blocks spacecraft communications.

"There is a very distinct advantage to getting this first mapping cycle right up front," Albee said. "The Martian dust storms occur roughly between February and August, so the atmosphere should be very clear in November when we begin mapping. An early start will also provide us with an excellent opportunity to obtain data before and after a Martian dust storm, since the dust storm period will resume in February 1994."

In addition to allowing Mars Observer to observe the birth and development of a global Martian dust storm — if one should occur this Martian year — the first 26-day mapping period will allow the scientific teams to become more familiar with operating their instruments from more than 200 million miles away. In its mapping orbit, the spacecraft will circle the planet at a speed of about 7,500 miles per hour (12,000 kilometers per hour) with respect to Mars and in an orbit that will take it close to both poles. On the day side of the planet, Mars Observer will be traveling from north to south. On each orbit, it will cross the equator at about 2 p.m. local Mars time. It will always see the surface of Mars on the daylit side as it appears at mid-afternoon. This "sun-synchronous" orbit puts the sun at a standard angle above the horizon in each image, with the mid-afternoon lighting angle casting shadows that will make surface features stand out. Solar conjunction begins on Dec. 20 and lasts through Jan. 3, 1994.

During this time, Mars' orbit around the sun will take the spacecraft behind the sun as seen from Earth. This planetary alignment will create a "command moratorium" for mission operations. "Ground-controllers will be out of contact with the spacecraft and will not transmit radio signals because of the noise caused by the sun's interference," Dallas said. "The spacecraft, however, will be fully deployed and in the mapping configuration, thereby minimizing the risk of problems occurring during this break in communications."

Experiment teams, who will control the spaceborne instruments from their home institutions, will also be out of contact with the spacecraft. But as two-way communication is restored in early January 1994, they will begin to command their experiments and access data daily using a computer network linking them to the Mars Observer operations center at JPL.

The Mars Observer mission is expected to yield more than 600 billion bits of scientific data, many times the amount of data returned by all previous missions, both American and Russian, to Mars. Mars Observer will examine the entire planet, from the ionosphere, an envelope of charged particles surrounding Mars, down through the atmosphere to the surface and deep into the interior. Scientists will glean valuable new information on daily and seasonal weather patterns, geological features, and the migration of water vapor over a complete Martian year.

As the primary mission winds down in late 1995, the Mars Observer science team will participate in a collaborative effort of international cooperation. The Mars Observer spacecraft carries a radio system supplied by the French Centre National d'Etudes Spatiales (CNES) that will periodically receive and relay data from small instrument packages landed on the Martian surface by the Russian Mars '94 mission, which will arrive at Mars in September 1995.

The Russian landed experiment packages will directly measure the Martian atmosphere and surface properties. Data will be relayed to both a Russian Mars '94 orbiter and the Mars Observer spacecraft. Once received, the data will be stored and processed for relay back to Earth. The Russian instrument packages are designed to operate for several years.

If Mars Observer is still operating as late as 1997, the spacecraft may support a second Russian mission, called the Mars '96 mission. The Mars '96 mission is being designed to release a balloon into the Martian atmosphere and possibly deploy landed stations or rovers that could explore the Martian surface either through remote control from Earth or under autonomous computer control. Following a launch in 1996, the Mars '96 spacecraft would reach Mars in 1997.

Glenn E. Cunningham of the Jet Propulsion Laboratory is the Mars Observer project manager. Dr. Arden Albee of the California Institute of Technology is the project scientist. Dr. Sam Dallas of the Jet Propulsion Laboratory is the mission manager. William C. Panter of NASA Headquarters is the program manager and Dr. Bevan M. French of NASA Headquarters is the program scientist. The spacecraft was built under contract to NASA and JPL by the Martin Marietta Astro-Space Division (formerly General Electric) in Princeton, N.J. NASA's Deep Space Network is supporting mission operations and tracking of the spacecraft throughout the primary mission. The Jet Propulsion Laboratory in Pasadena, Calif., manages the Mars Observer mission for the Solar System Exploration Division of NASA's Office of Space Science, Washington, D.C.

MARS OBSERVER ORBIT INSERTION MANEUVERS MOI 8/24/93

MOI is the first orbit insertion maneuver that brakes the spacecraft into the elliptical capture orbit. Once performed, the spacecraft will be in a three-day (75-hour) orbit around Mars. MOI occurs at 1:40 p.m. PDT on Aug. 24, 1993.

ECM-1 9/4/93 The first trim maneuver to the elliptical capture orbit is scheduled for 10.4 days after MOI. ECM-1 will put the spacecraft in a three-day "drift orbit" to distinguish it from the capture orbit.

ECM-2 9/15/93 The second trim maneuver is scheduled for 11.3 days after ECM-1 to lower the spacecraft into the one-day drift orbit. The spacecraft will be maintained in the drift orbit for the next 32 days, until the 2 p.m. solar orientation is nearly reached.

ECM-3 10/7/93 The third trim maneuver occurs 10 days before the TLO-1 maneuver. This is a small statistical maneuver to clean up errors in the drift orbit.

TLO-1 10/17/93 TLO-1 is the third braking maneuver, which lowers the spacecraft from the elliptical drift orbit to the 4.2-hour orbit. The 4.2-hour period of this intermediate orbit was selected to take out about half of the energy required to reach the mapping orbit.

TLO-2 10/28/93 TLO-2 is the fourth and last large braking maneuver to lower the spacecraft into the mapping orbit. It will occur 11 days after TLO-1. Because most of the bipropellant is being expended, this maneuver has the highest acceleration level of the mission.

OCM-1 11/8/93 The Orbit Change Maneuver is the final orbit maneuver to correct any errors in the mapping orbit. OCM-1 will be performed 11 days after TLO-2.

12/20/93 -1/3/94 Solar conjunction.

MARS OBSERVER SCIENCE OBJECTIVES

The Mars Observer mission will study the geology, geophysics, weather and climate of Mars. The mission's primary objectives are to identify and map the chemical and mineral composition of the surface; measure the topography of surface landforms; define the gravitational field; and search for a planetary magnetic field. The mission will also determine the distribution, abundance, sources and destinations of carbon dioxide, water and dust over a seasonal cycle, and will measure the atmospheric temperature and the water and dust content in the atmosphere.

Since all of these objectives involve global mapping, the mission will provide scientists with a global portrait of Mars as it exists today through a scientifically focused set of experiments. The spacecraft's instruments are similar to those now used to study the Earth. During its 687-day mapping mission, Mars Observer will return more than 600 billion bits of scientific data — more than that returned by all previous missions to Mars.

MISSION DESIGN

The near-polar orbit that was chosen for the Mars Observer mission is low enough to allow close-range study of Mars, but high enough so that the atmosphere does not drag excessively on the spacecraft. The orbit is also sun-synchronous; the spacecraft will pass over Mars' equator at the same local time during each orbit — about 2 p.m. on the day side and about 2 a.m. on the night side. This orbit is essential for a number of measurements, as it helps distinguish daily atmospheric variations from longer term, seasonal variations.

During the mission's mapping cycle, data reception from the spacecraft and command updates to the spacecraft and individual science instruments will be conducted on a daily basis.

The seven instruments on Mars Observer will gather data until November 1995. Near the end of its prime mission in the fall of 1995, Mars Observer will be joined at Mars by the Russian "Mars '94" spacecraft. Mars Observer is equipped to relay data to Earth from several penetrators and experiment packages that the Russians plan to place on the Martian surface. The radio-receiver equipment to accomplish this effort of international cooperation was supplied by the Centre National d'Etudes Spatiales (CNES) in France.

Once the primary mission is completed, the Mars Observer mission may be extended further — if the spacecraft and instruments are still in good condition and if there is enough fuel to control the spacecraft's altitude and orientation.

SPACECRAFT SCIENCE INSTRUMENTS

Mars Observer's battery of seven scientific instruments will make an intense and long-term examination of Mars from orbit. Collectively, the instruments cover much of the electromagnetic spectrum and form a complementary array. Each instrument produces sets of data that contribute to a wide variety of scientific investigations.

Gamma Ray Spectrometer (GRS)

The Gamma Ray Spectrometer will measure the chemical elements present on and near the surface of Mars with a surface resolution of a few hundred kilometers. The data will be obtained by measuring the intensities of gamma rays that emerge from the Martian surface.

These high-energy rays are created from the natural decay of radioactive elements in Martian surface materials and are also produced by the interaction of cosmic rays with the atmosphere and surface. By observing the number and energy of these gamma rays, it is possible to determine the chemical composition of the surface, element by element.

Mars Observer Camera (MOC)

The Mars Observer Camera system will photograph the Martian surface with the highest resolution ever accomplished by an orbiting civilian spacecraft. (Resolution is a measure of the smallest object that can be seen in an image.)

Low-resolution global images of Mars will also be acquired each day using two wide-angle cameras operated at 7.5-kilometer (4.7-mile) resolution per picture element. These same cameras will acquire moderate-resolution photographs at 240 meters (787 feet) per pixel.

A separate camera will acquire very-high-resolution images at 1.4 meters (4.6 feet) per pixel for features of special interest. Each of these camera systems uses a line array of several thousand detectors, and the motion of the spacecraft, to create the images.

The low-resolution camera system will capture global views of the Martian atmosphere and surface so that scientists may study the Martian weather and related surface changes on a daily basis. Moderate-resolution images will monitor changes in the surface and atmosphere over hours, days, weeks, months and years. The high-resolution camera system will be used selectively, because of the high data volume required for each image.

Thermal Emission Spectrometer (TES)

The Thermal Emission Spectrometer will measure infrared thermal radiation emitted from the Martian atmosphere and surface. From these measurements the thermal properties of Martian surface materials and their mineral content may be determined. When viewing the surface beneath the spacecraft, the spectrometer has six fields of view, each covering an area of 3 by 3 kilometers (1.9 by 1.9 miles).

The spectrometer (a Michelson interferometer) will determine the composition of surface rocks and ice and map their distribution on the Martian surface. Other capabilities of the instrument will investigate the advance and retreat of the polar ice caps, as well as the amount of radiation absorbed, reflected and emitted by these caps. The distribution of atmospheric dust and clouds will also be examined over the four seasons of the Martian year.

Pressure Modulator Infrared Radiometer (PMIRR)

This radiometer will measure the vertical profile of the tenuous Martian atmosphere by detecting infrared radiation from the atmosphere itself. For the most part, the instrument will measure infrared radiation from the limb, or above the horizon, to provide high-resolution (5-kilometer or 3-mile) vertical profiles through the atmosphere.

The measurements will be used to derive atmospheric pressure and determine temperature, water vapor and dust profiles from near the surface to as high as 80 kilometers (50 miles) above the surface. Using these measurements, global models of the Martian atmosphere, including seasonal changes that affect the polar caps, can be constructed and verified.

Mars Observer Laser Altimeter (MOLA)

The Mars Observer Laser Altimeter uses a very short pulse of light emitted by a laser to measure the distance from the spacecraft to the surface with a precision of several meters. These measurements of the topography of Mars will provide a better understanding of the relationship among the Martian gravity field, the surface topography and the forces responsible for shaping the large-scale features of the planet's crust.

Radio Science

The Radio Science investigation will use the spacecraft's telecommunication system and the giant parabolic (dish-shaped) antennas of NASA's Deep Space Network to probe the Martian gravity field and atmosphere. These measurements will help scientists determine the structure, pressure and temperature of the Martian atmosphere.

Each time the spacecraft passes behind the planet or reappears on the opposite side, its radio beam will pass through the Martian atmosphere briefly on it way to Earth. The way in which the radio waves are bent and slowed will provide data about the atmospheric structure at a much higher vertical resolution than any other Mars Observer experiment.

During that part of the orbit when the spacecraft is in view of Earth, precise measurements of the frequency of the signal received at the ground tracking stations will be made to determine the velocity change (using the Doppler effect) of the spacecraft in its orbit around Mars. These Doppler measurements, along with measurements of the distance from the Earth to the spacecraft, will be used to navigate the spacecraft and to study the planet's gravitational field.

Gravitational field models of Mars will be used along with topographic measurements to study the Martian crust and upper mantle. By the end of the mission, as a result of the low altitude of the orbit and the uniform coverage of Mars Observer, scientists will have obtained unprecedented global knowledge of the Martian gravitational field.

Magnetometer and Electron Reflectometer (MAG/ER)

Mars is now the only planet in the solar system, aside from Pluto, for which a planetary magnetic field has not yet been detected. In addition to searching for a Martian planetary magnetic field, this instrument will also scan the surface material for remnants of a magnetic field that may have existed in the distant past. The magnetic field generated by the interaction of the solar wind with the upper atmosphere of Mars will also be studied.

Mars Relay Experiment

The spacecraft also carries a radio system supplied by the French Centre National d'Etudes Spatiales (CNES) to support robotic missions being implemented by the Russians. The present Russian plan is to launch two spacecraft, one in 1994 and the other in 1996. The first would deploy penetrators into the surface of Mars and land small experiment packages on the surface for direct sampling of both the atmosphere and the surface. The associated Mars Observer relay equipment consists of a transmitter/receiver that will periodically receive and relay to Earth scientific and engineering data from these landed packages. In 1996 the Russians plan to launch instrument packages, a balloon and, perhaps, a surface rover that would relay information back to Mars Observer if the spacecraft is still operational in late 1997.

THE SPACECRAFT SYSTEM

The Mars Observer spacecraft uses, where possible, existing Earth-orbiting satellite component designs. The craft's main body is shaped like a box and is about 1.1 meters (3.25 feet) high, 2.2 meters (7.0 feet) wide, and 1.6 meters (5.0 feet) deep. Mars Observer was built by Martin Marietta (formerly General Electric) Astro Space in Princeton, N.J.

With its fuel, the spacecraft and its science instruments weigh about 2,565 kilograms (5,655 pounds). The spacecraft has a three-year design lifetime and is equipped with one large solar array, consisting of six 183 x 219 x 9.1-centimeter (6 x 7.2 x 0.3-foot) solar panels.

At launch, the spacecraft's main communication antenna, instrument booms and solar array were folded close to the spacecraft. During the cruise phase these structures were partially extended. The two 6-meter (20 foot) instrument booms carry two of Mars Observer's seven scientific instruments: the Magnetometer and Electron Reflectometer and the Gamma Ray Spectrometer.

After the Mars Observer spacecraft reaches its mapping orbit at Mars, the solar array and instrument booms will be fully unfolded. The main communication antenna — a 1.5-meter (5-foot) diameter parabolic antenna — will be raised on a 5.3-meter (17- foot) boom and rotated to have a clear view of Earth. The spacecraft will then power its instruments to begin conducting the mission experiments.

MAPPING CYCLE

In its near-circular mapping orbit, the Mars Observer spacecraft will rotate once per orbit in order to keep the instruments pointed at the planet. This will allow all the instruments to view the planet continuously and uniformly during the entire Martian year. The spacecraft, instruments and mission were designed so that sufficient resources, especially of power and data rate, are available to power all instruments as they collect data simultaneously and continuously on both the day and night sides of the planet. The camera system takes photos only on the day side and will acquire additional images every three days during real-time radio transmissions to the Deep Space Network. The rotation and orientation of the spacecraft are controlled by horizon sensors, a star sensor, gyroscopes and reaction wheels, as is common on Earth-orbiting satellites. The horizon sensors, adapted from a terrestrial design, continuously locate the horizon, providing control signals to the spacecraft. The star sensor will be used for attitude control during the 11-month cruise and as a backup to the horizon sensors during the mapping orbit.

Once during each 118-minute orbit, the spacecraft will enter the shadow of Mars and rely on battery power for about 40 minutes. The battery is charged by the spacecraft's large solar panel, which generates more than a kilowatt of power when it is in the sunlight to operate spacecraft subsystems and instruments.

Control of the spacecraft and instruments is accomplished through the use of onboard microprocessors and solid-state memories. Scientific and engineering data are stored on tape recorders for daily playback to Earth. Additional data operations will allow information to be returned in real-time from selected instruments whenever Earth is in view.

The lifetime of the spacecraft will most likely be determined by the supply of attitude-control fuel and the condition of the batteries. The following table summarizes some of the important performance features of the Mars Observer spacecraft.

SPACECRAFT STATISTICS

GENERAL

Design Life 3 years — Mapping Orbit Mars polar, nearly circular — Altitude Above Mars 378 km (234 miles) nominal — Key Features Seven science instruments (two mounted on 6-m booms) — Bi- and monopropulsion systems — Three-axis control system (highly stabilized) — Semiautonomous operation (stores up to 2000 commands) — Reliability Redundancy used to avoid single-point failures — Weight at launch Dry (with payload) 1,125 kg (2,480 lb.) — Fuel 1,440 kg (3,175 lb.) — Total Weight 2,565 kg (5,655 lb.) — Size (launch configuration): Length 1.6 m (5.0 ft) Width 2.2 m (7.0 ft) Height 1.1 m (3.25 ft)

COMMUNICATIONS Command Rate 12.5 commands/s (max)— Uplink Data Rate 500 bits/s (max) — Downlink Data Rate 85.3 kilobits/s (max) — Antennas 1.5-m-diam. high-gain parabolic articulating (on 5.3-m boom) three low-gain downlink RF — Power 44 watts — Tape Recorders 1.38×10^9-bit capacity

PROPULSION Bipropellant System Monomethyl hydrazine and nitrogen tetroxide — Monopropellant System Hydrazine Thrusters (20 total) (4) 490 N / (4) 22 N / (8) 4.5 N (orbit trim) / (4) 0.9 N (momentum unloading and steering) — Total Propellant Weight 1,346 kg (2,961 lb.)

ATTITUDE AND ARTICULATION CONTROL Pointing Accuracy Control: 10 mrad — Knowledge: 3 mrad — Pointing Stability 1 mrad (for 0.5 s) 3 mrad (for 12 s)

ELECTRICAL POWER Solar Array 6 panels, each 3.7 x 7 meters — Array Output Power 1,130 watts — Batteries 42-amp-hr NiCd (2) — Electronics Bus voltage regulation — Definitions: mrad = milliradian (~ 0.057 degree) / N = newton (~ 0.225 lb. force)

SCIENCE OPERATIONS

The Mars Observer mission operations at the Jet Propulsion Laboratory will be supported by NASA's Deep Space Network and the JPL Advanced Multimission Operations System. The 34-meter (111- foot), high-efficiency subnetwork, the newest of the Deep Space Network antenna subnets, will provide daily uplink and downlink communications with the spacecraft at X-band frequencies of 8.4 gigahertz. The 70-meter (230-foot) antenna network will also provide periodic very-long-baseline interferometry and real-time, high-rate telemetry and radio science support to the mission. The instrument scientists will remain at their home institutions, from which they will access Mars Observer data via a project database at JPL. Using workstations and electronic communications links, scientists will also be connected to the mission planning activities at JPL. In the same way, data products returned to the JPL database from the home institution for each of the instruments will be sent electronically to other investigators at their home institutions. This will allow scientists to have ready access to science data without moving to JPL for the duration of the mission.

More than 80 workstations will be connected to the project database at JPL, a centralized repository for downlink science and engineering telemetry data, ancillary data including navigation data, and uplink command and sequence data. This database, with about 30 gigabytes of on-line storage, will be electronically available to the science instrument investigators via NASCOM data links. The database will allow investigators to analyze their instrument data at their home institutions. During the mapping phase, the instrument investigations will return processed science data products to the database at JPL for access by the interdisciplinary scientists and the other investigation teams. The Mars Observer mission is expected to return more than 600 billion bits of scientific data to Earth — more than that returned by all previous missions to Mars and, in fact, roughly equal to the total amount of data returned by all planetary missions since the beginning of planetary exploration.

SOME SALIENT FACTS ABOUT MARS OBSERVER

Speed in Earth orbit (with respect to Earth) 7.73 km/s (17,300 mph) — Speed at TOS burnout (with respect to Earth) 11.5 km/s (25,700 mph) — Average speed during cruise (with respect to Sun) 25.0 km/s (56,000 mph) — Speed before Mars orbit insertion maneuver (with respect to Mars) 5.28 km/s (11,800 mph) Speed after Mars orbit insertion maneuver (with respect to Mars) 4.56 km/s (10,200 mph) — Speed in mapping orbit (with respect to Mars) 3.35 km/s (7,500 mph) — Distance traveled between Earth and Mars 7.24×10^8 km (450 million miles) — Distance from Earth at Mars arrival 3.4×10^8 km (210 million miles) — Distance from Earth during Min: 10^8 km (62 Mmi) mapping phase Max: 3.7×10^8 km (230 Mmi) — Time for command to reach spacecraft Min: 5.5 minutes during mapping phase Max: 20.5 minutes — Maximum acceleration on spacecraft (postlaunch) 0.1 G (occurs during transfer to low orbit) — Navigation target diameter at Mars 480 km (300 miles) (less than 1/10 of planet diameter)

MARS OBSERVER INVESTIGATORS

Gamma Ray Spectrometer (GRS)

TEAM LEADER: William V. Boynton, University of Arizona, Tucson, Ariz. James R. Arnold, University of California at San Diego, San Diego, Calif. Peter Englert, San Jose State University, San Jose, Calif. William C. Feldman, Los Alamos National Laboratory, Los Alamos, New Mexico Albert E. Metzger, Jet Propulsion Laboratory, Pasadena, Calif. Robert C. Reedy, Los Alamos National Laboratory, Los Alamos, New Mexico Steven W. Squyres, Cornell University, Ithaca, New York Jacob L. Trombka, NASA Goddard Space Flight Center, Greenbelt, Md. Heinrich Wanke, Max Planck Institute, Mainz, Germany Johannes Bruckner, Max Planck Institute, Mainz, Germany Darrell M. Drake, Los Alamos National Laboratory, Los Alamos, New Mexico Larry G. Evans, Computer Sciences Corporation, Lanham-Seabrook, Md. John G. Laros, Los Alamos National Laboratory, Los Alamos, New Mexico Richard D. Starr, Catholic University, Washington, D.C. Yuri A. Surkov, Vernadsky Institute, Russia

Mars Observer Camera (MOC)

PRINCIPAL INVESTIGATOR: Michael C. Malin, Malin Space Science Systems, Inc., San Diego, Calif. G. Edward Danielson Jr., California Institute of Technology, Pasadena, Calif. Andrew P. Ingersoll, California Institute of Technology, Pasadena, Calif. Laurence A. Soderblom, U.S. Geological Survey, Flagstaff, Ariz. Joseph Veverka, Cornell University, Ithaca, New York Genry A. Avenesov, Space Research Institute, Russia Merton E. Davies, The RAND Corporation, Santa Monica, Calif. William K. Hartmann, Planetary Science Institute, Tucson, Ariz. Philip B. James, University of Toledo, Toledo, Ohio Alfred S. McEwen, U.S. Geological Survey, Flagstaff, Ariz. Peter C. Thomas, Cornell University, Ithaca, New York

Thermal Emission Spectrometer (TES)

PRINCIPAL INVESTIGATOR: Philip R. Christensen, Arizona State University, Tempe, Ariz. Donald A. Anderson, Arizona State University, Tempe, Ariz. Stillman C. Chase, consultant, Santa Barbara, Calif. Roger N. Clark, U.S. Geological Survey, Denver, Colo. Hugh H. Kieffer, U.S. Geological Survey, Flagstaff, Ariz. Michael C. Malin, Malin Space Science Systems, Inc., San Diego, Calif. John C. Pearl, NASA Goddard Space Flight Center, Greenbelt, Md. Todd R. Clancy, University of Colorado, Boulder, Colo. Barney J. Conrath, NASA Goddard Space Flight Center, Greenbelt, Md. Ruslan O. Kuzmin, Vernadsky Institute, Russia Ted L. Roush, San Francisco State University, San Francisco, Calif. Arnold S. Selivanov, Institute for Space Devices Engineering, Russia

Pressure Modulator Infrared Radiometer (PMIRR)

PRINCIPAL INVESTIGATOR: Daniel J. McCleese, Jet Propulsion Laboratory, Pasadena, Calif. Robert D. Haskins, Jet Propulsion Laboratory, Pasadena, Calif. Conway B. Levoy, University of Washington, Seattle, Wash. David A. Paige, University of California at Los Angeles, Los Angeles, Calif. John T. Schofield, Jet Propulsion Laboratory, Pasadena, Calif. Fredric Taylor, Oxford University, England Richard W. Zurek, Jet Propulsion Laboratory, Pasadena, Calif. Michael D. Allison, NASA Goddard Space Flight Center, Greenbelt, Md. Jeffrey R. Barnes, Oregon State University, Corvallis, Ore. Terry Z. Martin, Jet Propulsion Laboratory, Pasadena, Calif. Peter L. Read, Oxford University, England

Mars Observer Laser Altimeter (MOLA)

PRINCIPAL INVESTIGATOR: David E. Smith, NASA Goddard Space Flight Center, Greenbelt, Md. Herbert V. Frey, NASA Goddard Space Flight Center, Greenbelt, Md. James B. Garvin, NASA Goddard Space Flight Center, Greenbelt, Md. James W. Head, Brown University, Providence, Rhode Island Duane Muhleman, California Institute of Technology, Pasadena, Calif. Gordon H. Pettengill, Massachusetts Institute of Technology,

Cambridge, Mass. Roger J. Phillips, Washington University, St. Louis, Mo. Sean C. Solomon, Carnegie Institute, Washington, D.C. Maria T. Zuber, NASA Goddard Space Flight Center, Greenbelt, Md. H. Jay Zwally, NASA Goddard Space Flight Center, Greenbelt, Md. W. Bruce Banerdt, Jet Propulsion Laboratory, Pasadena, Calif. Thomas C. Duxbury, Jet Propulsion Laboratory, Pasadena, Calif.

Radio Science (RS)

TEAM LEADER: G. Leonard Tyler, Stanford University, Stanford, Calif. Georges Balmino, Centre National d'Etudes Spatiales (CNES), France David Hinson, Stanford University, Stanford, Calif. William L. Sjogren, Jet Propulsion Laboratory, Pasadena, Calif. David E. Smith, NASA Goddard Space Flight Center, Greenbelt, Md. Richard Woo, Jet Propulsion Laboratory, Pasadena, Calif. Effraim L. Akim, Keldysh Institute of Applied Mathematics, Russia John W. Armstrong, Jet Propulsion Laboratory, Pasadena, Calif. Michael F. Flasar, NASA Goddard Space Flight Center, Greenbelt, Md. Richard A. Simpson, Stanford University, Stanford, Calif.

Magnetometer and Electron Reflectometer (MAG/ER)

PRINCIPAL INVESTIGATOR: Mario H. Acuna, NASA Goddard Space Flight Center, Greenbelt, Md. Kinsey S. Anderson, University of California at Berkeley, Berkeley, Calif. Sigfried Bauer, University of Graz, Austria Charles W. Carlson, University of California at Berkeley, Berkeley, Calif. Paul Cloutier, Rice University, Houston, Texas John E. P. Connerney, NASA Goddard Space Flight Center, Greenbelt, Md. David W. Curtis, University of California at Berkeley, Berkeley, Calif. Robert P. Lin, University of California at Berkeley, Berkeley, Calif. Michael Mayhew, National Science Foundation, Washington, D.C. Norman F. Ness, University of Delaware, Newark, Del. Henri Reme, University of Paul Sabatier, France Peter J. Wasilewski, NASA Goddard Space Flight Center, Greenbelt, Md. Michel Menvielle, University of Paris, Paris, France Diedrich Mohlmann, Deutsche Luft und Raumfahrt Gesellschaft, Germany James A. Slavin, NASA Goddard Space Flight Center, Greenbelt, Md. Alexander V. Zakharov, Space Research Institute, Russia

INTERDISCIPLINARY SCIENTISTS

Raymond E. Arvidson, Washington University, St. Louis, Mo. Bruce Fegley Jr., Washington University, St. Louis, Mo. Michael H. Carr, U.S. Geological Survey, Menlo Park, Calif. Alexander T. Bazilevsky, Vernadsky Institute, Russia Matthew P. Golombek, Jet Propulsion Laboratory, Pasadena, Calif. Harry Y. McSween Jr., University of Tennessee, Knoxville, Tenn. Andrew P. Ingersoll, California Institute of Technology, Pasadena, Calif. Howard Houben, NASA Ames Research Center, Mountain View, Calif. Bruce M. Jakosky, University of Colorado, Boulder, Colo. Leonid V. Ksanfomality, Space Research Institute, Russia Aaron P. Zent, Search for Extraterrestrial Intelligence (SETI) Institute, Mountain View, Calif. James B. Pollack, NASA Ames Research Center, Mountain View, Calif. Robert M. Haberle, NASA Ames Research Center, Mountain View, Calif. Vasily I. Moroz, Space Research Institute, Russia Laurence A. Soderblom, U.S. Geological Survey, Flagstaff, Ariz. E. Ken Herkenhoff, Jet Propulsion Laboratory, Pasadena, Calif. Bruce C. Murray, California Institute of Technology, Pasadena, Calif.

MARS OBSERVER POST LAUNCH MISSION REPORT

NASA
National Aeronautics and Space Administration
Washington, D.C. 20546

TO: A/Administrator
FROM: S/Associate Administrator for Space Science and Applications
SUBJECT: Mars Observer (MO) Post Launch Mission operations Report (MOR)

The MO spacecraft was successfully launched from Cape Canaveral Air Force Station on-board a Titan III/Transfer Orbit Stage on September 25, 1992, at 1:05 p.m. EDT. The spacecraft is in a stable planned configuration on the proper trajectory to Mars.

The enclosed MO Post Launch MOR is herein submitted as required by Headquarters Management Instruction 8610.1C dated November 26,1991.

L.A. Fisk Enclosure

FOREWORD

MISSION OPERATION REPORTS are published for the use of NASA Senior Management, as required by NASA Headquarters Management Instruction HQMI 8610.1C, effective November 26, 1991. The purpose of these reports is to provide a documentation system that represents an internal discipline to establish critical discriminators selected in advance to measure mission accomplishment, provide a formal written assessment of mission accomplishment, and provide an accountability of technical achievement.

Prelaunch reports are prepared and issued for each flight project just prior to launch. Following launch, updating (Post Launch) reports are issued to provide mission status and progress in meeting mission objectives.

Primary distribution of these reports is intended for personnel having program/project management responsibilities.

PREPARED BY: Solar System Exploration Division (Code SL) NASA Headquarters

PUBLISHED AND DISTRIBUTED BY: Executive Support Office (Code JA) Office of Management Systems and Facilities NASA Headquarters

SUMMARY

Mars Observer (MO) was successfully launched onboard a Titan III/Transfer Orbit Stage (TOS) from the Cape Canaveral Air Force Station at 1:05 p.m. EDT on September 25, 1992. MO is in a stable, planned configuration cruising normally on its way to Mars with all spacecraft subsystems performing well.

DISCUSSION

The spacecraft executed a normal countdown sequence with no anomalies. The Titan III countdown was delayed by two power failures in the Vehicle Integration Building (VIB), where the Titan III launch is controlled, and by delays in clearing personnel from the launch pad. Although it didn't impact the countdown, rollback of the Mobil Service Tower (MST) was delayed approximately 30 minutes due to the threat of rain. In addition, the predictions (based on low level winds which abated near the launch window opening) that a nitrogen tetroxide toxic plume would drift over populated areas in the event of a near-pad destruct of the Titan III caused a potential launch hold for most of the morning. Upper-level winds also restricted the launch azimuth to less than 105°, which reduced the planned 2-hour launch window to 1 hour and 41 minutes.

The Titan III performance as monitored real-time by the down-range stations was good. However, following separation from the Titan, neither the Advanced Range Instrumentation Aircraft (ARIA) nor the down-range stations received the TOS S-band carrier or telemetry, although a bright light (presumably the TOS solid motor burn) was observed by one of the ARIA during the proper interval, providing some real-time assurance of burn occurrence. The Canberra DSN stations, which acquired the spacecraft's X-band signal on schedule approximately 93 minutes after launch - confirmed a good TOS burn. The TOS telemetry data, although not received in real time, were recorded on the MO spacecraft's recorders, and were recovered at the Canberra Deep Space Network on October 7, 1992.

The Titan III/TOS injection accuracy estimate of the predicted state at the Target Interface Point (TIP) was 7.8 m/s at TIP +10 days. This error is within the maximum targeting error requirement of 9 m/s. The Figure of Merit (FOM) for this trajectory is 31.4 m/s at TIP +10 days, which is within the maximum FOM requirement of 35 m/s. The trans-Mars injection corresponds to a 1.8 sigma for the three-dimensional injection error.

When the spacecraft x-band signal was acquired, all indications, including solar array deployment, were nominal, with three exceptions: (1) a momentum unload firing with momentum continuing to build up about the x-axis, (2) the High-Gain Antenna (HGA) had not latched, and (3) a sun sensor was not functioning. Within a few minutes, based on the decreasing bus temperatures, it was evident that the HGA was not in its stowed position, which would have blocked the radiator panel. This obviated the need for any emergency time-critical commands. After about 5½ hours, the HGA telemetry deployment indication was received, indicating that the boom, although slow to deploy had deployed fully and latched. At the same time as the HGA latched, the momentum buildup began to stabilize; the HGA deployment direction was about the x-axis. The sun sensor redundancy management software switched sides after the appropriate time period, and operation was nominal. More recent data have shown that the sensor was shaded by the solar panels at that time and both sun sensors are operational.

The two science booms were extended to their cruise positions on September 29, 1992. On October 2, 1992, the Gamma Ray Spectrometer (GRS) cooler door and the Pressure Modulator Infrared Radiometer (PMIRR) vent and cooler doors were opened. The Electron Reflectometer (ER) cover was opened on October 16, 1992. This completed all spacecraft deployments until after insertion into the final Mars mapping orbit in late 1993.

The cruise configuration is demonstrated in Figure 1.

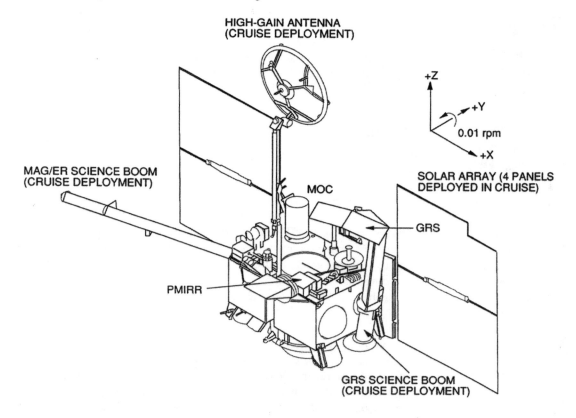

Figure 1. Cruise Configuration

All scientific instruments have been turned on; initial checkouts indicate that all are functioning properly. The inner cruise timeline is shown in Figure 2. An extended series of Radio Science (RS) characterizations during the month of November will complete preparations for the Gravity Wave Experiment early next year (see Figure 3).

The 1.8 sigma Titan III/TOS injection required an initial Trajectory Correction Maneuver (TCM- 1) velocity change of 50 m/s. On October 10, 1992, two of the four 490 Newton engines were ignited for 2 minutes and 13 seconds to achieve the desired velocity change. Three more TCM's are planned for next year (see Figure 3) to precisely trim the trajectory for Mars Orbit Insertion (MOI). Although the baseline mission plan is for a 4 month transition from MOI to mapping orbit, fuel budgets, based on the mid-window launch and nominal injection, are adequate for a faster transition period. Activity has been initiated to design the faster "power-in" transition sequence utilizing approximately 3 months versus the 4-month baseline.

Table 1 indicates the major spacecraft mission activities that are planned between now and end-of-mission.

Table 1. Significant MO Mission Events

Launch	September 25, 1992
Solar Array Deployed to Cruise	September 25, 1992
HGA Deployed to Cruise	September 25, 1992
GRS/MAG Booms Extended to Cruise	September 29, 1992
TCM-I	October 10, 1992
Complete Initial Instrument Checkout	October 10, 1992
Complete MOC Bakeout	December 25, 1992
Transition to HGA	January 4, 1993
TCM-2	January 8, 1993
TCM-3	February 8, 1993
Gravity Wave Experiment	April 11, 1993
TCM-4	August 9, 1993
MOI	August 24, 1993
Achieve Mapping Orbit	November 29, 1993
Deploy Solar Array to Mapping	December 6, 1993
Deploy HGA to Mapping	December 8, 1993
Deploy MAG Boom to Mapping	December 12, 1993
Start Data Acquisition	December 16, 1993
Deploy GRS Boom to Mapping	January 9-March 7, 1994
Start MBR Operations	September 1, 1995
Complete Data Acquisition	November 3, 1995
Complete MBR Operations	February 3, 1996

Note: All dates past MOI are baseline dates subject to "Power-in" design completion.

W. C. Panter
Mars Observer Program Manager
Solar System Exploration Division
Office of Space Science and Applications

REFERENCES

Luther, Joe L., "Titan III/TOS Injection Accuracy Report for Mars Observer, JPL IOM 312/92.6-516, 27 October 1992.

NASA Office of Space Science and Applications. Mars Observer *Mission Operation Report*, S-838-92-01, September 1992.

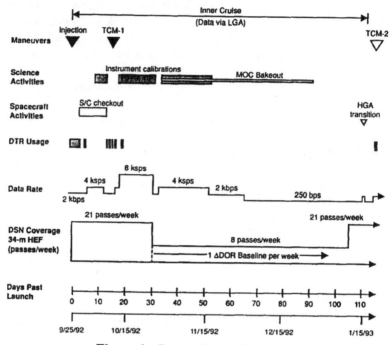

Figure 2. Inner Cruise Timeline

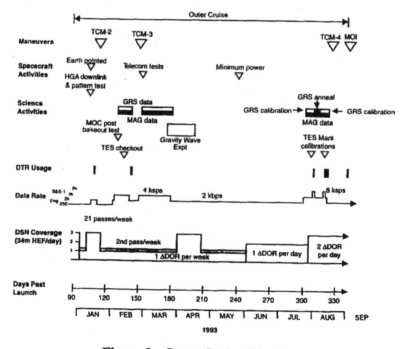

Figure 3. Outer Cruise Timeline

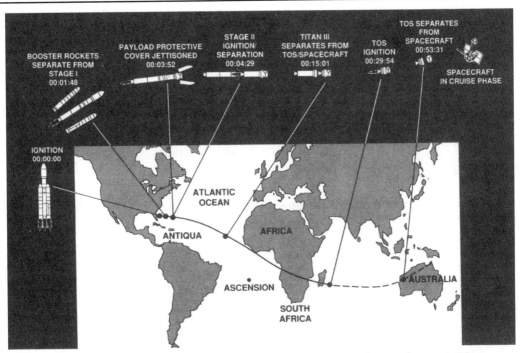

Launch sequence sends Mars Observer into orbit over two continents, then on to Mars.

MARS OBSERVER
INVESTIGATION REPORT

NASA
National Aeronautics & Space Administration
NEWS RELEASE

The final report by the independent investigation board on the failure of the Mars Observer spacecraft was delivered today (January 5, 1994) to NASA Administrator Daniel S. Goldin by Dr. Timothy Coffey, Chairman of the board. Dr. Coffey is Director of Research at the Naval Research Laboratory, Washington, D.C.

The Mars Observer spacecraft was to be the first U.S. spacecraft to study Mars since the Viking missions 18 years ago. The Mars Observer spacecraft fell silent just 3 days prior to entering orbit around Mars, following the pressurization of the rocket thruster fuel tanks.

Because the telemetry transmitted from the Observer had been commanded off and subsequent efforts to locate or communicate with the spacecraft failed, the board was unable to find conclusive evidence pointing to a particular event that caused the loss of the Observer.

However, after conducting extensive analyses, the board reported that the most probable cause of the loss of communications with the spacecraft on Aug. 21, 1993, was a rupture of the fuel (monomethyl hydrazine (MMH)) pressurization side of the spacecraft's propulsion system, resulting in a pressurized leak of both helium gas and liquid MMH under the spacecraft's thermal blanket. The gas and liquid would most likely have leaked out from under the blanket in an unsymmetrical manner, resulting in a net spin rate. This high spin rate would cause the spacecraft to enter into the "contingency mode," which interrupted the stored command sequence and thus, did not turn the transmitter on.

Additionally, this high spin rate precluded proper orientation of the solar arrays, resulting in discharge of the batteries. However, the spin effect may be academic, because the released MMH would likely attack and damage critical electrical circuits within the spacecraft.

The board's study concluded that the propulsion system failure most probably was caused by the inadvertent mixing and the reaction of nitrogen tetroxide (NTO) and MMH within titanium pressurization tubing, during the helium pressurization of the fuel tanks. This reaction caused the tubing to rupture, resulting in helium and MMH being released from the tubing, thus forcing the spacecraft into a catastrophic spin and also damaging critical electrical circuits.

Based on tests performed at the Jet Propulsion Laboratory (JPL) Pasadena, Calif., the board concludes that an energetically significant amount of NTO had gradually leaked through check valves and accumulated in the tubing during the spacecraft's 11-month flight to Mars.

In addition, the report listed other possible causes of the loss of the spacecraft as:
 · failure of the electrical power system, due to a regulated power bus short circuit;
 · NTO tank over-pressurization and rupture due to pressurization regulator failure;
 · the accidental high-speed ejection of a NASA standard initiator from a pyro valve into the MMH tank or other spacecraft system.

Other concerns noted by the board included:
 · a need to establish a policy to provide adequate telemetry data of all mission-critical events;
 · the lack of post-assembly procedures for verifying the cleanliness and proper functioning of the propellant pressurization system;
 · a current lack of understanding of the differences between the characteristics of European Space Agency and NASA pyro-initiators;
 · the potential for power bus short circuits, due to single component or insulation failure;
 · the potential for command and data handling control systems to be disabled by single-part failure;
 · the lack of fault protection external to the redundant crystal oscillator (RXO) should one of its two outputs fail;
 · the absence of information, in the telemetry, on the actual state of the RXO's backup oscillator;
 · deficiencies in systems engineering/flight rules;
 · too much reliance placed on the heritage of spacecraft hardware, software and procedures for near-Earth missions, which were fundamentally different from the interplanetary Mars Observer mission; and
 · the use of a firm fixed-price contract restricted the cost-effective and timely development of the unique and highly specialized Mars Observer Spacecraft.

Dr. Coffey notes, "We were challenged to conduct an extraordinarily complex investigation in which we had no hard evidence to examine nor communications with the spacecraft. However, after an extensive analysis covering every facet of the mission, operations and hardware, I believe that we are justified in arriving at the conclusions we have. If our findings will help to ensure that future missions won't suffer a similar fate, we feel we will have achieved our purpose."

Dr. Coffey also expressed his appreciation for the support provided to the investigation board by the six technical teams, other NRL and Air Force Phillips Laboratory contributors, NASA representatives, the JPL Project Team and Investigation Board, and the Martin Marietta Astro Space technical teams.

"I commend Dr. Coffey and his team for the thoughtful and thorough research into the tragic loss of the Mars Observer," said Dr. Wesley Huntress, Jr., Associate Administrator for NASA's Office of Space Science, Washington, D.C. "Their work will help and guide us in formulating a corrective action plan to help ensure future success as we plan for recovering our Mars science exploration objectives."

MARS PATHFINDER
PRESS KIT

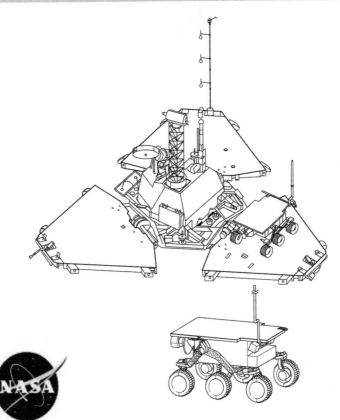

NATIONAL AERONAUTICS AND SPACE ADMINISTRATION

Mars Pathfinder Landing Press Kit July 1997
RELEASE: 96-207

Contacts

Douglas Isbell
Policy/Program Management
Headquarters,
Washington, DC
202/358-1753

Franklin O'Donnell
Mars Pathfinder Mission
Jet Propulsion Laboratory,
Pasadena, CA
818/354-5011

Mars Pathfinder Newsroom (June 30 to July 11) 818/354-8999

MARS PATHFINDER SET FOR INDEPENDENCE DAY LANDING

NASA's Mars Pathfinder mission — the first spacecraft to land on Mars in more than 20 years and the first ever to send a rover out to independently explore the Martian landscape — is set for touchdown July 4, initiating a new era of scientific exploration that will lead eventually to human expeditions to the red planet.

Mars Pathfinder is one of the first of NASA's Discovery class of missions, designed to foster rapidly developed, low-cost spacecraft with highly focused science objectives. Pathfinder's purpose is to demonstrate an innovative way of placing an instrumented lander on the surface of the planet. The lander will also carry a free-ranging robotic rover as a technology experiment. Landers and rovers of the future will benefit from the heritage of this pioneering mission.

Pathfinder's atmospheric entry and landing on the Martian surface are the centerpiece of the mission. Once the spacecraft hits the upper atmosphere at 10 a.m. Pacific Daylight Time on July 4, it will begin a 4½ minute, completely automated sequence of events to achieve its landing on the surface of the planet. After a fiery entry, the spacecraft releases a large, billowing parachute to slow its descent through the thin Martian atmosphere. Then a giant cocoon of airbags will inflate seconds before landing to cushion the spacecraft's impact. Along its descent to the ground, Pathfinder will be collecting engineering and atmospheric science data to help scientists profile the Martian environment.

"This is a new way of landing a spacecraft on a planet," said Brian Muirhead, Mars Pathfinder flight system manager at NASA's Jet Propulsion Laboratory, "and the first time a U.S. mission will use airbags to absorb the shock of landing and protect the lander from the rough, rocky terrain."

"The science investigations carried by Mars Pathfinder are going to give us unique insights into the planet's atmosphere and how it varies, and our first detailed understanding of the precise composition of its surface rocks and soils," said Joseph Boyce, Mars program scientist at NASA Headquarters, Washington, DC. "This knowledge is the key to helping unlock many interrelated mysteries about the history and evolution of Mars." Diving directly into the thin atmosphere at about 26,460 kilometers per hour (16,600 miles per hour), Pathfinder will release its parachute, then jettison the heat shield that protected it from the heat of entry. A tether will be deployed to lower the spacecraft from its backshell; airbags will then inflate about eight seconds before landing. Deceleration rocket engines will fire to nearly halt the craft for an instant in mid-air just before impact.

Once the rockets have been fired, Pathfinder will free-fall up to 30 meters (100 feet) before hitting the ground at a speed of up to 90 kilometers per hour (55 miles per hour). At impact, the spacecraft, now looking like a huge (5-meter-diameter (15-foot)) beach ball, will bounce many times, possibly as high as a 10-story building, until all impact energy dissipates. The interval between initial impact and complete halt may take as long as several minutes.

The site will be in darkness when Pathfinder lands at 3 a.m. local Mars time (shortly before 10 a.m. PDT; signal would be received on Earth at 10:07 a.m. PDT). After the spacecraft has come to a halt, its first task will be to deflate the airbags and open its petals. "The airbags might begin to deflate immediately after landing if they have been torn by the impact," said Robert Manning, Pathfinder flight system chief engineer at JPL. "If not, each of the bags has vents which will be opened to speed up the process of deflation."

As the Sun rises over the landing site at 12:45 p.m. PDT, Pathfinder will switch over from battery power to solar power. The lander will use its low-gain antenna to transmit critical data on the state of the spacecraft starting at about 2 p.m. PDT.

"By around 3 p.m. Pacific time, we'll have the critical data that we need to determine whether we have a basically healthy spacecraft in reasonably good condition, or whether we'll need to start thinking about contingency operations," Muirhead said. "Frankly, we will be very surprised if everything goes just right, since there are so many conditions that are unknown until we actually arrive at the landing site."

If no significant errors have been detected in the data and the spacecraft is healthy, a sequence will be sent to Pathfinder commanding it to unlock the imager camera head. The imager will first look for the Sun and, if found, the lander will use the location of the Sun to determine its orientation on the surface of Mars. The lander will then autonomously point the high-gain antenna toward Earth. Images of the spacecraft and the region around the rover petal will be taken and sent back to Earth through the high-gain antenna. If the camera, known as the Imager for Mars Pathfinder (IMP), does not successfully locate the Sun, then the mission will continue using the low-gain antenna. If this happens, fewer images will be sent back to Earth because of the antenna's much lower data rate.

Once images of the spacecraft and the rover on its petal have been received on Earth, the flight team will decide whether to deploy the rover ramps. This could take place on the first or second day of the mission. Once either or both ramps are unrolled, images of them and the terrain around the ends of the ramps will be taken and sent to Earth. If conditions are safe, the rover will be commanded to stand up and proceed down a ramp, either forward or backward. The rover should be deployed sometime during the first three days of the mission.

In addition to rover deployment images, Pathfinder's camera will take a panoramic image of its surroundings. If the high-gain communications link is operational, these images will begin to be transmitted as early as the first day of the mission. If the low-gain antenna is being used, these images will be sent down much later. The lander also will be transmitting science data on the temperature, atmospheric pressure and winds on Mars.

Once the rover sets out to explore Mars, it will rely on a tool kit of miniature instruments to study the composition of rocks and take close-up photos of Martian surface features. The rover is named Sojourner, after American civil rights crusader Sojourner Truth.

"Starting with the lander camera stereo images, we will use special goggles to view the terrain in three dimensions, and look for safe paths to travel along in order for the rover to reach specific rocks and regions to conduct science and technology experiments," said Brian Cooper, the primary rover driver on the mission. "Once the path is decided, we will drive the rover using a set of software instructions that will be uplinked to the rover each day."

The landing site, Ares Vallis, was chosen because scientists believe it is a relatively safe surface to land on and contains a wide variety of rocks washed down into this flood basin during a catastrophic flood. During its exploration of the surface, Sojourner will rely on the lander primarily for communications with Earth and for imaging support.

"Ares Vallis is particularly interesting to geologists because it drains a region of ancient, heavily cratered terrain that dates back to early Martian history, similar in age to the meteorite Allan Hills 84001, which contains scientific evidence suggesting life may have begun on Mars billions of years ago," said Dr. Matthew Golombek, Pathfinder project scientist. "By examining rocks in this region, Pathfinder should tell scientists about the early environment on Mars, which is important in evaluating the possibility that life could have begun there."

Pathfinder's instruments and mobile rover are not designed to provide an answer to the question of life on Mars. They are designed to provide an in-depth portrait of Martian rocks and surface materials over a relatively large landing area, thereby giving scientists an immediate look at some of the crustal materials that make up the red planet. Pathfinder data also will be used to verify observations that are made from space when an orbiter, called Mars Global Surveyor, arrives at Mars in September and later, in March 1998, begins its two-year mapping mission.

The Mars Pathfinder mission, along with Mars Global Surveyor, mark the beginning of a new era in Mars exploration and an ambitious new initiative by the United States to send pairs of spacecraft to the red planet every 26 months in a sustained program of robotic exploration extending well into the next century. This program of robotic exploration will expand scientists' knowledge of Mars in three important areas of investigation: the search for evidence of past life on Mars; understanding the Martian climate and its lessons for the past and future of Earth's climate; and surveying the geology and resources that could be used to support future human missions to Mars. The program will culminate in a robotic sample return mission to be launched as early as 2005.

Mars Pathfinder is the second in NASA's Discovery program of low-cost spacecraft with highly focused science goals. The Jet Propulsion Laboratory, Pasadena, CA, developed and manages the Mars Pathfinder mission for NASA's Office of Space Science, Washington, DC.

[End of General Release]

Media Services Information

NASA Television Transmission

NASA Television is broadcast on the satellite GE-2, transponder 9C, C Band, 85 degrees west longitude, frequency 3880.0 MHz, vertical polarization, audio monaural at 6.8 MHz. The schedule for television transmissions during the Mars Pathfinder landing period will be available from the Jet Propulsion Laboratory, Pasadena, CA; Johnson Space Center, Houston, TX; Kennedy Space Center, FL; and NASA Headquarters, Washington, DC.

Status Reports

Status reports on mission activities for Mars Pathfinder will be issued by the Jet Propulsion Laboratory's Public Information Office. They may be accessed online as noted below. Daily audio status reports are available by calling (800) 391-6654 or (818) 354-4210.

Pathfinder Newsroom

A newsroom will be operated at the Jet Propulsion Laboratory, Pasadena, CA, from June 30 to July 11, 1997. From June 30 to July 11, the newsroom telephone will be (818) 354-8999. Before that date, media may call (818) 354-5011 for information on credentialing.

Briefings

A pre-landing briefing on the missions and science objectives of Mars Pathfinder will be held at JPL at 10 a.m. PDT on July 1, 1997. Depending on the Space Shuttle launch schedule, this briefing may be videotaped for later replay on NASA Television. Daily news briefings will continue through the end of the Pathfinder rover's prime mission on July 11. Multiple briefings will be scheduled on landing day, July 4, and as required during the mission.

Image Releases

Images returned by the Mars Pathfinder lander and rover will be released to the news media in electronic format only during the mission. Images will be available in a variety of file formats at the web address http://www.jpl.nasa.gov/marsnews. This site will include files offering the highest spatial and color resolution of images returned by the Pathfinder lander and rover. Images will also be carried on NASA Television during daily Video File broadcasts.

Internet Information

Extensive information on Mars Pathfinder, including an electronic copy of this press kit, press releases, fact sheets, status reports and images, is available from the Jet Propulsion Laboratory's World Wide Web home page at http://www.jpl.nasa.gov/marsnews. The Mars Pathfinder Project also maintains a home page at http://mpfwww.jpl.nasa.gov.

Quick Facts

Lander

Spacecraft dimensions: Tetrahedron, three sides and base, standing 0.9 meter (3 feet) tall
Weight: 895 kilograms (1,973 pounds) at launch, fueled; 801 kilograms (1,766 pounds) dry
Science instruments: imager; magnets for measuring magnetic properties of soil; wind socks; atmospheric structure instrument/meteorology package.
Power: 160 watts peak power, up to 1,200 watt-hours per day from solar panels, batteries

Rover

Rover dimensions: 65 cm (2 feet) long by 48 cm (1.5 feet) wide by 30 cm (1 foot) tall
Weight: 10.6 kilograms (23 pounds)
Science instruments: alpha proton x-ray spectrometer, 3 cameras (also technology experiments)
Power: 16 watts peak power, up to 100 watt-hours per day from solar panels, batteries

Mission

Launch: December 4, 1996, at 1:58 a.m. EST from Cape Canaveral Air Station, FL, on a Delta II Launch vehicle
Mars landing: July 4, 1997, at approximately 1700 UTC (10 a.m. PDT)
Speed at atmospheric entry: 26,460 kilometers per hour (16,600 miles per hour)
Speed at surface impact: 70 to 90 kilometers per hour (45 to 55 miles per hour)
Landing site: Ares Vallis, approx. 19.4 degrees north latitude, 33.1 degrees west longitude
Sunrise at Martian landing site: 12:45 p.m. PDT
Sunset at Martian landing site: 1:45 a.m. PDT
One-way light time from Earth to Mars: 10 minutes, 35 seconds on July 4; 10 minutes, 40 seconds on July 5; 10 minutes, 44 seconds on July 6
Earth-Mars distance on landing day: 191 million kilometers (119 million miles)
Total distance traveled from Earth to Mars: 497 million kilometers (309 million miles)
Primary rover mission: 7 days
Primary lander mission: 30 days

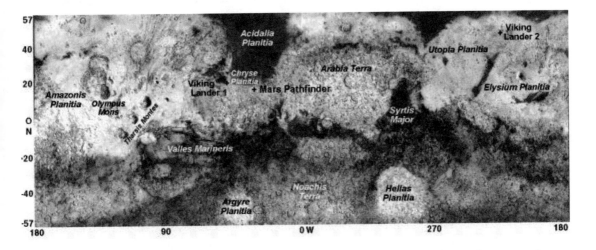

Mars Landing Sites

Mars at a Glance

General
* One of 5 planets known to ancients; Mars was Roman god of war, agriculture and the state
* Reddish color; at times the 3rd brightest object in night sky after the Moon and Venus

Physical Characteristics
* Average diameter 6,780 kilometers (4,217 miles); about half the size of Earth, but twice the size of Earth's Moon
* Mass 1/10th of Earth's; gravity only 38 percent as strong as Earth's
* Density 3.9 times greater than water (compared to Earth's 5.5 times greater than water)
* No magnetic field detected to date

Orbit
* Fourth planet from the Sun, the next beyond Earth
* About 1.5 times farther from the Sun than is Earth
* Orbit elliptical; distance from Sun varies from a minimum of 206.7 million kilometers (128.4 million miles) to a maximum of 249.2 million kilometers (154.8 million miles); average distance from Sun, 227.7 million kilometers (141.5 million miles)
* Revolves around Sun once every 687 Earth days
* Rotation period (length of day in Earth days) 24 hours, 37 min, 23 sec (1.026 Earth days)
* Poles tilted 25 degrees, creating seasons similar to Earth's

Environment
* Atmosphere composed chiefly of carbon dioxide (95.3%), nitrogen (2.7%) and argon (1.6%); only trace oxygen
* Surface atmospheric pressure less than 1/100th that of Earth's average
* Surface temperature averages -53 C (-64 F); varies from -128 C (-199 F) during polar night to 17 C (63 F) at equator during midday at closest point in orbit to Sun

Features
* Highest point is Olympus Mons, a huge shield volcano more than 27 kilometers (16 miles) high and 600 kilometers (370 miles) across; covers about the same area as Arizona
* Canyon system of Valles Marineris is largest and deepest known in solar system; extends more than 4,000 kilometers (2,500 miles) and has 5 to 10 kilometers (3 to 6 miles) relief from floors to tops of surrounding plateaus
* "Canals" observed by Giovanni Schiaparelli and Percival Lowell about 100 years ago were a visual illusion in which dark areas appeared connected by lines. The Viking missions of the 1970s, however, established that Mars has channels probably cut by ancient rivers

Moons
* Two irregularly shaped moons, each only a few kilometers wide
* Larger moon named Phobos ("fear"); smaller is Deimos ("terror"), named for attributes personified in Greek mythology as sons of the god of war

Historical Mars Missions

Mission, Country, Launch Date, Purpose, Results

Mars 1, USSR, 11/1/62, Mars flyby, lost at 106 million kilometers (65.9 million miles)
Mariner 3, U.S., 11/5/64, Mars flyby, shroud failed
Mariner 4, U.S. 11/28/64, first successful Mars flyby 7/14/65, returned 21 photos
Zond 2, USSR, 11/30/64, Mars flyby, failed to return planetary data
Mariner 6, U.S., 2/24/69, Mars flyby 7/31/69, returned 75 photos
Mariner 7, U.S., 3/27/69, Mars flyby 8/5/69, returned 126 photos

Mariner 8, U.S., 5/8/71, Mars flyby, failed during launch
Mars 2, USSR, 5/19/71, Mars orbiter/lander arrived 11/27/71, no useful data returned
Mars 3, USSR, 5/28/71, Mars orbiter/lander arrived 12/3/71, some data and few photos
Mariner 9, U.S., 5/30/71, Mars orbiter, in orbit 11/13/71 to 10/27/72, returned 7,329 photos
Mars 4, USSR, 7/21/73, failed Mars orbiter, flew past Mars 2/10/74
Mars 5, USSR, 7/25/73, Mars orbiter, arrived 2/12/74, some data
Mars 6, USSR, 8/5/73, Mars orbiter/lander, arrived 3/12/74, little data return
Mars 7, USSR, 8/9/73, Mars orbiter/lander, arrived 3/9/74 little data return
Viking 1, U.S., 8/20/75, Mars orbiter/lander, orbit 6/19/76-1980, lander 7/20/76-1982
Viking 2, U.S., 9/9/75, Mars orbiter/lander, orbit 8/7/76 - 1987, lander 9/3/76 - 1980; combined, the Viking
    orbiters and landers returned 50,000+ photos
Phobos 1, USSR, 7/7/88, Mars/Phobos orbiter/lander, lost 8/88 en route to Mars
Phobos 2, USSR, 7/12/88, Mars/Phobos orbiter/lander, lost 3/89 near Phobos
Mars Observer, U.S., 9/25/92, orbiter, lost just before Mars arrival 8/22/93 (8/21/93 PDT/EDT)
Mars Global Surveyor, 11/7/96, orbiter, en route to orbit insertion 9/12/97 (9/11/97 PDT/EDT)
Mars 96, Russia, 11/16/96, orbiter and landers, failed during launch
Mars Pathfinder, U.S., 12/4/96, en route to landing 7/4/97

Mission Timeline

All times for events on the spacecraft are given as the time signal would be received on Earth in Pacific Daylight Time (i.e. spacecraft event time plus one-way light time, which is approximately 10 minutes, 40 seconds). All operations events on Earth are in Pacific Daylight Time. Pacific Daylight Time is Universal Time minus 7 hours.

June 30:

12 a.m.: Mars Pathfinder is approximately 2 million kilometers (1.3 million miles) from Mars, traveling at a velocity of about 19,080 kilometers per hour (12,000 miles per hour) with respect to Mars.

July 1:

12 a.m.: Mars Pathfinder is about 1.6 million kilometers (982,000 miles) from Mars, traveling at a velocity of about 19,080 kilometers per hour (12,000 miles per hour) with respect to Mars.

July 2:

12 a.m.: Mars Pathfinder is about 1.1 million kilometers (696,000 miles) from Mars, traveling at a speed of about 19,080 kilometers per hour (12,000 miles per hour) with respect to Mars.

July 3:

12 a.m.: Mars Pathfinder is about 658,000 kilometers (408,000 miles) from Mars, traveling at a speed of about 19,080 kilometers per hour (12,000 miles per hour) with respect to Mars.

July 4:

12 a.m.: Mars Pathfinder is about 195,000 kilometers (121,000 miles) from Mars, traveling at a velocity of about 26,460 kilometers per hour (16,600 miles per hour) with respect to Mars.

9:32 a.m.: Cruise stage separation.

10:02 a.m.: Pathfinder enters the upper atmosphere of Mars at 26,460 kilometers per hour (16,600 miles per hour) and begins the sequence of events that will land the spacecraft on the surface. From this point on, the only likely signal from the spacecraft will be the carrier wave, a single frequency radio wave.

The shifting frequency of the carrier, known as the Doppler shift, will provide an indication of the decelerations occurring during entry and parachute deployment. The spacecraft is also designed to send back a frequency-keyed signal following certain key events; this signal is called a semaphore. The semaphore is very weak, and is not expected to be received in real time. However, careful analysis after-the-fact of the broad frequency spectrum recording of the radio signal will give the operations team considerable information on how events unfolded during the rapid descent to the surface.

Entry, descent and landing (EDL) takes approximately 4.5 minutes and follows the sequence below:

* Spacecraft rapidly decelerates in the atmosphere using the heat-shield
* Parachute deploys
* Heat shield separates
* Lander releases from backshell, descends on bridle
* Radar altimeter returns information on altitude
* Airbags inflate
* Rocket-assisted deceleration (RAD) engines fire
* Bridle cable is cut

10:07 a.m.: Landing on surface of Mars in Ares Vallis. Transmitter turned off shortly after landing to save power.

After touchdown, the following sequence will occur:

* Lander bounces and rolls to a stop
* Airbags deflate and are retracted up against the petals
* Petals open

These events of the entry, descent and landing phase will be complete between 11:32 a.m. and 12:33 p.m. PDT. A semaphore signaling the end of this phase may be received via the lander's low-gain antenna.

12:45 a.m.: Sunrise at the landing site. Operations begin for Sol 1 (a Sol is a Mars day, or 24 hours, 40 minutes).

1:56-3:13 p.m.: Transmitter is turned on, and the spacecraft signals Earth through the low-gain antenna. This communications session will contain telemetry from all engineering subsystems including the rover, and the first science data about the atmosphere taken during descent. Carrier is received at 1:55 p.m.; following ground processing, actual first information will probably be received by flight controllers at approximately 2:09 p.m.

Nominal Mission Scenario

If all spacecraft systems are normal, the mission will proceed on its "nominal" plan. On this plan, the following events will occur:

3:20 p.m.: The camera on the lander is released and begins searching for the Sun. The high-gain communications antenna is deployed and pointed toward Earth.

4:13-5 p.m.: First high-gain antenna downlink session. First engineering images of lander, airbags and the region around the lander. The very first image frame will be of a small region including part of the lander and an airbag. Assessment of these first images will tell the operations team about the condition of the spacecraft, the airbags and whether the rover ramps can be deployed. First color images of the region around the rover petal will be sent.

Low-Gain Antenna Communications Scenario

In all likelihood, there will be some condition or conditions of the spacecraft that will be different than the ideal case — for example, an unusually tilted orientation of the lander due to larger than anticipated rocks, or an airbag draping a solar panel, or some damaged hardware due to a harder than expected landing. At this point the mission team may enter a contingency mode where it uses commands and prepared sequences to further evaluate the health of the lander and improve its ability to continue the mission. Under such circumstances, the highest priority will be to assure the safety of the spacecraft and rover, and to insure enough power for operations and to recharge the battery.

Another possible contingency situation is loss of data due to a spacecraft or ground problem that would require using one of two remaining downlink sessions to retransmit data. Such a situation also will result in replanning the rest of the first day's activities.

Both the lander imager's Sun search and the high-gain antenna deployment must be completed successfully for the images described above to be received. If either activity is not completed fully, the team will intentionally go to a less complex plan of events using the lander's low-gain antenna. The low-gain antenna does not require knowledge of the spacecraft orientation on Mars or active pointing to Earth. This is a contingency scenario that has been well practiced and would proceed on the following timeline (attempts to find the Sun and point the high-gain antenna at Earth would normally resume on Sol 2).

6:06-7:51 p.m.: Low-gain antenna downlink session, including compressed rover ramp deployment images (black-and-white with 80-to-1 compression). Approximately 12 images will be sent. The rover team will evaluate the feasibility of ramp deployment based on these images.

7-8:15 p.m.: During this window, a decision will be made to deploy one or both rover ramps and command the rover to stand up. If more imaging is needed to make this decision, it will be requested at this time.

8:44 p.m.: If the decision is made to deploy either or both ramps, this will occur at this time in the following sequence: activate ramp deploy sequence; release the rover's alpha proton X-ray spectrometer instrument; release the rover from its stowed position; deploy the rover ramps. The rover will then stand up. A semaphore would be transmitted to Earth indicating that the command was received to begin the sequence.

In this low-gain antenna scenario, this is the end of Sol 1 because no more telemetry would be received. The downlink capability ends as the Earth sets to about 30 degrees above the Mars horizon.

Nominal Mission Scenario

If, on the other hand, the high-gain antenna is pointed toward Earth, the following timeline will be followed. It should be remembered that unexpected events can occur at any time which may change this timeline. As always, the highest priority will be to assure the safety of the spacecraft and rover, and to insure enough power for operations and to recharge the battery.

5:40-5:55 p.m.: Command conference to decide whether to deploy the rover ramp.

6 p.m.: Assuming that the rover and project team judge it safe, the ramp deployment sequence will begin at about this time.

6:55-7:57 p.m.: Second high-gain antenna downlink with rover deploy images (black-and-white), showing the ramps deployed. Engineering data, more detailed entry and weather data will be sent. Part of a black-and-white panorama image will be transmitted to Earth.

7:30-8:50 p.m.: Rover and project teams decide whether to deploy rover, based on the position of the ramp(s) on the surface and the expected ability of the rover to safely traverse the area immediately off the end of the ramp(s).

8:58 p.m.: If all conditions are judged acceptable, the rover deploy sequence will be activated and the rover will drive off the lander petal, down a ramp (either forward or backward), and roll out onto the surface of Mars. The alpha proton X-ray spectrometer will be lowered onto the soil to prepare for deployment.

9:24-10:26 p.m.: Third high-gain antenna downlink session. Images should show the rover on the surface of Mars. Based on this imaging, the rover team may decide to deploy the alpha proton X-ray spectrometer. Other images may include a black-and-white 360-degree panorama of the landing site.

10:30 p.m.: Sun sets at landing site, rover goes to sleep. If the alpha proton X-ray spectrometer was deployed, it will be taking measurements of rock and soil composition and storing data all night long.

July 5:

Low-Gain Antenna Communications Scenario

Resuming this scenario in the event that the high-gain antenna is not deployed on the first day:

Night of Sol 1: The flight team processes images of radiometric calibration target, and develops an estimate of the Sun's position. This information may then be used to estimate the lander orientation on the surface and allow the team to manually point the high-gain antenna at Earth on Sol 2. A set of commands will then be sent to the lander on the morning of Sol 2 to update the on-board orientation estimate.

2:20-2:50 p.m.: The first downlink session is conducted using the low-gain antenna. This communication session includes spacecraft health data taken at night. It also includes images acquired following ramp deployment at the end of Sol 1. The lander will then try a brief session with the high-gain antenna using the new pointing information uplinked early on Sol 2. If this communications demonstration is successful, the team will use the high-gain antenna for the second and final downlink on Sol 2. If not, a second low-gain antenna session will occur between 6:30 and 7:30 p.m. After the post-ramp deploy images are received on the ground, the rover team will make an assessment to determine if the rover can be deployed onto the surface. If conditions allow, the rover deploy sequence will be uplinked to the spacecraft and the rover will deploy at about 6:15 p.m. A final set of images of the rover sitting on the surface will then be acquired and down-linked during the final transmit session.

Nominal Mission Scenario

If the mission is on the high-gain communications scenario and the rover was deployed on Sol 1, the following is the sequence of events for Sol 2:

2:20-2:50 p.m.: The first downlink session on the high-gain antenna is conducted. This communication session includes night data and data from the alpha proton X-ray spectrometer.

Key activities on Sol 2 include obtaining and partially returning a color stereo panorama image and performing an extended rover traverse. The rover will conduct several experiments with soil mechanics during this traverse, and may attempt a second measurement with the alpha proton X-ray spectrometer at the end of the day. Additional transmit sessions may occur depending on available power; nominal time for these sessions are 8:20-9:20 p.m. and 10:30-11:20 p.m. Data expected during these sessions include engineering telemetry, weather observations, image data from the stereo color panorama and images acquired by the rover.

Why Mars?

Of all the planets in the solar system, Mars is the most like Earth and the planet most likely to support eventual human expeditions. Earth's Moon and Mercury are dry, airless bodies. Venus has suffered a runaway greenhouse effect, developing a very dense carbon dioxide atmosphere that has resulted in the escape of all of its water and the rise of torrid, inhospitable surface temperatures of nearly 500 degrees Celsius (about 900 degrees Fahrenheit). Mars, on the other hand, has all of the ingredients necessary for life, including an atmosphere, polar caps and large amounts of water locked beneath its surface. Mars, in fact, is the only other terrestrial planet thought to have abundant water that could be mined and converted into its liquid form to support human life.

Compared to Earth, Mars is about 6,800 kilometers (4,200 miles) in diameter, about half the diameter and about one-eighth the volume of Earth. Mars turns on its axis once every 24 hours, 37 minutes, making a Martian day — called a "sol" — only slightly longer than an Earth day. The planet's poles are tilted to the plane of its orbit at an angle of 25 degrees — about the same amount as Earth, whose poles are titled at 23.3 degrees to the ecliptic plane. Because of its tilted axis, Mars has Earth-like seasonal changes and a wide variety of weather phenomena. Although its atmosphere is tenuous, winds and clouds as high as 25 kilometers (about 15 miles) above the surface can blow across stark Martian deserts. Low-level fogs and surface frost have been observed by spacecraft. Spacecraft and ground-based observations have revealed huge dust storms that often start in the southern regions and can spread across the entire planet.

Early Mars may have been like early Earth. Current theories suggest that, early in its history, Mars may have once been much warmer, wetter and enveloped in a much thicker atmosphere. On Earth, evidence for life can be found in some of the oldest rocks, dating from the end of Earth's heavy bombardment by comets and meteors around 4 billion years ago. Surfaces on Mars that are about the same age show remains of ancient lakes, which suggests that liquid water flowed on the surface at one time and the climate was both wetter and substantially warmer. If this proves to be true, then further exploration may reveal whether life did develop on Mars at some point early in its history. If it did not, scientists will want to know why it didn't. Or perhaps they will be able to determine whether life that began early on in Mars' evolution could still survive in some specialized niches such as hydrothermal systems near volcanoes.

Mars is the most accessible planet on which to begin answering fundamental questions about the origin of life. Scientists want to know if we are alone in the universe. Is life a cosmic accident or does it develop anywhere given the proper environmental conditions? What happened to liquid water on Mars? Could life have begun on Mars and been transported to Earth?

Exploring Mars also will provide us with a better understanding of significant events that humankind may face in the future as Earth continues to evolve. What are the factors involved in natural changes in a planet's climate, for instance? On Earth, one of the most important questions now being studied is whether or not human activities are contributing to possible global warming. Could these climate changes bring about negative environmental changes such as sea level rise due to the melting of the ice caps? Mars provides a natural laboratory for studying climate changes on a variety of time scales. If Mars in the past was warmer and wetter, and had a thicker atmosphere, why did it change?

Layered deposits near the Martian polar caps suggest climatic fluctuations on a shorter time scale. If scientists can learn about the important factors controlling climatic changes on another planet, they may be able to understand the consequences of natural and human-induced changes on Earth.

Mars is an excellent laboratory to engage in such a study. Earth and Venus are active environments, constantly erasing all traces of their evolution with dynamic geological processes. On Mercury and on Earth's Moon, only relatively undisturbed ancient rocks are present. Mars, by contrast, has experienced an intermediate level of geological activity, which has produced rocks on the surface that preserve the entire history of the solar system. Sedimentary rocks preserved on the surface contain a record of the environmental conditions in which they formed and, consequently, any climate changes that have occurred through time.

The Search for Life

After years of exhaustive study of the data returned by the Viking spacecraft from their biology experiments, most scientists concluded that it is unlikely that any life currently exists on the surface of Mars. Centuries of fascination about the possibility of intelligent life on the red planet seemed to fade.

Since that time more than 20 years ago, however, much has been learned about the origins of life on Earth. Biologists learned that the most primitive single-celled microscopic organisms had sprung from hot volcanic vents at the very bottom of Earth's oceans. They found that the most fundamental carbonaceous organic material appear to demonstrate cell division and differentiated cell types, very similar to other fossils and living species. Geologists learned that these organisms could exist in regions along the floors of oceans in environments akin to pressure-cookers, at extremely hot temperatures devoid of light and prone to extreme pressures that no human being could survive. With new technologies and sophisticated instruments, they began to measure the skeletons of bacteria-like organisms lodged deep within old rocks.

Then, in August 1996, a NASA-funded team of scientists announced its findings of the first fossil evidence thought to be from Mars. The findings reignited the age-old question: Are we alone in the universe?

The two-year investigation by a team led by scientists from NASA's Johnson Space Center, Houston, TX, revealed evidence that strongly suggested primitive life may have existed on Mars more than 3.6 billion years ago. Researchers discovered an igneous meteorite in Earth's Antarctica that had been blasted away from the surface of Mars in an impact event; the rock was dated to about 4.5 billion years old, the period when Mars and its terrestrial neighbors were forming. According to scientists on the team, the rock contains fossil evidence of what they believe may have been ancient microorganisms.

The team studied carbonate minerals deposited in the fractures of the 2-kilogram (about 4-pound), potato-shaped Martian meteorite. They suggested living organisms deposited the carbonate — and some remains of the microscopic organisms may have become fossilized — in a fashion similar to the formation of fossils in limestone on Earth. Then, 16 million years ago, a huge comet or asteroid struck Mars, ejecting a piece of the rock from its subsurface location with enough force to escape the planet. For millions of years, the chunk of rock floated through space. It encountered Earth's atmosphere 13,000 years ago and fell in Antarctica as a meteorite.

In the tiny globs of carbonate, researchers found a number of features that can be interpreted as having been formed by possible past life. Team members from Stanford University detected organic molecules called polycyclic aromatic hydrocarbons (PAHs) concentrated in the vicinity of the carbonate. Researchers from NASA Johnson found iron mineral compounds commonly associated with microscopic organisms and the possible microscopic fossil structures.

Most of the team's findings were made possible only because of very recent technological advances in high-resolution scanning electron microscopy and laser mass spectrometry. Just a few years ago, many of the features that they reported were undetectable. Although past studies of the meteorite in question designated ALH84001 — and others of Martian origin failed to detect evidence of past life, they were generally performed using lower levels of magnification, without the benefit of the technology used in this research. In addition, the recent suggestion of extremely small bacteria on Earth, called nano-bacteria, prompted the team to perform this work at a much finer scale than had been done in the past.

The findings, presented in the August 16, 1996, issue of the journal Science, have been put forth to the scientific community at large for further study. The team was co-led by Johnson Space Center planetary scientists Dr. David McKay, Dr. Everett Gibson and Kathie Thomas-Keprta of Lockheed Martin, with the major collaboration of a Stanford University team headed by chemistry professor Dr. Richard Zare, as well as six other NASA and university partners. A variety of papers have been published in the months since that announcement that have argued for and against the claims that the evidence is suggestive of ancient life.

Whether or not the evidence stands up to scientific scrutiny, the suggestion alone has renewed interest in exploring the planets, stars and galaxies outside of the Milky Way galaxy. The questions resound: Does life exist elsewhere in the universe? And why does it exist at all? Did life as we know it originate on Earth or did it spring from other planets, only to be transported to Earth, where it found the most advantageous niche for continuing evolution?

In the year 2005, NASA plans to send to Mars a sample return mission, a robotic spacecraft that will be able to return soil and rock samples to Earth for direct study much as the Apollo astronauts returned hundreds of pounds of lunar rocks to Earth. Additional debate and scientific experimentation with Martian meteorites in the next several years may bring about an answer that may become a turning point in the history of civilization.

The Multi-Year Mars Program

Launch of the two 1996 missions to Mars — Mars Pathfinder and Mars Global Surveyor — ushered in a continuing U.S. program of Mars exploration. The program is designed to send low-cost spacecraft to Mars every 26 months well into the next decade.

Although they were launched within a month of each other in late 1996, Mars Global Surveyor and Mars Pathfinder have their roots in two separate NASA programs. Mars Pathfinder was approved as a stand-alone project under NASA's Discovery program, which was created in 1992 to fund low-cost solar system missions. Mars Global Surveyor, on the other hand, is the first in a multi-year series of missions under the Mars Surveyor program. After 1996, current plans call for two Mars Surveyor spacecraft to be sent to Mars during each launch opportunity in 1998, 2001 and 2003, and a single sample return spacecraft in 2005. The program is expected to continue beyond 2005 on a direction set by results obtained from earlier flights. In addition to the science goals stated, the purpose of these missions is to pave the way for human exploration some time early in the next century.

By 2005, NASA will have had a fleet of small spacecraft with highly focused science goals probing and watching the planet, setting in place a new way of exploring the solar system. Based on the space agency's philosophy of bringing faster, better and cheaper missions to fruition, combinations of orbiters and landers will take advantage of novel microtechnologies — lasers, microprocessors and electronic circuits, computers and cameras the size of a gaming die — to deliver an ingenious armada of miniaturized robotic payloads to Earth's planetary neighbor.

U.S. missions to Mars at this point are listed below. (Note: Projected costs are for spacecraft development only and do not include launch vehicles, mission operations after the first 30 days or spacecraft tracking.)

1996.

* Mars Pathfinder (Discovery mission). Demonstrate low cost-entry and landing system, and rover mobility; initiates mineralogy studies; continue study of surface characteristics and Martian weather. Cost: $171 million (capped at $150 million in fiscal year 1992 dollars), plus $25 million for rover.

* Mars Global Surveyor. Perform global reconnaissance of physical and mineralogical surface characteristics, including evidence of water; determine global topography and geologic structure of Mars; assess atmosphere and magnetic field during seasonal cycles; provide backup communication relay for the Mars Surveyor '98 lander and communication relay for the Mars Surveyor '98 microprobes. Cost: $148 million.

1998:

* Mars Surveyor '98 Orbiter. Launch scheduled December 10, 1998. Characterize the Martian atmosphere, including definition of atmospheric water content during seasonal cycles. Provide primary communication relay for the Mars Surveyor '98 lander.

* Mars Surveyor '98 Lander. Launch scheduled January 3, 1999. Access past and present-day water reservoirs on Mars; study surface chemistry, topology and mineralogy; continue weather studies. The spacecraft also will deliver two innovative soil microprobes developed under NASA's New Millennium program. Combined cost of both 1998 missions: $187 million, plus $26 million for the microprobes.

2001:

* Mars Surveyor '01 Orbiter. Characterize mineralogy and chemistry of surface, including identification of near surface water reservoirs.

* Mars Surveyor '01 Lander and Rover. Characterize terrain over tens of kilometers at site selected from

MGS and Mars Surveyor '98 orbital observations. Select and gather samples for possible later return. Characterize dust, soil and radiation conditions as they pertain to eventual human exploration. Test components of in-situ propellant production plant. Combined development cost of both 2001 missions: approximately $250 million.

2003:

* Mars Surveyor '03 Lander and Rover. Characterize terrain over tens of kilometers at a site chosen using earlier orbital observations; select and gather samples for possible later return. Other objectives, related to eventual human exploration, are expected to be added to both 2003 missions.

* Mars Surveyor '03 Orbiter: Provide communications and navigation facilities for 2003 and later missions on the Martian surface. Combined development cost of both 2003 missions: approximately $220 million.

2005:

* Sample Return Mission. Return a sample from one of the two rovers launched in 2001 and 2003. Development cost: approximately $400 million.

Beyond 2005:

* To be determined. Plans will depend on results of earlier missions.

International Cooperation

International collaboration on all Mars missions will be an important aspect of exploration in the next decade. Many space agencies around the world are considering participation in the planning stages of future missions, including those of Russia, Japan and many European countries. Scientists from the United States are consulting with international partners on the best ways to combine their efforts in Mars exploration. This may result in new proposals for cooperative missions in the first decade of the 21st century.

Among the ongoing programs taking shape is one called "Mars Together," a concept for the joint exploration of Mars by Russia and the United States. The program was initiated in the spring of 1994 and bore its first fruit in the summer of 1995. A Russian co-principal investigator and Russian hardware were incorporated into one experiment, the Pressure Modulator Infrared Radiometer, to be flown on NASA's Mars Surveyor 1998 orbiter. Dr. Vassili Moroz of the Russian Academy of Sciences Space Research Institute (IKI) in Moscow will co-lead the experiment with Dr. Daniel McCleese of NASA's Jet Propulsion Laboratory. The Russian institute also will provide the optical bench for the radiometer. In addition, IKI will furnish a complete science instrument, the LIDAR (Light Detection and Ranging) Atmospheric Sounder, for the 1998 Mars Surveyor lander.

Under Mars Together, NASA is discussing possible collaboration with Russia on a mission in 2001. This possible arrangement involves an additional rover launched and operated by Russia that also would select and gather samples for possible later return. The Mars Surveyor '01 orbiter would provide the communications relay for this rover.

Japan also is building an orbiter, called Planet B, to study the Martian upper atmosphere and its interaction with the solar wind. The spacecraft, to be launched in August 1998, will carry a U.S. neutral mass spectrometer instrument to investigate the upper atmosphere, in addition to a variety of Japanese instruments.

The nations of Europe are considering a mission in 2003 called Mars Express. The tentative plan includes an orbiter carrying one or more small landers and remote-sensing instruments that would study topography and surface minerals. A final decision on this mission is expected before the end of 1998.

Mission Overview

Mars Pathfinder will deliver a lander and small robotic rover, Sojourner, to the surface of Mars. The primary objective of the mission is to demonstrate a low-cost way of placing a science package on the surface of the red planet using a direct entry, descent and landing, with the aid of small rocket engines, a parachute, airbags and other techniques. Landers and free-ranging rovers of the future will share the heritage of Mars Pathfinder designs and technologies first tested in this mission. In addition, Pathfinder is studying ancient rocks to understand the nature of the early environment on Mars and the processes that have led to features that exist today.

Launch and Cruise

Mars Pathfinder was launched December 4, 1996, at 1:58 a.m. EDT atop a Delta II 7925 expendable launch vehicle from launch complex 17B at Cape Canaveral, FL. A PAM-D upper stage booster was used to inject the spacecraft on its interplanetary trajectory. While en route to Mars, the spacecraft spins at a rate of 2 rpm.

During the cruise to Mars, the spacecraft will have completed a total of four trajectory correction maneuvers to refine its flight path, on January 9, February 3, May 6 and June 25, 1997.

During the approach phase of the final 45 days before Mars arrival, spacecraft activities are focused on preparation for entry, landing and descent. The spacecraft continues in a 2 rpm spin-stabilized mode with the spin axis oriented toward Earth. Continuous coverage by NASA's Deep Space Network was required during this phase to support planning and execution of the fourth and final trajectory correction maneuver, and to support final entry preparations.

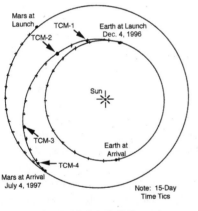

Mars Pathfinder's Earth-Mars trajectory

A final health and status check of the instruments and rover was conducted June 19, some 15 days before Mars entry. All onboard systems were operating normally. In addition, the Pathfinder rover was sent a "wake up" call; it responded normally and accepted minor software changes in preparation for landing. On June 23, the flight team began to load the 370 command sequences required to carry out Pathfinder's entry, descent and landing and initial surface operations.

Four days before Mars entry, on June 30, the spacecraft is scheduled to be turned about 7 degrees to orient it for entry. Commands will be issued to the spacecraft to initiate the software that controls the spacecraft during entry, descent and landing.

A fifth and final trajectory correction maneuver may be performed during one of two time windows, either 12 hours or six hours before entry into the Martian atmosphere, to insure that the spacecraft lands within its 100- by 200-kilometer (60- by 120-mile) target ellipse. A decision on this maneuver will not be made until a day or two before arrival.

Entry, Descent and Landing

The entry, descent and landing phase begins 1½ hours before Mars arrival and ends when the lander petals are fully deployed. Key activities during this phase include cruise stage separation, entry, parachute deployment, radar altimeter operations, airbag inflation, rocket-assisted deceleration burns, impact, airbag retraction and petal deployment. Real-time communications with the flight system will be possible through impact and, possibly, until the lander petals are deployed, depending on the landing orientation.

The entry trajectory for Mars Pathfinder is a ballistic, direct entry with an initial velocity of 26,460 kilometers per hour (16,600 miles per hour) and a mean flight path angle of 14.2 degrees (downward angle relative to

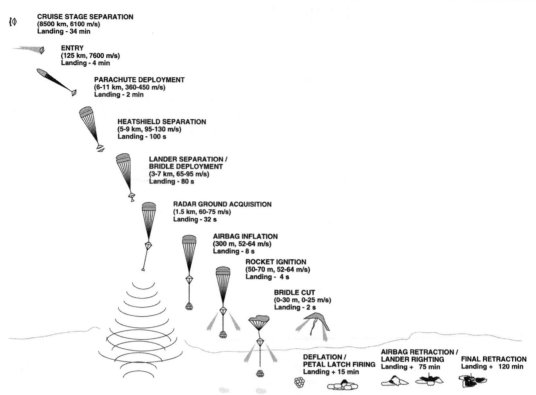

Mars Pathfinder entry, descent and landing

the surface of Mars' atmosphere). The entry velocity is approximately 80 percent faster than that of the Viking landers in the 1970s; the Vikings descended from Martian orbit, whereas Mars Pathfinder will enter the atmosphere directly from its interplanetary trajectory.

The peak aerodynamic deceleration during entry is about 20 g's, and occurs about 70 seconds after entry into the atmosphere (one g equals the normal force of gravity on Earth). A parachute will be deployed between 135 and 190 seconds after entry, at an altitude of between 6 and 10 kilometers (about 4 to 6 miles). The parachute will be deployed by firing a mortar to push the chute out of its canister. Once the parachute is deployed, the flight path angle will begin to bend until the vehicle is descending nearly vertically.

The heat shield will be released by a timer signal 20 seconds after parachute deployment to provide sufficient time for the chute to inflate and stabilize. Twenty seconds after heat shield release, the lander will be released and lowered from the backshell on a 20-meter (65-foot) bridle.

The radar altimeter begins measuring the distance to the surface at an altitude of about 1.5 kilometers (1 mile) above the surface. The spacecraft's airbags will inflate two seconds prior to ignition of the rocket-assisted deceleration rockets, and the rockets will fire about four seconds before impact. The total burn time of the rockets is approximately 2.2 seconds, but the bridle is cut prior to the end of the burn to allow enough extra thrust to carry the backshell and parachute away from the lander. This will prevent the backshell and parachute from falling onto the spacecraft. The lander will then free-fall the remaining distance to the ground.

Landing will occur about 4½ minutes after entry into the atmosphere. The imprecision of this time is caused by uncertainties about the altitude at the landing site and possible navigation targeting errors.

The lander could hit the ground in almost any orientation as a result of the rocket burn and bridle cut. At impact, the lander will bounce, roll and tumble until all impact energy dissipates. The interval between initial impact and the spacecraft's complete halt could be as long as several minutes. The airbags completely enclose the lander, so subsequent bounces should not result in high deceleration. Each face of the spacecraft's tetrahedron has a single six-lobed airbag, and energy is dissipated through vents in between the lobes.

Post-Landing

After the lander comes to a complete stop, the next key activities are deflation and retraction of the four airbags, and opening of the spacecraft's petals. Airbag deflation may begin to occur almost immediately after landing due to leaks in the bags. Each of the airbags has deflation patches which will be opened to speed up the process. These patches are opened by Kevlar cords inside the bags which are connected to a retraction motor. Additional cords are attached to other points inside each bag so that the airbags can be retracted after landing.

Flight software will control how the airbags are retracted. In general, the three airbags on the sides facing away from the ground will be retracted first. Once those bags have been retracted, the petals will be partially deployed so that the lander stands itself right side up. The final airbag on the side originally facing the ground will then be retracted before the petals are fully deployed. If the lander comes to rest on a rock, the entire lander may be tilted, but further maneuvering of the petals can be performed during surface operations to lower the overall tilt of the lander.

Telecommunications during entry should provide significant information about the behavior of the entry, descent and landing subsystem. Digital data will not be acquired, however, because of the extremely weak signal.

The amplitude and frequency of the spacecraft will be observed in real-time during entry and descent, and may be seen during petal deployment, depending on the lander's orientation once it comes to a stop on the Martian surface. Changes in amplitude are expected at cruise stage separation, parachute deployment, surface impact and during the airbag retraction and petal deployment. Changes in frequency reflect changes in the spacecraft's speed and will be most pronounced during the period of peak deceleration.

The spacecraft also will deliberately change the frequency of the sub-carrier to signal other key events. These include heat shield separation, bridle deployment, crossing the threshold altitude of 600 meters (about 2,000 feet) above the surface, completion of airbag retraction and completion of the petal deployment sequence. These planned frequency changes — called "semaphores" — are not likely to be detected in real-time, but can be extracted by post-processing the recorded data. In addition, key spacecraft telemetry data will be recorded and played back after landing.

Other key data to be transmitted to Earth include accelerometer measurements and selected atmospheric structure instrument measurements. The Deep Space Network's 70-meter (230-foot) and 34-meter (110-foot) antennas in Madrid, Spain, will be used to support entry communications.

Prime Mission

Mars Pathfinder's primary mission begins when its lander petals have been fully unfolded and the lander switches to a sequence of computer commands that will control its functions. The spacecraft lands about 2½ hours before sunrise on Mars and will spend the time in darkness retracting its airbags, standing itself upright and opening the petals so that solar panels can be powered up after sunrise.

The lander's first task will be to transmit engineering and science data collected during its descent through the thin atmosphere of Mars. If no errors are detected in these data and the spacecraft is basically healthy, a real-time command will be sent from Earth instructing the lander to unlock the imager camera head, deploy and point the high-gain antenna. If conditions are different than expected, which is not unlikely, the operation team will execute a contingency plan that has been placed onboard the spacecraft in expectation of such conditions.

In the normal plan, the lander's camera will begin taking images — including a panoramic view of the Martian landscape — and will begin transmitting the data directly to Earth at 2,250 bits per second. The first images of the Martian landscape will tell engineers whether the airbags are fully retracted and whether the rover's exit ramp can be safely and successfully deployed. Once either or both ramps are deployed, additional images will be acquired to show the terrain beyond the ramps so that engineers can decide on the safest exit route. If the high-gain antenna is not available, data will be sent over the lander's low-gain antenna at a much lower rate of 40 bits per second. In this case, only a few, highly compressed images will be sent.

Once a decision on the route has been made, commands will be sent to deploy the rover. Sojourner will spend about a quarter of an hour exiting its ramp. The rover should be deployed within the first three days after landing.

Driving off onto the floor of an old outflow channel, Sojourner will explore the surface at the command of Earth-based operators, who will rely on lander-based images to select a path and target for the rover. The six-wheeled Sojourner travels at 1 centimeter (0.4 inch) per second, performing mobility tests, imaging its surroundings and deploying an alpha proton x-ray spectrometer designed to study the elemental composition of rocks. During its prime mission, the rover will likely range a few tens of meters (yards) from the lander.

Also mounted on the lander are wind sensors, wind socks and high and low-gain antennas. Instruments will be used to measure the pressure, temperature and density of the Martian atmosphere. Magnets mounted on the lander will collect magnetic specimens of Martian dust and soil as small as 100 microns (about 1/250th of an inch).

Extended Missions

The primary mission lasts seven Martian days, or "sols," for the rover, and 30 Martian days, or "sols," for the lander. The rover could carry out an extended mission beyond that period, depending on how long its power sources and electronics last; engineers expect that the most probable reason for it to stop functioning is hot-cold cycling of its onboard electronics between Martian day and night.

Sojourner's extended mission activities would include repeating soil mechanics experiments on various soils; additional spectrometer measurements of both rocks and soil; obtaining images of selected areas with the rover camera, including close-ups of the lander; obtaining images of the lander's landing and tumbling path; and traveling longer distances, with the possibility of going over the horizon, up to hundreds of meters (yards).

For the lander, an extended mission lasting up to one year after landing is possible. Lander activities in the extended mission would include continued use of the lander camera to obtain images of the terrain and atmosphere, collection of key engineering telemetry and continued collection of meteorology data.

Mission Operations

All operations for Mars Pathfinder will be conducted at JPL, where the operations and science teams reside. Science data, both raw and processed, will be transferred after a period of validation to NASA's Planetary Data System archive for access and use by the planetary community at large and the general public. The Planetary Data System home page is at http://pds.jpl.nasa.gov/pds/home.html. Images from planetary missions are also available via the web from NASA's Planetary Photojournal at http://photojournal.jpl.nasa.gov.

Spacecraft

At launch the Mars Pathfinder spacecraft weighed about 895 kilograms (1,973 pounds), including its cruise stage, heat shield and backshell (or aeroshell), solar panels, propulsion stage, medium and high-gain antennas and 94 kilograms (207 pounds) of cruise propellant. The cruise vehicle measures 2.65 meters (8.5 feet) in diameter and stands 1.5 meters (5 feet) tall. The lander is a tetrahedron, a small pyramid standing about 0.9 meter (3 feet) tall with three triangular-shaped sides and a base.

When Pathfinder is poised to enter the Martian atmosphere, its main components are the aeroshell, folded lander and rover, parachute, airbag system and three rocket engines. Combined, the spacecraft's mass is about

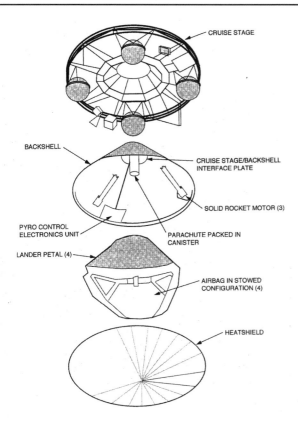

Mars Pathfinder flight system

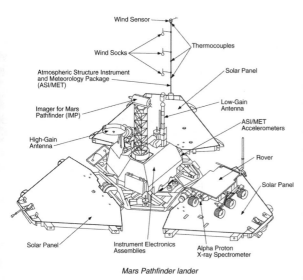

Mars Pathfinder lander

570 kilograms (1,256 pounds) at entry.

Once it has landed and its airbags have been deflated, Pathfinder's mass will be about 360 kilograms (793 pounds). Subsystems contributing to its landed weight include the opening/uprighting mechanism, lander cabling and electronics, instruments and rover. When it is unfolded and lying flat on the surface, the spacecraft will measure 2.75 meters (9 feet) across with a mast-mounted camera standing up about 1.5 meters (5 feet) from the ground.

The lander is controlled by a derivative of the commercially available IBM 6000 computer. This processor and associated components are radiation-hardened and mounted on a single electronics board. The computer has a 32-bit architecture which executes about 20 million instructions per second. The computer will store flight software as well as engineering and science data, including images and rover information, in 128 megabytes of dynamic random access memory.

During interplanetary cruise, the spacecraft requires 178 watts of electrical power, provided by 2.5 square meters (27 square feet) of gallium arsenide solar cells.

The lander has three solar panels, with a total area of 2.8 square meters (30 square feet) and supplying up to 1,200 watt-hours of power per day on clear days. At night, the lander will operate on rechargeable silver zinc batteries with a capacity at the beginning of the Mars surface mission of more than 40 amp-hours.

The Pathfinder lander carries a camera on a mast to survey its immediate surroundings. The camera has two optical paths for stereo imaging, each with a filter wheel giving 12 color bands in the 0.35 to 1.1 micron range; exposures through different filters can be combined to produce color images. The camera's field-of-view is 14 degrees in both horizontal and vertical directions, and it will be able to take one frame (256 by 256 pixels) every two seconds.

Sojourner Rover

Sojourner, the small rover onboard Mars Pathfinder, is named after an African-American crusader, Sojourner Truth, who lived during the tumultuous era of the American Civil War and made it her mission to "travel up and down the land" advocating the rights of all people to be free. The name was chosen in July 1995 by a panel of judges from the Jet Propulsion Laboratory and the Planetary Society following a year-long, worldwide

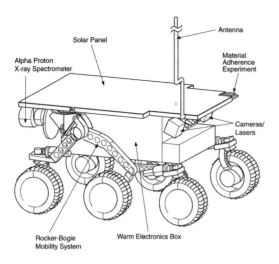

Mars Pathfinder's Sojourner rover

competition in which students up to 18 years old were invited to select heroines and submit essays about their historical accomplishments. The winning essay was submitted by Valerie Ambroise, now 14, of Bridgeport, CT. Sojourner Truth was shortened to Sojourner, which also means "traveler."

Sojourner with its mounting and deployment equipment weighed about 15.5 kilograms (34.2 pounds) at launch. Once it is mobile and operating on the Martian surface, it will weigh a mere 10.6 kilograms (23 pounds). The vehicle travels 1 centimeter (0.4 inch) per second and is about 65 centimeters (2 feet) long by 48 centimeters (1.5 feet) wide by 30 centimeters (1 foot) tall. During the cruise to Mars, it was folded in its stowage space and measured only 18 centimeters (7 inches) tall.

Equipped with three cameras — a forward stereo system and rear color imaging system — the Sojourner rover will take several images of the lander to assess the lander's health. The cameras are used in conjunction with a laser system to detect and avoid obstacles.

Sojourner is powered by a 0.2-square-meter (1.9-square-foot) solar array, sufficient to power the rover for several hours per day, even in the worst dust storms. As a backup and augmentation, lithium thionol chloride D-cell-sized batteries are enclosed in the rover's thermally protected warm electronics box. Thermal insulation is provided by a nearly weightless material called silica aerogel. Three radioisotope heater units (RHUs) — each about the size of a flashlight C-cell battery — contain small amounts of plutonium-238 (about 2.6 grams (less than 1/10th ounce) each) which gives off about 1 watt of heat each to keep the rover's electronics warm.

The rover's wheels and suspension use a rocker-bogie system that is unique in that it does not use springs. Rather, its joints rotate and conform to the contour of the ground, providing the greatest degree of stability for traversing rocky, uneven surfaces. A six-wheel chassis was chosen over a four-wheel design because it provides greater stability and obstacle-crossing capability.

Six-wheeled vehicles can overcome obstacles three times larger than those crossable by four-wheeled vehicles. For instance, one side of Sojourner could tip as much as 45 degrees as it climbed over a rock without tipping over. The wheels are 13 centimeters (5 inches) in diameter and made of aluminum. Stainless steel tread and cleats on the wheels provide traction and each wheel can move up and down independently of all the others. Three motion sensors along Sojourner's frame can detect excessive tilt and stop the rover before it gets dangerously close to tipping over. Sojourner is capable of scaling a boulder on Mars that is more than 20 centimeters (8 inches) high and keep on going.

The rover also will perform a number of technology experiments designed to provide information that will improve future planetary rovers. These experiments include: terrain geometry reconstruction from lander/rover imaging; basic soil mechanics by studying wheel sinkage; path reconstruction by dead reckoning and track images; and vision sensor performance.

In addition, Sojourner experiments also will determine vehicle performance; rover thermal conditions; effectiveness of the radio link; and material abrasion by sensing the wear on different thickness' of paint on a rover wheel. All rover communications is via the lander.

Scientists will study adherence of Martian airborne material by measuring dust accumulation on a reference solar cell that has a removable cover, and by directly measuring the mass of accumulated dust on a quartz crystal microbalance sensor.

The rover's control system calls for the human operator to choose targets and for the rover to autonomously control how it reaches the targets and performs tasks. The onboard control system is built around an Intel 8OC85 processor, selected for its low cost and resistance to upsets from space radiation. It is an 8-bit processor which runs at about 100,000 instructions per second.

Sojourner also carries an alpha proton X-ray spectrometer which is placed in contact with rocks or soil to measure the elemental composition of the material being studied.

Science Objectives

Mars Pathfinder carries a suite of instruments and sensors to accomplish a focused set of science investigations. These investigations include: studying the form and structure of the Martian surface and its geology; examining the elemental composition and mineralogy of surface materials, including the magnetic properties of airborne dust; conducting a variety of atmospheric science investigations, examining the structure of the atmosphere, meteorology at the surface, and aerosols; studying soil mechanics and properties of surface materials; and investigating the rotational and orbital dynamics of the planet from two-way ranging and Doppler tracking of the lander as Mars rotates.

In the first few days of the mission, the lander's stereo color imager will take several panoramic photographs of the Martian landscape. Scientists will use the imaging system to study Martian geologic processes and interactions between the surface and atmosphere. The imaging system will be able to observe the general physical geography, surface slopes and rock distribution of the surface so that scientists can understand the geological processes that created and modified Mars. Panoramic stereo imaging will take place at various times of the day, before and after the imager is deployed on its pop-up mast. In addition, observations over the life of the mission will reveal any changes in the scene over time that might be caused by frost, dust or redistribution of sand, erosion or other surface-atmosphere interactions.

The rover also will take close-up images of the terrain during its travels. A basic understanding of soil mechanics will be obtained by rover and lander imaging of rover wheel tracks, holes dug by rover wheels and any surface disruptions that have been caused by airbag bouncing and/or retraction.

The rover's alpha proton X-ray spectrometer along with spectral filters of the lander's imaging system and close-up images from the rover will measure the elemental composition of rocks and surface soil and infer their mineralogy. These data will provide a "ground truth" for orbital remote-sensing observations being obtained overhead by Mars Global Surveyor. Results will help scientists understand more about the crust of Mars, how it evolved into different feature types, and how weathering has affected surface features. The magnetic properties of airborne dust can also be investigated using a series of magnetic targets on the spacecraft.

During Pathfinder's entry and descent, an atmospheric instrument will profile the pressure, temperature and density of the atmosphere at various altitudes, beginning at 120 kilometers (about 75 miles) above the surface and continuing all the way down to the ground. After landing, a meteorology package records the weather. These weather data will be compared with the last data taken 20 years ago by the two Viking landers. Wind speed and direction will be determined by a wind sensor on top of the lander mast, along with three wind socks on the mast, to reveal more information about the forces present in the Martian atmosphere which act on small surface particles and draw them into the wind. Imaging of the atmosphere can be used to determine properties of aerosol particles such as their size, shape and distribution at different altitudes, as well as the abundance of atmospheric water vapor.

Orbital and rotational dynamics will be studied using two-way X-band ranging and Doppler tracking of the Mars Pathfinder lander by NASA's Deep Space Network. Ranging which is achieved by sending radio signals from Earth to the lander and back to Earth, then measuring the amount of time it takes to receive the returned signal — will provide an accurate measurement of the distance from a tracking station on Earth to Ares Vallis. After a few months of such tracking, scientists expect to know the location of the Pathfinder lander to within a few meters (yards) of accuracy.

Once the location of the lander is known, the pole of rotation of the planet can be determined. Knowledge of the orientation of Mars' pole of rotation will allow scientists to calculate the planet's precession — the gradual gyration of the planet's rotational axis over the course of many centuries that causes its north pole to point to different locations in space. Such information compared with the same measurement made by Viking should validate or disprove theories about Mars' interior, such as whether the planet has a metallic core, and shed new light on the forces which cycle volatiles such as water and carbon dioxide between the Martian atmosphere and its poles.

The Landing Site

NASA selected an ancient flood plain on Mars as the Mars Pathfinder landing site. Called Ares Vallis, the rocky plain was the site of great floods when water flowed on Mars eons ago. The site — at 19.4 degrees north latitude, 33.1 degrees west longitude — is 850 kilometers (about 525 miles) southeast of the location of Viking Lander 1, which in 1976 became the first spacecraft to land on Mars. Pathfinder will be the first craft to land on Mars since the twin Viking landers arrived more than 20 years ago. The spacecraft will land in Ares Vallis at the mouth of an ancient outflow channel chosen for the variety of rock and soil samples it may present.

Some constraints on the location were the result of engineering considerations. Since the Mars Pathfinder lander and Sojourner rover are solar-powered, the best site would be one with maximum sunshine; in July 1997, the Sun is directly overhead at 15 degrees north Martian latitude. The location's elevation had to be as low as possible so the descent parachute would have sufficient time to open and slow the lander to the correct terminal velocity. The landing will be within a 100- by 200-kilometer (60- by 120-mile) ellipse around the targeted site due to uncertainties in navigation and atmospheric entry.

The importance of the landing site on the potential scientific return was the driving factor in the scientific community's choice of a landing site. In 1994, more than 60 scientists from the United States and Europe participated in a workshop to recommend a landing site for Pathfinder. More than 20 individual landing sites were proposed before Ares Vallis was chosen.

A number of potential sites were considered. Among them were Oxia Palus, a dark highlands region that contains highland crust and dark wind-blown deposits; Maja Valles Fan, a delta fan which drained an outflow channel; the Maja Highlands, just south of Maja Valles;

Isidis Planitia, a lowlands site; and Tritonis Lacus, in Elysium Planitia. All of the sites were studied using Viking orbiter and Earth-based radar data.

Ares Vallis met several general criteria. First, it was a "grab bag" location, set at the mouth of a large water outflow channel in which a wide variety of rocks would be potentially within reach of the rover. Even though the exact origins of the samples would not be known, since many rocks washed onto the plain from highlands in ancient floods, the chance to sample a variety of rocks in a small area could reveal much about Mars. In addition, scientists wanted to choose a site that contained highland rocks because they make up two-thirds of the crust of Mars. With highland samples, they would be able to address questions about the early evolution of both the crust and the Martian environment. This was of particular interest to exobiologists, who are interested in beginning their search for evidence of life — extinct or existing — by first surveying the rock types in the highlands to find out if Mars had an early environment that was suitable for the beginnings of life.

Once the site was selected, scientists fanned out until they found a geological site very similar to Ares Vallis on Earth which they could study firsthand. In September 1995, planetary scientists traveled to the Channeled Scabland, near the cities of Spokane and Moses Lake, in central eastern Washington State to examine landforms and geologic features created by one or more giant, catastrophic floods which swept through the area as the North American continent thawed from an ice age.

The Scabland formed when waters in glacial Lake Missoula with the volume of Lake Erie and Lake Ontario combined broke through a glacial dam and flooded the region in just two weeks. The flooding carved landforms and geologic features similar to those on Mars' Ares Vallis. The site was an ideal Earth-based laboratory for studying landing site conditions and testing rover mobility.

Mars Pathfinder is the first mission to characterize the rocks and soils in a landing area over hundreds of square meters (yards) on Mars. The new information will provide a calibration point or "ground truth" for remote-sensing observations taken by orbiters that are surveying the planet's surface.

Planetary Protection Requirements

The United States is a signatory to the United Nation's Treaty of Principles Governing the Activities of States in the Exploration and Use of Outer Space, Including the Moon and Other Celestial Bodies (12/19/66). Known as the "Outer Space Treaty," this document states in part that exploration of the Moon and other celestial bodies shall be conducted "so as to avoid their harmful contamination and also adverse changes in the environment of the Earth resulting from the introduction of extraterrestrial matter."

NASA policy establishes basic procedures to prevent contamination of planetary bodies. Different requirements apply to different missions, depending on which solar system object is targeted; the spacecraft or mission type (flyby, orbiter, lander, sample-return and so on); and the importance of the object to the study of the origins of life. For some bodies such as the Sun, Moon and Mercury, there are no outbound contamination requirements. Current requirements for the outbound phase of missions to Mars, however, are particularly rigorous. Planning for planetary protection begins during pre-mission feasibility planning.

Planetary protection requirements called for Pathfinder's surfaces to contain a maximum of 300 spores per square meter (about 250 spores per square yard) and no more than 300,000 spores total for the entire spacecraft. To meet this requirement, the spacecraft was cleaned to the same level as the Viking landers before they were sterilized. Also by requirement, the final assembly of the spacecraft and its pre-launch processing were performed under special clean room conditions. Technicians continually cleaned the spacecraft throughout development by rubbing down surfaces with ethyl alcohol. Large surface areas, such as the airbags, thermal blankets and the parachute, had to be baked for about 50 hours at 110 degrees C (230 F). The spacecraft was checked constantly during processing at NASA's Kennedy Space Center in Florida and was given a final planetary protection inspection just before its integration with the Delta II launch vehicle.

The result of this effort was an exceptionally clean spacecraft. With the microbiological sampling and assays, the cleaned surfaces had a spore burden density of 43 spores per square meter. By analysis, the partially sterilized surfaces had even lower spore counts. The final total spore count by direct assay and by analysis was less than 24,000 spores.

Science Experiments

Mars Pathfinder and its rover carry several science instruments that will image terrain in the vicinity of the landing site, explore the composition of rocks and make measurements of the Martian atmosphere. The payload of science instruments includes:

* **Imager.** The camera on the lander is a stereo imaging system with color capability provided by a set of selectable filters for each of the two camera channels. It has been developed by a team led by the University of Arizona at Tucson, with contributions from the Lockheed Martin Corp., Max Planck Institute for Aeronomy in Katlenberg-Lindau, Germany, Technical University of Braunschweig in Germany and the Orsted Laboratory, Niels Bohr Institute for Astronomy, University of Copenhagen, Denmark. Principal investigator is Peter Smith, University of Arizona, Tucson, AZ.

The imager consists of three physical subassemblies: a camera head with stereo optics, filter wheel, charge-coupled device and pre-amplifier, mechanisms and stepper motors; an extendable mast with electronic cabling; and three plug-in electronics cards which plug into slots in the warm electronics box within the lander. Full panoramas of the landing site are acquired during the mission using the stereo baseline provided by the camera optics. Multispectral images of a substantial portion of the visible surface may be acquired with as many as 13 spectral bands.

A number of atmospheric investigations will be carried out using the imager.

Aerosol opacity is measured periodically by imaging the Sun through two narrow-band filters. Dust particles in the atmosphere are characterized by observing Phobos, one of Mars' moons, at night, as well as the Sun during the day. Water vapor abundance is measured by imaging the Sun through filters in the water vapor absorption band and in the spectrally adjacent continuum. Images of wind socks located at several heights above the surrounding terrain are used to assess wind speed and direction.

A magnetic properties investigation is included as part of the imaging investigation. A set of magnets of differing field strengths will be mounted to a plate and attached to the lander. Images taken over the duration of the lander mission are used to determine the accumulation of magnetic species in the wind-blown dust. Multispectral images of these accumulations may be used to differentiate among likely magnetic minerals.

The imaging investigation also includes the observation of wind direction and speed using wind socks that are located at various heights on a 1-meter-tall (40-inch) mast. The wind socks will be imaged repeatedly by the imager; orientations of the wind socks will be measured in the images to determine the wind velocity at three different heights above the surface. This information can then be used to estimate the aerodynamic roughness of the surface in the vicinity of the lander, and to determine the variation in wind speed with height. Because the Viking landers had wind sensors at only one height, such a vertical wind profile has never been measured on Mars.

This new knowledge will help to develop and modify theories for how dust and sand particles are lifted into the Martian atmosphere by winds, for example. Because erosion and deposition of wind-blown materials has been such an important geologic process on the surface of Mars, the results of the wind sock experiment will be of interest to geologists as well as atmospheric scientists.

* **Alpha Proton X-ray Spectrometer.** This instrument is designed to determine the elements that make up the rocks and soil on Mars. It is a derivative of instruments flown on the Russian Vega and Phobos missions and identical to the unit that flew on the Russian Mars '96 landers, which were lost shortly after launch. Thanks to the mobility provided by the Mars Pathfinder rover, the alpha proton X-ray spectrometer can take not only spectral measurements of the Martian dust but, more importantly, may be moved to distinct rock outcroppings, permitting analysis of the native rock composition for the first time. The alpha and proton portions are provided by the Max Planck Institute for Chemistry, Mainz, Germany. The X-ray spectrometer portion is provided by the University of Chicago. Principal investigator is Dr. Rudolph Rieder of the Max Planck Institute for Chemistry; co-investigators are Dr. Thanasis Economou of the University of Chicago and Dr. Henry Wanke of the Max Planck Institute for Chemistry.

The instrument can measure the amounts of all elements present except hydrogen, as long as they make up more than about 1/10th of 1 percent of the mass of the rock or soil. The spectrometer works by bombarding a rock or soil sample with alpha particle radiation — charged particles that are equivalent to the nucleus of a helium atom, consisting of two protons and two neutrons. The sources of the particles are small pieces of the radioactive element curium-244 onboard the instrument. In some cases, the alpha particles interact with the rock or soil sample by bouncing back; in other cases, they cause X-rays or protons to be generated. The "backscattered" alpha particles, X-rays and protons that make it back into the detectors of the instrument are counted and their energies are measured. The number of particles counted at each energy level is related to the abundance of various elements in the rock or soil sample, and the energies are related to the types of elements present in the sample. A high-quality analysis requires about 10 hours of instrument operation while the rover is stationary; it may be done at any time of day or night.

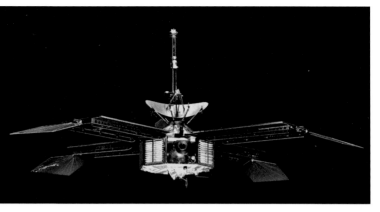

The design used for Mariners 3 & 4. Mariner 4 became the first spacecraft to photograph another planet at close-range in 1965. (above)

The launch of the first successful US Mars probe, Mariner 4, from Complex 12, 9:22 a.m. November 28 1964. The Atlas Agena D launch vehicle performed flawlessly.
The following July Mariner 4 flew past Mars and fulfilled its mission with 22 still pictures of the Red Planet. (above)

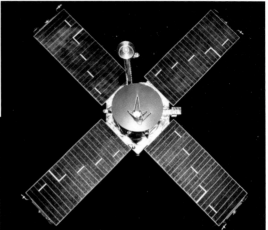

(Above) The Mariner '69 Mars flyby probe. This design was successfully flown on the Mariner 6 and Mariner 7 missions in 1969. Mariner 7 (below) and Mariner 6 (left) both successfully utilised the Atlas-Centaur launch vehicle.

An example of the beautiful pictures taken by the Hubble telescope from Earth orbit. (right). Before Hubble the quality of Earth-bound pictures was considerably worse than those taken by Mariner 6 and 7. (left)

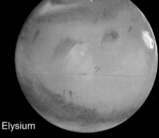

Acidalia

Elysium

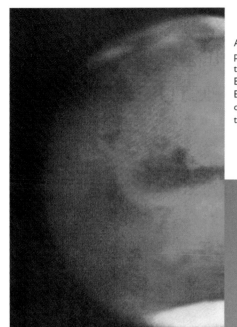

Mars as seen by Mariner 7 in 1969. Each Mars mission improved on its predecessor. This was one of the best color pictures of Mars available in 1969. (above)

Mariner 9 is launched 6:23 pm EDT, May 30th 1971 (right) atop Atlas-Centaur 23. At that time Mars was 63 million miles from Earth.

Tharsis

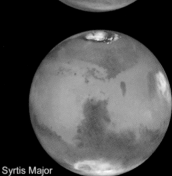

The highly successful Mars orbiter — Mariner 9. After 313 days in orbit around Mars it had transmitted nearly 7000 pictures of Mars back to Earth. Its identical sister ship, Mariner 8 failed shortly after launch.

Syrtis Major

The Viking lander as it would have appeared after fully deploying its science package on the surface of Mars. (Above)

Viking 1 is launched atop a Titan/Centaur rocket from Complex 41 at Cape Canaveral. August 20th 1975. (Left)

Viking 1 is encapsulated in its payload shroud at the KSC Assembly Facility July 11th 1975. (Below right)

The Viking Orbiter as it would have appeared in orbit around Mars in 1976 (Below left)

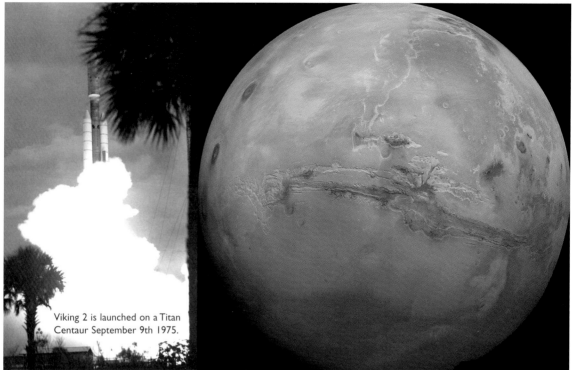

Viking 2 is launched on a Titan
Centaur September 9th 1975.

Mosaic of the Valles Marineris hemisphere of Mars. (above right) The view is 2,500 kilometers from the surface of the planet. The mosaic is composed of 102 Viking Orbiter images of Mars. The center of the picture shows the entire Valles Marineris canyon system, over 3,000 kilometers long and up to 8 kilometers deep. The three Tharsis volcanoes, each about 25 kilometers high, are visible to the west.

The Viking Lander 1 site Chryse Planitia. (below) The large rock just left of center is about 2 meters wide. This rock was named "Big Joe" by the Viking scientists. The top of the rock is covered with red soil. Many of the rocks around the two Viking Lander sites and the Pathfinder site were given names so that scientists could discuss the rocks without the need for images.

The ill-fated Mars Observer (left) is launched atop a Titan III with a Transfer Orbit Stage (right) September 25th 1992. An investigation concluded that the vehicle fell victim to a fuel leak.

The Viking Lander 2 site Utopia Planitia. The rounded rock in the center foreground is about 20 cm wide. There are two trenches that were dug in the regolith to the right of the rounded rock, as well as one behind and slightly to the left.

The Martian horizon, three kilometers distant, as seen by Viking 2 in September 1976. Viking's surface sampler housing is at left, the antenna for receiving commands from Earth at right.

VIKING LANDER 2   CAMERA 2   CE LABEL 22A003/000
DIODE   BGI/T   STEP SIZE 0.12 CHANNEL/MODE 2/1
VIKING LANDER 2   CAMERA 2   CE LABEL 22A016/002
DIODE   BGI/T   STEP SIZE 0.12 CHANNEL/MODE 2/1
VIKING LANDER 2   CAMERA 2   CE LABEL 22A018/002
DIODE   BGI/T   STEP SIZE 0.12 CHANNEL/MODE 2/1
COLOR MOSAIC OF RADCAM OUTPUT SPEC MIN 0. MAX 4.5 *
LABCAT
SAR - LGEOM
MASKVL
                   SEGMENT 1 OF 1
        IPL PIC ID 76/09/14/125832   HDR/11473BX

The astonishingly successful Sojourner rover deployed on Mars in 1997 by the Pathfinder lander. The two vehicles proved many new technologies such as using airbags for landing (above foreground) as well as returning a wealth of new scientific information from the Martian surface. Cruise phase (left)

Mars Pathfinder is launched atop a Delta II launch vehicle December 4th 1996 (far left). The lander and rover performed flawlessly for weeks on the Martian surface returning such spectacular images as the so-called "Presidential" panorama (bottom)

An oblique, color image of central Valles Marineris. The photograph is a composite of Viking high-resolution images in black and white and low-resolution images in color. Ophir Chasma on the north is approximately 300 km across and as deep as 10 km. (below)

Mars Global Surveyor is launched atop a Delta II 7925A on November 7th 1996. (left) As of Summer 2000 the probe was still relaying massive amounts of data to Earth about Mars climate and structure.

A Viking Orbiter mosaic showing the caldera of the enormous volcano Olympus Mons. The mountain stands nearly 27 km above the surrounding terrain. The image above is approximately 40 by 60 km.

The volcanoes of the Tharsis region (left) as seen by the Mars Global Surveyor spacecraft (below). Olympus Mons dominates the upper left corner—it is nearly 550 km (340 miles) wide. The white features are clouds. The four largest volcanoes are more than 15 km (9 mi) high.

Courtesy NASA/JPL/Malin Space Science Systems

The last unmanned missions to Mars of the 20th century were destined to both fail.
Mars Climate Orbiter (right — December 11th 1998) and Mars Polar Lander
(left—January 3rd 1999) were both launched on Delta II 7425 rockets.

Polar Lander (left) failed during the final phases of entering Mars atmosphere, both the lander and the accompanying Deep Space Two penetrator probes were lost. Climate Orbiter (right) was the victim of a mix-up in measurements between Imperial and Metric standards. It met a fiery end when it burned up in the Martian atmosphere.

Most of the instrument's electronics are located on the rover in a container called the warm electronics box. Cables run from that box to the instrument sensor head, which contains the radioactive sources and particle detectors. The instrument sensor head is held by a robotic arm attached to the back of the rover. This arm has a flexible "wrist" and can place the sensor head in contact with rocks and soil at various angles depending on how rough the rocks or soils might be. Sensors on a bumper ring attached to the sensor head indicate to the rover when adequate contact has been made with the sample rock or soil. When the sensor head is in position, it analyzes an area of rock or soil within a circle 5 centimeters (2 inches) across. Additional information about the rock or soil can be obtained by taking pictures of it using a small color camera on the back of the rover, or by rotating the rover and imaging it with stereo cameras on the front of the rover.

* **Atmospheric Structure Instrument/Meteorology Package.** The atmospheric structure instrument and meteorology package — or ASI/MET — is an engineering subsystem which acquires atmospheric information during the descent of the lander through the atmosphere and during the entire landed mission. It is implemented by JPL as a facility experiment, taking advantage of the heritage provided by the Viking mission experiments. Dr. Alvin Seiff of San Jose State University, San Jose, CA, was the instrument definition team leader. The science team that will use the data acquired by the package is led by Dr. John T. "Tim" Schofield of JPL.

Data acquired during the entry and descent of the lander permit reconstruction of profiles of atmospheric density, temperature and pressure from altitudes in excess of 100 kilometers (60 miles) from the surface.

The accelerometer portion of the atmospheric structure instrument depends on the attitude and information management subsystem of the lander. It consists of sensors on each of three spacecraft axes. The instrument is designed to measure accelerations over a wide variety of ranges from the micro-g accelerations experienced upon entering the atmosphere to the peak deceleration and landing events in the range of 30 to 50 g's.

The ASI/MET instrument hardware consists of a set of temperature, pressure and wind sensors and an electronics board for operating the sensors and digitizing their output signals. Temperature is measured by thin wire thermocouples mounted on a meteorological mast that is deployed after landing. One thermocouple is placed to measure atmospheric temperature during descent; three more are located to monitor atmospheric temperatures at heights of 25, 50 and 100 centimeters (about 10, 20 and 40 inches) above the Martian surface after landing. Pressure is measured by a Tavis magnetic reluctance diaphragm sensor similar to that used by Viking, both during descent and after landing. The wind sensor employs six hot wire elements distributed uniformly around the top of the mast. Wind speed and direction 100 centimeters (about 40 inches) above the Martian surface are derived from the temperatures of these elements.

What's Next?

One month before the launch of Mars Pathfinder, another Mars spacecraft called Mars Global Surveyor was launched toward the red planet. Mars Global Surveyor will reach Mars on September 12, 1997, and be captured in orbit after a 10-month journey to the planet.

At first, the spacecraft will be in a highly elliptical orbit and spend four months dipping into Mars' upper atmosphere using a technique called aerobraking to bring it into a low-altitude, nearly circular mapping orbit over the poles. This critical phase of the mission will be flown in a modified aerobraking configuration to accommodate a solar panel that is not fully deployed. The panel, one of two 3.5-meter (11-foot) wings, is tilted 20.5 degrees from its fully deployed position.

Shortly after launch, ground controllers discovered that a small damper arm — part of the solar array deployment mechanism at the joint where the inboard panel is attached to the spacecraft — broke during the panel's initial rotation during the first day of flight. The piece of metal became lodged in a 5-centimeter (2-inch) space in the shoulder joint at the edge of the solar panel.

After completion of a series of diagnostic activities in January and February 1997 to characterize the situation, the Surveyor flight team turned its attention to analyzing the possibility of carrying out the mission in the solar array's current configuration. In late April 1997, the JPL flight team, in collaboration with NASA Headquarters and its partners at Lockheed Martin Astronautics in Denver, decided to perform the aerobraking phase with the tilted solar panel rather than attempt some slight maneuvers to jiggle the debris free and allow the panel to lock in place.

Analysis of the situation indicates that the array, in its current state, will pose little risk to the aerobraking operations or the science goals of the mission. Using a two-axis gimbal drive assembly at the base of the solar array wing, the solar panel can be adjusted to the proper position for aerobraking and mapping. During aerobraking, the panel will be rotated 180 degrees so that the side with the solar cells faces into the direction of air flow as the spacecraft dips repeatedly into the Martian atmosphere.

Once in its mapping orbit, Mars Global Surveyor will complete one orbit around Mars about every two hours. Each new orbit will bring the spacecraft over a different part of Mars. As the weeks pass, the spacecraft will create a complete global portrait of Mars, capturing the planet's ancient cratered plains, huge canyon system, massive volcanoes, channels and frozen polar caps. During its mission, Mars Global Surveyor will pass over the terrain where the two U.S. Viking landers — separated by more than 6,400 kilometers (4,000 miles) — have rested for 22 years. The spacecraft also will be passing over the Mars Pathfinder lander and rover, which likely will no longer be operating by then.

By March 1998, Surveyor will be ready to begin data collection, compiling a systematic database as it surveys the Martian landscape with multi-spectral measurements and high-resolution photographs of unique features, such as giant volcanoes, deep canyons, changing polar caps and Mars' network of sinuous, intertwining river channels. In addition, the spacecraft's altimeter will fire laser pulses that will measure the heights of Martian surface features.

Mapping will begin on March 15, 1998 and last until January 31, 2000 — a period of one Martian year or 687 Earth days (almost two Earth years). The spacecraft will transmit its recorded data back to Earth once a day during a single 10-hour tracking pass by antennas of the Deep Space Network. During mapping operations, the spacecraft will return more than 600 billion bits of scientific data to Earth — more than that returned by all previous missions to Mars and, in fact, roughly equal to the total amount of data returned by all planetary missions since the beginning of planetary exploration, with the exception of the Magellan mission to Venus.

Mars Global Surveyor, the first in NASA's decade-long program of robotic exploration, will study Mars' early history, geology and climate. The spacecraft, which will orbit Mars in a near-polar orbit that will take it over most of the planet, carries a suite of sophisticated remote-sensing instruments designed to create a global portrait of Mars by mapping its morphology, mineral composition, topography, magnetism and atmosphere. Some of the instruments are flight spares from experiments flown on Mars Observer, which was lost shortly before Mars arrival in 1993. With this comprehensive archive of the red planet, scientists will be able to address a multitude of questions surrounding the evolution of Mars.

Program/Project Management

The Mars Pathfinder mission is managed by the Jet Propulsion Laboratory for NASA's Office of Space Science, Washington, DC. At NASA Headquarters, Dr. Wesley T. Huntress is associate administrator for space science. Joseph Boyce is Mars program scientist and program scientist for Mars Pathfinder. Kenneth Ledbetter is director of the Mission and Payload Development Division.

At the Jet Propulsion Laboratory, Norman Haynes is director for the Mars Exploration Directorate. Donna Shirley is manager of the Mars Exploration Program. For Mars Pathfinder, JPL's Anthony Spear is project manager and Brian Muirhead is flight system manager and deputy project manager. Richard Cook is mission manager and Dr. Matthew Golombek is project scientist.

MARS PATHFINDER MISSION STATUS REPORTS

PUBLIC INFORMATION OFFICE
JET PROPULSION LABORATORY
CALIFORNIA INSTITUTE OF TECHNOLOGY
NATIONAL AERONAUTICS AND SPACE ADMINISTRATION
PASADENA, CALIF. 91109.
TELEPHONE (818) 354-5011 http://www.jpl.nasa.gov

Mars Pathfinder mission Status July 5, 1997 6 p.m. Pacific Time

A communications problem between the Mars Pathfinder rover and lander has been solved, the Mars Pathfinder operations team reported at a 5 p.m. press briefing. Engineers are ready to move ahead with deployment tonight of the rover's ramps and the rover itself.

Although the operations team was not able to pinpoint a specific event that could explain the communications fix, information from a 3:20 p.m. downlink session suggested that it might have been the lander's flight computer which, for unknown reasons, reset itself at the end of the first day's activities.

"The spacecraft is fine," said Richard Cook, Mars Pathfinder mission manager. "It essentially operated all night, just as it was supposed to and we got all the data back showing that all subsystems were fine, but we're a little perplexed as to what happened (with the flight computer). We'll be getting back more diagnostic data later and we should be able to figure it out."

Telecommunications engineers discovered last night that the Sojourner rover, which is programmed to communicate with the lander as frequently as every minute, was not "completing full sentences" in its transmissions to the lander. The rover team thought the faulty communications might have been the result of a software synchronization problem between the rover's modem and the lander. With new information from today's high-gain downlink session, however, they discovered that the flight computer had reset itself at the end of the first day of activities at about 10:30 p.m. PDT on July 4. The automatic reset may have overwritten previous software timeout commands. As more telemetry becomes available, the team will be able to identify and understand the problem. In the meantime, they were not concerned about a repeat performance since the problem is solvable.

New images also were returned during the mid-afternoon downlink session. Immediately after the press briefing, Mars Pathfinder camera team leader Dr. Peter Smith, of the University of Arizona, and six of his team members rolled out a banner 360- degree mosaic of the Ares Vallis landing site made from all of the image frames received to date.

The rover team sent commands to Pathfinder late today to acquire more imagery of the ramps and Martian terrain around the regions where the ramps would touch the surface. They expected to deploy both ramps by about 7 p.m. PDT, with rover deployment to follow two or three hours later. If the rover rolls onto the surface of Mars by the end of Sol 2 tonight, it would be instructed to place its alpha proton X-ray spectrometer on the ground and take measurements of the soil overnight.

NASA TV will continue to carry coverage of events as they unfold this evening, including ramp and rover deployment. A final wrap-up briefing, scheduled at 9 p.m. PDT, may be canceled if no new developments have occurred by that time. The next scheduled briefing will be held at 10 a.m. PDT on Sunday, July 6, in JPL's von Karman Auditorium.

Mars Pathfinder Mission Status July 6, 1997 9 p.m. Pacific Daylight Time

With a perfectly healthy lander and rover on the surface of Mars, scientists on the Mars Pathfinder team burned the midnight oil last night to design their first day of activities for the robust, 23-pound rover named Sojourner Truth.

The rover spent its first night on Mars very near the end of the rear ramp from which it exited the lander last night, taking measurements of the Martian soil with its alpha proton X-ray spectrometer. At 3:45 p.m. PDT today, the operations team at JPL "woke up" the rover by playing "Final Frontier," the theme song from the television program "Mad About You," in keeping with the traditional wake-up songs used to arouse the astronauts during space shuttle missions.

Two experiments for Sol 3 were radiated to Pathfinder earlier today. The first is a soil mechanics experiment, in which the rover will lock five of its wheels, then turn the sixth wheel in both directions. Scientists will observe the depth of the wheel tracks and the movement of a top layer of fine-grain material. The second experiment will send Sojourner to its first rock, nicknamed "Barnacle Bill," for its very rough, barnacle-like surface.

This traverse to Barnacle Bill will involve two maneuvers for the rover: the first to turn its wheels approximately 70 degrees in the direction of the rock; the second maneuver to move backward with its fully deployed spectrometer until it touches the rock. The rover will travel approximately 36 centimeters (1.2 feet) to reach the rock. Once the spectrometer has been placed against Barnacle Bill, Sojourner will spend the night gathering data on its composition.

"With any luck, we will get a picture of Sojourner holding hands with Barnacle Bill," said Brian Muirhead, deputy project manager, at a 6 p.m. press briefing.

Among the most anticipated data expected to be returned tonight are 12-color, high resolution images that will be pieced together like a mosaic to create a 360-degree, color panorama, or "monster pan," of the landing site. The color variations and higher resolution will help scientists identify more geological features worthy of exploration in this very rocky flood plain. The first picture of the Pathfinder lander taken by the rover is also expected to be returned tonight.

Scientists pointed out other interesting rocks, soil deposits and features on the horizon in this evening's press briefing. A pair of mountain peaks, nicknamed "Twin Peaks," revealed ribbons of different colored rock. Scientists noted that the horizontal bands could be sedimentary layers or terraces cut by erosion. Angular rocks appearing in the foreground, all leaning in the same direction, suggest they were ejected from a nearby impact crater. A variety of smooth round rocks suggested that they were transported by water in Mars' early evolution.

"In the initial analysis of these images, we see multiple episodes of flooding, not just one catastrophic event, but many," said Dr. Ronald Greeley, of Arizona State University, a co-investigator on the Imager for Mars Pathfinder (IMP) team.

"We have a view of Mars that we have never seen before," added Dr. Matthew Golombek, project scientist at JPL. "We really have a grab bag of rocks here, varying in color, texture, fabric, sizes and shapes. They are completely different from the Viking landing sites and from each other too."

Scientists expect to begin reporting results of the first day of science activities at a 10 a.m. PDT press briefing at JPL. Meanwhile, all instruments and spacecraft systems continue to perform exceptionally well. The operations team will be increasing Pathfinder's data rate to more than 8,000 bits per second tomorrow in order to maximize the return of science measurements.

Mars Pathfinder Mission Status July 7, 1997 1 p.m. Pacific Daylight Time

Moderate weather yesterday, temperatures hovering around minus 76 degrees Fahrenheit, pressure about 6.8 millibars, steady light winds blowing from the southeast. Afternoon temperatures reached about 10 degrees Fahrenheit. The forecast for today: 10 degrees Fahrenheit, cooling overnight to about minus 105 degrees Fahrenheit.

A little extreme for an Earthly weather report? Perhaps, but with that, scientists on the Mars Pathfinder mission today presented the first weather report from Ares Vallis, an outflow channel on the surface of Mars.

Four days into surface operations, the Mars Pathfinder lander, rover and instruments are performing perfectly and returning a wealth of new data on the rocks, soils and atmosphere of Mars.

"The site is everything we hoped it would be," said Dr. Matthew Golombek, Pathfinder project scientist, at a 10 a.m. PDT press briefing. "We are finding more and more surprises as we look in detail at the rocks and terrain."

Images presented this morning included the first photograph of the lander taken by the rover. The image showed final retraction of the airbags in a very high, puffy clump that blocked most of the lander from view.

Meanwhile, the lander's Imager for Mars Pathfinder (IMP) camera has provided a new perspective on rocks and hills on the Martian horizon now that it is deployed on its mast and photographing the site at an elevation of 1.0 meters (3.2 feet) above the lander, said Dr. Peter Smith, IMP principal investigator from the University of Arizona.

Another new image presented this morning showed Sojourner Truth, the 23-pound rover that has begun to explore rocks around the landing site, as it was gathering data overnight on "Barnacle Bill." This rock, which was about 36 centimeters (1.2 feet) from the rover after it exited the lander, is thought to be about 8- to-10-inches tall, Smith said, and has a very distinctive surface that looks almost as if it is covered with barnacle-shaped objects.

"Here we have proof that Sojourner sort of nestled up and kissed Barnacle Bill," Golombek said as the photograph was presented.

"We have also received data from the rover's first soil experiment. The APXS (alpha proton X-ray spectrometer) is working perfectly," Golombek continued. "However, because we started taking data earlier in the day than we originally planned, the temperatures on Mars were warmer than the detectors liked and we have a bit of noise in the spectra. The team needs an extra day to try to figure out how to subtract that noise out."

The science team said a full chemical analysis of both the Martian soil and Barnacle Bill would be reported at tomorrow's 11 a.m. PDT press briefing. Meanwhile, Sojourner will travel to a larger rock later today, called "Yogi," and study the composition of the soil around it using the alpha proton X-ray spectrometer. Several scientists have commented that a smooth depression of soil around the rock resembles a moat.

Looking south at a pair of sloping hills, called "Twin Peaks," that are about a mile away, Smith pointed out new observations made possible by the fully deployed IMP camera. A depression in the landscape in front of the peaks suggests the presence of a channel. "This is actually a channel back behind those rocks, we're on the edge of a channel," he said.

A high resolution close-up of the Martian soil near the base of the lander also revealed a texture perfectly preserved in the Martian environment. Dr. Jim Bell of Cornell University explained the calibration targets that are used to achieve the true color of the Martian landscape. Color variations allow scientists to identify different types of minerals that are present in the environment. The bright reddish color of the soil, for example, points to the presence of oxidized iron in surface materials.

"The surface of Mars is rusting," Bell said. "We don't know when or how fast it's rusting, but we hope to find these things out. Not all of the surfaces are the same, though. There's lots of diversity and variation in the landscape. We can see some surfaces that are much less red, for example, and more consistent with volcanic rocks."

Building on comments made yesterday by Dr. Ronald Greeley (Arizona State University) about the evidence for floods in this region, Dr. Michael Malin, an interdisciplinary scientist, said the floods were so catastrophic that they would have filled up the Mediterranean basin here on Earth. Evidence, he said, can be seen in the variety of rocks, sediments and "puddles" left in the Martian soil that materials from the highlands were swept into this flood basin.

A full color, 360-degree panorama of the Pathfinder landing site will be presented at tomorrow's 11 a.m. press briefing, as will data about the composition of the Martian soil and Barnacle Bill.

Briefings are carried live on NASA TV, which is available on GE-2, transponder 9C at 85 degrees west longitude, vertical polarization, with a frequency of 3880 MHz, and audio of 6.8 MHz.

Mars Pathfinder Mission Status August 18, 1997 4 p.m. Pacific Daylight Time

Daily communications with the Mars Pathfinder lander and rover have resumed after an interruption on Saturday, Aug. 16, that was caused by an automatic reset of the lander's flight computer. The cause of the reset is not known, said Mars Pathfinder Flight Director Rob Manning.

The flight team was able to reactivate the lander by sending it instructions to reinitialize high-gain pointing and then begin its scheduled downlink session on Sunday evening. The downlink session began right on time at 10:09 p.m. PDT Aug. 17.

New images indicated that the rover had stopped its traverse to a rock nicknamed Shark after partially climbing up a rock called Wedge. The rover's hazard avoidance software is designed to stop the vehicle when it begins to tilt too much. Wedge is one of many small rocks forming a gateway to the Rock Garden, which is Sojourner's next destination. This portion of the landing site is more challenging than other regions because it is much rockier than any terrain explored to date by the rover.

The flight team will instruct Sojourner to continue its trek to Shark tonight. Shark is of interest to scientists because it is a large, smooth rock, which is relatively dust-free and, therefore, an ideal candidate for the next spectrometer analysis. Shark is part of a cluster of large rocks standing straight up from the ground, but leaning slight to the left. Scientists have theorized that these large, angular rocks may have been tilted by a huge flood which swept through Ares Vallis early in Mars' history.

Twenty-six megabits of data were received last night, on Sol 44 of the Pathfinder mission. The data included rover health status data, meteorological data and three additional sections of the "super" panorama.

Mars Pathfinder Mission Status July 8, 1997 3 p.m. Pacific Daylight Time

The first in-situ chemical measurements ever obtained of a rock on Mars — nicknamed Barnacle Bill for its rough, barnacle-like surface — surprised scientists and raised questions about the duration of volcanic activity occurring on Mars in its early formation.

Dr. Rudolph Rieder, of the Max Planck Institute for Chemistry, Germany, and principal investigator on the Alpha Proton X-Ray Spectrometer (APXS) team, reported that Barnacle Bill, an 8-to-10-inch tall rock near the Mars Pathfinder lander, was unusually rich in silicon, which is more characteristic of Earth rocks than Martian rocks. On Earth, volcanic rocks contain significant amounts of free silica in the form of quartz. The rich silicon content puts Barnacle Bill in one of the most common categories of volcanic rocks on Earth, known as "andesites."

"It turns out this rock has some rather peculiar chemical characteristics, which make it very unlike the other SNC meteorites," said Dr. Hap McSween, University of Tennessee, who is a participating scientist on the APXS team. (The SNC meteorites are those found on Earth that are believed to be of Martian origin.)

"In particular, it has a very high content of silicon or silicon dioxide (quartz)," McSween said. "It appears that Barnacle Bill falls into a category called 'andesites,' which are among the most common volcanic rocks on Earth."

Andesites are mixtures of very fine crystalline and other minerals that are formed through a process known as differentiation. Differentiation is the process by which crustal materials deep within a planet's interior are repeatedly melted and remelted, thereby shaping and reshaping the surface of the planet. Mars today has very few volcanoes and no continental plates like those found on Earth to suggest it was internally active for very long. Barnacle Bill's chemical signature may throw that theory into question.

Today's weather report was similar to yesterday's: at 3 p.m. local Mars time, it was about 5 degrees Fahrenheit, pressure was about 6.74 millibars, with very light winds out of the northwest.

"The weather on Mars is pretty boring," said Dr. Jeffrey Barnes, Oregon State University, who is a member of the atmospheric/meteorological experiment. "Northern summer in the subtropics on Mars is pretty much the same from day to day. Fifty or 60 days from now, we'll start to see dramatic changes with fall."

Atmospheric opacity — or how clear the sky is according to Pathfinder's atmospheric experiment — showed that Mars is moderately dusty up to about 40 kilometers (25 miles) above the surface. The dust appears to be uniformly distributed, and is expected to increase as Mars approaches its dusty season in the fall, Barnes said. The visibility on Mars was estimated to be about 32 kilometers (20 miles) or more, roughly equivalent to a moderately smoggy day in Los Angeles.

The rover's next task later today will be to perform a chemical analysis of the soil around a large rock named "Yogi." Once the soil measurements are taken, Sojourner will then back up to the left side of the rock and begin a chemical analysis using the APXS instrument.

On the fifth day of surface operations since Pathfinder's historic July 4 landing, all spacecraft and rover systems continue to operate extremely well. Pathfinder is returning data at an unprecedented rate of more than 8,500 bits per second and has returned 1,575 images of the Martian surface to date. A 360- degree, color panorama of the Ares Vallis landing site is expected to be released within the next few days.

Mars Pathfinder Mission Status July 9, 1997 3 p.m. Pacific Daylight Time

Six days after landing in an ancient outflow channel called Ares Vallis, the Mars Pathfinder lander and rover continue to operate extremely well, returning unprecedented amounts of data during daily downlink sessions.

Yesterday, Pathfinder returned 85 megabits of data on the Martian atmosphere, weather, soil and a rock called "Barnacle Bill," the first rock on Mars ever to be studied up close and personal. Additional rover and lander imaging was also returned.

Tonight the operations team will perform a low-gain antenna session from 6:30 p.m.- 7 p.m. PDT to acquire data on the health of the lander and rover. A three-hour high-gain transmission will begin later this evening, at 10 p.m. - 1:30 a.m. PDT, at the higher data rate.

The rover has completed its soil analysis of the smooth, moat-like terrain around a large boulder named "Yogi." After completing the analysis, the rover retracted the alpha proton X-ray spectrometer, then conducted a wheel abrasion experiment in which it dug into the soil and disturbed the crusty material as it was turning its wheels. This soil abrasion test is one of many technology and mobility experiments planned for the rover to help engineers understand soil dynamics on the Martian surface for future generations of rovers.

"We used the rover as sort of a bulldozer to push this rock and crusty material up," said Dr. Matthew Golombek, Pathfinder project scientist at a 1 p.m. PDT press briefing. "Next the rover moved slightly to the left and imaged Yogi with its front cameras, then turned around and imaged the lander with its rear camera. After that, the rover will photograph Yogi at close range. That data will be returned tonight."

Further preliminary analysis of "Barnacle Bill" showed that its texture seems to be consistent with volcanic "andesites," the second most common volcanic rock on Earth, said Dr. Jeff Johnson, of the U.S. Geological Survey in Flagstaff, AZ, who is on the Imager for Mars Pathfinder (IMP) camera team.

Scientists will use reflectance spectra collected by the lander and rover cameras to determine whether the rock, which measures about 40 centimeters (1.3 feet) across and 1.1 to 1.5 centimeters (8-to-10-inches) tall, is a sedimentary rock composed of many different rock fragments, or whether it is "homogenous," which would be consistent with scientists' first impression that it is a volcanic rock.

On a lighter note, Dr. Peter Smith, principal investigator of the IMP team, shared some of his personal insights on what it's like to be living on local Mars time, which means working on a 24-hour, 37-minute clock each day.

"When you say good morning, and the sun is setting, now that's living on Martian solar time…When your sunglasses start looking like this (holding up the red-and-blue stereo glasses used to view images in 3-D), that's living on Martian time…When you start admiring strange-looking rocks and giving them names, then telling your friends, that's living on Martian time…When your days are called Sols, and your nights are called days, that's living on Martian time…But when you start laughing at the engineers' jokes, you know you're living on Martian time."

Next on the rover's schedule of investigations are two rocks that appear white or very light in color: "Casper" and "Scooby Doo," located off to the left of the Pathfinder lander. Among the many images planned in the next week are shots of the Martian sunset and sunrise; pictures of the Martian moons Phobos and Deimos; and pictures of "Twin Peaks," two sloping hills that are about 800 meters (about half a mile) away from the landing site.

The next scheduled press briefing will be held at 12:30 p.m. Pacific Daylight Time on July 10 in JPL's von Karman Auditorium.

Mars Pathfinder Mission Status July 10, 1997 2 p.m. Pacific Daylight Time

Seven days into surface activities on the Mars Pathfinder mission, all spacecraft systems and instruments are continuing to perform well. The rover remains in excellent health and appears to be driving a little bit faster when left to its own devices than when it receives instructions from Earth.

"Basically the rover overshot its target rock, Yogi, by a little bit last night," explained Dr. Justin Maki, of the University of Arizona, who is a member of the Imager for Mars Pathfinder (IMP) team. Maki showed a movie of Sojourner as it approached the large boulder and began to climb up its side with one wheel. In this type of dead reckoning, the rover performed just as it should have, which was to back off the rock once it knew the rock was in the way, then turn and move away from the object. Although the rover travels about 1 centimeter per second (about 2 feet per minute), it appeared to be moving a little bit faster on its own.

The science team targeted the left side of Yogi for alpha proton X-ray spectrometer study because it appears to be dark and free of Martian dust. However, that side turned out to be tricky for the rover because of the rock's uneven contours and the slight depression in the soil beneath the rock. The rover team will instruct Sojourner to attempt instrument placement again tonight. Multiple attempts to position the science instrument were anticipated, making this repeat attempt nothing out of the usual.

The navigation team also announced the Ares Vallis landing site coordinates today as 19.33 degrees north latitude, 33.55 degrees west longitude.

Dr. Carol Stoker of NASA Ames Research Center showed some of the virtual reality products that her team is beginning to produce from the Pathfinder data during today's press briefing. Data from the lander camera's stereo images are overlain with terrain models to create the three-dimensional perspective, which can then be rotated in any direction on any plane on a computer screen. The 3-D perspective will be very useful to the science team in planning rover traverses and in analyzing data.

Dr. Julio Magalhaes, also of NASA Ames Research Center, a member of the atmospheric structure instrument/meteorology package (ASI/MET) on board the Pathfinder lander, reported that upper atmospheric temperatures are extremely cold. Temperatures at an altitude of 80 kilometers (50 miles) above the surface were 171 Celsius (minus 275 degrees Fahrenheit). In the lower atmosphere, between 60 km to 13 km (37 to 8 miles) above the surface, the temperatures are warmer and very close to those recorded by the Viking landers of the mid-1970s.

The next scheduled press briefing will be held at 12:00 Noon Pacific Daylight Time on July 11 in JPL's von Karman Auditorium.

Mars Pathfinder Mission Status July 11, 1997 11:15 a.m. Pacific Daylight Time

Commands for the next day of activities for Mars Pathfinder were not sent last night because the Pathfinder spacecraft's receiver had not been turned on in advance of the uplink session.

NASA's Deep Space Network conducts a routine frequency sweep before uplink sessions each day. The Goldstone, CA station initiated this sweep yesterday at about 1:35 p.m. PDT, when it came online. Because Pathfinder's receiver is only turned on at specific times each day to conserve power, it was not scheduled to be turned on until 1:46 p.m., an 11-minute miscalculation. Therefore the planned command link to the spacecraft was not established.

The operations team did not discover the problem until it was ready to begin its downlink session at 9:12 p.m. PDT last night. That 30-minute downlink would have been followed by a later downlink of data at 10:30 p.m. to 12:20 a.m.

The lost transmission session did not impact the mission in any way, except to delay the rover and lander activities. The operations team will retransmit the same set of sequences tonight during the 8 p.m. PDT session.

Activities planned for today will repeat the tasks not completed yesterday, including backing Sojourner down from Yogi and repositioning its science instrument against the rock. A full color panorama is also planned.

Meanwhile, all spacecraft and rover systems are performing well. Today is Sol 8 of the Mars Pathfinder mission.

Mars Pathfinder Mission Status July 11, 1997 2 p.m. Pacific Daylight Time

After determining Pathfinder's landing site coordinates yesterday, the Mars Pathfinder navigation team today reconstructed the spacecraft's novel entry, descent and landing at a 12 noon Pacific briefing. The team has been analyzing data acquired in the last week to come up with this preliminary landing profile.

Pathfinder was "right on the money," within a kilometer (6/10ths of a mile) of the target landing site, said Dr. Sam Thurman, one of the entry, descent and landing team members.

The spacecraft's terminal velocity as it parachuted to the ground was about 60 meters per second (134 miles per hour). An algorithm onboard the spacecraft that controlled the retro rockets recorded Pathfinder's speed at about 61.5 meters per second (140 miles per hour) at the time the RAD (rocket-assisted deceleration) rockets fired.

One issue of great importance to the Mars Global Surveyor team was Pathfinder's performance during descent, while it was subjected to the forces of the Martian environment. The Pathfinder navigation team reported that the spacecraft did indeed pick up some horizontal wind velocity on the order of about 13 meters per second (20 to 25 miles per hour), which was still well within the limits of the descent and landing design. That information will be very useful to the Mars Global Surveyor flight team when its spacecraft begins aerobraking through the upper atmosphere of Mars in order to circularize the spacecraft's orbit.

Pathfinder next fired its retro rockets at about 98 meters (321 feet) above the ground, just slightly higher than the 90-meter (295-foot) predicted elevation target, but also well within the parameters of the landing strategy. The 65-foot bridle was cut at about 21 meters (65 feet) above the ground, four seconds before impact.

Pathfinder's airbags — a new component of the spacecraft never before tested for a semi-hard landing — hit the ground at a speed of about 18 meters per second (40 miles per hour) and bounced horizontally across the landscape at about 12.5 meters per second (28 miles per hour). Pathfinder bounced about 15 meters (50 feet) high after impact, then bounced about 14 or 15 times more before coming to a stop. In all, the spacecraft bounced and rolled for about 2.5 minutes and traveled about 1 kilometer (6/10ths of a mile) before coming to a halt.

Activities for Sol 8 of Pathfinder's nearly flawless mission will include a set of commands to drive Sojourner off the large boulder, named Yogi, that it began to climb yesterday before automatically stopping itself. The rover team will send the rover new commands to reposition itself near the rock and attempt to place the alpha proton X-ray spectrometer against the rock again. The imaging team, meanwhile, released the famous "monster pan" today, a full 360-degree, three-dimensional color panorama of the Ares Vallis landing site.

Mars Pathfinder Mission Status July 12, 1997 12:45 a.m. Pacific Daylight Time

Flight controllers reestablished radio contact with the Mars Pathfinder lander tonight and repositioned the Sojourner rover after an initial communications session did not take place because the spacecraft's computer reset itself.

Beginning at about 3 p.m. Pacific Daylight Time Friday, the flight team sent commands to the Pathfinder lander instructing the Sojourner rover to back away from a rock nicknamed Yogi. The rover had been lodged against the rock since Wednesday evening; it was believed that commands sent Thursday to move the rover were not received by the lander because of an error in the timing of a radio uplink session.

The flight team received a signal from the Pathfinder lander's low-gain antenna at 6:47 p.m. Friday confirming that the lander received Friday's commands and was beginning to execute them. The team then expected to receive data over the lander's high-gain antenna beginning at 8:47 p.m. However, no signal was received at that time.

The team then sent a command to the Pathfinder lander via its low-gain antenna at about 9:45 p.m. instructing the lander to send a signal back to Earth. This signal was received. Flight controllers concluded that the lander's computer must have reset itself sometime between 6:47 and 8:47 p.m. PDT Friday evening. The commands to move the rover would not have been executed, because they were scheduled to take place later.

Commands were then sent at about 11 p.m. instructing the lander to point its high-gain antenna at Earth and begin a half-hour downlink session sending engineering data reporting on the status of the lander and rover. Commands were also sent to back the rover away from Yogi. Another command instructed the lander to take and transmit an image of the rover confirming that the repositioning had been completed.

At 12:15 a.m. PDT, the Deep Space Network station near Canberra, Australia, acquired a signal from the Pathfinder lander at the beginning of a half-hour downlink session sent via the lander's high-gain antenna. Data sent during this session indicated that the rover did in fact receive and execute commands to back it away from the Yogi rock and reposition itself on Thursday. It reexecuted these commands when they were resent Friday. In addition, an image was received at about 12:25 a.m. which showed the rover backed away from Yogi.

Friday evening marks the second time that the Pathfinder lander's computer has reset itself since its landing July 4. The flight team is not certain why the resets are taking place, but engineers noted that both incidents occurred during periods of heavy communication between the lander and rover.

Telemetry indicated that all spacecraft and rover systems are performing normally. Today is Sol 8 of the Mars Pathfinder mission.

Mars Pathfinder Mission Status July 13, 1997 7 a.m. Pacific Daylight Time

Mars Pathfinder's lander transmitted to Earth one-third of a sweeping color panorama image, as well as engineering data to help flight controllers fine-tune spacecraft operations, during the Martian day that just ended.

The flight team expects the lander to take and transmit image frames for the other two-thirds of the color panorama tonight. Much as landscape painters often work at the same time of day over a number of days, the exposures of the Martian panorama are divided over successive days so that lighting conditions are consistent in various portions of the 360-degree image.

Pathfinder's lander also transmitted 3 megabytes of engineering data, including all data available on why the lander's computer reset itself Friday evening. The flight team is continuing work to understand and correct the problem, but it appears to be related to the behavior of software that manages interaction between activities running on the computer. The team has already sent an initial set of commands aimed at preventing future resets.

The team also sent commands for the Sojourner rover, instructing it to stand by for another day. Mission managers wanted to receive the color image panorama and engineering data before resuming rover exploration of the landing site. The team will decide later today whether to initiate any rover activities tonight.

The downlink session began at 11:15 p.m. Pacific Daylight Time Saturday night and lasted about an hour. Earth set at the landing site was at 4:27 a.m. PDT, and sunset was at 6:56 a.m. — concluding Pathfinder's ninth day on Mars, or Sol 9.

Mars Pathfinder Mission Status July 14, 1997 10 a.m. Pacific Daylight Time

Mars Pathfinder's lander sent about an hour's worth of data to Earth last night — including portions of a 360-degree color panorama image — before the lander's computer appeared to reset itself, terminating the downlink session.

Engineers are continuing to debug the reset problem, which appears to be related to software that manages how the lander's computer handles a number of different activities simultaneously. "Saturday night, we 'serialized' activities by having the lander do one thing at a time, whereas last night the lander was handling a number of activities when the reset occurred," said Brian Muirhead, Mars Pathfinder flight system manager. "Tonight we will return to a 'serialized' approach to try to avoid the possibility of a reset." The reset occurred at 1:06 a.m. Pacific Daylight Time (PDT), about halfway through a planned two-hour downlink session.

Data received during last night's downlink session indicated that the Sojourner rover is positioned against the rock nicknamed Yogi, with its alpha proton X-ray spectrometer (APXS) instrument in place to study the rock's elemental composition. Data from the APXS should be received tonight; the science team will then decide whether to move the rover.

After rover data, the next highest priority for tonight is to complete the 360-degree color panorama image.

Last night's downlink session was during Sol 10, or Pathfinder's 10th Martian day. On Sol 10, Earth rise occurred at 3:27 p.m. PDT Sunday and sunrise was at 6:36 p.m. PDT. Earth set was at 5:07 a.m. Monday, followed by sunset at 7:35 a.m. Tonight the flight team expects to hear from the Pathfinder lander during downlink sessions beginning at about 9 p.m. and 11 p.m. PDT.

Mars Pathfinder Mission Status July 15, 1997 12 noon Pacific Daylight Time

The Mars Pathfinder flight team today reported on a very successful night of data transmission, receiving an unprecedented 90 megabits of data on the chemical makeup of a boulder nicknamed Yogi, atmospheric measurements and nearly all remaining portions of a 360-degree color panorama image of the landing site.

Last night's downlink sessions contained detailed information on the chemistry of Yogi taken by the rover after a second attempt to position its alpha proton X-ray spectrometer against the rock. The new data also included measurements of the aerosol content of the Martian atmosphere, which was used in parallel with new Hubble Space Telescope images of Mars to characterize changes in regional and global weather patterns in the last three weeks.

Recent incidents in which the Pathfinder lander's computer reset itself were discussed by Glenn Reeves, flight software team leader. According to Reeves, computer resets have occurred a total of four times during the mission — on July 5, 10, 11 and 14. The flight team has attempted to avoid future resets by instructing the computer to handle one activity at a time — "serializing" activities — rather than juggling a number of activities at once.

The team continues to troubleshoot the problem by testing all of the sequences leading up to reset in JPL's Mars Pathfinder testbed; considering changes in the flight software that would allow for immediate recovery if the flight computer were to reset itself; and modifying operational activities to minimize data loss if a reset should occur again. "In a sense, the reset itself is not harmful because it brings us back into a safe state," said Reeves. "But it does cause a disruption of the operational activities."

Among the science highlights, the Pathfinder mineralogy team presented new information about Barnacle Bill, a very roughly textured rock, and Yogi, a much larger boulder nearby, which was successfully measured last night.

Yogi, low in quartz content, appears to be more primitive than Barnacle Bill, "having not gone through the cooking that Barnacle Bill and other andesites have gone through," said Dr. James Greenwood, University of Tennessee, a member of the mineralogy science team. Although these observations are very preliminary, Yogi appeared to be more like the common basalts found on Earth. The next rock to be studied is "Scooby Doo," followed by others, including "Half Dome," "Wedge," "Shark" and "Flat Top," all located in a different region of the landing site. Some are near the lander petal on which Sojourner flew to Mars.

Observations from the Earth-orbiting Hubble Space Telescope revealed a lot of surface-atmospheric transport activity. A dust storm detected in Vallis Marineris just prior to Pathfinder's landing, for instance, had all but vanished within two weeks according to new Hubble images, noted Dr. Steven Lee, University of Colorado, a Hubble investigator. Some of the dust from that regional storm had diffused to the Pathfinder landing site, which was consistent with recent Pathfinder atmospheric opacity measurements.

In observations taken between May 18 and July 11, the amount of dust near the Pathfinder landing site had nearly tripled. "There's obviously a lot of very rapid transport going on here, with some of the dust diffusing toward the landing area," Lee said. "This is consistent with Pathfinder observations on the surface."

The increase in atmospheric dust appears to be diminishing the amount of cloudiness, Lee added. Clouds observed near the southern polar hood had begun to decrease in the most recent Hubble images as the dust diffused throughout the southern hemisphere. The Hubble team estimated that these clouds were relatively low, hovering around 15 to 16 kilometers (9 to 10 miles) above the surface, because the tips of some Martian volcanoes could be seen peeking through the cloud tops.

On Pathfinder's 11th Martian day — or Sol 11 — Earth rise at the Martian landing site was at 4:07 p.m. Pacific Daylight Time on Monday, July 14, followed by sunrise at 7:16 p.m. PDT. The flight team radioed commands to the Pathfinder lander beginning at 7:40 p.m. PDT. Data were downlinked from Pathfinder's lander from 9:02 to 9:35 p.m. PDT using the lander's low-gain antenna; this session included the spectrometer data on Yogi. A second downlink session, on the lander's high-gain antenna, began at 1:20 a.m. and ran until 5:10 a.m. PDT, with a half-hour break in the middle while the antenna was adjusted; this session included the new portions of the color panorama image. Earth set was at 5:46 a.m. and sunset was at 8:15 a.m. PDT.

Mars Pathfinder Mission Status July 16, 1997 9 a.m. Pacific Daylight Time

The Sojourner rover moved away from the rock nicknamed Yogi and headed toward a rock dubbed Scooby Doo during Mars Pathfinder's 12th Martian day, or Sol 12, which just concluded.

The rover moved a total of 3.6 meters (about 12 feet), and has about 2 to 3 meters (7 to 10 feet) to go to reach Scooby Doo. On the way between the two rocks, the rover's alpha proton X-ray spectrometer (APXS) instrument will be taking readings of Martian soil. The science team expects for the APXS instrument to take readings of Scooby Doo on Sol 14 — equivalent to Thursday night and Friday morning, July 17-18.

Also during the past Martian day, the lander's Imager for Mars Pathfinder (IMP) instrument captured pictures of a Martian sunrise, atmospheric opacity and the lander's windsocks. The imager is also expected to take pictures shortly of a sunset and Mars' moon Phobos.

During Sol 12, Pathfinder's lander sent a total of 65 megabits of data to Earth. All systems are functioning normally. On this Martian day, Earth rise was at 4:47 p.m. Pacific Daylight Time on Tuesday, July 15; sunrise was at 7:56 p.m. PDT; Earth set was at 6:26 a.m. PDT Wednesday, July 16; and sunset was at 8:54 a.m.

Mars Pathfinder Mission Status July 17, 1997 11 a.m. Pacific Daylight Time

Mars Pathfinder engineers reported a day of flawless operations of the lander and Sojourner Rover on Mars with the end of the mission's 13th day on Mars this morning, and also noted that they have found and are in the process of fixing a software bug that had caused the lander's computer to reset itself four times in recent days.

"The resets on the lander computer were caused by a software task that was unable to complete the task in the allotted time," said Flight Director Brian Muirhead. "We found that the task was being cut short because it had not been given high enough priority to run through to completion. Basically, we just need to add one instruction to the computer software to raise the priority of that task."

The problem was reproduced and isolated in testing at JPL. Further tests and verification will be completed today and tomorrow, with radio transmission of a software "patch" to change the lander's software scheduled for Saturday, Muirhead said.

Overnight, the Pathfinder team received all of the planned 58 megabits of data expected from the lander, along with the first of eight image sectors that will be combined to create a so-called "super-pan" high-resolution color panorama of the Martian terrain surrounding the spacecraft. The rest of the images will be transmitted back to Earth over the next several days.

A new "rover movie" created from time-lapse images taken by the lander was returned overnight. It shows Sojourner moving 2.5 meters (about 8 feet) and closing in on the whitish rock dubbed Scooby Doo. During the next Martian day, Sol 14, Rover drivers at JPL will bring the vehicle closer to the rock so Sojourner's alpha proton X-ray spectrometer can be placed against the rock.

On this Martian Day, Sol 13, Earth rise was at 5:27 p.m. yesterday, sunrise was at 9:35 p.m., Earth set was at 7:06 a.m. and sunset was at 9:33 a.m.

Mars Pathfinder Mission Status July 18, 1997 10:15 a.m. Pacific Daylight Time

The Pathfinder lander and Sojourner rover concluded their 14th successful day of operations on the surface of Mars today, JPL engineers reported.

Highlights of the scientific data returned overnight include a series of images that show Sojourner approaching the white rock Scooby Doo and deploying the rover's X-ray spectrometer instrument to study the rock's surface. Imaging data from the lander's camera also included a high-resolution view of the northernmost of the Twin Peaks seen on the horizon from the landing site.

Pathfinder engineers said all subsystems on both the lander and rover performed flawlessly and that no resets of the lander's computer were detected.

On this Martian Day, Sol 14, Earth rise was at 6:07 p.m. PDT yesterday, sunrise was at 9:15 p.m. PDT, Earth set was at 7:46 a.m. PDT and sunset was at 10:12 a.m. PDT. The day's total data return from the Mars station was 58 megabits.

Mars Pathfinder Mission Status July 19, 1997 10 a.m. Pacific Daylight Time

Last night's receipt of scientific data from Mars Pathfinder was delayed until tonight due to minor ground station problems that interfered with capturing all of Pathfinder's radio transmissions, mission engineers said today.

A short downlink opportunity of just 1½ hours and a problem with ground station computers combined to prevent most of Pathfinder's scientific data from being received last night. But engineering data from the rover and lander show that both remain in excellent health as they completed the first day of their third week on the surface of Mars.

"All the telemetry from the lander and rover continue to show that we have two very healthy spacecraft," said project manager Brian Muirhead. "We successfully completed the rover's seven-day prime mission and have finished the first week of its extended mission, and then we are half-way through the lander's 30-day prime mission. Everything looks good for continued operations with outstanding science return from both lander and rover," he said.

Last night's scheduled science data return will be retransmitted during the next Mars day, Sol 16, which begins tonight. Engineers also plan to send a new software patch to remove the software bug that had caused the lander's computer to reset itself earlier in the mission. The next downlink session is scheduled to include images of the Martian moon Phobos, along with observations of early morning fog, measurements of the rock Scooby Doo and images of various features around the lander.

Mission engineers said that overnight, Sojourner had successfully executed commands to move its wheels to scrape off the top layer of dust from the rock Scooby Doo. The rover's spectrometer was to have then repositioned its sensor to measure the newly revealed surface of the rock. The extended sensor head, however, apparently overshot the edge of the rock and did not make contact. Engineers will analyze data on the position of the rover and its spectrometer and plan to reposition the instrument tonight.

On this Martian Day, Sol 15, Earth rise was at 6:07 p.m. PDT yesterday, sunrise was at 9:55 p.m. PDT, Earth set was at 8:25 a.m. PDT and sunset was at 10:51 a.m. PDT. The day's total data return from the Mars station was 2 megabits.

Mars Pathfinder Mission Status July 20, 1997 1 p.m. Pacific Daylight Time

The Pathfinder mission operations team commanded the lander early this morning and did obtain a carrier signal over the high-gain antenna starting at 3:14 a.m. PDT for the normal period of about 66 minutes, but the signal strength was below expected levels and no scientific data was received.

"This told us that the spacecraft was basically healthy but that there was a problem with the telecommunications link," said Project Manager Brian Muirhead. A later attempt to communicate with the lander through its high-gain antenna from 7:03 to 7:27 a.m. PDT was not successful.

"The flight team is assessing the possible causes of the communication problem, said Muirhead. "This morning's problem may be related to some extent to configuration problems between the spacecraft and the Deep Space Network, but more data is needed to fully assess the problem. We are trying to troubleshoot a problem with very little information," he said.

The flight team is preparing sequences for a low-gain antenna communications session for about midnight tonight (July 20, Pacific Time). A communications session with the high-gain antenna is planned for about 3:30 a.m. PDT tomorrow, July 21.

"Since we have only limited windows to communicate with the spacecraft we must wait patiently for our next opportunity, Muirhead said. "We will go through the usual steps that have worked for us before, and then we will get to the bottom of the problem as we have before." The telecommunications problem is not thought to be related to the reset problem previously experienced by the lander's computer.

The rover remains safely at the rock called Scooby Doo. Earth will rise over the Sagan Memorial Station at 8:07 p.m. PDT today July 20, and sunrise will be at 11:15 p.m. Earth set is at 9:45 a.m. July 21.

An audio update on Pathfinder's status can be heard by calling 1-800-391-6654.

Mars Pathfinder Mission Status July 21, 1997 10 a.m. Pacific Daylight Time

The Mars Pathfinder flight team successfully reestablished contact with the Pathfinder lander and rover early this morning, completing several communications sessions using both the low-gain and high-gain antennas.

"What a difference a day makes," said Brian Muirhead, Pathfinder project manager. "The project team has successfully regained full communication capability on both the low-gain and high-gain antennas. The team is extremely pleased with our current status."

Most of the communications problem experienced over the weekend was associated with ground operations, not with the spacecraft on Mars, Muirhead said. "We'll be working to eliminate the cause of these problems in the coming days, as we return to a more normal mode of operations."

The flight team successfully initiated its first low-gain communications session of the Martian day at 10:38 p.m. Pacific Daylight Time on July 20, then began a second low-gain session at 1:36 a.m. July 21. Both sessions were returning data at the low data rate of 40 bits per second. At 3:22 a.m. PDT, the team conducted a third, brief low-gain session at a slightly higher data rate of 150 bits per second.

"All sessions worked perfectly, and we gained all of the basic engineering and telemetry data that had been stored onboard," Muirhead reported. "We verified that all spacecraft subsystems were healthy."

At 4:50 a.m. PDT, the team conducted a brief high-gain antenna session to make sure the high-gain antenna was pointed at Earth. A full high-gain antenna session at 8,200 bits per second was later performed beginning at 6:43 a.m. PDT. The team acquired all data on lander and rover health and completed acquisition of all of the spacecraft engineering data. They also sent a software update to correct sequences onboard the flight computer which have caused it to automatically reset itself.

Tonight's science activities will include downlinking measurements of a white-colored rock named Scooby Doo and continuing to acquire data from a full resolution color panoramic photograph of the landing site.

On this Martian day, Sol 17, Earth rose over the newly named Sagan Memorial Station at 8:07 p.m. PDT yesterday July 20. Sunrise was at 11:15 p.m. July 20 and Earth set occurred this morning at 9:45 a.m. July 21.

An audio update on Pathfinder's status can be heard by calling 1-800-391-6654.

Mars Pathfinder Mission Status July 22, 1997 12 Noon Pacific Daylight Time

Two-and-a-half weeks after landing in an ancient Martian flood basin known as Ares Vallis, Mars Pathfinder has fulfilled all of its primary science goals and continues to operate nearly flawlessly, the flight team reported at today's press briefing.

More than 300 megabits of data have been returned just in the last week, said Dr. Matthew Golombek, Pathfinder project scientist. The rover continues to follow an aggressive series of maneuvers to study rocks and soils identified by the science teams for their interesting features. In addition, the rover's wheel tracks and soil abrasion experiments are beginning to yield new information about the Martian soil, which appears to be finer than talcum powder.

Worldwide interest in the mission has peaked, with more than 400 million hits reported on the Internet today, said Kirk Goodall, Mars Pathfinder web engineer. Goodall and David Dubov, Mars Pathfinder webmaster, constructed 20 Pathfinder mirror sites prior to landing day to service the public. The most hits received in a single day — 46 million — occurred on July 8, Goodall said, which is more than double the number of hits received in a single day during the 1996 Olympic Games in Atlanta, Georgia.

A communications problem experienced last weekend has been resolved, reported Richard Cook, Mars Pathfinder mission manager. The problem was associated with ground operations, which has been required to reconfigure equipment and software on a daily basis, and the necessity of establishing communications links only during the short periods of time each day when the lander's transmitter is on.

Scientists are beginning to learn more about the Martian soil by studying the rover's wheel tracks, asking it to perform soil abrasion experiments and measuring the material properties of dust and soil through these wheel-soil interactions. Dr. Henry Moore, a rover scientist with the U.S. Geological Survey in Menlo Park, CA, likened the Martian soil to a very fine-grained silt that could be found in Nebraska. The Martian particles are less than 50 microns in diameter, which is finer than talcum powder.

Dust coverage on some of the spacecraft instruments is accumulating at a very low rate of about a quarter of a percent per day, added Dr. Geoffrey Landis, NASA Lewis Research Center in Cleveland, which is very close to the team's original predictions. These measurements also indicated that the dust was not moving toward the Martian poles right now. Additional study of dust patterns in the Martian environment may shed more light on the ways in which dust leaves the Martian atmosphere.

Dr. Peter Smith, University of Arizona, who is principal investigator of the lander camera, described more about the Martian landscape, pointing out a shallow riverbed crossing through the landing site and rocks in the distance that were washed into this outflow channel from the Martian highlands. About four distinct impressions left by the airbags were evident in the images presented today, noted Dr. Tim Parker, a science team member at JPL. The disturbed soil suggested that the spacecraft was nearly rolling, rather than bouncing, by the time it came to a stop. Parker estimated that the spacecraft bounced 15 to 20 times over a kilometer (6/10ths of a mile) of the landing site before stopping.

Science activities tonight will take the rover through the "cabbage patch," an area of soil in between Scooby Doo and a light-colored rock named Lamb. The rover will conduct a soil experiment, then turn and move toward Lamb. Scientists will take measurements of the dark soil near that rock before moving Sojourner close enough to place its spectrometer against the rock.

On this Martian day, Sol 18, Earth rose over the Sagan Memorial Station at 8:47 p.m. PDT yesterday, July 21. Sunrise was at 11:54 p.m. July 21 and Earth set occurred this morning (July 22) at 10:25 a.m. PDT.

An audio update on Pathfinder's status can be heard by calling 1-800-391-6654.

Mars Pathfinder Mission Status July 23, 1997 1:30 p.m. Pacific Daylight Time

The Mars Pathfinder lander and rover continue to operate flawlessly on the surface of Mars, 19 days after landing in an ancient outflow channel called Ares Vallis.

Pathfinder's 1-foot-tall roving geologist — named Sojourner — continues to collect data on crustal materials and rocks in the immediate vicinity to provide scientists with new information on the geology of this region. The Pathfinder lander, on the other hand, has become a virtual weather station, using its wind socks, wind sensors and image magnets around-the-clock now to profile the pressure, temperature, density and opacity of the Martian atmosphere.

Two downlink sessions were successfully completed by 11 a.m. today, using the 70-meter (230-foot) antenna of NASA's Deep Space Network facility in Madrid, Spain, reported David Gruel, Mars Pathfinder flight director for Sol 19. The flight team retrieved a total of 45 megabits of data over night, most of which was imaging data from the ongoing science experiments.

"The lander and rover are in excellent health and continue to operate flawlessly," Gruel said. "Meteorological data are being gathered around the clock."

First on Sojourner's list of activities tonight is a wheel abrasion experiment, in which the 10.5-kilogram (23-pound) vehicle will turn and dig some of its wheels into the fine Martian sand to measure material properties of the surface. Next the rover will position its alpha proton X-ray spectrometer face-down in the soil next to a rock called "Lamb" and make measurements of the rock's chemical composition.

On this Martian day, Sol 19, Earth rose over the Sagan Memorial Station at 9:30 p.m. PDT yesterday, July 22. Sunrise was at 12:30 a.m. July 23, and Earth set occurred at 11:04 a.m. PDT today.

Mars Pathfinder Mission Status July 24, 1997 2:30 p.m. Pacific Daylight Time

All communications sessions between the Pathfinder lander and rover were successfully completed today, one day short of the mission's three-week anniversary on the surface of Mars.

Sol 20 began when the Earth rose over Mars' horizon at 10:30 p.m. Pacific Daylight Time last night (July 23), enabling the flight team to initiate communications with the spacecraft. The Sun later rose at 1:15 a.m. PDT this morning, supplying the lander and rover with the energy needed to carry out specific tasks.

Communications were carried out using the 70-meter (230- foot) antenna of NASA's Deep Space Network facility in Madrid, Spain. Forty-seven megabits of data during two downlink sessions were returned on Sol 20.

The data indicated that both the lander and rover remain in excellent health and are continuing to operate masterfully. Flight Director Dave Gruel reported that no further flight software resets have occurred since the team sent modified flight software three sols, or days, ago.

Today's data included numerous images taken for ongoing science experiments. The Imager for Mars Pathfinder (IMP) also completed another section of the 12-color super panorama image of the landing site, then imaged the rover to add to an ongoing "rover movie" that is being assembled. IMP took a final, end-of-the-day photo of Sojourner following completion of its activities.

Sojourner traveled a total of 7/10ths of a meter (2.3 feet) today and performed another soil mechanics experiment that involved staging a "wheely." The last of its activities was to lower the alpha proton X-ray spectrometer onto the soil near the rock named Lamb. Presently, because it is night on Mars, the rover is powered down and using only its battery to operate the spectrometer and gather data on the Martian soil near Lamb. That data will be transmitted to Earth via the lander during the next Martian day, Sol 21, which begins when Earth rises over Mars tonight at 8:48 p.m. PDT.

Activities for Sol 21 will include another rover soil mechanics test, some more autonomous driving and repositioning of Sojourner's spectrometer against the side of Lamb in preparation for data-gathering the following night.

The lander's meteorological experiment reported highs today of minus 2 degrees Celsius (28 degrees Fahrenheit) and morning low temperatures of minus 73 degrees Celsius (minus 99 degrees Fahrenheit). The weather detectors also recorded large fluctuations of 3/10ths millibars in total pressure on the surface of Mars.

On this Martian day, Sol 20, the Earth set at 11:45 a.m. PDT, ending spacecraft communications with Earth for the day. The Sun set at 1 p.m. PDT.

Mars Pathfinder Mission Status July 25, 1997 1:30 p.m. Pacific Daylight Time

Mars Pathfinder celebrated its three-week anniversary on the surface of Mars today, with all spacecraft systems, science instruments and rover activities continuing to go exceptionally well.

On this Martian day, Sol 21, Earth rose at 10:48 p.m. PDT July 24 and Sunrise occurred at 1:53 a.m. PDT today.

The science team finished analyzing alpha proton X-ray spectrometer data from the rock nicknamed "Scooby Doo," the third rock measured by the rover since rolling off its ramp on July 5. ("Barnacle Bill" and "Yogi" were the first two rocks to be measured.) "Scooby Doo," of interest to scientists because of its light color, has a chemical signature very similar to other soils measured at the Pathfinder landing site. However, initial analysis shows that it contains slightly higher amounts of calcium and silicon.

Data returned during successful communications sessions last night indicated that the lander and rover remain in excellent health, reported Guy Beutelschies, Pathfinder flight director for Sol 21.

Sojourner performed a "self-guided" traverse today, receiving a minimum of instructions from Earth before driving off to find its own way to the next rock. Up until now, Sojourner has relied on detailed instructions and "way points," or X-axis and Y-axis coordinates, to find its way to the next rock target.

Today's 3-meter (10-foot) excursion, however, involved only two sets of way point instructions and an additional command to "find the rock." Sojourner used its own hazard avoidance system to locate the two way points, as it usually does, but then relied only on its laser light beams to find the next rock and line up with it. By 11 a.m. PDT, Sojourner had stopped just 25 centimeters (10 inches) in front of "Souffle," the next rock to be studied.

The rover will begin making measurements of "Souffle" on Sol 22, using its alpha proton X-ray spectrometer.

Meanwhile, atmospheric and meteorological data on the temperatures and density of the Martian atmosphere continue to be received during daily telecommunications sessions. Data stored onboard Pathfinder last week, while the flight computer was automatically resetting itself, were returned on Sol 21.

The lander camera snapped images of the disturbed soil near the rock called "Lamb," and photographed three more rocks: "Half Dome," "Shark" and "Pumpkin."

The Earth set today — Sol 21 — at 12:24 p.m. PDT. The Sun set at 2:46 p.m. PDT.

Mars Pathfinder Mission Status July 26, 1997 3 p.m. Pacific Daylight Time

The Mars Pathfinder lander and rover remain healthy and are continuing to carry out science experiments on this Martian day, Sol 22. The Earth rose over Mars at 11:28 p.m. PDT July 25. The sun rose today at 2:33 a.m. PDT.

Sojourner's self-guided journey to the rock "Souffle" was interrupted briefly by a software sequencing error, which was identified and corrected immediately. A sequencing error is easily corrected by modifying the numerical coding in the program responsible for executing the command, just as a computer user would modify coding in a program that runs the main menu or desktop functions of a personal computer.

"The problem was corrected immediately and a new sequence was radiated to the rover during the second downlink session," said Becky Manning, flight director for Sol 22. "By the end of that session, ground controllers had received confirmation that the rover had successfully received and was executing the instructions to continue its traverse to Souffle."

Sojourner will leave Souffle on Sol 23 and circumnavigate the lander. When that journey has been completed, the rover will be in the vicinity of three new rocks named "Baker Bench," "Desert Princess" and "Marvin."

The Mars Pathfinder lander imager (IMP) returned more data from the "insurance" panorama and "super" panorama today. It is preparing to take the standard end-of-the-day photograph of the rover before surface operations conclude in 30 minutes.

On Sol 22, the Earth set over Mars at 1:04 p.m. today. The sun will set in about 25 minutes, at 3:25 p.m. PDT.

An audio briefing on Mars Pathfinder activities is updated at the conclusion of each Martian day by calling 1-800-391-6654.

Mars Pathfinder Mission Status July 27, 1997 4:30 p.m. Pacific Daylight Time

All lander and rover systems and science instruments continue to operate well on Sol 23 of the Mars Pathfinder mission. Earth rise at the landing site occurred at midnight PDT July 26; sunrise followed at 3:13 a.m. PDT today.

In keeping with the tradition of playing wake-up songs for the space shuttle astronauts, the rover and flight teams were awakened on Sol 23 to the music of the Blues Brothers' version of "Raw Hide," a 1960s television western. The song was chosen to match "a long day of driving" for the rover.

Flight Director Jennifer Harris reported that start-of-the-day images showed the rover had begun to climb up the side of the rock named "Souffle," but was not able to position its science spectrometer against the rock. Consequently, no alpha proton X-ray spectrometer data were acquired today.

"However, this did not deter the rover from executing a long traverse which took it past the lander, through the 'Rock Garden' and past a rock named 'Casper,' before coming to a stop near the rocks 'Desert Princess' and 'Baker's Bench,'" Harris said. In all, Sojourner traveled six meters (nearly 20 feet) to complete the traverse, the longest excursion it has taken yet.

Images of the traverse, as well as routine beginning and end-of-day images, were taken by the Imager for Mars Pathfinder (IMP) camera. These images will go into a "rover movie," which is being compiled by the imaging team.

The IMP imaged sunrise on Mars, Phobos, one of Mars' two small moons, and the next portion of the super panorama, Harris said. The flight team also completed its downlink of the IMP stowed-position "insurance pan," which will enable them to begin downloading another portion of the super panorama.

Tomorrow's activities will include sending the rover to a way point beyond the rocks "Calvin" and "Hobbes." There it will be instructed to turn toward a rock named "Mini Matterhorn," take a picture of it and then image the lander.

On this Martian day, Sol 23, the Earth set at 1:43 p.m. PDT and the sun set at 4:04 p.m. PDT.

Mars Pathfinder Mission Status July 28, 1997 3 p.m. Pacific Daylight Time

The Mars Pathfinder lander and rover remain healthy and are continuing science experiments on the surface of Mars. The Earth rose over Mars on this Martian day — Sol 24 — at 12:48 a.m. PDT. The sun rose at 3:53 a.m. PDT.

The Imager for Mars Pathfinder (IMP) camera focused its lens on the sky today to photograph dust in the upper atmosphere and to search for clouds. IMP also imaged the wind socks onboard the Pathfinder lander to give scientists more information on wind direction and strength. Also included in today's photography session were images of Phobos, one of Mars' two small moons, and plans to image the Martian sunset later today.

The flight team continued to downlink data for the super panorama of the landing site, which is being assembled by the IMP team, said Flight Director Guy Beutelschies.

Sojourner was awakened this morning with the pop song "Radar Love," and executed a 7-meter (23-foot) traverse, the longest trip yet to be completed. The rover began its journey near the rock "Souffle" and ended it near the rock called "Mini Matterhorn." Next the rover imaged the rock and then the lander. Plans for tomorrow (Sol 25) call for more imaging of "Mini Matterhorn," after which the rover will begin a new traverse toward a rock called "Mermaid."

The Earth set at 2:23 p.m. PDT today and the sun will set at 4:43 p.m. PDT.

Mars Pathfinder Mission Status July 29, 1997 4:30 p.m. Pacific Daylight Time

Imaging the atmosphere of Mars — how clear or dusty it is and whether there are traces of water vapor — was the focus of science activities on the surface of Mars today.

The Mars Pathfinder imaging team also photographed the lander's wind socks, three small socks attached at different heights to a 1-meter mast. Visual images of these small socks provide scientists with information on wind strength and direction.

Temperatures on Sol 25 were typical, ranging from highs near minus 12 degrees Celsius (10 degrees Fahrenheit) and lows of minus 79 degrees Celsius (minus 110 degrees Fahrenheit). Today the Earth rose over Mars at 1:28 a.m. PDT and the Sun rose at 4:32 a.m. PDT.

The Atmospheric Science Instrument/Meteorology Package (ASI/MET) instrument team reported a very successful day of data return, said Flight Director Jennifer Harris, receiving more information than ever before on the pressure of the Martian atmosphere. Also included in the downlink sessions was more imaging data for the high-resolution "super panorama" of the landing site. In all, a total of 48 megabits of data was successfully returned.

A sequencing transmission error prevented the rover from executing its daily traverse, Harris said. The situation was quickly corrected and the rover was able to complete an accelerometer diagnosis sequence, which involved making a 120- degree turn in place. Sojourner will complete its traverse to the rock nicknamed Mini Matterhorn tomorrow and then turn to image the lander.

The Earth set today at 3:03 p.m. PDT and the Sun will set at 5:22 p.m. PDT.

Mars Pathfinder Mission Status July 30, 1997 4:30 p.m. Pacific Daylight Time

Pathfinder's 10.5-kilogram (23-pound) rover called Sojourner stalled today during the last stretch of its journey to a rock nicknamed Mermaid, but was quick to recover and prepare for completion of the traverse tomorrow.

Data returned this morning from the Sagan Memorial Station indicated that the rover's left front wheel stalled during the third of four waypoint maneuvers. Waypoints are navigational instructions — consisting of x- and y-axis coordinates — used by the rover to travel from one rock to another. To complete today's traverse to the rock nicknamed Mermaid, for instance, the rover had to make four short trips based on four sets of waypoint coordinates. Its wheel jammed during the third segment of the journey.

"This stall was probably caused by a small rock becoming jammed in one of the rover wheel's cleats," said Flight Director David Gruel. "Once the rover detected the stall on its own, she was able to autonomously clear the problem by backing up a short distance. Since the stall did not exist after the backup was performed, there's a high probability that Sojourner is ready to continue the drive around the lander tomorrow."

Approximately 55 megabits of engineering and science data were returned today. All data indicated the lander and rover are healthy and the lander's battery continues to power the craft through the subfreezing nights on Mars. The rover returned a "spectacular" image today of the rear of the lander and the rock nicknamed Mini Matterhorn.

Temperatures on Mars today ranged from a balmy minus 13 degrees Celsius (8 degrees Fahrenheit) at 5:35 p.m. local solar time to minus 79 degrees Celsius (minus 105 degrees Fahrenheit) at 5:30 a.m. local time. Winds were light and from the west.

On this Martian day, Sol 26, Earthrise occurred at 2:09 a.m. PDT and sunrise followed at 5:12 a.m. PDT. The Earth later set at 3:43 p.m. PDT and Pathfinder will observe its 26th sunset at 6 p.m. PDT.

Mars Pathfinder Mission Status July 31, 1997 4:30 p.m. Pacific Daylight Time

The Mars Pathfinder flight team has completed all of its science and engineering goals, four days before the primary mission draws to a close, said Dr. Matthew Golombek, Pathfinder project scientist, at today's press briefing.

Atmospheric-surface interactions were the focus of today's presentation. To set the stage, Dr. Mark Lemmon, a member of the Imager for Mars Pathfinder (IMP) camera team from the University of Arizona, presented new images of the Martian sunrise and sunset. True to color, the dawn images revealed pale pink sunrises and clouds floating overhead. The reddish tint is the result of Martian dust, composed of oxidized iron, which is present in the atmosphere. The sunset images — color-enhanced to bring out structural detail in the atmosphere — showed a sky darkening to salmon-colored hues.

These spectacular images of the Martian summer are possible by return of an unprecedented amount of science and engineering data — on the order of 400 megabits just in the last nine days — Golombek pointed out.

Temperature highs and lows at the landing site have not varied much, said Dr. Robert Haberle, a participating scientist from the NASA Ames Research Center. They range from highs of about minus 12 degrees Celsius (8 degrees Fahrenheit) to lows near minus 76 Celsius (minus 105 Fahrenheit). Frozen water-ice clouds are evident in the Martian sky during the early morning hours, but evaporate once temperatures rise. "We expect late night and early morning clouds, but we expect those clouds will burn off fairly rapidly with sunrise, giving way to a dusty Martian day," Haberle said. Although there has not been much variation in these weather conditions since Pathfinder arrived, they are expected to begin changing in about a month, as fall arrives and ushers in the dust storm season.

Atmospheric pressures, on the other hand, are fluctuating dramatically, sometimes peaking two, three or four times a day, Haberle noted. Pressure oscillations are indicative of a global scale thermal tidal system that is moving dust, water-ice or vapor clouds and other volatiles through the atmosphere. On Mars, these atmospheric variations are sizable, whereas on Earth they almost never occur.

Since data-gathering began, the maximum change in pressure over the course of a day has been 0.3 millibars, which is about 4.5 percent of the average pressure on Mars. On Earth, pressures that low might occur during a severe hurricane. A better understanding of these pronounced pressure oscillations will help scientists understand the processes by which volatiles enter and escape the Martian atmosphere, and may shed more light on the rise of regional and global dust storms.

Wind speeds have been increasing with altitude, reported Dr. Robert Sullivan of Arizona State University. And temperatures will vary dramatically with elevation. When ground temperatures are 16 to 21 degrees Celsius (60 to 70 degrees Fahrenheit), they can drop to minus 23 to 27 degrees Celsius (minus 10 to minus 15 degrees Fahrenheit) just five and a-half feet above the ground.

Images of the Martian landscape also revealed a shiny object about 1,200 meters (7/10ths of a mile) away from the lander. Dr. Michael Malin, a participating scientist, said the object is about the same dimensions and is probably the spacecraft's discarded backshell, which separated just before the spacecraft landed.

Although the Pathfinder lander and rover remain healthy, engineers plan to recharge the lander's battery during a two-day hiatus beginning Sunday, Aug. 3. The lander will perform some science experiments during the day, but will use most of its solar energy to charge the battery. At night, the craft essentially goes to sleep.

The rover will continue its daily traverses and spectrometer studies, rolling off to a smooth, dark region of soil called Mermaid Dune tomorrow. After taking measurements of the soil, scientists will identify one of three large, dust-free rocks — Shark, Half Dome and Wedge — as the next target for study.

On this Martian day, Sol 27, Earthrise occurred at 2:49 a.m. PDT and sunrise followed at 5:52 a.m. PDT. The Earth later set at 4:23 p.m. PDT and the sun set at 6:41 p.m. PDT.

Mars Pathfinder Mission Status August 1, 1997 6 p.m. Pacific Daylight Time

A wealth of new information about Martian weather and atmosphere was acquired by the Pathfinder lander on the 27th and 28th days of surface operations in the Ares Vallis outflow channel on Mars.

Data are showing that daily temperatures do not vary much, but minute-by-minute temperatures can fluctuate dramatically, reported Flight Director Rob Manning. Atmospheric pressures can also change significantly within a matter of minutes. Scientists think some of these variations may be caused by small dust devils that can be whipped up by a gust of wind.

The flight team has been taking advantage of longer downlink sessions over the last several Martian days to acquire as much weather data as possible. Regular weather measurements will be limited in the days ahead due to the lander's limited battery power. A two-day shutdown of lander operations to recharge the battery will occur on Sunday and Monday, Aug. 3-4.

Summer temperatures remain in the same range. This morning's low was minus 75 degrees Celsius (minus 103 degrees Fahrenheit), while highs rose to minus 15 degrees Celsius (5 degrees Fahrenheit). Winds were blowing lightly from the south and swinging around from the northwest during the day.

The rover continued its trek south today after undergoing a temporary delay. Yesterday the vehicle detected a jam in its left front wheel and autonomously backed up to free the pebble from the cleats in its stainless steel wheels. Today Sojourner marched about 4 meters (13 feet) to the south.

"Using waypoints specified the day before, Sojourner stopped, spun about and carefully backed up and onto a small dune named Mermaid today," Manning said. "This afternoon she continued to back up onto the dune and then lowered the alpha proton X-ray spectrometer onto the dune. Tonight the rover will collect elemental spectra of this interesting feature."

The lander and rover returned more than 60 megabits of science and engineering data on Sols 27 and 28.

"This data included yet another rover traverse movie and a series of photometric strips that will allow scientists to precisely gauge the optical properties of key features near the lander," Manning added.

On this Martian day, Sol 28, Earthrise occurred at 3:29 a.m. PDT and sunrise followed at 6:32 a.m. PDT. The Earth set at 5:02 p.m. PDT and sunset occurred at 7:20 p.m. PDT.

Mars Pathfinder Mission Status August 2, 1997 5 p.m. Pacific Daylight Time

Mars Pathfinder lander and rover operations were curtailed today when the lander downlink session, scheduled to begin at 1:20 p.m. PDT, was not initiated. The cause of the problem is currently unknown, said Mars Pathfinder Mission Manager Richard Cook.

The flight team, led by Flight Director Carl Steiner, subsequently reestablished communications with the lander and received a brief carrier blip at 3:30 p.m. PDT. Additional downlink sessions were not attempted, however, because of the lack of time before the Earth set over the landing site at 5:42 p.m. PDT.

Data on the health of the lander and rover, in addition to other engineering telemetry, will be acquired tomorrow. The flight team expects Sol 30 operations to proceed normally.

Additional information will be posted on the JPL home page at http://www.jpl.nasa.gov as it becomes available. An audio update is also available by calling 1-800-391-6654.

On this Martian day, Sol 29, Earthrise occurred at 4:09 a.m. PDT and sunrise followed at 7:11 a.m. PDT. The sun set over the landing site at 7:59 p.m. PDT.

Mars Pathfinder Mission Status August 3, 1997 7 p.m. Pacific Daylight Time

Although the reason for yesterday's loss of downlink opportunities has not yet been identified, science activities proceeded normally today on the surface of Mars. Today, Sol 30, marks the end of the Mars Pathfinder primary mission, 30 days after the spacecraft landed in an ancient outflow channel called Ares Vallis.

The Imager for Mars Pathfinder (IMP) continued to image the thin Martian atmosphere, the lander's wind socks, the Sun and the rover as it roamed to another destination, said Carl Steiner, Mars Pathfinder flight director. Acting as a weather station, the Pathfinder lander — now called the Sagan Memorial Station — gathered weather data for the 12th consecutive day. Data from yesterday's surface operations had been stored onboard the lander and were downlinked today. Highs on Mars rose to minus 10 degrees Celsius (14 degrees Fahrenheit) today and dipped to minus 70 degrees Celsius (minus 94 degrees Fahrenheit).

The rover finished its soil analysis of Mermaid Dune before heading toward the Rock Garden. An onboard tilt protection circuit caused the rover to shut down after reaching 10 centimeters (0.3 feet) of motion.

"An especially noisy accelerometer had caused this problem in past, but had successfully guarded the rover against excessive tilt," Steiner said. "The rover team thought it prudent to activate this device, even with the possibility of inadvertent shut-down, because of the uneven path to the Rock Garden and the long traverse." Sojourner will resume this traverse on Sol 32, as Pathfinder's extended mission gets under way.

Three downlink sessions were successfully carried out today using the low-gain antenna once and the high-gain antenna for the next two sessions, Steiner said. The operations team, however, was unable to complete its planned downlink of an eighth (octant) of the so-called "super pan" before the end of the day and the beginning of a two-day sleep period for the lander.

"This will be the first time in nearly 240 days that the lander's electronics will be powered off," Steiner said. "At the conclusion of today's activities, all lander electronics, with the exception of a few computer chips that comprise the hibernate circuit, will be powered off to conserve energy through the Martian evening and prolong our waning battery."

The hibernate circuit has been programmed to wake up the lander at 7:30 a.m. local solar time tomorrow. A backup circuit will wake the lander at 8 a.m. if the lander is still asleep. Tomorrow's activities will focus on recharging the battery to the fullest capacity possible. No science experiments are planned.

On this Martian day, Sol 30, Earthrise occurred at 4:49 a.m. PDT and sunrise occurred at 7:51 a.m. PDT. The Earth set over the landing site at 6:22 p.m. PDT and the sun set at 8:39 p.m. PDT.

Mars Pathfinder Mission Status August 18, 1997 4 p.m. Pacific Daylight Time

Daily communications with the Mars Pathfinder lander and rover have resumed after an interruption on Saturday, Aug. 16, that was caused by an automatic reset of the lander's flight computer. The cause of the reset is not known, said Mars Pathfinder Flight Director Rob Manning.

The flight team was able to reactivate the lander by sending it instructions to reinitialize high-gain pointing and then begin its scheduled downlink session on Sunday evening. The downlink session began right on time at 10:09 p.m. PDT Aug. 17.

New images indicated that the rover had stopped its traverse to a rock nicknamed Shark after partially climbing up a rock called Wedge. The rover's hazard avoidance software is designed to stop the vehicle when it begins to tilt too much. Wedge is one of many small rocks forming a gateway to the Rock Garden, which is Sojourner's next destination. This portion of the landing site is more challenging than other regions because it is much rockier than any terrain explored to date by the rover.

The flight team will instruct Sojourner to continue its trek to Shark tonight. Shark is of interest to scientists because it is a large, smooth rock, which is relatively dust-free and, therefore, an ideal candidate for the next spectrometer analysis. Shark is part of a cluster of large rocks standing straight up from the ground, but leaning slight to the left. Scientists have theorized that these large, angular rocks may have been tilted by a huge flood which swept through Ares Vallis early in Mars' history.

Twenty-six megabits of data were received last night, on Sol 44 of the Pathfinder mission. The data included rover health status data, meteorological data and three additional sections of the "super" panorama.

Mars Pathfinder Mission Status August 27, 1997

Images of the Martian sunrise and sunset, with water ice clouds floating through the atmosphere, were unveiled today at a Mars Pathfinder press briefing, held on Sol 53 of the mission, at NASA's Jet Propulsion Laboratory.

Today's collection of photographs included one portion of the super panorama view looking to the north-northeast from the Sagan Memorial lander. The super panorama of the landing site, which is being constructed from high resolution color images taken by the Imager for Mars Pathfinder (IMP) instrument, will be comprised of about 3,000 images when it is completed in about eight weeks, said Dr. Matthew Golombek, Pathfinder project scientist at JPL. This mammoth color and stereo data set, which is now about 65 percent finished, will be used to derive high quality topographic maps of the Martian surface and detailed shapes of rocks and other surface features. Scientists will also be able to examine subtle chemical, mineralogical and textural variations in rocks and soils from this panorama.

Temperatures on Mars today remained in roughly the same temperature range. Today's low was minus 75 degrees Celsius (-103 degrees Fahrenheit) and the high was minus 10 degrees Celsius (14 degrees Fahrenheit). The highest pressure measurements seen yet on Mars were recorded yesterday (Sol 52) at 6.8 millibars, said Dr. Tim Schofield, atmospheric structure/meteorology package team leader at JPL. In addition to temperature and pressure measurements, Pathfinder has observed a total of 12 dust devils, small swirls of dust kicked up by winds blowing down through the canyons in the Ares Vallis landing zone.

Scientists are finding that Martian temperatures are cooler at higher altitudes (about 80 kilometers or 50 miles) than on the ground. They think that ice clouds forming about 10 kilometers to 15 kilometers (6 to 9 miles) above the surface are responsible for this cooling trend higher in the atmosphere. The clouds are thought to be made of very small ice particles, about one-tenth the size of Martian dust or one-thousandth the thickness of a human hair.

Dr. Mark Lemmon, a member of the lander camera imaging team at the University of Arizona, noted color variations in some of the sunset pictures. The blue color is not caused by clouds of water ice but by Martian dust in the atmosphere, Lemmon said. The dust absorbs blue light, giving the sky its red color, but it also scatters some of the blue light into areas that looked very blue around the Sun. The blues only show up near sunrise and sunset, when the light has to pass through the largest amount of dust.

Sojourner, which remains in excellent health, began exploring the Rock Garden yesterday, after spending about a week en route to the region. The Rock Garden is an assemblage of several large boulders and many smaller rocks near the lander. After conducting a chemical analysis yesterday of the rock nicknamed Shark, the rover moved toward another rock called Half Dome today, but climbed too high up on the rock and automatically shut itself off.

Tomorrow the rover team will instruct the rover to back down the rock and reposition the alpha proton X-ray spectrometer against the side of Half Dome. Chemical analyses of all of the rocks studied so far indicate that at least two types of rocks are present in the Pathfinder landing zone: those with high levels of silicon and those with high levels of sulfur, reported Dr. Tom Economou, co-investigator of the alpha proton X-ray spectrometer team at the University of Chicago.

Soil mechanics experiments using the rover's wheels and cleats to dig below the surface have revealed different layers of material, Howard Eisen, principal investigator on the soil mechanics technology experiment at JPL, pointed out. Soil surfaces differ near the lander, where the soil contains a mixture of pebbles, fine-grained sand and clods, from regions a bit farther out. There, the surface is covered with a bright drift material, Eisen said. Using the rover's cleats to dig below the surface, scientists have discovered that cloddy material was present underneath the drift.

After traveling a total of about 80 meters (263 feet) around the landing site, Sojourner will continue to explore the Rock Garden for the next several days, taking as many chemical analyses as possible of the large boulders in the vicinity. After these rocks have been studied, the rover will head back to the ramp on which it exited the lander and study a dust sample that has been accumulating on a magnet, Golombek said. This study may provide new information about magnetic properties that might be present in the Martian soil. Longer range plans for the rover may take it much farther away from the lander, so that it may peer over the rim of what appears to be a shallow riverbed, and photograph a region that cannot be seen by the lander.

The Earth rose over Mars on Sol 53 at 8:15 p.m. Pacific Daylight Time and the Sun rose at 11:05 p.m. PDT yesterday (Aug. 26). The Earth set this morning at 9:55 a.m. PDT and the Sun set at 11:35 a.m. PDT.

Images and comprehensive updates on Pathfinder science results are available on the Internet at http://mpfwww.jpl.nasa.gov or via JPL's home page at http://www.jpl.nasa.gov/marsnews/ . Daily audio updates are also available by calling 1-800-391-6654.

Mars Pathfinder Mission Status October 1, 1997

After experiencing difficulties in communicating with the Mars Pathfinder spacecraft for the past three days, the operations team was able to reestablish a brief two-way communications session Tuesday using the lander's auxiliary transmitter. Receipt of this beacon signal indicated that the spacecraft is still operational.

The team began having communications problems with the spacecraft on Saturday, Sept. 27. These problems could be related to degradation of the spacecraft's battery. The last successful data transmission cycle from Pathfinder was completed at 3:23 a.m. Pacific Daylight Time on Sept. 27, which was Sol 83 of the mission.

No signal was received from the spacecraft on the next Martian day, Sol 84, which began in the evening of Sept. 27. The team's transmission session began at 11:15 p.m. PDT. The lack of a signal, at that time, was thought to be caused by a possible computer reset incident, ground system problem or low voltage condition. A reset or a low voltage condition, caused by the aging of the battery, would cause the spacecraft sequence to automatically stop and not execute its planned communication with Earth.

The team attempted to communicate with the spacecraft again on Sept. 29 (Sol 85) and Sept. 30 (Sol 86) with no success.

Tonight, on Sol 88 of the mission, the team will use the auxiliary transmitter again to attempt to acquire engineering data that will help them assess the cause of the communications problem. Meanwhile, the rover, which receives its instructions from Earth via the lander, is currently running a contingency program which has instructed it to stand still rather than begin its trek around the lander.

The team will repeat these activities on subsequent days and attempt to receive telemetry that will give them more information about the health of the lander and rover.

If Pathfinder operations do not return to normal tonight, a Mars Pathfinder team representative will provide an update on the situation at the beginning of the planned Mars Global Surveyor science news briefing at 9 a.m. PDT on Thursday, Oct. 2.

Mars Pathfinder Mission Status October 7, 1997

The Mars Pathfinder operations team reestablished communications with the lander today, on Sol 92 of the mission, after four days of silence from the spacecraft. The team received a transmission from the spacecraft's main transmitter. The signal was detected using the Madrid, Spain 34-meter antenna.

No data was received, but receipt of a spacecraft signal indicates that the lander is operational and the battery is off-line. Meanwhile, the rover, which is programmed to begin a contingency sequence when it has not heard from the lander for five days, started that activity on Sol 90. In this mode of operation, the rover is instructed to return to the lander and begin circling it.

The Mars Pathfinder operations team will repeat commands tomorrow night, on Sol 93, to verify two-way communications with the lander's main transmitter and attempt to return engineering data on the health of the lander and rover. If successful, that information would be returned the following day, on Sol 94 of the mission.

Contact: Diane Ainsworth

FOR IMMEDIATE RELEASE: October 9, 1997

PATHFINDER TEAM PAINTS AN EARTH-LIKE PICTURE OF EARLY MARS

Mars is appearing more and more like a planet that was very Earth-like in its infancy, with weathering processes and flowing water that created a variety of rock types and a warmer atmosphere that generated clouds, winds and seasonal cycles.

Those observations, along with new images taken by the Mars Pathfinder rover and lander, and an update on the condition of the spacecraft, were presented at an Oct. 8 press briefing originating from NASA's Jet Propulsion Laboratory.

"What the data are telling us is that the planet appears to have water-worn rock conglomerates, sand and surface features that were created by liquid water," said Dr. Matthew Golombek, Mars Pathfinder project scientist at JPL. "If, with more study, these rocks turn out to be made of composite materials, that would have required liquid water flowing on the surface to round the edges in pebbles we see on the surface or explain how they were embedded in larger rocks. That would be a very important finding."

Golombek also stressed the amount of differentiation — or heating, cooling and recycling of crustal materials — that appears to have taken place on Mars. "We're seeing a much greater degree of differentiation — the process by which heavier elements sink to the center of the planet while lighter elements rise to the surface — than we previously thought, and very clear evidence that liquid water was stable at one time in Mars' past.

"Water, of course, is the very ingredient that is necessary to support life," he added, "and that leads to the $64,000 question: Are we alone in the universe? Did life ever develop on Mars? If so, what happened to it and, if not, why not?"

Despite recent communications problems with Earth, the Mars Pathfinder lander and rover are continuing to operate during the Martian days, when they can receive enough energy to power up spacecraft systems via their solar panels. The mission is now into Sol 94, or the 94th Martian day of operations, since landing on July 4.

"Everything that we have seen over the last 10 days (with respect to communications) is like a twisty little maze with passages all alike," said Jennifer Harris, acting mission manager. "I am happy to report that we have made contact with the spacecraft using its main transmitter. We were able to confirm that we could send a command to the spacecraft to turn its transmitter on and then turn it off.

"We don't know yet whether we are receiving that signal over the low-gain or high-gain antenna," she added, "but we should be able to determine this over the next few days."

The Mars Pathfinder team began having communications problems with the spacecraft on Saturday, Sept. 27. After three days of attempting to reestablish contact, they were able to lock on to a beacon signal from the spacecraft's auxiliary transmitter on Oct. 1, which meant that the spacecraft was still operational.

At that time they surmised that the communications problems were most likely related to depletion of the spacecraft's battery and uncertainties in the onboard clock. The last successful data transmission cycle from Pathfinder was completed at 3:23 a.m. Pacific Daylight Time on Sept. 27, which was Sol 83 of the mission.

Since then, efforts have been made during each Martian day to reestablish contact with both the primary and auxiliary transmitter and obtain engineering telemetry that would tell the team more about the health of the lander and rover. On Oct. 7, the team was able to lock on to Pathfinder's signal, via NASA's Deep Space Network 34-meter-diameter (112-foot) dish antenna in Madrid, Spain, for about 15 minutes, using the main transmitter. However, in repeating the process on Oct. 8, they did not receive a signal.

The rover, which receives its instructions from Earth via the lander, is currently running a contingency software program that was preprogrammed to start up if the vehicle did not hear from the lander after five Martian days. That program was powered on Oct. 6, on Sol 92 of the mission. In this contingency mode, the rover is instructed to return to the lander and begin circling it. This precaution is designed to keep Sojourner close to the lander in the event that the spacecraft was able to begin communicating with it again.

If normal communications are reestablished, the rover team will send new commands to Sojourner to halt the contingency circling and begin a traverse to a specific location.

Dr. William Folkner, an interdisciplinary scientist at JPL, presented data on the rotation and orbital dynamics of Mars, which are being obtained from two-way ranging and Doppler tracking of the lander as Mars rotates. Measurements of the rate of change in Mars' spin axis have important implications for learning more about the density and mass of the planet's interior. Eventually, scientists may be able to determine whether Mars' core is presently molten or fluid. The size of the core also can be used to characterize the thickness, or radius, of Mars' mantle.

"By measuring the spin axis of Mars, we can learn something about the interior of the planet, because the speed of the change in its orientation is related to how the mass is distributed inside," Folkner said. "If the core is fluid, its spin and the way in which the planet wobbles slightly will be different from the spin and wobble of a planet with a solid core.

"If Mars' core is solid, then it can't be less than about 1,300 kilometers (807 miles) in radius, out of the planet's total radius of 3,400 kilometers (2,112 miles)," Folkner added. "If the core is made up of something less dense than iron, if it's a mixture of, say, iron and sulfur, then the core would be bigger, but it couldn't be bigger than about 2,000 kilometers (1,242 miles) in radius."

New close-up images of dunes around the landing site are showing some scientists clear evidence that there is sand on the surface of Mars. Identification of sand, as opposed to dust or pebbles, is a significant factor in establishing that weathering processes such as erosion, winds and flowing water all contributed to Mars' present landscape.

"We've made significant progress in establishing that water was a dominant agent in forming the surface, and now we can say that there is another agent at work, and that is the wind, that has created and modified some of the landforms on a smaller and medium scale," said Dr. Wes Ward of the U.S. Geological Survey, Flagstaff, AZ, a member of the Imager for Mars Pathfinder team. "And because the water is no longer there, wind probably is the dominant agent shaping the Martian surface at this moment."

Ward showed images of Ares Vallis, taken by the rover and Viking 1 orbiter images to point out the structural difference in these surface features. While Viking 1 surface features around a rock nicknamed "Big Joe" showed drifts, the dune-like surfaces in the Ares Vallis flood basin resemble sand that has been blown southwest over the landing site. The presence of sand also points to the likely presence of liquid water, needed to create these small, 1-millimeter-diameter granules, and weathering agents such as wind to blow them into small ridges and moats present around the Ares Vallis rocks.

"The wind is quite an active agent," Ward said. "Sand is the smoking gun, and as far as I'm concerned, the gun is smoking and has Colonel Mustard's prints all over it. We are seeing sand at the landing site."

Dr. Greg Wilson, of Arizona State University, who is on the Pathfinder atmospheric experiment team, reported increases in the pressure of the Martian atmosphere and a drop in surface temperatures.

"We expect to see a continued increase in pressure and decrease in temperatures as the dust season approaches and winds begin to lift more dust into the Martian atmosphere," he said. "The dust season on Mars usually begins in the next few weeks."

Additional information, images and rover movies from the Mars Pathfinder mission are available on JPL's Mars news media web site at http://www.jpl.nasa.gov/marsnews or on the Mars Pathfinder project's home page at http://marsweb.jpl.nasa.gov . Images from Mars Pathfinder and other planetary missions are available at NASA's Planetary Photojournal web site at http://photojournal.jpl.nasa.gov.

The Mars Pathfinder mission is managed by the Jet Propulsion Laboratory for NASA's Office of Space Science, Washington, DC. The mission is the second in the Discovery Program of fast-track, low-cost spacecraft with highly focused science goals. JPL is a division of the California Institute of Technology, Pasadena, CA.

Mars Pathfinder Mission Status October 16, 1997

The Mars Pathfinder operations team is continuing its efforts to attempt to send commands to the Pathfinder lander while, at the same time, investigating possible scenarios to explain what might be occurring onboard the spacecraft. The last signal received from the spacecraft was on Sol 93, which was Tuesday, October 7, at 7:21 a.m. Pacific Daylight Time.

There is no indication at this time that the spacecraft is no longer operating. The difficulty in communications is thought to be related to the degradation of the spacecraft battery. In the "no battery" mode of operations, the spacecraft cannot keep track of time accurately and will also be powered on for a smaller portion of each day. As time progresses, spacecraft hardware will become colder. In regular operations, by turning on the transmitter, the spacecraft hardware warms up sufficiently to operate normally. It is possible that because the team has not been able to turn on the transmitter for long periods of time over several days that the spacecraft temperatures are colder than normal. Those lower temperatures could cause the spacecraft hardware to operate differently than expected.

In their attempts to communicate with the spacecraft recently, the operations team is focusing on both the non-operational battery scenario and the colder temperatures at which the spacecraft is probably operating. They are also experimenting with the timing of their commands in the belief that the spacecraft may be waking up later than normal due to the faulty onboard clock.

The team will continue its daily uplink sessions with Mars Pathfinder. Daily audio updates are available by calling 1-800- 391-6654.

Mars Pathfinder Mission Status October 22, 1997

The Mars Pathfinder operations team is continuing its efforts to reestablish communications with the Pathfinder lander. Although they are experiencing communications difficulties, the team is confident that the spacecraft is still operating on the surface of Mars, according to Mission Manager Richard Cook. The last time they were able to send a command to the Pathfinder lander instructing it to transmit a signal back to Earth was on Sol 93, which was Tuesday, October 7, at 7:21 a.m. Pacific Daylight Time.

Team members suspect that the spacecraft may not be receiving commands from Earth properly because the lander's hardware has become much colder than normal. In regular operations, when the lander's transmitter is turned on, spacecraft hardware warms up sufficiently to operate normally. Since the transmitter has not been on for several days, engineers suspect that temperatures within the lander are considerably colder than normal. Predicted internal temperatures drop to as low as -50 C (-58 F) in the early morning and only rise to about -30 C (-22 F) in the late afternoon. These temperatures are about 20 C (38 F) colder than the coldest previous operational temperatures.

The lower temperatures cause the spacecraft radio hardware to operate outside the range of radio frequencies that ground controllers have used in the past. During the past three weeks the operations team has been transmitting to the spacecraft at a lower frequency and sweeping through a wider frequency range, a technique that has been used on other missions to attempt to cause the spacecraft receiver to lock on to the transmitted signal. Once ground controllers finish this, they send commands instructing the lander to turn on its transmitter and send a signal back to Earth.

To be certain that they investigate all possibilities, team members are also consulting with experts knowledgeable about the radio and other key elements of the spacecraft. They have identified some new scenarios that are being pursued to regain communications. These recommendations include doing more testing of the engineering model hardware in the laboratory to better understand how the spacecraft might be behaving. Another recommendation has suggested shifting and increasing the range of frequencies being swept through much more than previously attempted.

According to Project Manager Brian Muirhead, the possibility exists that an unrecoverable problem may have occurred. Team members expected that, once the lander's onboard battery died, cold and thermal cycling could result in a failure of some other element of Pathfinder and thereby end the mission. "However, the team will continue to do everything possible to reestablish communications until all options have been exhausted," Muirhead said. The mission has already exceeded all of its goals in terms of spacecraft lifetime and data return.

The science team, meanwhile, continues to process and analyze the large volume of data sent back by Pathfinder's lander and rover. Further science products are planned and new results will continue to be presented as they develop.

The team will continue its daily uplink sessions with Mars Pathfinder. Daily audio updates are available by calling (800) 391-6654.

Mars Pathfinder Mission Status October 29, 1997

Mars Pathfinder's operations team is continuing daily efforts to reestablish communications with the lander. Over the last month the team has been working through all credible problem scenarios and taking a variety of actions to try to recover the link with Pathfinder. They plan to continue sending commands to the spacecraft for another week before shifting to a contingency plan of less frequent commanding and listening.

During the past month, the team has investigated a variety of scenarios that could explain why the Pathfinder lander has not sent telemetry to Earth since September 27. Since that time, ground stations have detected a carrier signal from the lander on two occasions, but on each attempt following the receipt of the carrier signals they were not able to reestablish a link, and therefore no digital data was received to enable determination of the spacecraft's condition.

The team initially investigated the possibility that the lander's battery had failed. This scenario would have resulted in spacecraft clock uncertainties and unknown spacecraft power conditions due to the lander only operating on solar power. They then investigated the possibility that, because the lander's transmitter had been turned off for many days, the lander's temperature had dropped to a range between -50 and -30 C (-58 to -22 F), some 20 to 40 degrees C (about 35 to 70 degrees F) colder than previous operating temperatures, causing its radio receiver to shift away from its normal frequency range.

Currently the team is sending commands to the lander to investigate the possibility that the spacecraft's flight computer is not operating normally. "Under this scenario, the thought is that perhaps the computer is not booting up fully," said Mission Manager Richard Cook. "The team is sending resets to the lander at various times of the day before we attempt to send other commands."

All scenarios are expected to have been fully investigated by end of day on Tuesday, November 4. If the team does not reestablish contact by then, said Project Manager Brian Muirhead, they plan on shifting to a contingency strategy of sending commands to the lander only periodically, perhaps once a week or once per month. "The normal extended mission would be over, but there is still a small chance of reestablishing a link, so we'll keep trying at a very low level," Muirhead said. "Of course the science team will continue to process, catalog and understand the large volume of science data we have received, which will keep us very busy for several months."

Although the true cause of the loss of lander communications may never be known, recent events are consistent with predictions made at the beginning of the extended mission in early August. When asked about the life expectancy of the lander, project team members predicted that the first thing that would fail on the lander would be the battery; this apparently happened after the last successful transmission September 27. After that, the lander would begin getting colder at night and go through much deeper day-night thermal cycles. Eventually, the cold or the cycling would probably render the lander inoperable. According to Muirhead, it appears that this sequence of events may have taken place. The health and status of the rover is also unknown, but since initiating its onboard backup operations plan three weeks ago, it is probably in the vicinity of the lander attempting to communication with the lander.

At the time the last telemetry from the spacecraft was received, Pathfinder's lander had operated nearly three times its design lifetime of 30 days, and the Sojourner rover operated 12 times its design lifetime of seven days. Since its landing on July 4, 1997, Mars Pathfinder has returned 2.6 billion bits of information, including more than 16,000 images from the lander and 550 from the rover, as well as more than 15 chemical analyses of rocks and extensive data on winds and other weather factors. The only remaining objective was to complete the high-resolution 360- degree image of the landing site called the "Super Pan," of which 83 percent has already been received and is being processed.

Daily audio updates on Mars Pathfinder's status are available by calling (800) 391-6654.

Mars Pathfinder Mission Status Tuesday, March 10, 1998

The long goodbye to NASA's Mars Pathfinder lander and the Soujourner rover ended today when the lander failed to respond to the final command to communicate with controllers at NASA's Jet Propulsion Laboratory. The Pathfinder mission, which operated three times longer than its original 30-day planned lifetime on the Martian surface, is acknowledged as one of NASA's most successful endeavors as a dramatic example of the space agency's new style of "faster, better, cheaper" planetary exploration.

Today's last-ditch effort to listen for a signal from Pathfinder effectively ends the mission, said project manager Brian Muirhead. No further attempts will be made to communicate with Pathfinder, he added.

Pathfinder flight controllers Ben Toyoshima and Rob Smith at JPL spent nearly four hours today alternately commanding the lander to turn on its transmitter, then listening for a response via NASA's Deep Space Network's 34-meter antenna at Goldstone, California, in the Mojave Desert. One-way radio communications to Mars from Earth take nearly 20 minutes.

The final Pathfinder telecommunications session ended at 1:21 p.m. PST when no transmissions had been detected from Pathfinder.

A description of today's efforts to reestablish contact with Pathfinder can be found at the following URL: http://mars.jpl.nasa.gov/readme.html

Mars Global Surveyor Arrival Press Kit

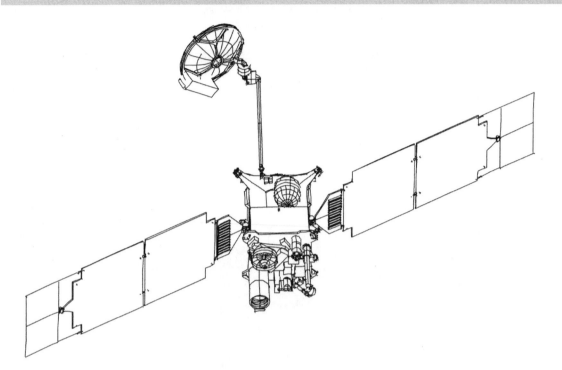

NATIONAL AERONAUTICS AND SPACE ADMINISTRATION
RELEASE: September 1997

Contacts
Douglas Isbell
Policy/Program Management
Headquarters,
Washington, DC
202/358-1753

Franklin O'Donnell
Mars Global Surveyor Mission
Jet Propulsion Laboratory,
Pasadena, CA
818/354-5011

MARS GLOBAL SURVEYOR ON TARGET FOR ARRIVAL

NASA's Mars Global Surveyor, the first in a new series of spacecraft destined to explore the red planet, is preparing to intercept the orbit of Mars and begin a two-year mapping mission after a 10-month, 700-million-kilometer (435-million-mile) interplanetary journey.

The orbiter will fire its main engine beginning at 01:17 Universal Time on September 12 (6:17 p.m. Pacific Daylight Time September 11) for 22 minutes to slow its speed enough to be captured in orbit around the planet.

Mars Global Surveyor is a global mapping mission, carrying a suite of science instruments designed to study the entire Martian surface, atmosphere and interior. Measurements will be collected beginning in March 1998 from a low-altitude, nearly polar orbit 378 kilometers (234 miles) above the Martian surface over the course of one complete Martian year, the equivalent of nearly two Earth years.

"Throughout its primary, two-year mission, Mars Global Surveyor will gather information on the geology, geophysics and climate of Mars," said Glenn E. Cunningham, Global Surveyor project manager at NASA's Jet Propulsion Laboratory, Pasadena, CA.

"The mission will provide a global portrait of Mars as it exists today," he said. "This new view will help planetary scientists to better understand the history of Mars' evolution, and will provide clues about the planet's interior and surface evolution. With this information, we will have a better understanding of the history of all of the inner planets of the solar system, including our home planet, Earth."

Mars Global Surveyor continues NASA's long exploration of the red planet, which began more than 30 years ago with the Mariner 4 spacecraft that produced the first pictures of the planet's cratered surface. Following the successful landing of the Mars Pathfinder lander and rover on July 4, 1997, Mars Global Surveyor is the first in a multi-year series of missions called the Mars Surveyor program that will lead to eventual human expeditions to the red planet.

Mars Global Surveyor was launched at 12:00:49 p.m. Eastern Standard Time on November 7, 1996, atop a three-stage Delta II launch vehicle from launch pad 17A at Cape Canaveral Air Station, FL. The third-stage Star 48B solid rocket later propelled the spacecraft out of Earth orbit and on its way to Mars.

Once on course for the cruise to Mars, the spacecraft deployed its two solar panels to begin generating solar power. One of the solar panels did not fully deploy and is tilted about 20 degrees from its intended position; this is not, however, expected to pose a significant risk to the mission. The low-gain antenna was used for initial spacecraft communications, until the spacecraft was far enough away from Earth in early January 1997 to begin using its 1.5-meter-diameter (5-foot) high-gain antenna.

Mars Global Surveyor's six science instruments — the thermal emission spectrometer, laser altimeter, magnetometer/electron reflectometer, ultra-stable oscillator, camera and radio relay system — were calibrated during the cruise to Mars. Three trajectory correction maneuvers were performed to fine-tune the spacecraft's flight path. All spacecraft systems and the instrument payload performed well as Mars Global Surveyor headed for its destination, according to Joe Beerer, Global Surveyor flight operations manager.

When Surveyor reaches Mars, its 660-newton main engine will fire to slow the spacecraft's speed by more than 973 meters per second (2,176 miles per hour) with respect to Mars and allow the craft to be captured by Mars' gravity.

"Mars Global Surveyor will be flying over the north pole when it enters orbit around Mars," said Wayne Lee, Mars Global Surveyor mission planner. "The spacecraft will spend the first six days in this highly elliptical orbit around the planet, completing one orbit around Mars in about 45 hours, or just less than two days."

Instrument calibrations and some science measurements will take place during the elliptical orbit phase, said

Dr. Arden Albee, Mars Global Surveyor project scientist.

"The spacecraft will be passing in and out of the planet's magnetic field, if indeed Mars has one, during the early and larger elliptical orbits around the planet," he said. "Global Surveyor will be able to make unique observations of interactions of magnetic field lines with the solar wind. In addition, it will make calibrations of the magnetometer and electron spectrometer that would not be possible from the lower-altitude mapping orbit.

"The thermal emission spectrometer and the camera will obtain initial observations on the surface and atmosphere of Mars," Albee continued. "These will provide valuable insight into changes in the atmosphere that might affect the safety of the spacecraft during aerobraking operations."

Six days after Mars arrival, the spacecraft will begin an innovative braking process, called aerobraking, to lower itself into a low-altitude mapping orbit. Aerobraking allows a spacecraft to use the drag of a planet's atmosphere to lower its orbit without relying on propellant. The technique was first tested in the summer of 1993, using the Magellan spacecraft orbiting Venus.

During each of its orbits shortly after Mars arrival, Mars Global Surveyor will pass through the upper fringes of the Martian atmosphere each time it reaches periapsis, the point in its orbit closest to the planet. Friction from the atmosphere will cause the spacecraft to be slowed slightly and lose some of its momentum during each orbit. Each time the spacecraft dips into the atmosphere, its one tilted solar panel will be rotated 180 degrees to protect it from folding up. As the spacecraft loses momentum, the apoapsis, or the point in its orbit farthest from Mars, will also be lowered.

Trimming its orbit from the highly elliptical, 45-hour orbit to a nearly circular, two-hour orbit will take about four months. Five engine burns will accomplish the first orbital adjustments, lowering the periapsis (or closest point to Mars) from about 250 kilometers (156 miles) to about 112 kilometers (69 miles) above the surface.

Next, Mars Global Surveyor will spend about three months adjusting the farthest part of its orbit from 54,000 kilometers (33,480 miles) to about 2,000 kilometers (1,240 miles). As the spacecraft's orbit is trimmed, the time it takes to make one complete revolution around Mars will diminish to less than three hours.

In the final three weeks of aerobraking, Global Surveyor will raise the closest part of its orbit once again, until it is circling Mars in a 350 by 410-kilometer (217- by 254-mile) orbit, very close to the final mapping orbit. With this adjustment, the spacecraft will be orbiting Mars about once every 118 minutes, and crossing Mars' equator at about 2 p.m. local solar time each orbit.

As the spacecraft continues to circularize its orbit, the 34-meter-diameter (112-foot) antennas of NASA's Deep Space Network will be used to begin a navigation and radio science experiment, measuring small shifts in the spacecraft's velocity that will tell scientists more about the planet's gravity field.

All of the spacecraft's instruments will be turned on around March 10, 1998, and the mapping mission will begin on March 15. Data from this final checkout phase will allow the spacecraft to obtain one complete global map of the surface — a process that will take seven days — before the dust storm season begins in the spring.

"In 1998 the Martian dust storms occur roughly between February and August, so the atmosphere may be disturbed when mapping begins," Albee said. "But we may have an excellent opportunity to obtain data on the spread of a global Martian dust storm, should one occur next year."

In its final mapping orbit, Mars Global Surveyor will circle Mars at a speed of about 3.4 kilometers per second (7,600 miles per hour) in an orbit that will take it close to both poles. On the day side of the planet, Global Surveyor will be traveling from north to south. On each orbit, it will cross the equator at about 2 p.m. local mean solar time. The spacecraft will always see the surface of Mars on the daylit side as it appears at mid-afternoon. This "sun-synchronous" orbit puts the Sun at a standard angle above the horizon in each image.

Experiment teams will control their spaceborne instruments from their home institutions. Each 24 hours worth of data will be transmitted to Earth during daily, 10-hour tracking passes performed by NASA's Deep Space Network.

The Mars Global Surveyor mission is expected to yield more than 700 billion bits of scientific data, more than the amount of data returned by all previous Mars missions, and exceeded only by the Magellan Venus mission.

Mars Global Surveyor will examine the entire planet — from the ionosphere, an envelope of charged particles surrounding Mars, down through the atmosphere to the surface and deep into Mars' interior. Scientists will glean valuable new information on daily and seasonal weather patterns, geological features and the migration of water vapor from hemisphere to hemisphere over a complete Martian year.

As the primary mission winds down in late 1999, Global Surveyor will be used to relay data from microprobes penetrators beneath the surface of Mars that will have been deployed before arrival by the 1998 Mars Surveyor, and will be used as a backup relay platform for data from the Mars Surveyor '98 lander itself. Depending on its lifetime, Global Surveyor may also serve as a communications relay station for other spacecraft arriving at Mars.

Mars Global Surveyor is the first mission in a sustained program of robotic exploration of Mars, managed by the Jet Propulsion Laboratory, Pasadena, CA, for NASA's Office of Space Science, Washington, DC. JPL is a division of the California Institute of Technology.

[End of General Release]

Media Services Information

NASA Television Transmission

NASA Television is broadcast on the satellite GE-2, transponder 9C, C Band, 85 degrees west longitude, frequency 3880.0 MHz, vertical polarization, audio monaural at 6.8 MHz. Television coverage of the Mars Global Surveyor orbit insertion begins at 5 p.m. Pacific Daylight Time on Sept. 11 and concludes at approximately 8:30 p.m. PDT, following a live news briefing. The schedule for NASA TV programming is available from the Jet Propulsion Laboratory, Pasadena, CA; Johnson Space Center, Houston, TX; Kennedy Space Center, FL; and NASA Headquarters, Washington, DC.

Status Reports

Status reports on Mars Global Surveyor will be issued by the Jet Propulsion Laboratory's Public Information Office. They may be accessed online as noted below. Daily audio status reports are available by calling (800) 391-6654.

Briefings

A pre-arrival mission and science overview briefing, originating from the Jet Propulsion Laboratory, will be carried live on NASA TV at 10 a.m. Pacific Daylight Time on Tuesday, September 9. Status briefings will be held at 10 a.m. PDT Wednesday and Thursday, September 10-11. A post-arrival briefing will be held at 7:40 p.m. PDT Thursday, September 11. A briefing presenting initial science findings is tentatively scheduled the week of September 22-26.

Image Releases

Images returned by Mars Global Surveyor will be released to the news media in electronic format only during the mission. The images will be available in a variety of file formats from the JPL home page at http://www.jpl.nasa.gov . Images will also be distributed from the Mars Orbiter Camera principal investigator's web site at http://www.msss.com.

Internet Information

Extensive information on Mars Global Surveyor — including an electronic copy of this press kit, as well as news releases, fact sheets, status reports, spacecraft and science data and images — is available from the Jet Propulsion Laboratory's World Wide Web home page at http://www.jpl.nasa.gov. The Mars Global Surveyor Project also maintains a home page at http://marsweb.jpl.nasa.gov. Data from all of the mission's science instruments will be available on the home pages of each instrument's principal investigator; these addresses are linked to the Mars Global Surveyor project's home page.

Quick Facts

Spacecraft

Dimensions: Main structure 1.2 by 1.2 by 1.8 meters (4 by 4 by 6 feet); 12 meters (40 feet) across with fully deployed solar panels
Spacecraft mass at Mars arrival: 767 kg (1,691 lbs.)
Science instruments: thermal emission spectrometer; laser altimeter; magnetometer/electron reflectometer; ultra-stable oscillator; camera; radio relay system
Solar arrays: 2 panels, each 3.5 meters (11 feet) long; power 980 watts maximum; cells composed of gallium arsenide and silicon
Radio: 25-watt transmitter, X-band (8 GHz)
High-gain antenna diameter: 1.5 meters (4.9 feet)

Launch/Cruise

Launch: 12:00:49 p.m. Eastern Standard Time (17:00:49 Universal Time) November 7, 1996, from Cape Canaveral Air Station, FL
Cruise: 10 months

Mars Orbit Insertion

Orbit insertion burn: Begins 01:17:16 and ends at 01:39:33 Universal Time on Sept 12, 1997 (6:17 p.m. to 6:39 p.m. Pacific Daylight Time Sept 11, 1997); duration 22 min, 17 sec
One-way speed of light time on arrival day: 14 minutes, 6 seconds
Time burn observed on Earth: Begins 6:31 p.m. and ends 6:53 p.m. PDT Sept 11, 1997
Mars-Earth distance on arrival day: 254.6 million km (158 million mi)
Change in velocity: 973 m/sec (2,176 mph)
Velocity before burn (with respect to Mars): 5.09 km/sec (11,386 mph)
Velocity after burn (with respect to Mars): 4.40 km/sec (9,842 mph)
Average deceleration due to burn: 7/100ths of 1 Earth G
Martian seasons at arrival: Fall in northern hemisphere, spring in south

Aerobraking and Mapping Mission

Period of aerobraking: Begins September 17, 1997; continues for 4 months
Period of initial elliptical orbit: 45 hours, plus or minus 3 hours
Final mapping orbit: period 117 min, mean altitude 378 km (234 mi), polar, sun-synchronous
Primary mapping mission: March 15, 1998 to January 31, 2000; covers 1 Martian year (687 Earth days)
Martian seasons when mapping begins: Winter in northern hemisphere, summer in south

Program

Development time: 26 months
Costs: $148 million pre-launch development; $52.6 million launch; $46.4 million mission ops
Spacecraft industrial partner: Lockheed Martin Astronautics, Denver, CO

Mars at a Glance

General

* One of 5 planets known to ancients; Mars was Roman god of war, agriculture and the state
* Reddish color; at times the 3rd brightest object in night sky after the Moon and Venus

Physical Characteristics

* Average diameter 6,794 kilometers (4,219 miles); about half the size of Earth, but twice the size of Earth's Moon
* Mass 1/10th of Earth's; gravity only 38 percent as strong as Earth's
* Density 3.9 times greater than water (compared to Earth's 5.5 times greater than water)
* No magnetic field detected to date

Orbit

* Fourth planet from the Sun, the next beyond Earth
* About 1.5 times farther from the Sun than Earth is
* Orbit elliptical; distance from Sun varies from a minimum of 206.7 million kilometers (128.4 million miles) to a maximum of 249.2 million kilometers (154.8 million miles); average distance from Sun, 227.7 million kilometers (141.5 million miles)
* Revolves around Sun once every 687 Earth days
* Rotation period (length of day in Earth days) 24 hours, 37 min, 23 sec (1.026 Earth days)
* Poles tilted 25 degrees, creating seasons similar to Earth's

Environment

* Atmosphere composed chiefly of carbon dioxide (95.3%), nitrogen (2.7%) and argon (1.6%)
* Surface atmospheric pressure 8 millibars (less than 1/100th that of Earth's average)
* Surface temperature averages -53 C (-64 F); varies from -128 C (-199 F) during polar night to 27 C (80 F) at equator during midday at closest point in orbit to Sun

Features

* Highest point is Olympus Mons, a huge shield volcano more than 15,900 meters (52,000 feet) high and 600 kilometers (370 miles) across; has about the same area as Arizona
* Canyon system of Valles Marineris is largest and deepest known in solar system; extends more than 4,000 kilometers (2,500 miles) and has 5 to 10 kilometers (3 to 6 miles) relief from floors to tops of surrounding plateaus
* "Canals" observed by Giovanni Schiaparelli and Percival Lowell about 100 years ago were a visual illusion in which dark areas appeared connected by lines. The Viking missions of the 1970s, however, established that Mars has channels probably cut by ancient rivers

Moons

* Two irregularly shaped moons, each only a few kilometers wide
* Larger moon named Phobos ("fear"); smaller is Deimos ("terror"), named for attributes personified in Greek mythology as sons of the god of war

Historical Mars Missions

Mission, Country, Launch Date, Purpose, Results

- Mars 1, USSR, 11/1/62, Mars flyby, lost at 106 million kilometers (65.9 million miles)
- Mariner 3, U.S., 11/5/64, Mars flyby, shroud failed
- Mariner 4, U.S. 11/28/64, first successful Mars flyby 7/14/65, returned 21 photos
- Zond 2, USSR, 11/30/64, Mars flyby, failed to return planetary data
- Mariner 6, U.S., 2/24/69, Mars flyby 7/31/69, returned 75 photos
- Mariner 7, U.S., 3/27/69, Mars flyby 8/5/69, returned 126 photos
- Mariner 8, U.S., 5/8/71, Mars flyby, failed during launch
- Mars 2, USSR, 5/19/71, Mars orbiter/lander arrived 11/27/71, no useful data returned
- Mars 3, USSR, 5/28/71, Mars orbiter/lander, arrived 12/3/71, some data and few photos
- Mariner 9, U.S., 5/30/71, Mars orbiter, in orbit 11/13/71 to 10/12/72, returned 7,329 photos
- Mars 4, USSR, 7/21/73, failed Mars orbiter, flew past Mars 2/10/74
- Mars 5, USSR, 7/25/73, Mars orbiter, arrived 2/12/74, some data
- Mars 6, USSR, 8/5/73, Mars orbiter/lander, arrived 3/12/74, little data return
- Mars 7, USSR, 8/9/73, Mars orbiter/lander, arrived 3/9/74, little data return
- Viking 1, U.S., 8/20/75, Mars orbiter/lander, orbit 6/19/76-1980, lander 7/20/76-1982
- Viking 2, U.S., 9/9/75, Mars orbiter/lander, orbit 8/7/76-1987, lander 9/3/76-1980; combined, the Viking orbiters and landers returned 50,000+ photos
- Phobos 1, USSR, 7/7/88, Mars/Phobos orbiter/lander, lost 8/88 en route to Mars
- Phobos 2, USSR, 7/12/88, Mars/Phobos orbiter/lander, lost 3/89 near Phobos
- Mars Observer, U.S., 9/25/92, orbiter, lost just before Mars arrival 8/22/93 (8/21/93 EDT)
- Mars Global Surveyor, 11/7/96, orbiter, en route to orbit insertion 9/12/97 (9/11/97 EDT)
- Mars 96, Russia, 11/16/96, orbiter and landers, failed during launch
- Mars Pathfinder, U.S., 12/4/96, landed 7/4/97, lander and rover, fulfilled all science objectives

(Editor's note: To save space several pages have been omitted from the MGS Press Kit at this point. These pages were general information about Mars and can be seen verbatim in the Pathfinder Press Kit — pages 214-218.)

Mission Overview

Mars Global Surveyor is a global mapping mission, designed to gather data on the atmosphere, surface and interior of Mars. Global data sets will enable scientists to determine Mars' current state and characterize its evolution. Among a myriad of science objectives, Mars Global Surveyor will study Mars' climate, surface topography and subsurface resources, and map the entire globe. Information from the mission will serve as a foundation for planning future robotic and human missions to Mars.

Launch

Mars Global Surveyor was launched at 12:00:49 p.m. Eastern Standard Time on November 7, 1996, atop a Delta II 7925A launch vehicle from launch pad 17A at Cape Canaveral Air Station, FL. The launch was delayed one day due to clouds and upper level winds on the first day of the launch period.

After liftoff, the first stage of the three-stage Delta rocket and the nine solid-fuel strap-on boosters lifted the spacecraft to an altitude of about 115 kilometers (70 miles) above Earth, and the Delta's second stage then boosted the payload into a circular parking orbit 185 kilometers (115 miles) above Earth. After achieving the parking orbit, the booster and spacecraft coasted for about 35 minutes, then fired to raise the apogee, or high point, of the parking orbit.

Small rockets were used to spin up the Delta's third stage and spacecraft to 60 rpm. The second stage was jettisoned and the third stage, a Star 48B solid rocket, was ignited to complete the trans-Mars injection burn and send Global Surveyor on its way to the red planet.

Solar panel deployment.

About an hour after launch, the spacecraft's two 3.5-meter (11-foot) solar arrays were unfolded and a piece of metal called the "damper arm" on one of the panels was broken off during the deployment. (The damper arm is the part of the solar array deployment mechanism at the joint where the entire panel is attached to the spacecraft.) The metal fragment became trapped in the 50-millimeter (2-inch) space between the panel's shoulder joint and the edge of the solar panel, leaving the array about 19 degrees from its fully deployed position.

The damper arm connects the panel to a device called the "rate damper," which functions in much the same way as the hydraulic closer on a screen door acts to slow the speed at which the door closes. In Global Surveyor's case, the rate damper was used to slow the motion of the solar panel as it unfolded from its stowed position.

The operations team studied the problem with engineering data and computer-simulated models over the next two weeks. Two strategies for correcting the problem emerged. The first involved performing several slight maneuvers using the spacecraft's electrically driven solar array positioning actuators to try to gently shake the array and free the trapped debris. These maneuvers were carried out in January and February 1997, to no avail.

The second strategy was to reconfigure the solar panel during aerobraking so that the unlatched side of the panel would not be facing into the direction of the air flow and at risk of folding up on itself. The solar arrays are essential to the aerobraking technique and will be used to provide the drag surface that will slow the spacecraft's orbital speed, transforming its initial elliptical orbit into the final, circular mapping orbit. The technique allows Global Surveyor to carry considerably less fuel to Mars and take advantage of the planet's atmospheric drag to lower itself into the correct orbit.

After testing and analysis, the flight team determined that aerobraking could be safely accomplished by rotating the panel 180 degrees, turning the panel's solar-cell side into the flow of wind each time the spacecraft dips into the Martian atmosphere. By turning the panel's solar-cell side into the direction of the air flow, force will be exerted on the debris that is lodged in the hinge. The force of the air flow on the opposite side of the panel will insure the panel does not close up, and may possibly exert enough force to snap the panel into the latched position.

Cruise

The spacecraft spent 309 days en route to Mars, following what navigators call a Type 2 trajectory. This type of flight path took the spacecraft more than 180 degrees around the Sun and, compared with other types of trajectories, is a slower way to reach Mars. Because the spacecraft has been traveling at a slower velocity, however, it will require less propellant to slow down once it is ready to be captured in orbit around the destination planet.

On the first part of its flight, all spacecraft communications with Earth occurred through Global Surveyor's broad-beam, low-gain antenna. The dish-shaped, narrow-beam high-gain antenna sat on the spacecraft in a stowed, body-fixed orientation during cruise, making communications with Earth through the high-gain antenna impossible unless the spacecraft was turned to point the high-gain antenna directly at Earth.

Outer cruise began on January 9, 1997, when the spacecraft switched from the low-gain to the high-gain antenna for communications with Earth. The switch-over became feasible when the angle between the Sun and Earth as seen from the spacecraft fell to a level low enough to allow the solar panels to collect adequate power while pointing the antenna at Earth.

Three trajectory correction maneuvers were performed along the trip to Mars to fine-tune the spacecraft's flight path. These thruster firings were performed on November 21, 1996, March 20, 1997, and August 25, 1997. Another maneuver had been scheduled for April 21, 1997, but was not necessary and, hence, not performed.

During the last 30 days of approach to Mars, the flight team focused on final targeting of the spacecraft to the proper aim point, and preparations for orbit insertion. On August 19 and 20, the spacecraft's camera took a series of eight images of Mars, at a distance of 5.5 million kilometers (3.4 million miles). With a resolution of about 20 kilometers (12.5 miles) per picture element, these images were processed to create a movie of the planet as it rotated.

Mars Orbit Insertion

Mars Global Surveyor will perform an attitude reorientation maneuver once it reaches the orbit of Mars to turn the spacecraft's main engine toward the direction of its motion, or toward Mars. Then the spacecraft will fire its 660-newton main engine for approximately 22 minutes, 17 seconds, to slow down. The burn will begin at 01: 17:16 and conclude at 01: 39:33 Universal Time (UT) September 12, 1997 (6:17 to 6:39 p.m. September 11, 1997 Pacific Daylight Time (PDT). Because it takes 14 minutes, 6 seconds for radio signals to travel from Mars to Earth on arrival day, the beginning and end of the burn will be detected on Earth at 6:31 and 6:53 p.m. PDT, respectively.

Telecommunications with the spacecraft will cease 9 minutes into the burn as Mars Global Surveyor passes behind Mars as seen from Earth. This occultation will last 14 minutes. NASA's Deep Space Network tracking facilities in California and Australia will regain communications with Global Surveyor at 01:57:00 UT (6:57 p.m. PDT), when the spacecraft reemerges from behind Mars.

By completion of the burn, the spacecraft will have slowed down by about 973 meters per second (2,176 miles per hour) with respect to Mars. Global Surveyor will be in a highly elliptical orbit, completing one revolution around Mars every 45 hours, plus or minus 3 hours.

Aerobraking

Selection of the less expensive Delta II booster in order to stay within program costs placed mass limitations on Mars Global Surveyor and prevented it from carrying enough propellant to Mars to directly achieve a low-altitude mapping orbit. An innovative, mission-enabling braking technique known as aerobraking was chosen instead to trim the spacecraft's initial, highly elliptical capture orbit and lower it to a nearly circular mapping orbit.

The Magellan spacecraft at Venus was the first planetary spacecraft to try aerobraking, as a demonstration, in the summer of 1993. The success of this demonstration cleared the way for its implementation in the Mars Global Surveyor mission design.

Global Surveyor's use of aerobraking will differ from that performed with the Magellan spacecraft in two important ways. First, Global Surveyor must aerobrake before it can start its primary mapping mission, whereas Magellan tested aerobraking as an engineering demonstration at the conclusion of its mission. Consequently, Global Surveyor's mission objectives are dependent on successfully aerobraking through the Martian atmosphere until the proper mapping orbit is achieved. Not only must aerobraking be successful, but it must be accomplished so that the spacecraft crosses the equator within a few minutes of 2 p.m. local solar time each orbit. This is called a "sun-synchronous" orbit. These two elements of the aerobraking phase make the Mars Global Surveyor's navigation by far the most challenging of the planetary missions to date.

Mars Global Surveyor Orbit Insertion Timeline

Thursday, September 11, 1997 All times are Earth-received, Pacific Daylight Time

- 12:40 p.m.   Deep Space Station 14 at Goldstone, CA, begins tracking
- 12:40   Deep Space Station 15 at Goldstone, CA, begins tracking
- 4:31   Gyro #2 turned on
- 4:35   Deep Space Station 43 in Australia begins tracking

- 4:35   Deep Space Station 45 in Australia begins tracking
- 5:01   Spacecraft begins loading maneuver control parameters
- 5:17   Deep Space Station 15 transmitter off
- 5:31   Spacecraft begins maneuver command block
- 5:55   Spacecraft transmitter switches from high-gain antenna to low-gain antenna #1
- 5:55   Spacecraft's telemetry turned off; transmits only carrier signal
- ~5:59   Deep Space Stations reacquire signal
- 6:01   Thrusters enabled for steering and attitude control during orbit insertion
- 6:14   Spacecraft starts turn to align rocket engine in burn direction
- 6:15   Solar arrays begin turning to orbit insertion orientation
- 6:29   Inertia measurement unit set to supply accelerometer data
- 6:30   Fuel and oxidizer valves enabled and armed
- 6:31   Main engine burn begins
- 6:40   Deep Space Station 45 transmitter on
- 6:43   Spacecraft passes behind Mars; radio signal lost
- 6:44   Closest approach to Mars (periapsis #1)
- 6:53   Main engine burn ends
- 6:55   Attitude control returned to reaction wheels
- 6:55   Spacecraft turns to Earth point and resumes array-normal-spin attitude
- 6:56   Solar arrays moved to array-normal-spin position
- 6:57   Spacecraft emerges from behind Mars
- ~6:57 - 7:03   Deep Space Stations acquire spacecraft carrier signal
- 7:10   Spacecraft transmitter switches from low-gain antenna to high-gain antenna
- 7:10   Spacecraft resumes transmitting telemetry (data)

During each of its orbits shortly after Mars arrival, the spacecraft will pass through the upper fringes of the Martian atmosphere each time it reaches periapsis, the point in its orbit closest to the planet. Friction from the atmosphere will cause the spacecraft to slow slightly and lose some of its momentum during each orbit. Loss of momentum will lower the spacecraft's apoapsis, or the point in its orbit farthest from Mars.

Aerobraking will take place over four months, beginning with an initial "walk-in" phase. After the spacecraft is captured in a 45-hour (plus or minus 3 hours) elliptical orbit around Mars, its apoapsis and periapsis will be gradually adjusted as scientists and engineers learn more about the density of Mars' upper atmosphere. The flight team will be able to gauge atmospheric density and variations from one orbit to another, while the spacecraft is tracked continuously by the 34-meter-diameter (112-foot) antennas of NASA's Deep Space Network, which are designed to both transmit and receive X-band signals.

In the initial walk-in phase, the spacecraft's closest pass over Mars will be lowered from the capture orbit of 250 kilometers (156 miles) above the surface to about 112 kilometers (69 miles) above the surface. This will be done with a series of five propulsive maneuvers, using the spacecraft's thrusters. The first of these propulsive burns, scheduled to take place on September 16, will be the largest and drop the spacecraft's periapsis altitude to 150 kilometers (93 miles). The next four burns, occurring on September 18, 20, 22 and 24, will lower the spacecraft gradually to the 112-kilometer (69-mile) aerobraking altitude.

After completion of the walk-in phase, the spacecraft will spend about three months in the main phase of aerobraking. During this time, Global Surveyor's apoapsis altitude of 54,000 kilometers (33,480 miles) will be dropped to about 2,000 kilometers (1,240 miles). As the spacecraft's orbit is trimmed, the time it takes to make one complete revolution around Mars will drop to less than three hours.

The final three weeks of aerobraking constitute the "walk-out" phase and will reduce the spacecraft's apoapsis to 450 kilometers (279 miles) above the surface of Mars. As Global Surveyor lowers its apoapsis, it will also use its thrusters to gradually raise its periapsis from 112 kilometers (69 miles) to 143 kilometers (89 miles) above the surface. By doing so, the spacecraft will be slowly "walking out" of the atmosphere during each closest approach to Mars. Adjustments in both the farthest and closest points in Mars Global Surveyor's orbit around the planet will be reshaping its flight path from highly elliptical to a nearly circular orbit.

Aerobraking will end with a termination burn performed on about January 18, 1998. This burn will raise Global Surveyor's periapsis one final time, from 143 kilometers (89 miles) to approximately 450 kilometers (279 miles) above the surface of Mars. The spacecraft will then be circling Mars in a 400- by 450-kilometer (248- by 279-mile) orbit, just slightly off from its final mapping orbit. By then Global Surveyor will be orbiting Mars about once every 118 minutes, and crossing Mars' equator at just about 2 p.m. local solar time each orbit.

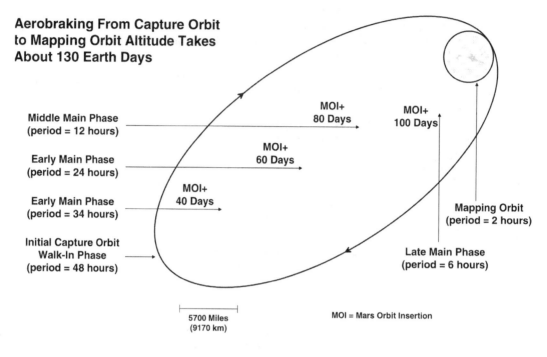

Aerobraking From Capture Orbit to Mapping Orbit Altitude Takes About 130 Earth Days

Middle Main Phase (period = 12 hours)

Early Main Phase (period = 24 hours)

Early Main Phase (period = 34 hours)

Initial Capture Orbit Walk-In Phase (period = 48 hours)

MOI+ 80 Days

MOI+ 100 Days

MOI+ 60 Days

MOI+ 40 Days

Mapping Orbit (period = 2 hours)

Late Main Phase (period = 6 hours)

5700 Miles (9170 km)

MOI = Mars Orbit Insertion

Mars Global Surveyor aerobraking orbits

Mars Global Surveyor will begin aerobraking in the southern spring, and the historical record of global dust storms suggests that this is the most likely season for such storms to occur. During these storms, the dust itself is not of concern, since there is no evidence that it reaches anywhere near the altitudes at which aerobraking will occur. Scientists are more concerned with the possible increase in atmospheric density at the aerobraking altitude that could be caused by increased heating due to a dustier atmosphere in general.

Data from the highly successful Mars Pathfinder mission provided valuable new information about the density of the Martian atmosphere all the way up to about 120 kilometers (75 miles), which will be used to update the atmospheric models before the start of the Mars Global Surveyor aerobraking phase. Pathfinder also recorded dust activity and pressure and temperature fluctuations during its primary mission in July 1997. If Pathfinder's lander and rover are still operating by the time Mars Global Surveyor arrives at Mars, surface temperature, pressure and opacity measurements will be monitored to help the Global Surveyor navigation team adjust the spacecraft's orbit.

Several measurements will also be made from the Hubble Space Telescope and from Earth. Hubble will take multiple images of Mars near the start of the aerobraking phase, before Mars gets too close to the Sun as viewed from Earth. Hubble images taken during the week before Pathfinder landed showed local dust activity near the landing site in Ares Vallis. When combined with Earth-based observations made at the Space Science Institute in Boulder, CO, using the National Radio Astronomy Observatory microwave antenna, these measurements will yield a global average atmospheric temperature profile from the ground to about 60 kilometers (35 miles) above the surface.

Scientists have been able to infer that moisture in the atmosphere usually confines Martian dust to lower altitudes, where it has a minimal impact on the densities at higher altitudes. As Mars gets closer to the Sun, atmospheric temperatures rise, along with water vapor content, and allow dust to circulate at higher altitudes. The Hubble and radio astronomy microwave observations will be used to keep a close eye on the Martian atmosphere during aerobraking, not only because the spacecraft's periapsis will have to be raised to survive the effects of a global dust storm, but also because the aerobraking phase coincides with the first half of the dust storm season.

Mars Global Surveyor will also be dipping repeatedly into the Martian atmosphere in a different configuration than was originally planned. Because one of the spacecraft's two solar panels did not fully deploy and latch into place after launch, it is tilted at about 20 degrees from its fully deployed position. After careful analysis, engineers determined that the only risk posed by the tilted panel was the possibility that the panel might fold up on itself if enough wind flow was exerted.

To minimize that risk, the operations team will rotate the unlatched panel 180 degrees each time the spacecraft encounters the strongest air flow, which will be at periapsis. By rotating the panel and turning its solar-cell side into the direction of the air flow, the latch will not be in danger of folding up. The air flow, when exerted on the opposite side of the latch, may, in fact, push the array into the fully locked position. Whether or not that is accomplished, Global Surveyor's new aerobraking configuration does not pose significant risk to the science objectives of the mission.

Science During Aerobraking

Early science measurements conducted during aerobraking will provide the navigation team with enough information to perform alternative aerobraking strategies should the situation arise. Scientists will benefit from new observations of Mars achieved with unique lighting geometry from Global Surveyor's aerobraking orbit. The spacecraft's low-altitude passes will allow scientists to record information from much closer to the planet than will be possible during the mapping period. This 4½ month transition will also give experiment teams time to calibrate their instruments and prepare for continuous mapping operations.

The Mars orbital camera will image the surface of Mars at low Sun elevations between 15 degrees north latitude and 45 degrees south latitude, as the spacecraft slowly unwinds from the aerobraking attitude. Image swaths taken over 100 orbital revolutions will yield more than 5,000 pictures at five times the surface resolution of Viking orbiter imaging.

Data from the thermal emission spectrometer will include 100 atmospheric profiles of temperature and dust content at an unprecedented altitude of about 100 kilometers (60 miles). This will be followed each orbit by mid-latitude infrared measurements with a resolution of 30 to 40 kilometers (19 to 25 miles). In addition, daily global coverage of thermal and compositional properties will be recorded at a resolution of about 300 to 400 kilometers (185 to 250 miles) resolution, in association with unique local times of day and seasons.

As the elliptical orbit shrinks, the magnetometer will complete its primary science objective of determining the existence of a global magnetic field and measuring solar wind interaction with Mars' magnetopause, the boundary beyond which Mars no longer influences space. The electron reflectometer portion of this experiment will observe electron density variations in the ionosphere at the lowest altitude during its two months of operation.

Pointing down at the planet, the laser altimeter will have a unique opportunity during the spacecraft's closest approach to the planet on its third orbit after arrival to obtain a 600-kilometer (373-mile) swath, centered at 30 degrees north latitude, with 200 meters (655 feet) of spatial resolution and 2 meters (6.5 feet) of altitude precision. This orbit also permits the only forward/aft viewing by the thermal emission spectrometer prior to mapping the planet at 5:30 p.m. local solar time.

Finally, as the spacecraft's orbit is adjusted to achieve a periapsis of 200 kilometers (124 miles) from the surface, the radio science team will see a four-fold improvement in the sensitivity of their measurements when Global Surveyor flies over features known to produce gravity variations, such as the south polar cap.

The Mapping Mission

The final mapping orbit will be nearly circular, at 350 by 410 kilometers (217 by 254 miles), or an average of 378 kilometers (234 miles) above the planet's surface.

After the mapping orbit is achieved, spacecraft systems will be deployed and instruments will be checked out over the next 10 days.

The primary mission begins once the spacecraft has reached its mapping orbit and is completing one orbit around Mars about every two hours. Each new orbit will bring the spacecraft over a different part of Mars. As the weeks pass, the spacecraft will create a global portrait of Mars — capturing the planet's ancient cratered plains, huge canyon system, massive volcanoes, channels and frozen polar caps. During its mission, Mars Global Surveyor will pass over the terrain where the two U.S. Viking landers — separated by more than 6,400 kilometers (4,000 miles) — have rested for 21 years, and over Ares Vallis, home of the Mars Pathfinder lander (or Carl Sagan Memorial Station) and the Sojourner rover.

The primary mapping mission will begin on March 15, 1998, and last until January 31, 2000 — a period of one Martian year or 687 Earth days (almost two Earth years). The spacecraft will transmit its recorded data back to Earth once a day during a single 10-hour tracking pass by antennas of the Deep Space Network. During mapping operations, the spacecraft will return more than 700 billion bits of scientific data to Earth — than that returned by all previous missions to Mars and, in fact, roughly equal to the total amount of data returned by all planetary missions since the beginning of planetary exploration with the exception of the Magellan mission to Venus.

As Mars rotates beneath the spacecraft, a suite of onboard instruments will record a variety of detailed information. Detectors will measure radiation — visible and infrared — from the surface to deduce the presence of minerals that make up Mars. These same instruments will record infrared radiation from the thin Martian atmosphere, gathering data about its changing pressure, composition, water content and dust clouds. By firing short pulses of laser light at the surface and measuring the time the reflections take to return, a laser altimeter will map out the heights of Mars' mountains and the depths of its valleys.

The camera system will use wide and narrow-angle lenses to record land forms and atmospheric cloud patterns. Another sensor will look for a Martian magnetic field. As the telecommunications subsystem transmits information back to Earth, engineers will use the signal of the orbiting spacecraft to derive data about the planet's atmosphere and gravitational field.

By the time global mapping operations are over, Mars Global Surveyor will have obtained an extensive record of the nature and behavior of the Martian surface, atmosphere and interior. Such a record will aid in planning more specialized explorations that might involve robots, scientific stations deployed to the Martian surface, sample return missions and perhaps even human landings. Just as importantly, this record will help scientists understand planet Earth and what the future might have in store for humanity.

Mission Operations

Throughout the two years of Mars Global Surveyor's mapping mission, principal investigators, team leaders and interdisciplinary scientists will have science operations planning computers at their home institutions. All will be electronically connected to the project database at the Jet Propulsion Laboratory in Pasadena, CA, giving them direct involvement in mission operations.

Their computers will be equipped with software that allows the science teams to remotely initiate most of the commands required by their instruments to conduct desired experiments. The teams will be able to access raw science data within hours of their receipt on Earth. This automated operation will provide "quick-look" science data and let investigators easily monitor the health and performance of their instruments.

Many images and other data will be immediately available to the public on the Internet. After a short period of data validation, science data, both raw and processed, along with supplementary processing information and documentation, will be transferred to NASA's Planetary Data System archive for access and use by the broader planetary science community and the general public.

Control and operation of Mars Global Surveyor will be performed by a team of engineers located at JPL and at Lockheed Martin Astronautics Inc. in Denver, CO. Engineers in Denver will be electronically linked to JPL, providing monitoring and analysis of Mars Global Surveyor based on telemetry received from the spacecraft through NASA's Deep Space Network. The team will also develop the sequence of commands that will be sent to the spacecraft via the Deep Space Network. The electronic networking eliminates the costs of relocating mission operations team members during the mission.

The Spacecraft

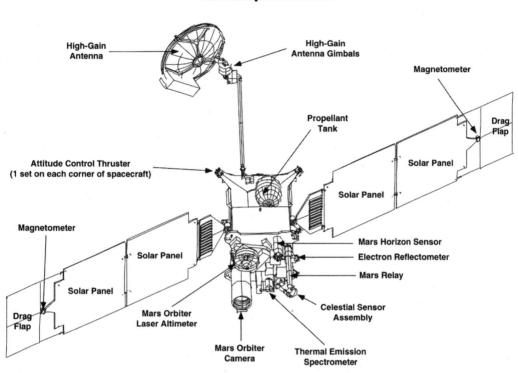

Mars Global Surveyor spacecraft

The main component of the Mars Global Surveyor spacecraft is a rectangular body, or bus, that houses computers, the radio system, solid-state data recorders, fuel tanks and other equipment. Attached to the outside of the bus are several rocket thrusters, which are fired to adjust the spacecraft's path during the cruise to Mars and to modify the spacecraft's orbit around the planet.

At launch, the spacecraft weighed 1,060 kilograms (2,337 pounds), including the science payload and fuel, and stands about 3 meters (10 feet) tall. The bus or main body of the spacecraft measures 1.2 by 1.2 meters (4 by 4 feet) and is 12 meters (40 feet) across from tip to tip when the solar panels are fully unfolded. The high-gain antenna, deployed in the cruise phase, measures 1.5 meters (4.9 feet) in diameter. The high-gain antenna will be deployed on a 2-meter-long (6-1/2-foot) boom.

To minimize costs, spare units left over from the Mars Observer mission were used in portions of the spacecraft's electronics and for some of the science instruments. The spacecraft design also incorporates new hardware — the radio transmitters, solid state recorders, propulsion system and composite material bus structure.

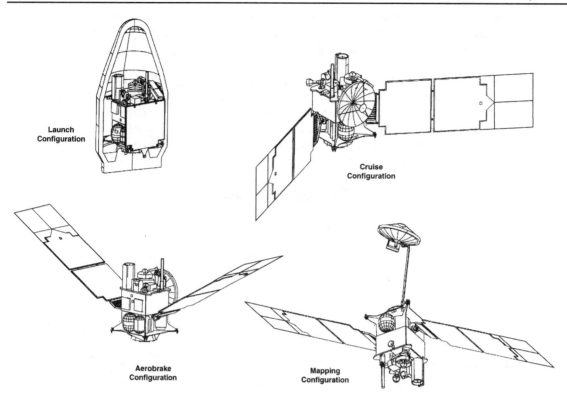

Mars Global Surveyor configurations

Mars Global Surveyor will orbit the planet so that one side of the bus, called the nadir deck, always faces the Martian surface. Of the six science instruments, four — the Mars orbiter camera, the Mars orbiter laser altimeter, the electron reflectometer and the thermal emission spectrometer — are attached to the nadir deck, along with the Mars relay radio system. The magnetometer sensors are mounted on the ends of the solar arrays.

The bus has two solar-array wings and a boom-mounted high-gain communications antenna. The solar arrays, which always point toward the Sun, provide 980 watts of electricity for operating the spacecraft's electronic equipment and for charging nickel hydrogen batteries. The batteries will provide electricity when the spacecraft is mapping the dark side of Mars. To maintain appropriate operating temperatures, most of the outer exposed parts of the spacecraft, including the science instruments, are wrapped in thermal blankets.

In its mapping configuration, the dish-shaped high-gain antenna is deployed on the end of a 2-meter (6½ foot) boom so that its view of Earth will not be blocked by the solar arrays as the spacecraft orbits Mars. Measuring 1.5 meters (4.9 feet) in diameter, this steerable antenna will be pointed toward Earth even though the spacecraft's position will be continuously adjusted during mapping to keep the nadir deck pointed toward Mars. The spacecraft's radio system, including the high-gain antenna, also will function as a science instrument. Researchers will use it in conjunction with NASA's Deep Space Network ground stations for the radio science investigations.

Spacecraft communications with Earth will always utilize X-band frequencies for radio tracking, return of science and engineering telemetry, commanding and the radio science experiments.

However, the spacecraft's telecommunications equipment also accommodates Ka-band downlink, which was furnished as an experiment to demonstrate its feasibility for future missions. Primary communications to and from the spacecraft occur through the 1.5-meter-diameter (4.9-foot) high-gain antenna.

From launch through the aerobraking operations, the high-gain antenna remains fixed to one side of the spacecraft, and the spacecraft must be slewed to point the antenna directly toward Earth for communications. Just before the start of mapping, the high-gain antenna will be deployed on the end of a 2-meter (6.5-foot) boom mounted to the same side of the spacecraft. This configuration will allow the antenna to automatically track Earth by using two gimbals that hold the antenna to the boom.

In addition to the high-gain antenna, the spacecraft also carries four low-gain antennas that could be used in the event ground-controllers lose the signal from the high-gain antenna. Two of these low-gain antennas function as transmit antennas, while the other two can receive signals from Earth. Placement of these four low-gain antennas insures that the spacecraft can receive commands and transmit downlink telemetry in a wide range of orientations in space.

The spacecraft's 25-watt radio frequency amplifiers allow Global Surveyor to transmit science and engineering telemetry at data rates between 21,333 symbols per second to 85,333 symbols per second. A symbol is a specially encoded bit. It takes approximately 1.147 bits of storage space to encode one bit of raw data with this particular encoding scheme.

Data rates for sending commands to the spacecraft will vary from as low as 7.8 bits per second using the low-gain antennas to as high as 500 bits per second using the high-gain antenna. The standard date rate is 125 bits per second. Global Surveyor can receive instructions from Earth at a maximum rate of 12.5 commands per second.

Science Objectives

Mars Global Surveyor is designed to provide a detailed, global map of Mars that will allow scientists to study its geology, climate and interior. Key science objectives are to:

* Characterize the surface features and geological processes on Mars.

* Determine the composition, distribution and physical properties of surface minerals, rocks and ice.

* Determine the global topography, planet shape and gravitational field.

* Establish the nature of the magnetic field and map the crustal remnant field. (A crustal remnant field is evidence of magnetism that is detected within the planet's crust or rocks, produced by the planet's own magnetic field at the time of formation.)

* Monitor global weather and the thermal structure of the atmosphere.

* Study interactions between Mars' surface and the atmosphere by monitoring surface features, polar caps that expand and recede, the polar energy balance, and dust and clouds as they migrate over a seasonal cycle.

Among the questions scientists wish to answer are those relating to Mars' early atmosphere and the dramatic climate changes which sent the planet into a deep freeze. All the ingredients necessary for life exist on Mars, including water, yet the surface of Mars is totally dry and probably devoid of life today.

Water is fundamental to the understanding of geological processes and climate change. But water cannot exist in liquid form at the low atmospheric pressures that currently prevail on the surface of Mars; it turns into water vapor or ice. Spacecraft have photographed numerous large and fine channels across the surface of Mars, with shapes and structures that indicate, almost beyond a doubt, that they were carved by running water. Where has the water gone? Only a tiny fraction is known to exist in the northern polar cap and in the atmosphere.

Some of it may have escaped into space, but scientists believe that most of it should have remained on Mars. They want to know if water could be hidden in permafrost — thick layers of ice-rock — beneath the surface, just as it is trapped in the polar regions on Earth. The origin and evolution of Mars are still a mystery. Thought to have formed 4.6 billion years ago, in much the same way as the other rocky planets of the inner solar system, Mars has two distinct hemispheres, roughly divided by the equator. The southern hemisphere is badly battered, perhaps the result of an intense bombardment by debris as the planet was forming. This part of Mars may be closest in history and age to the heavily cratered faces of the Moon and Mercury. Other regions of Mars may be widespread plains of volcanic lava, which erupted from within the planet over a long period of time. Similar eruptions spread across Earth's Moon to form the dark areas known as lunar maria, or "seas."

During the last 2 billion or 3 billion years, Mars also developed features that resemble those of Earth rather than the Moon. Geologic activity in the younger, northern hemisphere created huge, isolated volcanoes — most notably Olympus Mons and the other volcanoes along the Tharsis uplift — as the interior of Mars melted and lava rose to the surface. A huge canyon just below the Martian equator, called Vallis Marineris, would dwarf Earth's Grand Canyon, stretching 5,000 kilometers (3,100 miles) across the planet's surface. Many sinuous channels, apparently cut by running water that may have flooded regions of Mars hundreds of millions of years ago, also appear in the northern hemisphere.

Science Experiments

Mars Global Surveyor carries a complement of six scientific instruments which have been furnished by NASA centers as well as universities and industry. They are:

* **Thermal Emission Spectrometer.** This instrument will analyze infrared radiation from the surface. From these measurements, scientists can determine several important properties of the rocks and soils that make up the Martian surface: how hot and cold they get during the cycles of night and day; how well they transmit heat; the distribution of rock and grain sizes; and the amount of the surface covered by large rocks and boulders. Scientists will also be able to identify minerals in solid rocks and sand dunes, which will be key to understanding how Martian bedrock has weathered over millions of years and how it might be weathering today. The instrument can also provide information about the Martian atmosphere, especially the locations and nature of short-lived clouds and dust. Principal investigator is Dr. Philip Christensen, Arizona State University.

* **Mars Orbiter Laser Altimeter.** This experiment will measure the height of Martian surface features. A laser will fire pulses of infrared light 10 times each second, striking a 160-meter (525-foot) area on the surface. By measuring the length of time it takes for the light to return to the spacecraft, scientists can determine the distance to the planet's surface. Data from this instrument will give scientists elevation maps precise to within about 30 meters (100 feet) from which they will be able to construct a detailed topographic map of the Martian landscape. Principal investigator is Dr. David Smith, NASA Goddard Space Flight Center.

* **Magnetometer/Electron Reflectometer.** The magnetometer/electron reflectometer will search for evidence of a planetary magnetic field and measure its strength, if it exists. These measurements will provide critical tests for current speculations about the early history and evolution of the planet. The instrument will also scan the surface to detect remnants of an ancient magnetic field, providing clues to the Martian past when the magnetic field may have been stronger due to the planet's higher internal temperature. Principal investigator is Dr. Mario Acuna, NASA Goddard Space Flight Center.

* **Radio Science.** The radio science investigation will use data provided by the spacecraft's telecommunications system, high-gain antenna and an onboard ultra-stable oscillator, which is like an ultra precise clock, to map variations in the gravity field by noting where the spacecraft speeds up and slows down in its passage around Mars. From these observations, a precise map of the gravity field can be constructed and related to the structure of the planet. In addition, scientists will study how radio waves are distorted as they pass through Mars' atmosphere in order to measure the atmosphere's temperature and pressure. Dr. G. Leonard Tyler, of Stanford University, is the radio science team leader.

* **Mars Orbiter Camera.** Unlike cameras on spacecraft such as Galileo or Voyager, which take conventional, snapshot-type exposures, the Mars orbiter camera uses a "push-broom" technique that builds

up a long, ribbon-like image as the spacecraft passes over the planet. The camera will provide low-resolution global coverage of the planet every day, collecting images through red and blue filters. It will also obtain medium and high-resolution images of selected areas. The wide-angle lens is ideal for accumulating a weather map of Mars each day, showing surface features and clouds at a resolution of about 7.5 kilometers (4.6 miles). These global views will be similar to the types of views obtained by weather satellites orbiting the Earth. The narrow-angle lens will image small areas of the surface at a resolution of 2 to 3 meters (6.5 to 9.5 feet). These pictures will be sharp enough to show small geologic features such as boulders and sand dunes — perhaps even the Mars Pathfinder lander and the now-silent Viking landers — and may also be used to select landing sites for future missions. Principal investigator is Dr. Michael Malin, Malin Space Science Systems Inc., San Diego, CA.

* **Mars Relay System.** Mars Global Surveyor carries a radio receiver/transmitter supplied by the French space agency, Centre National d'Etudes Spatiales, which was originally designed to support the Russian Mars '96 mission, lost during launch. Now it will be used to relay data from the microprobes carried on the 1998 Mars Global Surveyor Lander mission, as well as to serve as a backup to relay data from the lander itself. Data relayed from the surface to Mars Global Surveyor will be stored in the large solid-state memory of the orbiter's camera, where it will be processed for return to Earth. This collaborative effort will maximize data collection. Following support of the Mars '98 mission, the Mars relay system is expected to provide multiple years of in-orbit communications relay capability for future international Mars missions.

Program/Project Management

Mars Global Surveyor is the first mission in a sustained program of Mars exploration called the Mars Surveyor program — which will send low-cost pairs of orbiters and landers to Mars every 25 months well into the next century.

The Mars Global Surveyor mission is managed by the Jet Propulsion Laboratory for NASA's Office of Space Science, Washington, DC. At NASA Headquarters, Dr. Wesley T. Huntress is associate administrator for space science. Kenneth Ledbetter is director for mission and payload development. Dr. William Piotrowski is Mars Global Surveyor program executive. Dr. Patricia Rogers is Mars Global Surveyor program scientist.

At the Jet Propulsion Laboratory, Norman Haynes is director of the Mars Exploration Directorate. Donna Shirley is manager of the Mars Exploration Program. Glenn E. Cunningham is Mars Global Surveyor project manager. Dr. Arden Albee of the California Institute of Technology, Pasadena, CA, is Mars Global Surveyor project scientist.

JPL's industrial partner is Lockheed Martin Astronautics, Denver, CO, which developed and operates the spacecraft. Navigation and ground data support are provided by JPL. Science operations will be carried out by NASA's Goddard Space Flight Center, Greenbelt, MD; Arizona State University, Tempe, AZ; Malin Space Science Systems, San Diego, CA; Stanford University, Palo Alto, CA; and NASA's Jet Propulsion Laboratory, Pasadena, CA.

Mars Global Surveyor Project Mission Plan

Final Version, Rev. B (MGS 542-405)

Jet Propulsion Laboratory
California Institute of Technology Pasadena CA, 91109

JPL D-12088
November 1996

National Aeronautics and Space Administration
Mars Global Surveyor Project
Mission Plan Document
Final Version, Rev. B (MGS 542-405, November 1996)

Prepared by:
Wayne Lee MGS/MSOP Mission Planner

Approved by:
Joseph Bearer MSOP Flight Operations Manager

Approved by:
Sam Dallas MGS Mission Design Manager
National Aeronautics and Space Administration

1. Overview

The Mission Plan for Mars Global Surveyor (MGS) describes a baseline strategy for successfully achieving the scientific objectives of the mission within the capabilities and constraints of the project systems, subject to the project policies on the use of these systems. This strategy will serve as a starting point for planning detailed event sequences, both before launch and during the mission. The Mission Plan also serves as the basic reference for a detailed description of the mission. It documents the major design features and options considered by the Project along with the rationale for major technical decisions.

1.1 Scope

The Mission Plan Document for Mars Global Surveyor responds to the high-level project policy and requirements documents. Accordingly, the Mission Plan does not originate new mission and system requirements. It describes a strategy for conducting the baseline mission which is consistent with the requirements, capabilities, and constraints defined in other project documentation. The baseline mission is the planned series of mission events that barring failure, normally proceed from launch to the end of the mission.

Section 2   Overview of Spacecraft Design, Science Instruments, and Baseline Mission

Section 3   Description of Launch Phase

Section 4   Description of Cruise Phase

Section 5   Description of Orbit Insertion Phase

Section 6   Description of Mapping Phase

Release Version and Release Comments

This November 1996 Mission Plan release is an update to the Final Version (released September 1995) and supersedes all previous releases. The following list summarizes the major changes between this Mission Plan and the previous release.

a)   Mission event calendar updated to be consistent with a 6 November 1996 launch date and new end of mission date on 1 August 2000. Delta-V budget also updated accordingly.
b)   Mission synopsis, spacecraft description, and payload description completely rewritten to be more clear for those not directly involved with the project's design.
c)   Gravity wave search campaign added to the mission schedule.
d)   Launch targets and lift-off times updated to be consistent with the final report from McDonnell Douglas.
e)   Launch sequence of events and post-separation sequence of events updated to match latest sequence design from Lockheed-Martin.
f)   Post AB1 burn (first aerobrake maneuver) periapsis altitude is now 150 km.
g)   Aerobrake walk-in profile updated to be consistent with "critical scale height" maneuver strategy.
h)   Aerobrake walk-out profile updated to be consistent with two-day orbit lifetime.
i)   Orbital elements for mapping updated using the Mars 50c gravity field model.
j)   Science campaign schedule updated and now includes new 29-day campaign at the start of the mapping phase. This campaign will primarily be dedicated to production of the MOLA global topography map.

k)   Data rate usage profile during mapping updated to reflect latest telecom performance profile numbers.

Disclaimer

The information presented in this document represents the final state of the project's mission design efforts and is current as of 31 October 1996. However, some design issues may be examined in more detail during the cruise phase of the mission. Any changes will be documented in change requests or in technical memorandums.

1.2 Relationship to Other Documents

The Mission Plan is consistent with and responsive to the requirements and objectives of the following project documents:

Mars Global Surveyor Project Plan	(542-010, December 1994)
Investigation Description and Science Requirements Document	(542-030, February 1995)
Mission Requirements Document	(542-400, September 1995)
Planetary Protection Policy	NASA NHB 80201.12B

Based on the best information available as of 31 July 1996, the Mission Plan is consistent with the capabilities and constraints of the project systems as described in the following documents:

Spacecraft Requirements	(542-200, October 1994)
Mission Operations Specification	(542-409, September 1995)
Navigation Plan	(542-406, September 1995)
Planetary Protection Plan	(542-402, October 1996)

Some more detailed aspects of the mission design for Mars Global Surveyor are described in the following documents:

Trajectory Characteristics Document	(542-410, September 1995)
Delta 2 Target Specification	(542-411, June 1995)
Mars Observer Planetary Constants and Models	(642-321, November 1990)
Mission Sequence Plan	(542-407, September 1995)
Detailed Mission Requirements on the DSN	(542-424, September 1996)

1.3 Update History

During the development phase of the mission, updates for the Mission Plan occurred as specified by the following schedule. Future revisions may be released during flight if deemed necessary by the Flight Operations Manager. However, there are no further releases scheduled at this time.

Draft Version August 1994
Preliminary September 1994
Draft for Final June 1995
Final Version September 1995
Final, Revision A July 1996
Final, Revision B November 1996

1.4 Acknowledgments

The authors of the mission plan are extremely grateful to the following members of the MGS team who provided information and inputs critical to the writing of the plan:

Joe Beerer, Bill Blume, Daren Casey, Dan Johnston, Dan Lyons   Mission Design (JPL)
Nick Smith, Jim Taylor, Wayne Sidney, Bill Willcockson   Mission Design (LMA)
John Callas, Mick Connally, Tom Thorpe   MGS Science Office (JPL)
Gene Bollman, Pat Esposito   Navigation (JPL)
Stan Butman, George Chen, Charles Whetsel   Spacecraft & Subsystems (JPL)

1.5 Questions or Comments?

General comments, corrections, suggestions, or inquires about this document may be submitted to the authors:

Wayne Lee Jet Propulsion Laboratory
Mail Stop 264-235 4800 Oak Grove Drive
Pasadena, CA 91109 (818) 354-8784
wayne@cranberry.jpl.nasa.gov

2. Mission Background

In November 1996, NASA and the Jet Propulsion Laboratory will begin America's return to Mars after a 20-year absence by launching the Mars Global Surveyor (MGS) spacecraft. The MGS mission will recover most of the lost objectives of the 1992 Mars Observer mission by delivering a single spacecraft to the red planet for a two-year study of Mars' surface, atmosphere, and gravitational and magnetic fields. Achieving the scientific objectives of the MGS mission will require placing the spacecraft in a low-altitude, near-polar, Sun-synchronous orbit around Mars and returning data over a complete Martian year. This document will provide an overview of the MGS mission plan, spacecraft, and science instruments.

2.1 Mission Synopsis

A Delta 2 (7925A) launch vehicle will boost the 1,062-kilogram Mars Global Surveyor spacecraft from Cape Canaveral Air Station (CCAS) during the November 1996 launch opportunity (see Figure 2-1 on for the overall mission timeline). The spacecraft will utilize a Type-2 transfer trajectory with a trans-Mars flight time

Figure 2-1: *General Timeline for MGS Mission*

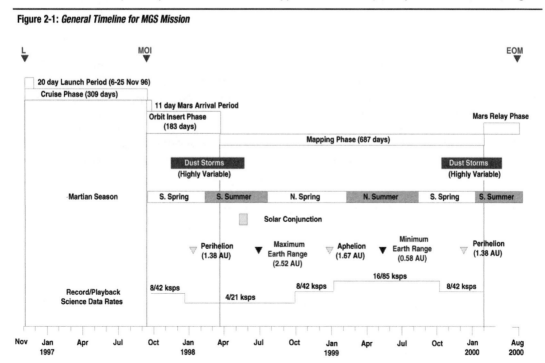

of about ten months. After arriving at the red planet in September 1997, MGS will be propulsively inserted into an initial, highly elliptical capture orbit with a period of 48 hours. Over the next five months, the spacecraft will be gradually lowered into the mapping orbit by the use of aerobraking. This technique works by dipping the spacecraft into Mars' upper atmosphere on every periapsis passage in order to slow down and lower apoapsis.

Mars Global Surveyor will utilize a Sun-synchronous mapping orbit at a 378-km index altitude, and with a descending node orientation of 2:00 p.m. with respect to the fictitious mean Sun. In this 92.9° inclination orbit, the MGS spacecraft will circle the red planet once every 117.65 minutes. Once every seven Martian days (sols), the spacecraft will approximately retrace its ground track. After each seven sol cycle (88 orbits), the ground track pattern will be offset eastward by 59 km from the tracks on the previous cycle. This orbit will provide a repeat cycle scheme that will allow 99.9% global coverage to be built up from repeated instrument swaths across the planet.

Repetitive observations of the planet's surface and atmosphere from the mapping orbit will be conducted over a time span of one complete Martian year (687 Earth days), from March 1998 to January 2000. Data returned from six prime experiments on the spacecraft will provide for a better understanding of the geology, geophysics, and climatology of Mars. Five of those six will utilize scientific instruments mounted to the spacecraft. The sixth investigation will collect data about Mars by analyzing the spacecraft's radio signal after it reaches the Earth.

Throughout the entire mapping period, the spacecraft will remain in an orientation with the scientific instruments nadir pointed. Because MGS lacks a scan platform, any scanning capability will be provided by the instruments. The normal sequence of collecting science data will involve recording continuously for 24 hours, and then playing it back through the Deep Space Network (DSN) during a 10-hour tracking pass once every Earth day. Approximately every third Earth day, an additional tracking pass will be scheduled to return high-rate, realtime data.

From the end of mapping until the end of mission on 1 August 2000, MGS is scheduled to support the Mars Exploration Program by relaying data from various landers and atmospheric vehicles back to the Earth through the spacecraft's Mars Relay antenna.

2.2 Mission Phases

Five mission phases have been defined to simplify the description of different periods of activity. These are the launch, cruise, orbit insertion, mapping, and relay. Several time epochs will be used throughout this document for defining activities and the boundaries of some of the mission phases and subphases. Launch (L) denotes the liftoff time, defined to be the instant of ignition of the Delta 2's first stage. Injection (TMI) represents the burn-out time of the Delta 2's third stage, used for the trans-Mars injection burn that places MGS onto a trajectory bound for the red planet. Mars Orbit Insertion (MOI) designates the time that the spacecraft begins the propulsive maneuver to slow down and enter Martian orbit after the completion of its interplanetary trajectory from Earth. End of Mission (EOM) marks the end of ground operations to control spacecraft activities and collect data. Table 2-1 summarizes the dates of the mission phases and some key mission events. The dates in Table 2-1 are specific to a mission that launches at the opening of the launch period on 6 November 1996. Under the current launch period strategy, MGS can launch as late as 25 November 1996.

2.2.1 Launch

After lift-off, the first stage of the three-stage Delta rocket will boost the spacecraft to an altitude of 115 km. From there, the second stage will take over and achieve a 185-km, circular, parking orbit at about L+ 10 minutes. After parking orbit insertion, the booster and spacecraft will coast for between 24 and 37 minutes (variable with launch date) until reaching a position over the eastern Indian or western Pacific ocean. At that time, stage two will re-start and thrust for nearly two minutes to raise the apogee of the parking orbit. Then,

Table 2-1: *Overview of Mission Phases*

Event	Date	Comments
Launch	6-Nov-1996 (early afternoon EST)	Launch period opens on 6-Nov and closes on 25-Nov-96
Inner Cruise Phase	6-Nov-1996 to 6-Jan-1997	Communications through LGA only because solar arrays must be pointed at a fixed angle to the Sun
TCM1	21-Nov-1996 (L+ 15 days)	Trajectory correct for injection errors, remove aim-point biasing introduced for Mars planetary quarantine
Outer Cruise Phase	7-Jan-1997 to 11-Sep-1997	Communications through HGA, phase begins when Earth-MGS-Sun angle falls below 60°
TCM2	21-Mar-1997 (TCM1+ 120 days)	Correct for execution errors from TCM1
TCM3	20-Apr-1997 (TCM2+ 30 days)	Correct for execution errors from TCM2
TCM4	22-Aug-1997 (MOI- 20 days)	Final adjustment to MOI aim point
Mars Orbit Insertion (MOI)	11-Sep-1997 (about 01:15 UTC)	MOI can vary from 11-Sep-97 to 22-Sep-97 depending on exact launch date
Orbit Insertion Phase	11-Sep-1997 to 14-Mar-1998	Begins at MOI, lasts 5 months to reach mapping orbit using aerobraking and propulsive maneuvers
Mapping Phase	15-Mar-1998 to 31-Jan-2000	Mars mapping operations for one Martian year, about 687 Earth days in duration
Relay Phase	1-Feb-2000 to 1-Aug-2000	About 6 months of support for future Mars missions

spin rockets will spin-up the third stage and spacecraft to 60 r.p.m., followed by third stage ignition. The Delta's third stage, a STAR 48B solid, will fire for 87 seconds to complete the trans-Mars injection (TMI) burn. After completing the TMI burn, but before third stage jettison, a yo-yo cable device will deploy from the STAR 48B to de-spin the spacecraft. Section 3 contains more detail.

2.2.2 Trans-Mars Cruise

Cruise covers the time of ballistic flight between Earth and Mars. The spacecraft will take between 301 and 309 days to reach the red planet on its Type-2 trajectory depending on the Earth departure date within the 20-day launch period. A launch at the open of the launch period on 6 November 1996 will correspond to a Mars arrival date of 11 September 1997, while a launch at the close of the period on 25 November 1996 will result in an arrival on 22 September 1997. During cruise, a set of four trajectory correction maneuvers (TCMs) will adjust the interplanetary trajectory to ensure that the spacecraft reaches the proper velocity and position targets prior to the Mars orbit insertion (MOI) burn.

Inner Cruise

During the first part of cruise, called inner cruise, initial deployment and checkout of the spacecraft and payload will be accomplished, and navigation tracking data will be taken to determine the flight path for the purpose of planning and executing the first of four planned trajectory correction maneuvers (TCMs). TCM1 is scheduled to occur 15 days after launch (L+ 15 days).

In inner cruise, all spacecraft communications with the Earth will occur though the low-gain antenna (LGA). The reason is primarily due to the spacecraft configuration and solar panel geometry. Because the high gain antenna (HGA) sits on the spacecraft in a stowed, body-fixed orientation during cruise, communicating with the Earth through the HGA will require turning the spacecraft to point the antenna directly at Earth. However, such an orientation would push the incidence angle of sunlight on the panels past acceptable levels for minimum power generation. Therefore, communications through the LGA represents the only feasible option.

Outer Cruise

Outer cruise will begin when the spacecraft switches from use of the low gain to the high gain antenna for communications with the Earth. The exact time when the switch becomes feasible depends on when the angle between the Sun and Earth as seen from the spacecraft (SPE) falls to a level low enough to allow good power while the spacecraft is oriented to point the HGA directly at Earth. This angle starts at about 120° at the time of launch and falls to less than 60° by 6 January 1997, assuming a launch on 6 November 1996.

Currently, the transition date to switch to the HGA from the LGA will occur on 6 January 1997 for the purpose of planning command sequences. However, this date will be subject to change during flight as the spacecraft team evaluates the telemetry. In the interest of maintaining the highest possible communications link margin with Earth, switch over will occur as early as possible.

Most of outer cruise will consist of minimal activity as the spacecraft transits to Mars. The vast majority of the events will involve acquiring navigation and tracking data to support the remaining TCMs. During the last 30 days of approach to Mars, the focus will be on final targeting of the spacecraft to the proper aim point, and preparations for orbit insertion. Some science observations during this "Mars approach" time period will also occur.

2.2.3 Mars Orbit Insertion

MOI will slow the spacecraft and allow Mars to capture it into an elliptical orbit. Before the burn, the spacecraft's velocity relative to Mars will measure approximately five kilometers per second. Near the periapsis of the inbound hyperbolic trajectory, the 659-N main engine will fire for between 20 to 25 minutes to provide a Delta-V of about 980 m/s. Burn ignition will occur about 10 minutes before periapsis. During the burn, the spacecraft will utilize a "pitch-over" maneuvering strategy to slew the spacecraft at a constant rate in an attempt to keep the thrust nearly tangent to the trajectory arc. After MOI-burn cut-off 10 minutes after periapsis, the spacecraft will orbit Mars on a highly elliptical orbit with a period of 48 hours and periapsis altitude of about 300 km (periapsis radius of 3,700 km).

2.2.4 Aerobraking

The MGS spacecraft will not carry enough propellant to propulsively reach the required low-altitude, Sun-synchronous mapping orbit due to the relatively low interplanetary injected mass capability of the low-cost Delta booster. Consequently, the spacecraft will rely on aerobraking, an innovative mission-enabling technique, to trim the initial, highly elliptical, capture orbit down to mapping orbit altitudes. During aerobraking, the spacecraft will pass through the upper fringes of the Martian atmosphere on every periapsis pass. Friction from the atmosphere during the drag pass will cause the spacecraft to lose a small amount of momentum and will cause the altitude on the next apoapsis pass to slightly decrease. The rate at which the apoapsis altitude decreases will be determined by the amount of drag generated. Aerobraking deeper in the atmosphere will provide greater drag and reduce the orbit faster, but will generate higher spacecraft temperatures and dynamic pressures.

Walk-in

Aerobraking will begin nine days after MOI with the first (AB1) of four to six maneuvers designed to lower the periapsis altitude into the Martian atmosphere in gradual steps. The AB1 burn will be the largest of the six and will lower the periapsis to 150 km. The next maneuvers (AB2, AB3, AB4, AB5 and AB6) will provide a further drop to about 112 km. This need for a gradual walk-in is due to both the large uncertainty in the atmospheric density model of Mars, and the time required by the navigation team to gauge the atmospheric density and its orbit-to-orbit variation. Consequently, the navigation team may decide to add or eliminate maneuvers as necessary during walk-in operations. All of these burns will be performed using the attitude control thrusters.

Main Phase

After completion of walk-in, the spacecraft will spend about three months in the main phase of aerobraking. During this phase, the apoapsis altitude will shrink in size from about 56,675 km down to 2,000 km. As needed, small propulsive maneuvers (ABMs) executed at apoapsis will maintain periapsis within a well-defined periapsis altitude corridor low enough to produce enough drag to reduce the orbit within the time constraints to reach the 2:00 p.m. node, yet high enough to avoid spacecraft heating and maximum dynamic pressure limits. Due to the oblateness of Mars and the fact that periapsis will be migrating northward toward the pole during main phase, the altitude of periapsis will tend to rise. Consequently, most of the ABMs will be in the down direction to lower the periapsis altitude into the control corridor.

Figure 2-2: *Schematic Diagram of Key Mission Phases (not to scale)*

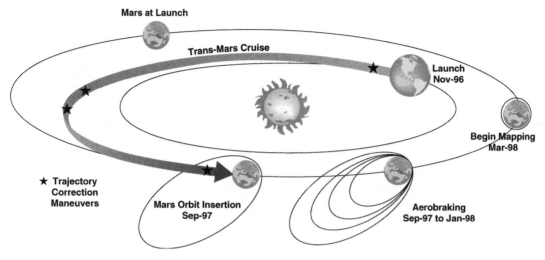

Walk-Out

The three weeks of aerobraking following the main phase will represent an extremely critical period as the spacecraft lowers its apoapsis toward the target finish altitude of 450 km. During this time, the spacecraft will slowly "walk-out" of the atmosphere by gradually raising its periapsis altitude to 143 km. Daily ABMs will be performed as necessary to maintain a guaranteed worst-case, two-day orbit lifetime. In other words, in the absence of ABMs due to unforeseen events that inhibit the ability of flight controllers to send commands, the spacecraft will always be at least two days from crashing into the surface.

Aerobraking will end with a termination burn (ABX) performed sometime during mid-January 1998. This burn will raise the orbit periapsis out of the atmosphere to an altitude of approximately 450 km. At this time, the spacecraft will be circling in a 400 x 450 km orbit with a period slightly under two hours. In addition, the descending node location will have regressed from its original MOI position at 5:45 p.m. with respect to the fictitious mean Sun to nearly 2:00 p.m.

2.2.5 Transition to Mapping

Transition-to-mapping will begin at the end of aerobraking and lasts until the final mapping orbit has been established and the spacecraft is declared ready to begin mapping operations. After the aerobrake termination burn (ABX), the spacecraft will circle Mars in a transition orbit for a month. During this waiting period, the oblateness of Mars will alter the orbit and cause the location of periapsis to drift to a position almost immediately above the Martian South Pole. At that time, currently scheduled for mid-February 1998, the transition-to-mapping-orbit (TMO) burn will be performed with the intent of "freezing" the periapsis location at the South Pole, and establishing the proper altitude for mapping operations. Throughout this entire transition period between ABX and TMO, the navigation team will conduct gravity calibrations to update the Martian gravity field model using in-flight navigation data returned from the red planet. This update will be crucial toward accurately executing the TMO and other future burns.

An orbit-trim maneuver (OTM) burn will be executed to refine the frozen orbit 12 days after TMO. The 12 days is driven by the navigation team's need for four days to track the spacecraft after the TMO burn and another eight days to plan the maneuver. After the OTM, a ten-day spacecraft deployment and checkout period will follow to allow the operations team time to configure the spacecraft and its instruments for mapping operations.

Figure 2-3: *View of the MGS Spacecraft in Mapping Configuration*

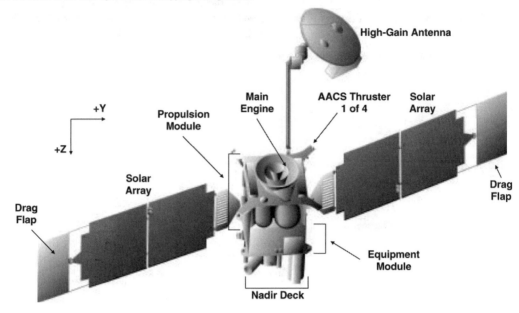

2.2.6 Mapping

Mapping phase represents the period of concentrated return of science data from the mapping orbit. This phase will start on 15 March 1998 and last until 31 January 2000, a time period of one Martian year (687 Earth days). These dates will remain fixed and are independent of the actual day of lift-off within the launch period. During this phase, the spacecraft will keep its science instruments (+Z panel of the spacecraft) nadir pointed to enable data recording on a continuous basis. On a daily basis, the spacecraft will transmit 24 hours of recorded data back to Earth during a single 10-hour Deep Space Network (DSN) tracking pass. An articulating high gain antenna (HGA) on the spacecraft will allow data recording to proceed while downlink to Earth is in progress.

2.2.7 Relay

The relay operations phase will begin at the end of mapping and continue for about six months. During this phase, the spacecraft will function as a relay satellite for various Mars landers and orbiters in support of the Mars Exploration Program. End of mission will occur on 1 August 2000.

Before relay operations begin, a quarantine orbit raise maneuver will elevate the spacecraft to a near-circular orbit with an average altitude of approximately 400 kilometers. This maneuver is needed to reduce the probability of spacecraft impact with Mars before L+ 20 years and L+ 50 years to the values required for planetary protection. The primary goal of the quarantine maneuver is to raise the orbit. If enough propellant remains after the mapping mission, a slight inclination change will also be performed to maintain the spacecraft's Sun-synchronous orientation. However, no project or program requirements exist with respect to repeat cycle characteristics for the relay orbit.

2.3 Spacecraft Description

Lockheed Martin Astronautics is currently building the Mars Global Surveyor spacecraft at their Denver facility. The design for this three-axis stabilized spacecraft was primarily derived from the Mars Observer spacecraft, with necessary modifications made to support aerobraking, to incorporate the action plans from the Mars Observer failure reports, and to fit onto the Delta 2 launch vehicle. When fully loaded with propellant at the time of launch, the Mars Global Surveyor spacecraft will weigh 1,062.1 kilograms under the current design and Delta-V budget. In order to meet this target mass, the spacecraft structure will consist of lightweight composite material divided into four sub-assemblies known as the equipment module, the propulsion module, the solar array support structure, and the high-gain antenna support structure (see Figure 2-3).

2.3.1 Equipment Module

For the most part, the equipment module houses the avionics packages and science instruments. The dimensions of this rectangular shaped module measure 1.221 x 1.221 x 0.762 meters in the X, Y, and Z directions, respectively. With the exception of the Magnetometer, all of the science instruments will be bolted to the nadir equipment deck, mounted above the equipment module on the +Z panel. The Mars Relay antenna is the tallest instrument and extends up 1.115 meters above the nadir equipment deck.

Flight Computers

Inside the equipment module, two redundant flight computers will orchestrate almost all of the spacecraft's flight activities. Although only one of the two units will control the spacecraft at any one time, identical software will run concurrently in the backup unit. Each computer control unit consists of a Marconi 1750A microprocessor, 128 K RAM for storing command sequences as long as six weeks in duration uploaded from Earth, and 20 K PROM that contains code to run basic survival sequences upon entry into fault protection mode due to anomalous conditions.

Additional storage space for science and engineering data will be provided by two solid-state recorders, each with a 1,500 Mbit capacity. The MGS mission will represent America's first interplanetary spacecraft to exclusively use RAM instead of a tape recorder for mass data storage. This technological improvement will dramatically reduce operational complexity, thereby reducing mission planning costs during flight.

Attitude Control

The equipment module also contains three reaction wheels mounted in orthogonal directions to provide spacecraft pointing control authority for all mission events, except for major propulsive maneuvers. A fourth wheel mounted in a direction skewed to the other three will serve as a redundant unit. Attitude information for the spacecraft will be provided by Sun sensors, an inertial measurement unit, Mars horizon sensors, and a star scanner. Table 2-2 describes these devices in more detail.

Table 2-2: *Spacecraft Attitude Knowledge Devices*

Device	Location	Comments
Sun Sensors	Solar Panels	Provides vector to the Sun, used to begin attitude re-initialization in the event of an anomaly.
Inertial Measurement Unit (IMU)	Inside Equipment Module	Contains gyroscopes and accelerometers that measure angular rates and linear accelerations. Used to determine yaw attitude during Mars mapping, and to aid spacecraft in pointing inertially for both fixed pointing and attitude slews during maneuvers.
Mars Horizon Sensor (MHSA)	Nadir Equipment Deck	Looks at atmospheric horizon during Mars mapping to define the nadir direction by sensing roll and pitch errors.
Star Scanner (CSA)	Nadir Equipment Deck	Complements IMUs by scanning distant stars to provide inertial attitude data.

2.3.2 Propulsion Module

The propulsion module serves as the adapter between the launch vehicle and contains the nitrogen tetroxide (NTO) and hydrazine tanks, main engine, propulsion feed system, and attitude control thrusters. This module bolts beneath the equipment module on the -Z panel and consists of a rectangular shaped box 1.063 meters on each side, with a 0.310 meter tall cylindrical shaped launch vehicle adapter extending from the bottom of the box. Each corner of the box portion of the module contains a small metal protrusion that houses attitude control thrusters. Including the length of these protrusions, the diagonal widths of the propulsion module measure 2.464 and 2.394 meters long.

The main engine, used for large maneuvers such as major trans-Mars trajectory corrections (TCMs) and Mars orbit insertion (MOI), will burn a bi-propellant combination of NTO and hydrazine, and will deliver an Isp of 315 to 318 seconds at a thrust level of 659 N. During main engine burns, four rocket-engine modules (REMs), each containing three 4.45 N thrusters (two aft facing, and one for roll control), will burn hydrazine in mono-propellant, pulse-on mode to provide attitude control. In addition, these mono-propellant thrusters will also be used in pulse-off mode for small trajectory corrections during cruise and orbit trim maneuvers at Mars, and for unloading momentum from the reaction wheels. This propulsion system differs from a conventional bi-propellant system in that the same hydrazine tank will serve both the main engine and attitude-control thrusters, rather than using a separate hydrazine tank for each system.

In total, the propulsion system will provide the spacecraft with a Delta-V potential of 1,281.6 m/s. This budget assumes a spacecraft dry mass of 673.73 kg and a total mass of 1,062.1 kg. Table 2-3 shows the Delta-V budget for each maneuver in the mission. "Translational" stands for the translational Delta-V that affects the trajectory parameters, "Rotational" stands for the rotational Delta-V used for attitude control during large maneuvers, and Delta-i stands for inclination change. All velocities are expressed in meters per second.

Table 2-3: *Mission ΔV Budget (Launch on 6 November 1996)*

Phase	Maneuver & Type	Translational	Rotational	Comments
Cruise	TCM 1 and 2 (bi)	36.0	0.3	Values based on 95% statistical confi-
	TCM 3 and 4 (mono)	2.0		dence level
Orbit Insertion	MOI (bi)	968.2	5.9	Pitch-over burn, ΔV includes finite burn loss, Rp = 3,700 km, T = 48 hours
Aerobrake Walk-in	AB1 (bi)	7.5	0.1	1.5 m/s reserved for Δi, Hp = 150 km
	AB2 - AB6 (mono)	2.5		Used to drop periapsis to about 112 km
Aerobrake Main Phase	ABM (mono)	5.0	5.0	Corridor control burns
Aerobrake Walk-out	ABM (mono)	20.0	30.0	Used to slowly raise periapsis
Transition to Mapping	ABX (bi)	60.2	0.4	Raise periapsis to about 400 km
	TMO (bi)	16.1	0.1	Establish mapping orbit
	OTM1 (bi)	7.5		Refine frozen orbit
Mapping	OTM Drag (mono)	3.9		Drag make-up
	Momentum Unload		40.0	Unload reaction wheels
Quarantine Raise	PQ (mono)	22.8		Performed at end of mapping
Relay	OTM Drag (mono)	1.0		Drag make-up
	Momentum Unload		10.0	Unload reaction wheels
Contingency	Pre-launch reserve (bi)	20.1		Used for aerobrake and other emer-
	AB Pop-up (bi/mono)	14.5 / 2.5		gency situations
Total (bi-propellant)		1130.1		Grand total is 1,281.6 m/s for the entire
Total (mono-propellant)		59.7	91.8	mission

2.3.3 Power

Two solar arrays, each 3.531 meters long by 1.854 meters wide, will provide energy for the MGS spacecraft. Each array mounts close to the top of the propulsion module on the +Y and -Y panels, near the interface between the propulsion and equipment modules. Including the adapter that holds the array to the propulsion module, the tips of the arrays extend 4.270 meters from the sides of the spacecraft. Rectangular shaped, metal drag "flaps" mounted onto the ends of both arrays add another 0.813 meters to the overall array structure. These "flaps" serve no purpose other than to increase the total surface area of the array structure to increase the spacecraft's ballistic coefficient during aerobraking.

Each array consists of two panels, an inner and outer panel comprised of gallium arsenide and silicon cells, respectively. Available power will start at 1,100 W immediately after launch. During mapping operations at Mars, this amount will vary from a high of roughly 980 W at Mars perihelion to about 660 W at aphelion. When the spacecraft moves into eclipse or turns away from the Sun, energy will flow from two nickel-hydrogen (NiH2) batteries, each with a capacity of 20 Amp-hours.

2.3.4 Communications

Spacecraft communications with Earth will always utilize X-band frequencies for radiometric: tracking, return of science and engineering telemetry, commanding, and radio science experiments. However, the spacecraft's telecommunications equipment also accommodates Ka-band carrier-only downlink for the purposes of providing a feasibility demonstration. Primary communications to and from the spacecraft will occur through the 1.5-meter diameter high-gain antenna (HGA). From launch until the start of mapping operations at Mars, the HGA will remain body fixed to the spacecraft on the +X side of the spacecraft. Consequently, using the HGA will require slewing the spacecraft to point directly at the Earth. During mapping operations, the HGA will be deployed and will sit at the end of a 2.0-meter boom mounted to the +X panel of the propulsion module. This configuration will allow the HGA to automatically track the Earth by means of two single-axis gimbals that hold the antenna to the boom.

In addition to the HGA, the spacecraft also carries four low-gain antennas (LGA) for emergency communications. Two of the LGAs will function as transmit antennas, while the other two will receive. Placement of these four LGAs will ensure that the spacecraft can receive commands and downlink telemetry over a wide range of attitude orientations. The primary transmit LGA is mounted on the HGA, while the backup is mounted on the +X side of the propulsion module. The two receive LGAs are mounted on the -X panel of the equipment module and the +X side of the propulsion module.

As shown in Table 2-4, the spacecraft's 25-Watt RF power amplifiers will provide the capability for downlink of science and engineering telemetry at data rates between 21,333 sps to 85,333 sps, depending on the varying Earth to Mars distance. "sps" stands for symbols per second, and a symbol is essentially a Reed-Solomon encoded (250:218 ratio) bit. Therefore, it takes approximately 1.147 bits of storage space to encode one bit of raw data with this encoding ratio.

Table 2-4: *Data Rate Modes*

Mode	Contents	Realtime Rate	Record Rate	Playback Rate
S&E1	Engineering and Science	4,000 sps	4,000 sps	21,333.33 sps
		8,000 sps	8,000 sps	42,666.67 sps
		16,000 sps	16,000 sps	85,333.33 sps
S&E2	Engineering and Science	40,000 sps	n/a	n/a
		80,000 sps	n/a	n/a
ENG	Engineering Only	2,000 bps	2,000 bps	8,000 bps
		250 bps	n/a	n/a
		10 bps	n/a	n/a

During flight, the spacecraft will generate telemetry using one of three different modes. Depending on the specific data mode (S&E1, S&E2, or ENG), the data can be recorded for later playback, returned in realtime, or both. Table 2-4 lists the different data modes and the data rates that correspond to those modes. As shown in the table, S&E1 is a dual purpose science and engineering data record or realtime mode, while S&E2 is a realtime only mode. Bit allocations for each science instrument will vary depending on the selection of S&E1 or S&E2, and the specific data rate chosen within the mode.

For example, if the S&E1 mode is chosen, data can be recorded and/or downlinked in realtime at rates of 4 ksps, 8 ksps, or 16 ksps. Data recorded for later playback will normally utilize the 21.3 ksps, 42.7 ksps, or 85.3 ksps rates, respectively. This S&E1 strategy corresponds to a 5.333:1 playback to record rate and will allow the spacecraft to return 24 hours of recorded data during a single 10-hour DSN pass.

Data rates for reception of commands from the Earth will vary from as low as 7.8125 bps (emergency situations over the LGA) to as high as 500 bps, with a normal rate of 125 bps. The maximum command reception rate amounts to about 12.5 commands received per second.

2.4 Spacecraft Operating Configurations

Throughout the mission, the Mars Global Surveyor spacecraft will utilize several different operating configurations that depend on the mission phase. Each configuration is characterized by a unique physical state of the spacecraft's appendages (solar array and high gain antenna) and philosophy of choice of attitude to balance power, thermal, and communications constraints. The main modes are launch, array normal spin, maneuver, drag pass, and mapping. Table 2-5 summarizes the configuration modes and associated spacecraft attitude by mission phase.

Table 2-5: Configuration Modes by Mission Phase

Mission Phase	Normal Configuration Mode	Spacecraft Attitude
Launch	Launch	Z axis pointed along Delta's longitudinal axis, +Z pointed toward top of rocket
Inner Cruise	ANS	+X axis pointed 60° off Sun, slow roll about +X
Outer Cruise	ANS	+X axis pointed directly at Earth, slow roll about +X
TCMs and MOI	Maneuver	Z axis aligned along inertial direction of thrust vector, +Z in the direction of the desired ΔV.
Orbit Insertion (MOI to mapping)	ANS	+X axis pointed directly at Earth, slow roll about +X
Aerobrake (drag pass only)	Aerobrake	-Z forward along velocity vector, +X nadir pointed
Mapping	Mapping	+X forward along velocity vector, +Z nadir pointed
Relay	Mapping	+X forward along velocity vector, +Z nadir pointed

2.4.1 Launch Configuration

Although the solar panels and HGA attach to the propulsion module, they will be initially stowed and folded upward against the rectangular equipment module at the time of launch. When attached to the Delta rocket, the XY plane of the spacecraft will lie perpendicular to the longitudinal axis of the booster. Since the instruments will be pointed in the spacecraft's +Z direction, they will point upward along the Delta's longitudinal axis, toward the top of the booster's payload fairing during flight (see Figure 2-4).

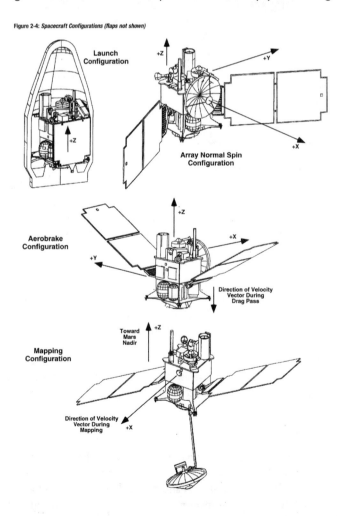

Figure 2-4: Spacecraft Configurations (flaps not shown)

About half a minute after the Delta jettisons its first stage, it will also jettison the payload fairing and expose the MGS spacecraft to the ambient environment. Because one of the mission flight rules dictates that the +Z axis (science instruments) of the spacecraft can never be pointed within 30° of the Sun, the Delta must fly an ascent trajectory, parking orbit, and trans-Mars injection profile that keeps the rocket's longitudinal axis at least 30° from the Sun.

Shortly before the Delta jettisons its second stage, it will spin the third stage and the spacecraft to a rate of 60 revolutions per minute along the booster's longitudinal axis for stabilization purposes. Therefore, the spacecraft will also spin about its +Z axis at the same rate. De-spin will occur several minutes after third stage burn-out. Under normal circumstances, the time during the third stage burn will be the only time that the spacecraft will spin at a rate faster than one revolution every 100 minutes.

2.4.2 Array Normal Spin (ANS)

In array normal spin (ANS), the solar arrays will be swept forward 30° above the Y axis in the +X direction, and the spacecraft will roll about the +X axis at the rate of one revolution every 100 minutes. This roll serves to maintain a thermal balance on the spacecraft and to constantly slew the star sensor (CSA) across the sky to maintain attitude reference (see Figure 2-4).

The spacecraft's attitude in this configuration will vary by mission phase and will balance power, thermal, and communications constraints. For example, during inner cruise, the +X axis will point to a position halfway between the Earth and Sun. This attitude represents a compromise between needing to point the +X axis directly at the Earth for maximum communications link margin, and needing to point the solar arrays at the Sun for adequate power generation. Communications with the Earth will always occur through the low gain antenna during inner cruise because the undeployed HGA on the +X axis must point directly at the Earth for use.

During all other mission phases that will utilize ANS (outer cruise, capture orbit, aerobrake non-drag pass periods, transition to mapping), the +X axis will point directly at the Earth. Consequently, the high gain antenna (HGA) will be usable because it will also point directly at Earth.

2.4.3 Maneuver Configuration

All maneuvers greater than 10 seconds in duration will use the main engine, located on the bottom (-Z panel) of the propulsion module. During the burns, the solar panels will be swept back 30° below the Y axis, toward the -Z direction. The reason for this choice is that the thrust direction of the main engine lies along the -Z axis. By sweeping the panels in that direction, as opposed to the +X direction, the spacecraft center of mass will be better aligned with the main engine's thrust axis. In all cases, the active side of the solar panels will also point in the -Z direction. The exact attitude of the spacecraft will be different for each maneuver depending on the location of the thrust vector. For main engine burns, the Z axis must point along the inertial thrust direction.

The spacecraft will use the maneuver configuration for all four trajectory correction maneuvers (TCMs) and Mars orbit insertion (MOI). In addition, this configuration will also be used for all major orbit change maneuvers at Mars such as OTMs, TMO, and ABx burns whether or not the burn is performed with the main engine or attitude control system.

2.4.4 Aerobrake Drag Pass Configuration

The normal spacecraft configuration during aerobrake drag passes will resemble the maneuver configuration, except the solar panels will point in the opposite direction. In this case, the panels will be swept 30° above the Y axis, toward the +Z direction, and the active side of the solar array will point in the +Z instead of the -Z direction (see Figure 2-4). Because the spacecraft will fly through the Martian atmosphere -Z axis (main engine end) forward during the drag pass, this orientation will maximize protection for the arrays by keeping their active side away from the oncoming airflow. In addition, the +X axis will remain nadir pointed throughout the duration of an aerobrake drag pass.

For all propulsive maneuvers and for the aerobraking drag passes, the spacecraft will turn under three-axis control to the proper attitude. Then, upon completion of the maneuver or drag pass, the spacecraft will return to the configuration required for the current mission phase. Usually, that configuration will be array normal spin.

2.4.5 Mapping Configuration

After insertion into the mapping orbit after the aerobrake phase of the mission, the MGS spacecraft will be configured for mapping operations. In this mode, the spacecraft will be three-axis controlled, using input from the horizon sensors to maintain the science instruments on the +Z panel pointed in the nadir direction. Also, the +X side of the spacecraft will point forward in the direction of orbital motion, and the spacecraft will complete one revolution about the Y axis once per orbit (see Figure 2-4).

Because the mapping orbit is Sun-synchronous with respect to 2:00 p.m. of the fictitious mean Sun, the Sun will always shine (except during eclipse periods) on the +Y spacecraft side at an angle that varies between 50° and 74° from the +Y axis (or 16° to 40° from +X), depending on the Martian time of year. The solar arrays gimbal drive control will be enabled to automatically track the Sun as the spacecraft progresses around the orbit. In addition, the HGA boom will be deployed and its gimbal drive control will allow the antenna to track the Earth around each orbit. This configuration will allow the HGA to point directly at Earth without rotating the entire spacecraft.

2.4.6 Contingency Configurations

These configurations will allow the spacecraft to regain inertial attitude knowledge in the event that an anomaly causes entry into safe or contingency mode. Upon entering one of these two modes, the spacecraft will automatically transition to a "Sun coning" mode where it will find the Sun, off point the +X axis from the Sun by a pre-selected angle, and then roll about the vector to the Sun at the rate of one revolution every 100 minutes. The key to this mode is that once attitude knowledge has been lost, the only inertial reference that the spacecraft can automatically obtain is the vector to the Sun as determined from the Sun sensors.

After transitioning to Sun coning, the spacecraft will wait until communications has been re-established with the ground. At that time, the control team can command the spacecraft to transition to SUN-STAR-INIT. This mode looks exactly the same as Sun coning except for the fact that the star sensors will scan the sky to reinitialize attitude reference as the spacecraft rolls.

During the cruise, orbit insertion, mapping, and relay phases of the mission, the +X axis offset angle from the Sun for coning will be zero degrees. Consequently, the spacecraft will spin Sun pointed along the +X axis. However, because the high gain antenna (HGA) also points along the +X axis, communications with the ground will occur only through the LGA during Sun coning and SUN-STAR-INIT as using the HGA requires pointing it directly at the Earth.

The only time the spacecraft will take on a different attitude during Sun coning and SUN-STAR-INIT will be immediately after separation from the Delta's third stage. At that time, the spacecraft will not know its inertial attitude because the third stage will have been spinning at a high rate that will saturate the gyroscopes in the spacecraft's inertial measurement units (IMUs). During this initial attitude initialization, the spacecraft will point its +X axis 60° off of the Sun. As the spacecraft spins around the Sun line, the +X axis will trace a path around an imaginary cone with a half-angle of 60° and longitudinal axis along the vector to the Sun. In this configuration, the spacecraft +X axis is said to be "coning 60° off the Sun."

2.5 Science Payload

Mars Global Surveyor will carry six of the original eight scientific instruments flown on the Mars Observer mission. These scientific devices were either Mars Observer flight spares or built from spare components. During 687 days of mapping operations at Mars, the science payload will generate more than 700 Gbits of raw data to support the fulfillment of the five basic goals of the mission: to characterize the surface morphology at high spatial resolution; to determine the global elemental, thermophysical, and mineralogical character of the surface material; to define the global topographical and gravitational field; to establish the nature of the magnetic field; and to monitor the global weather and thermal structure of the atmosphere to evaluate their seasonal impact on the polar caps, atmospheric dust, and clouds. In addition, the data will support mission and scientific planning for future Mars expeditions with special emphasis on the selection of possible future landing sites. See Figure 2-5 for photographs of the science instruments.

The names of the scientific instruments for MGS, the location of the principal investigator's home institution, and the measurement objectives for each instrument appear in Table 2-6. More detailed descriptions of each instrument follow in subsequent paragraphs in this section of the Mission Plan. In addition, Figure 2-6 shows the sizes and shapes of the instrument footprints on the surface of Mars.

Figure 2-5: *Science Instruments* (CCW from top left: MAG/ER, MOLA, MR, TES, USO, MOC)

Table 2-6: *Science Instruments Summary*

Acronym	Full Name	Lead Center	Objective
MAG / ER	Magnetometer and Electron Reflectometer	Goddard Space Flight Center (GSFC)	Intrinsic magnetic field and solar wind interaction with Mars
MOC	Mars Orbiter Camera	Malin Space Science Systems (MSSS)	Surface and atmospheric imaging
MOLA	Mars Orbiter Laser Altimeter	Goddard Space Flight Center (GSFC)	Surface topography and gravity field studies
MR	Mars Relay Radio System	Centre Nationale d'Estudies Spatiales (CNES, France)	Support for future Mars missions, both American and international
TES	Thermal Emission Spectrometer	Arizona State University (ASU)	Mineralogy, condensates, dust, thermal properties, and atmospheric measurements
USO (RS)	Ultra Stable Oscillator for Radio Science	Stanford University (team leader)	Gravity field determination and atmospheric refractivity profiles

2.5.1 Magnetometer and Electron Reflectometer (MAG/ER)

Currently, Mars is the only planet from Mercury to Neptune whose magnetic field has not yet been measured. The design of the solar-array-mounted magnetometers will allow for the in-situ measurement of the global magnetic field in three perpendicular directions over an extremely wide dynamic range between 16 and 65,536 nT. Given this data, the electron reflectometer will determine the local, near-surface magnetic fields by measuring the deflection of ambient electrons (energy range of 1 - 20 keV) by the local magnetic field. Because the paths of these electrons are significantly altered by the magnetic field, use of the electron reflectometer in conjunction with the magnetometer will provide for a resolution capability 10 to 100 times greater than using the magnetometer alone.

Depending on the ground track spacing and the amount of collected data, the Magnetometer will be able to identify surface magnetic features of sufficient strength with a resolution roughly equal to the orbit altitude. With careful calibration, the MGS spacecraft's contribution to the magnetic field can be subtracted from this data. Both the MAG and ER share a common data processing unit, and the data will be combined and reduced before downlink. Dr. M.H. Acuña of the NASA Goddard Space Flight Center in Greenbelt, MD leads the MAG/ER team.

2.5.2 Mars Orbiter Camera (MOC)

This instrument consists of two independent cameras mounted onto a single assembly. Both cameras are supported by a 32-bit microprocessor for data acquisition and compression, and a 12-Mbyte buffer for temporary image storage. The narrow-angle camera employs a 70-cm tall, f/10 Ritchey-Critien reflector with a focal length of 3.5 meters. Inside, two 2048-element charged couple device (CCD) line arrays sit on the focal plane and are mounted in a direction perpendicular to the spacecraft's velocity vector during mapping operations. Two dimensional images of the Martian surface at an unprecedented resolution of 1.4 m/pixel will be formed as the motion of the spacecraft sweeps the detectors forward.

Figure 2-6: *Science Instrument Footprints*

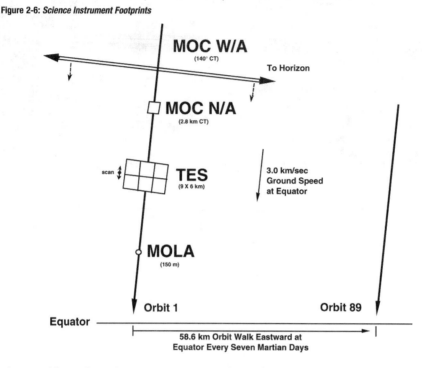

The color-capable, wide-angle camera consists of two f/6, 9.7 mm focal length, 140° field of view fish-eye lenses feeding into a single focal plane containing two 3456-element CCD line arrays. In order to produce color images, the two lenses use red (575 - 625 nm) and blue (400 - 500 nm) filters, respectively. Wide-angle images of the surface with a resolution of 250 m/pixel at nadir (2 km/pixel at the limb) will be produced in the same "motion swept forward" fashion as the narrow-angle images. These wide-angle pictures will contribute to the MOC's global monitoring mode, an experiment that will provide daily, full-planet observations of the Martian atmosphere and surface similar to the weather pictures of Earth shown during newscasts. Dr. M. Malin from Malin Space Science Systems in San Diego, CA leads the MOC team.

2.5.3 Mars Orbiter Laser Altimeter [MOLA)

The Mars Orbiter Laser Altimeter (MOLA) experiment will generate high-resolution topographic profiles of Mars for studies of geophysical and geological structures and processes. This goal will be accomplished by using a diode-pumped neodymium-yttrium aluminum-garnet laser that will fire 45-mj pulses of light at the Martian surface at a rate of 10 bursts per second. By recording the time that the pulse takes to reach the surface and bounce back to the instrument's 50-cm Cassegrain collecting mirror, the MOLA team will be able to compute the local altitude under the spacecraft along the ground track.

Each laser spot will measure about 160 meters in diameter on the surface with a spacing of about 300 meters between spots along the ground track. The accuracy in measuring relative topography will vary from one to 10 meters, with an absolute accuracy of about 30 meters. Ultimately, the absolute accuracy will depend on precise post-reconstruction of the spacecraft orbital position from navigation and radio-science data. Dr. D.E. Smith from the NASA Goddard Space Flight Center leads the MOLA team.

2.5.4 Mars Relay (MR)

The Mars Relay (MR) consists of a radio system and antenna designed to return measurements and imaging data from spacecraft deployed on the surface of the red planet. This instrument consists of a 1-meter tall helix antenna mounted on the nadir panel of the spacecraft and all of the associated electronics. Unlike the main X-band communications system, this device operates at UHF frequencies. The antenna pattern (-3 db) takes the form of a 65° cone emanating from the tip of the antenna, providing coverage with a 5,000 km effective range for a 8 kbps data rate, and a 1,300 km range for a 128 kbps rate. As the spacecraft orbits Mars, the MR will transmit a 1.3-W, 437.1-MHz beacon to the surface, indicating to the landers that the MGS spacecraft is currently in view. This beacon will serve as an indicator for the landers to begin transmitting their data. The MR is provided by the Centre Nationale d'Estudies Spatiales in France and will be used to relay data from Russian landers (1996 launch) and American landers (1998 launch).

2.5.5 Radio Science (RS)

Radio-science experiments, led by Dr. G.L. Tyler of Stanford University in Palo Alto, California, will advance two fields fundamental to the study of Mars. First, observations of distortions (frequency, phase, and amplitude) in the spacecraft's radio signal as it passes through the Martian atmosphere on the way to Earth will be used to derive high-resolution temperature profiles of the atmosphere with a vertical resolution of 200 meters. Second, by using Doppler tracking to carefully monitor small changes in the frequency of the radio signal from the spacecraft as it orbits Mars, the radio-science team will be able to reconstruct the Martian gravity field to an unprecedented level of accuracy, perhaps higher than a 50 x 50 field.

Figure 2-7: Mars Orbiter Camera (top) and Laser Altimeter (bottom)

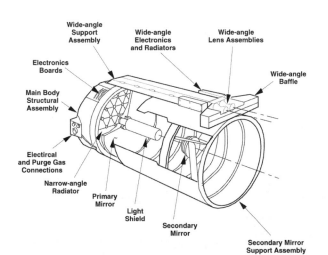

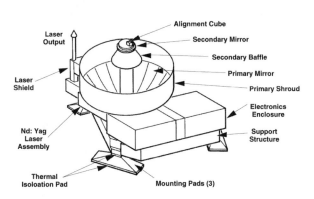

Both radio-science experiments will require precise tracking of the spacecraft's radio signal by the antennas of NASA's Deep Space Network. In order to facilitate this requirement, the spacecraft will employ an ultra-stable oscillator (USO) to provide an extremely stable frequency reference for the X-band telecommunications system. This high-quality, low-noise oscillator resonates at 19.143519 MHz and will have a long-term frequency variation limit of less than 1.0×10^{-10} Hz.

2.5.6 Thermal Emission Spectrometer (TES)

The Thermal Emission Spectrometer (TES) will function as a combined infrared spectrometer and radiometer designed to measure heat energy radiated from surface and atmosphere of Mars. The investigation team, led by Dr. P.R. Christensen of Arizona State University in Tempe, will use data from TES to determine the thermal and mineralogical properties of the surface, and to learn about Martian atmospheric properties, including cloud type and dust opacity.

Figure 2-8: *Thermal Emission Spectrometer (TES)*

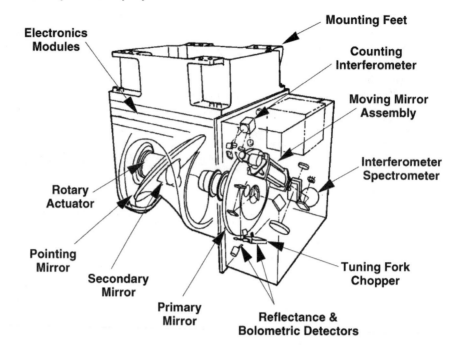

Table 2-7: *DSN Tracking Requirements*

Time Period	Antenna	Tracking Coverage	Data Types
Launch to L+ 30 days (includes TCM1 at L+ 15 days)	34m HEF	Continuous	2-way coherent Doppler and ranging, angular data from launch to L+ 1 day, acquire tracking data as soon as possible after launch.
L+ 30 days to MOI- 90 days	34m HEF	10 hours/pass 1 pass/day	2-way coherent Doppler, 3-way Doppler and ranging.
TCM2 (TCM1+ 120 days) TCM3 (TCM2+ 30 days)	34m HEF	Continuous for a period of 3 days before TCM to 3 days after TCM	2-way coherent Doppler and ranging.
Gravity Wave Campaign (14-Apr-97 to 5-May-97)	34m HEF	Continuous	2-way coherent Doppler and ranging
MOI- 90 days to MOI- 30 days	34m HEF	10 hours/pass 2 pass/day	2-way coherent Doppler, 3-way Doppler and ranging.
MOI- 30 days to start of mapping (includes TCM4 at MOI- 20 days)	34m HEF	Continuous	2-way coherent Doppler, 3-way Doppler and ranging.
MOI- 24 hours to MOI+ 24 hours	70m	Continuous	1-way Doppler and ranging
Routine Mapping Operations (15-Mar-98 to 31-Jan-00)	34m HEF	10 hours/pass 4 pass/3 days	2-way coherent Doppler and ranging, 3-way Doppler and ranging, open loop recording during atmospheric occultations.
Science Campaigns A: 15-Mar-98 to 13-Apr-98 B: 29-Jun-98 to 6-Jul-98 C: 26-Oct-98 to 2-Nov-98 D1: 5-Jan-99 to 12-Jan-99 D2: 20-Jan-99 to 27-Jan-99 D3: 3-Feb-99 to 10-Feb-99 D4: 18-Feb-99 to 25-Feb-99 E: 3-May-99 to 10-May-99 F: 4-Oct-99 to 11-Oct-99 G: 13-Dec-99 to 20-Dec-99	34m HEF	Continuous	Same as during routine mapping
Diametric Occultations Edge-on Orbital Configuration (28-days centered on 28-Oct-98 and 19-Feb-99)	34m HEF	10 hours/pass 2 pass/day (w/ 2 hour overlap)	During the overlap period, simultaneous 2-way coherent Doppler and 3-way Doppler. Otherwise, same as during routine mapping operations.
Communications Relay Phase (1-Feb-00 to 1-Aug-00)	34m HEF	10 hours/pass 1 pass/day	2-way coherent Doppler and ranging

This instrument primarily consists of two nadir-pointing telescopes. The larger of the two is a 15.24 cm diameter Cassegrain design that feeds a two-port Michelson interferometer spectrometer with a spectral range from 6.25 to 50 μm. The smaller of the two serves two bolometric channels (0.3 to 3.9 μm and 0.3 to 100 μm) and takes the form of an off-axis parabola shaped telescope. Each telescope utilizes six detectors, each with an 8.3 x 8.3 mrad field of view. Together, the six form a rectangular grid three frames wide (cross-track) and two frames deep (down-track). Although the instrument will normally remain nadir pointed, a rotatable scan mirror will allow the TES' telescopes to view Mars at any arbitrary oblique angle from horizon to horizon.

2.6 DSN Utilization

The 34-meter high-efficiency (34m HEF) antennas of the Deep Space Network (DSN) will provide almost all of the tracking coverage for the mission. This type of antenna was selected for its capability to both transmit and receive X-band signals. Project use of other antenna types, such as the 34-meter beam waveguide (34m BWG), will only be accepted on a negotiated case by case basis.

During periods of normal operation in cruise, mapping, and relay phase, the project's requirement for DSN support are modest at one 10-hour track per day. However, critical operation such as launch, maneuvers (TCMs and MOI) and aerobraking operations will require continuous coverage. In addition, the project will also require continuous coverage for special, week-long science campaigns during mapping to take advantage of special orbit geometry conditions or to observe unique seasonal changes on Mars.

Table 2-7 provides a profile of the tracking support required by the Mars Global Surveyor Project. Stations other than, but equivalent to a 34m HEF, such as a 34m BWG or 70m with comparable up and downlink performance, may only be substituted with negotiation from the project.

3. Launch Phase

Launch of the Mars Global Surveyor spacecraft will occur aboard a McDonnell Douglas Delta 2 (7925A) launch vehicle from Space Launch Complex 17A (SLC17A) at the Cape Canaveral Air Station (CCAS) in Florida. In the past, Delta rockets have primarily been used by NASA and commercial organizations to launch small to medium sized payloads to low Earth or geosynchronous orbits. The current Delta rocket design represents the latest in a long line of highly-reliable and successful family of launch vehicles. This mission will represent the first use of the low-cost Delta to send a spacecraft to another planet, thus producing a savings for NASA of nearly $300 million as compared to Titan launch vehicles used for previous Mars missions.

The launch period for this mission spans a 20-day period from 6 November 1996 through 25 November 1996. Launch phase will start at the beginning of launch countdown and last until separation of the spacecraft from the Delta's third stage following the trans-Mars injection burn.

3.1 Brief Launch Vehicle Description

Delta 7925A launch vehicles consist of five major components that include the three main stages, nine

Figure 3-1: *Parts of the Delta Launch Vehicle*

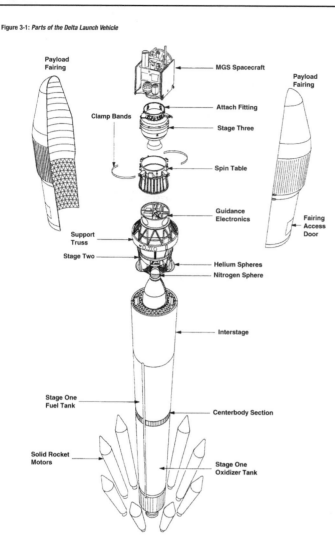

solid rocket motors that attach to the first stage, and the payload fairing. The Delta measures 38.2 meters tall from the tip of the payload fairing to the bottom of the first stage's body. Including the spacecraft (assumed to weigh 1,060 kg for launch vehicle planning purposes), the Delta will weigh approximately 231,325 kg at the time of launch.

3.1.1 First Stage

Stage one employs a Rocketdyne RS-27A main engine with a 12:1 expansion ratio. This engine is a single start, liquid, bi-propellant rocket that will provide nearly 890,000 N of thrust at the time of lift-off. Its propellant load (95,655 kg) consists of RP-1 fuel (thermally stable kerosene) and liquid oxygen (LOX) for oxidizer. The RP-1 fuel tank and liquid oxygen tank on the first stage are separated by a center body section that houses control electronics, ordnance sequencing equipment, a telemetry system, and a rate gyro. In addition, stage one also employs two Rocketdyne vernier engines. They will provide roll control during the main engine burn, and attitude control between main engine cutoff (MECO) and second stage ignition.

3.1.2 Solid Rocket Motors

A set of nine solid-propellant graphite epoxy motors (GEMs), each fueled with approximately 12,000 kg of hydroxyl-terminated polybutadiene (HTPB) solid propellant, attach to the first stage to provide augmentation thrust. Each GEM will provide an average thrust of 446,000 N. Six of the nine GEMs, the main engine, and the vernier engines will ignite at the time of lift-off, producing a total thrust of about 2,850,000 N. The remaining three GEMs will ignite 65 seconds into flight, shortly after the initial six burnout.

3.1.3 Second Stage

The Delta second stage uses a re-startable, liquid, bi-propellant Aerojet AJ10-118K engine that consumes a combination of Aerozine-50 fuel (a 50/50 mix of hydrazine and un-symmetric dimethyl hydrazine) and nitrogen tetroxide (N_2O_4) oxidizer. Since this propellant combination is hypergolic, no catalyst or igniter in the engine thrust chamber is required. In total, the second stage will burn nearly 6,000 kg of propellant at an average thrust of 43,370 N. A set of hydraulically activated engine gimbals will provide pitch and yaw control during powered flight, and a nitrogen cold gas jet system will provide the roll authority. In addition, the nitrogen jets will also provide attitude control for the coast phases.

This stage will provide the thrust needed to boost the launch vehicle and spacecraft into low Earth orbit after first stage jettison. Also, the second stage will impart some of the velocity needed to send the spacecraft to an Earth-escape trajectory.

3.1.4 Third Stage and Payload Fairing

A spin-stabilized third stage will provide most of the velocity required to boost the MGS spacecraft from low-Earth orbit to a trans-Mars trajectory This stage consists primarily of a Thiokol Star-48B solid motor. The engine on the third stage will provide an average thrust of 66,370 N and will burn about 2,000 kilograms of ammonium perchlorate propellant.

During launch and ascent through the lower atmosphere, a 2.9 meter diameter payload fairing will protect the spacecraft and Delta third stage from aerodynamic forces. The fairing will be jettisoned from the launch vehicle at an altitude of approximately 129 kilometers (L+ 286 seconds), shortly after second stage ignition.

3.2 Pre-Launch Activity Overview

Delivery of the spacecraft to Cape Canaveral is currently scheduled for 15 August 1996, after completion of system testing Lockheed-Martin's Waterton facility near Denver, Colorado. The spacecraft will be flown to Florida on a C-17 cargo jet. After arrival, initial inspection and checkout will take place at the KSC Payload Hazardous Servicing Facility (PHSF) where a DSN end-to-end compatibility test via the MIL-71 complex will be conducted. Technicians at KSC will then load the spacecraft with propellants and mate it to the Delta third stage. Following mating at the PHSF, initial interface verification testing will be performed.

Stacking of the Delta first and second stage will occur at Pad-17A in parallel with the activities at the PHSE. Roughly three weeks before launch, the coupled spacecraft and third stage will be moved to Pad-17A for mating with the bottom two stages of the Delta booster. Following final interface testing and close-out activities at the launch pad, the payload fairing will be installed, and Delta launch readiness will be verified.

3.3 Launch Strategy

Designing a launch strategy for Mars Global Surveyor provided many challenges due to the presence of many aspects not commonly found on interplanetary launches. Some of these include near-instantaneous launch windows and fixed azimuth flight profiles during boost. These constraints were dictated by the operational characteristics of the Delta combined with a "heavy" spacecraft relative to the performance capability of the launch vehicle. Nevertheless, the launch strategy developed for this mission will provide a probability of lift-off within the 20-day launch period of greater than 97%.

3.3.1 Launch Window

Delta rockets do not possess the capability to automatically "re-target" their launch azimuth in realtime at the launch pad, nor does the first stage possess the capability to yaw-steer during ascent. Once the guidance software has been loaded, the booster must fly at a single, pre-determined azimuth. Switching azimuths will require manually loading new guidance targets. Consequently, traditional launch windows (typically two hours in duration for the Mars Observer launch) are not possible. Instead, the booster must launch within a very small time period (about ±1 second) when the launch site rotates precisely into the plane containing the parking orbit and the departure asymptote. These extremely short windows are referred to as "instantaneous launch windows" or "instantaneous launch opportunities."

Despite the implementation of the instantaneous launch window, the need to avoid potential orbiting debris in the Delta's flight path may require the booster to launch up to 30 seconds before the calculated lift-off time. If this situation occurs, the Delta will fly exactly the same boost profile as if a normal launch had occurred. The error incurred in the spacecraft's Earth departure trajectory will be removed at TCM 1 with a penalty of approximately 5 m/s.

3.3.2 Launch Period

The current mission baseline calls for a launch period that opens on 6 November 1996 and closes 20 days later on 25 November 1996. This time period is bounded on both ends by trans-Mars trajectory C3 values too high for the Delta to achieve, given the mass of the spacecraft. Over the duration of the launch period, the declination of the departure asymptote (DLA) will vary from a minimum of 20.8° at the open, and 36.4° at the close. Geometrical constraints of the interplanetary injection problem dictate that the parking orbit prior to trans-Mars injection must lie at an inclination greater than or equal to the DLA. In order to satisfy this constraint, minimize the number of parking orbits in the launch vehicle targeting specification, and maximize the probability of launch, the Delta will fly at one of three different launch azimuths.

For the first half of the launch period (6 November to 15 November 1996), the Delta will launch at an azimuth of either 93° or 99.89° and fly to a low-Earth parking orbit of 28.47° and 29.82°, respectively. During the second half of the launch period (16 November to 25 November 1996), the declination of the hyperbolic departure asymptote will exceed the latitude of Cape Canaveral Air Station, and the Delta will launch at a 110° azimuth (with a dog-leg ascent) to reach a 36.5° parking orbit to compensate. Although a booster performance penalty normally exists for reaching higher inclination orbits, they will be in part offset by the lower C3 requirements (minimum of 8.89 km^2/s^2 on 18 November) during the second half of the launch period.

Utilization of two launch azimuths during the first half of the launch period will give the Delta two discrete, instantaneous launch opportunities each day. If the booster misses the first due to minor hardware problems or short term weather violations, the launch team can reload the guidance targets and make a second attempt at lift-off on the same day. This scheme will increase the overall probability of liftoff within the launch period and will specifically increase the probability of a lift-off near the open of the period.

The choice of the two specific launch azimuths for the first half of the launch period was driven by constraints imposed by Delta launch operations. Specifically, the two lift-off time solutions corresponding to the two different azimuths must be separated by no less than 64 minutes and no more than 73 minutes. The former represents the minimum time required for the ground team to reload the guidance targets, and the latter corresponds to the maximum time that the liquid oxygen can remain in the first stage tanks without freezing vital parts of the Delta's propellant feed system.

The choice of the launch azimuth for the second half of the launch period resulted from range safety constraints. Normally, rockets launched from the Cape must fly at an azimuth several more degrees south of east than the chosen 110° in order to reach an inclination of 36.5° directly. Instead, the Delta first and second stages will each perform a "dog-leg" maneuver to reach the proper parking orbit inclination to avoid violating the range safety constraints.

3.3.3 Launch Opportunity

For any given launch azimuth, two opportunities exist every day for a rocket to launch and inject its payload onto the proper Earth escape trajectory. The primary difference between the two, called the long and short coasts, is the length of time that the Delta must wait in its low-Earth parking orbit before reaching the proper location to perform the trans-Mars injection burn. On any given day (relative to the time when the launch site on Earth rotates through the departure asymptote), the long coast launch opportunity always occurs first.

During the wait in low Earth orbit for the trans-Mars injection burn, the spacecraft will rely on its batteries for power because the solar panels will have not yet been deployed. Preliminary analysis shows that on the long coast, the spacecraft will need to rely on battery power for up to 91 minutes. Although this length of time is undesirably long, battery depth of discharge does not represent the limiting factor in the choice between the two launch opportunities.

The major constraint involves Sun avoidance. An MGS flight rule specifies that science instruments (located on the +Z axis of the spacecraft) must always remain pointed at least 30° away from the Sun under normal (non-safe-mode) conditions. On this mission, the long coast requires a dawn launch from the Cape. Because launches occur generally in the eastward direction, the science instruments will be pointed almost directly at the Sun at the time of payload fairing jettison. For this reason, the MGS mission will utilize the short coast launch opportunity for all three launch azimuths.

3.3.4 Probability of Commanded Shutdown (PCs)

Probability of commanded shut-down (PCS) for the Delta second stage plays a key role in determining the duration of the MGS launch period. The reason is that the second burn of the second stage provides part of the energy required to place the spacecraft on the trans-Mars trajectory. PCS defines the probability that the second stage engine will complete its burn before the propellant supply depletes. By accepting a lower PCS value, it is possible to achieve a higher injected mass. Current project policy dictates a minimum PCS value of 95% in order to achieve the necessary C3 for a given launch day. However, under the current MGS launch design, the PCS does not fall below 97% until the end of the launch period.

Predicting the net effect of a PCS violation during flight is difficult. If the propellant supply of the second stage runs out during the stage's second burn, the third stage will still ignite at the proper time. In this scenario, the total Delta-V imparted by the combination of second burn of the second stage and the third stage burn will fall short of the required amount to achieve the proper C3 target. If propellant depletion occurs very late during the second burn of the second stage, it is conceivable that the velocity deficit could be corrected at the first trajectory correction maneuver (TCM) or with a contingency TCM. However, such a scheme would result in a significant reduction in total mission Delta - V capability and would prohibit flying the mission according to the current baseline plan.

3.4 Boost Profile and Injection

In general, the exact mission elapsed times for key events depend on the orientation and location of the Earth departure asymptote (variable with each launch day), and the launch azimuth of the booster (either 93°, 99.89° or 110°). However, in all cases, the mission elapsed event times for the first stage boost profile will always remain constant. Table 3-1 lists the times for critical events in the boost sequence.

Table 3-1: *Major Events During Ascent*

Event	Time (93° Azimuth)	Time (99.89° Azimuth)	Time (110° Azimuth)
Lift-Off	0.000 seconds	0.000 seconds	0.000 seconds
Mach 1	32.248	32.248	32.248
Maximum Dynamic Pressure	49.410	49.410	49.410
Solid Motor Burn-Out (6 of 9)	63.120	63.120	63.120
Solid Motor Ignition (3 of 9)	65.500	65.500	65.500
Solid Motor Jettison (6 of 9)	66.500	66.500	66.500
Solid Motor Burn-Out (3 of 9)	128.820	128.820	128.820
Solid Motor Jettison (3 of 9)	131.500	131.500	131.500
Stage 1 Main Engine Cut-Off	260.664	260.664	260.664
Stage 1 Jettison	268.664	268.664	268.664
Stage 2 Ignition	274.164	274.164	274.164
Payload Fairing Jettison	286.000	286.000	286.000
Stage 2 First Cut-Off	576.592	577.324	583.573
Stage 2 Restart	2420 / 2812	2213 / 2557	2027 / 2550
Stage 2 Second Cut-off	2548 / 2932	2340 / 2677	2147 / 2672
Stage 2 Jettison	2601 / 2985	2394 / 2731	2200 / 2725
Stage 3 Ignition	2638 / 3022	2431 / 2768	2238 / 2762
Stage 3 Burn-Out	2725 / 3110	2518 / 2855	2325 / 2849
Yo-yo Deploy and Despin	3003 / 3387	2796 / 3132	2602 / 3127
Stage 3 Jettison	3008 / 3392	2801 / 3137	2607 / 3132

In Table 3-1, the appearance of two time values for a single event indicates that the event time varies with the launch date. For the 93° and 99.89° flight azimuths, the first and second times correspond to the 6 November and 15 November launch dates, respectively. For the 110° azimuth, the first and second times correspond to the 16 November and 25 November launch dates, respectively.

During the launch phase, the booster will not provide the spacecraft with any power or telemetry capabilities. The spacecraft will launch with power for the computer, receiver, and attitude control sensors supplied from the batteries. Switch-over from launch-pad power to internal spacecraft power will occur at T- 4 minutes prior to launch.

3.4.1 Stage One

Lift-off will occur from SLC-17A at Cape Canaveral Air Station. This time will vary from as late as 13:37 to as early as 9:30 EST depending on launch azimuth and launch date. At the time of lift-off, the main engines and six of the nine solid rocket motors will ignite. Approximately 63 seconds into flight, the solids will burn-out and be jettisoned. Then, the remaining three solids will ignite and burn for another 63 seconds before being ejected. Main engine cut-off (MECO) will occur 261 seconds after lift-off at a sub-orbital altitude of 115.2 km, 541.3 km down range from the launch site. The vernier engines will continue to burn for another six seconds, and first stage jettison will occur two seconds later at L+ 269 seconds.

3.4.2 Stage Two

At L+ 274 seconds, roughly five seconds after stage one jettison, stage two will ignite. Twelve seconds later, the free molecular heating rate on the vehicle will have dropped to below 1135.0 W/m², allowing the Delta to jettison its payload fairing and expose the spacecraft to the vacuum of space. In total, stage two will thrust for approximately five minutes to boost the spacecraft from its sub-orbital state at MECO to a circular, low-Earth parking orbit at an altitude of 185 km. Second stage cut-off (SECO1) will occur between L+ 577 seconds to L+ 584 seconds, depending on the exact date of launch and choice of flight azimuth.

3.4.3 Trans-Mars Injection

After parking orbit insertion, the booster and spacecraft will coast for between 24 to 37 minutes (variable with launch date) until they reach the proper position to begin the two-burn trans-Mars injection sequence. First, stage two will re-start and thrust for roughly 120 seconds to raise the apogee of the parking orbit. After cut-off of the second stage engine (SECO2), the Delta will coast for 53 seconds before jettisoning the second stage. During that time, small rockets on the spin table (attached to the bottom of stage three) will fire to spin stage three and the spacecraft to a rate of 60 r.p.m. for spin-stabilization purposes.

Once 37 seconds have elapsed after stage two jettison (90 seconds after SECO2), stage three will ignite and burn for 87.3 seconds to complete the trans-Mars injection sequence. At the end of the burn, the spacecraft will be on an Earth escape trajectory. For the baseline, short-coast option, trans-Mars injection will almost always occur in darkness, somewhere over the Indian Ocean.

The combination of the second burn of the second stage and the third-stage burn will provide the Delta-V needed for trans-Mars injection. During every day of the launch period, the third stage will impart the same amount of Delta -V to the spacecraft. The burn time and Delta-V of the second stage's second burn will vary depending on the specific C3 requirements of the given launch day.

3.4.4 Despin and Spacecraft Separation

At the time of stage three burn-out, the Delta and MGS spacecraft will still be spinning at 60 r.p.m. This rotation rate must be nullified because the spacecraft functions on three-axis stabilization. In order to despin the spacecraft, a yo-yo cable device on the third stage will deploy approximately 278 seconds after burn-out (365 seconds after ignition). The device works by transferring the angular momentum of the third stage and spacecraft to the cable in a fashion similar to how a spinning figure skater slows her spin by extending her arms.

Five seconds after yo-yo deploy, pyrotechnic devices will fire to sever the connection between the spacecraft and third stage. A set of four springs will then uncoil to impart a relative separation velocity of between 0.6 and 2.4 m/s between the third stage and spacecraft. The 278 second wait after burn-out for separation is

Figure 3-2: *Timeline of Launch Events*

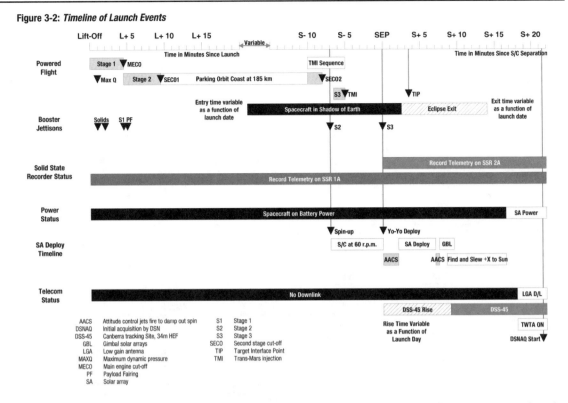

designed to allow adequate time for residual thrust from the third stage to tail-off and to ensure that the stage will not collide with the spacecraft after separation. During this waiting period, a set of thermal blankets located on the third stage will protect the spacecraft from thermal soakback.

The actual mission elapsed time of separation depends on the length of time that the spacecraft spends in the low-Earth park orbit before trans-Mars injection and will vary with each launch day. However, separation will always occur 370 seconds after third stage ignition. The choice of 370 seconds is in part driven by the standard cascaded event timers that McDonnell Douglas installs on the third stage.

3.5 Launch Targets

Launch vehicle targets represent the state that the Delta must deliver the spacecraft to in order to place the spacecraft on the proper trans-Mars trajectory. The targets are defined as osculating C_3, DLA, and RLA achieved at the target interface point, defined as 10 minutes after third stage ignition. Table 3-2 lists the launch targets for each launch date. In the table, two values for each targeting parameter are listed for launch dates between 6 November 1996 and 15 November 1996, inclusive. The top number corresponds to the 93.0° launch azimuth, while the bottom one corresponds to the 99.89° launch azimuth. For the rest of the launch dates, the Delta will fly at an 110° azimuth. All of the departure states listed in the table have been biased to satisfy Mars quarantine requirements. The following definitions and assumptions apply to the launch vehicle departure targets listed in Table 3-2

C_3	Departure energy or hyperbolic excess velocity squared (km²/s²)
LO	Lift-off time (hh:mm:ss UTC)
TOF	Time from lift-off to target interface point (hh:mm:ss)
DLA	Declination of the departure asymptote vector (degrees, EME2000)
RLA	Right ascension of the departure asymptote vector (degrees, EME2000)

Table 3-2: *Departure Launch Targets*

Launch Date	Lift-Off Time	Time of Flight	Departure C₃	Departure DLA	Departure RLA
6-Nov-96	17:11:17	0:53:58	10.1656	21.2815	173.2868
	18:15:44	0:50:31	10.1551	21.3062	173.2849
7-Nov-96	17:00:50	0:54:18	9.9846	21.8151	173.3121
	18:05:56	0:50:49	9.9747	21.8430	173.3104
8-Nov-96	16:48:11	0:54:43	9.7984	22.4474	173.1808
	17:54:09	0:51:11	9.7890	22.4782	173.1780
9-Nov-96	16:35:20	0:55:11	9.6412	23.0918	173.1528
	17:42:19	0:51:36	9.6324	23.1246	173.1488
10-Nov-96	16:20:05	0:55:44	9.4837	23.8150	172.9559
	17:28:23	0:52:05	9.4755	23.8502	172.9506
11-Nov-96	16:04:47	0:56:20	9.3549	24.5143	172.8633
	17:14:37	0:52:35	9.3473	24.5515	172.8567
12-Nov-96	15:46:14	0:57:03	9.2314	25.2983	172.6072
	16:58:15	0:53:12	9.2246	25.3382	172.5995
13-Nov-96	15:24:50	0:57:56	9.1257	26.1236	172.3150
	16:39:57	0:53:55	9.1198	26.1667	172.3063
14-Nov-96	15:02:16	0:58:55	9.0466	26.8873	172.1402
	16:15:21	0:54:55	9.0287	27.1824	171.7576
15-Nov-96	14:29:58	1:00:22	8.9811	27.7543	171.8299
	15:47:27	0:56:08	8.9725	28.1834	171.3326
16-Nov-96	18:36:33	0:47:18	8.9162	28.7341	171.5825
17-Nov-96	18:21:17	0:47:52	8.8939	29.6635	171.2614
18-Nov-96	18:07:04	0:48:25	8.8888	30.4871	171.0732
19-Nov-96	17:52:40	0:48:58	8.9003	31.2739	170.9072
20-Nov-96	17:34:27	0:49:42	8.9336	32.2347	170.6113
21-Nov-96	17:17:37	0:50:23	8.9721	33.0381	170.4474
22-Nov-96	16:59:17	0:51:08	9.0213	33.8373	170.2947
23-Nov-96	16:33:04	0:52:16	9.1017	34.8534	170.0216
24-Nov-96	16:05:44	0:53:27	9.1758	35.6671	169.8820
25-Nov-96	15:09:45	0:56:02	9.2576	36.4706	169.7374

4. Cruise

Trans-Mars cruise covers the time of ballistic flight between Earth and Mars. The spacecraft will take between 301 and 309 days to reach the red planet on its Type-2 trajectory depending on the Earth departure date within the 20-day launch period. A launch at the open of the launch period on 6 November 1996 will correspond to a Mars arrival date of 11 September 1997, while a launch at the close of the period on 25 November 1996 will result in an arrival on 22 September 1997.

During cruise, a set of four trajectory correction maneuvers (TCMs) will adjust the interplanetary trajectory to ensure that the spacecraft reaches the proper velocity and position targets prior to the Mars orbit insertion (MOI) burn. Other primary activities during cruise will include daily monitoring of the subsystems and science instrument checkout and calibration activities. Figure 4-1 shows an overall timeline for the cruise phase based on a 6 November 1996 launch date.

4.1 Initial Deployment and Acquisition

The cruise phase of the mission will start immediately after spacecraft separation from the third stage of the Delta 2. Separation from the booster will trigger a command sequence that will initiate deployment of the solar arrays and turn on the spacecraft's transmitter to allow for acquisition from the tracking antennas of the Deep Space Network (DSN). This sequence represents the most critical mission period for two reasons. First, spacecraft survival will critically depend on proper deployment of the solar arrays because the spacecraft can only operate on battery power for a limited amount of time after launch.

Second, initial acquisition of the spacecraft's signal by the DSN must occur within about 100 minutes after separation because the X-band acquisition-aid antenna can only receive the spacecraft's signal to a range of about 40,000 km.

The acquisition-aid antenna is a small, wide-beam dish mounted on the 26-meter antenna at the Canberra tracking site. This small antenna will initially direct pointing of the larger, narrow-beam, 34-meter high-efficiency tracking antenna (34m HEF) that will normally be used to track the spacecraft. Not acquiring the spacecraft's signal will prevent the navigation team from gathering the initial two-way Doppler data necessary to determine the spacecraft's trajectory for the purpose of producing an accurate ephemeris. In turn, lack of an accurate ephemeris will dramatically hamper the ability of the 34m HEF to find the spacecraft.

Figure 4-1: Cruise Timeline

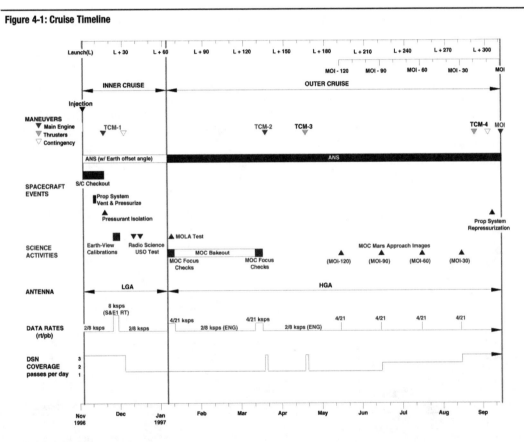

4.1.1 Initial Deployment

Before the spacecraft separates from the Delta third stage, a yo-yo cable device attached to the stage will deploy to reduce the spin rate from 60 r.p.m. down to zero with a plus or minus two r.p.m. uncertainty. The device works by transferring the angular momentum of the third stage and spacecraft to the cable. During the despin process, redundancy management will be disabled to prevent the computer logic from incorrectly identifying possibly-saturated gyros as failed in the unlikely event that rotational rates exceed 9° per second (1.5 r.p.m.) after despin.

Five seconds after yo-yo release, pyrotechnic devices will fire to sever the connection between the spacecraft and third stage. A set of four springs will then uncoil to impart a relative separation velocity of between 0.6 and 2.4 m/s between the third stage and spacecraft. The actual mission elapsed time of separation will vary as a function of launch day between the values of 43 and 56 minutes, but will always occur 370 seconds after third stage ignition. For a launch at the open of the period on 6 November 1996, separation will occur about 50 minutes after launch. Table 4-1 shows a listing of the post-separation events.

A set of breakwires attached to the Delta's third stage will provide a signal to the spacecraft that separation has occurred. Once this signal triggers, the post-separation command sequence will activate. First, the attitude-control thrusters will be armed and enabled in the first two seconds after separation detection, allowing the launch despin control mode to be activated as soon as possible. The despin control software will "snapshot" the current attitude and fire the appropriate thrusters to despin the spacecraft and hold attitude. This process will take no more than two minutes under normal circumstances. After this two minute time period, the thrusters will be disarmed in preparation for solar array deployment.

Following thruster disarming, the two folded solar arrays will deploy one after the other by a release of each array's outer and inner panel in that order. Once released, the solar arrays will unfold to their fully deployed configuration by four pairs of spring-driven hinges. After each hinge rotates approximately 180°, a latch will engage and lock the hinge and panel in place. In the timeline, five minutes has been allocated for the solar array deployment process. In addition, another 20 seconds has been allocated for the attitude control thrusters to remove any incidental spacecraft rotation introduced by the release of the solar panels.

At this time the spacecraft will begin to acquire the attitude required for DSN initial acquisition. Because the spacecraft will be spinning when it separates from the Delta, it will not have a three-axis attitude reference. The only attitude knowledge the spacecraft will possess is the location of the Sun as determined from the Sun sensors (SSA). For DSN initial acquisition, the spacecraft will enter the Sun acquisition and coning mode, called SUN-COM-POWER, in which the +X axis will be aligned to the measured Sun vector. In the timeline, five minutes has been allocated for the spacecraft to slew until the SSA detects the Sun, and an additional three minutes has been allocated to point the +X axis at the Sun. Under normal circumstances, the spacecraft will enter the DSN initial acquisition attitude within 18 minutes of spacecraft separation. At that time, the spacecraft will begin to spin about its +X axis at a rate of one revolution every 100 minutes.

Table 4-1: *Post-Separation Timeline*

Event Description	Time From Separation	Comments
Third stage burn out	-370 seconds	Actual time varies by launch date (between L+ 39 to 52 minutes)
Yo-yo deploy	-5.0	Despin spacecraft and third stage from 60 r.p.m.
Third stage jettison	0.0	Breakwires attached to third stage provide separation signal to spacecraft
Arm thruster string 1 & 2	0.1	Thruster enable commands already issued during parking orbit coast
Begin thruster firings	0.6	Remove any residual rotation not removed by yo-yo device
Disarm thrusters	126.6	Prevents external torques during solar panel deployment
Initiate solar panel deploy	127.6	Five minutes allocated for deploy
Exit eclipse (penumbra)	159.3	Value for 6-Nov, maximum value of 835 seconds for 25-Nov launch
Outer solar array gimbal begin	428.6	Gimbal from -85° to +30° at 0.7°/sec
Inner solar array gimbal begin	471.5	Gimbal from +90° to -15° at 0.7°/sec, begin when outer gimbal at -55°
Turn on star sensor (CSA)	479.7	CSA turned on 30 seconds prior to use
Arm thrusters	478.6	Delay from array deploy begin time to keep panels out of thruster plume
Begin null attitude rates	488.6	Two minutes allocated to remove spin due to solar array deployment
Initiate payload Sun avoidance	509.6	Enable autonomous Sun avoidance logic for payload protection
Begin Sun acquisition	509.7	Eight minutes allocated to find Sun and point +X axis at Sun
End Sun acquisition	989.7	Roll at 0.06°/sec, will occur later if eclipse exit has not yet occurred
Turn on TWTA filament	1050.9	Allow four minutes for warm-up
Turn on TWTA beam	1290.9	Beam power will use 55 Watts of power
Turn on transponder (MOT)	1291.9	Begin DSN initial acquisition period of two hours
Move to Sun coning orientation	8491.9	Begin slew to +X axis 60° off vector to Sun
Begin Sun coning	8691.9	Begin coning at 0.06°/sec (100 min/rev)
Begin SUN-STAR-INIT	8899.0	Use CSA to establish attitude reference (takes about 200 minutes)

During the transition from the post-despin attitude to the DSN initial-acquisition attitude, the solar panels will automatically be commanded to their cruise orientation by rotating the outboard (azimuth) gimbals by 120° and the inboard (elevation) gimbals by 90°. In this configuration, the solar arrays will be swept forward toward the Sun, 30° above the Y axis in the +X direction. Approximately 12 minutes after separation, four minutes prior to the spacecraft achieving the DSN initial-acquisition attitude, the filament of the TWTA configured to the primary transmit low-gain antenna (LGA) will be activated for warm-up.

4.1.2 DSN Initial Acquisition

Initial acquisition will begin about 18 minutes after the spacecraft separates from the Delta's third stage. At this time, the post-separation command sequence will command the spacecraft to begin transmitting realtime engineering data over the LGA at a rate of 2,000 bps. This transmission rate will allow the ground-control team to instantly determine whether the spacecraft entered safe mode prior to initial acquisition. The reason is that if safe mode entry occurs during launch, then the first transmission will appear at the slower, safe mode utilized rate of 10 bps.

Use of the short coast launch trajectory will place the spacecraft over the Canberra tracking site for initial acquisition. The DSN estimates that using the listen only, wide-beam, X-band acquisition-aid (ACQ-AID) antenna, they will "lock-up" on the carrier portion of the signal within a few minutes after the spacecraft begins transmitting. After detection of the carrier, the DSN will listen to the downlink signal for about TBD minutes to "fine tune" the best uplink frequency with which to establish a coherent, two-way lock.

In order to establish two-way, the DSN will track the spacecraft downlink with the ACQ-AID and use its pointing data to point the narrow-beam, 34m HEF antenna. Once the 34m HEF locks onto the spacecraft's signal, the ACQ-AID will no longer be needed. The DSN estimates that under normal circumstances, establishing a coherent, two-way lock with the spacecraft will require up to 30 minutes after detecting the signal on the ground.

Under all circumstances, signal lock-up with the 34m HEF must occur within 105 minutes of the time that the spacecraft begins to transmit telemetry to the ground. After 105 minutes elapse, the spacecraft's range to the Canberra tracking site will exceed 40,000 km, a distance greater than the ACQ-AID's specified "listen range" given the spacecraft's transmission link margin. If 105 minutes elapse and lock-up has not yet occurred, the 34m HEF can perform initial acquisition. However, such a task will be extremely difficult because of the HEF's narrow beamwidth.

During the initial acquisition period, the spacecraft will maintain its orientation of +X axis pointed directly at the Sun, solar panels swept forward 30° above the Y axis in the direction of +X, and roll rate of one revolution every 100 minutes about the +X axis. The spacecraft will continue to hold this attitude for a total of two hours starting from the time that LGA begins transmitting. This time period exists primarily to allow the navigation team to collect at 1.5 hours of two-way, coherent Doppler data for orbit prediction purposes. In addition, the ground operations team will examine the realtime engineering telemetry to assess the health and status of the spacecraft.

The DSN multi-mission navigation team will require at least 1.5 hours of coherent, two-way Doppler data in order to guarantee an orbit prediction accurate enough to point the 34m HEF antenna at the spacecraft to within a half beamwidth. In the current DSN initial acquisition timeline, the first 30 minutes that elapse after the transmitter turns on has been allocated to establishing a two-way lock on the spacecraft. Therefore, the view geometry from the LGA to the ground must support a viable downlink until two hours after the transmitter turns on (2 hours, 22 minutes after separation).

4.1.3 Attitude Initialization

Two hours after the start of the DSN initial-acquisition period (142 minutes after separation), the spacecraft will begin its attitude-initialization sequence. In order to establish a 3-axis attitude reference, the spacecraft must scan its celestial star sensor (CSA) around the sky to identify known stars. This attitude initialization process will be accomplished by pointing the spacecraft +X axis somewhere along the edge of an imaginary cone with a half-angle of 60° and longitudinal axis along the vector to the Sun. In this orientation, the spacecraft will roll so that its +X axis traces a path around this 120° Sun exclusion cone once every 100 minutes. Therefore, the spacecraft +X axis is said to be "coning 60° off the Sun." The coning motion will last for several 100-minute revolutions to allow the star sensor (CSA) enough time to acquire a three-axis attitude reference.

While the spacecraft cones around the Sun to perform attitude initialization, communications with the Earth will periodically fade in and out at 100 minute intervals. The reason is that during this time, the angle between the Sun and Earth, as seen from the spacecraft, will measure between 65° to 105° (depending on the launch day), and the extremes of the spacecraft's +X axis during the coning will place the axis plus or minus 60° from the Sun. Because the LGA sits on the rim of the high gain antenna (HGA) and will point in the +X direction while the HGA sits in its stowed position, the LGA will cycle through positions of pointing almost directly at the Earth to pointing 125° or more away from the Earth. Current analysis shows that if downlink is not guaranteed when the LGA points more than 90° from the Earth, then downlink will only be available during 45 minutes out of each 100-minute coning revolution.

After attitude initialization has been completed, the ground team will command the spacecraft to inner-cruise orientation. In this mode, called array normal spin (ANS), the solar arrays will lie in the same position relative to the spacecraft body (30° swept forward toward +X axis) as during Sun coning. However, the difference is that the +X axis will point 60° off the Sun vector in the direction of the Earth, and the spacecraft will roll at a rate of one revolution every 100 minutes around the +X axis.

Spinning 60° off Sun represents a compromise between needing to point the +X axis directly at the Earth for maximum communications link margin, and needing to point the solar arrays at the Sun for adequate power generation. Communications with the Earth will always occur through the low gain antenna during inner cruise because the undeployed HGA on the +X axis must point directly at the Earth for use.

4.2 Trajectory Correction Maneuvers

During cruise, a set of four trajectory correction maneuvers (TCMs) will adjust the interplanetary trajectory to ensure that the spacecraft reaches the proper velocity and position targets prior to the Mars orbit insertion burn. In general, the TCMs will predominantly be statistical maneuvers to correct for injection errors from the Delta third stage, orbit determination errors, unmodelled forces, and slight execution errors from previous TCMs. In addition, the TCMs will remove the Mars aim-point biasing introduced by the Delta for planetary-protection purposes. This launch biasing intentionally aims the spacecraft away from Mars by about 50,000 km to guarantee a sufficiently-low probability of the Delta third stage impacting the planet. Figure 4-2 shows the locations of the TCM maneuvers during cruise.

Table 4-2: *TCM Schedule and Burn Magnitudes*

Maneuver / (type)	Time	Mean / (95%) Value	Comments
TCM1 (main engine - biprop)	L+ 15 days (21-Nov-96)	15.0 m/s (95% magnitude of 33.0 m/s)	Correct for most of the injection errors, remove most of the launch biasing due to planetary quarantine purposes
	L+ 15 days (10-Dec-96)	21.1 m/s (95% magnitude of 41.7 m/s)	
TCM2 (main engine - blow down)	TCM1+ 120 days (21-Mar-97)	4.1 m/s (95% magnitude of 5.7 m/s)	Correct for execution errors from TCM1, remove remaining launch injection errors and planetary quarantine biasing
	TCM1+ 105 days (25-Mar-97)	5.6 m/s (95% magnitude of 6.0 m/s)	
TCM3 (AACS - monoprop)	TCM2+ 30 days (20-Apr-97)	0.087 m/s (95% magnitude of 0.151 m/s)	Correct for errors from TCM2
	TCM2+ 30 days (24-Apr-97)	0.197 m/s (95% magnitude of 0.359 m/s)	
TCM4 (AACS - monoprop)	MOI- 20 days (22-Aug-97)	0.26 m/s (95% magnitude of 0.53 m/s)	Final adjustment to MOI aim point
	MOI- 20 days (2-Sep-97)	0.26 m/s (95% magnitude of 0.50 m/s)	
Total for TCMs		Launch on 6-Nov-96 (95% magnitude of 36.98 m/s) Launch on 25-Nov-96 (95% magnitude of 47.90 m/s)	Values for the TCM total represent the statistically combined total for the entire cruise phase at a 95% magnitude, not an algebraic sum. Also includes allowance for 95% launch vehicle PCS and revised spin and nutation time constants (59 r.p.m., 86 s)

Table 4-2 shows the TCM schedule along with the expected magnitudes for each of the four burns. Current project policy calls for budgeting the TCMs at a 95% confidence level instead of a 99% level due to tight Delta-V budget situation. In the table, each row corresponds to a different TCM and contains two sets of values for the maneuver dates and burn magnitudes. The top set of numbers in each row reflects values for a launch on 6 November 1996, while the bottom set corresponds to values for a 25 November 1996 launch.

Figure 4-2: *Trans-Mars Cruise Trajectory Diagram and Location of TCMs*

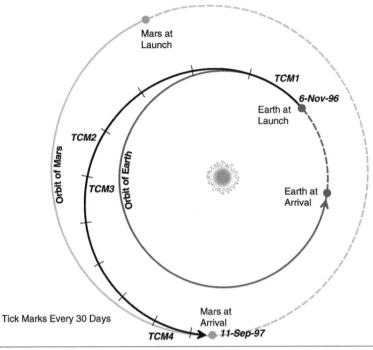

The first and largest of the trajectory correction maneuvers, called TCM1, will always occur 15 days after launch. This maneuver will primarily correct for injection errors introduced by the Delta third stage and remove most of the launch biasing introduced into the trajectory for Mars planetary quarantine purposes. TCM2, scheduled for between 105 and 120 days after TCM1, will correct for errors in TCM1 and remove the remaining planetary-quarantine bias from the trajectory. Both of the first two TCMs will be performed with the main engine. However, the second TCM will be performed in "blow-down" mode with the residual pressure level remaining in the tanks after the isolation of the high-pressure helium line sometime after TCM1.

The last two correction maneuvers, called TCM3 and TCM4, will each consist of small burns designed to correct for slight execution errors in the previous maneuvers. In addition, TCM4 at 20 days prior to Mars orbit insertion (MOI) will precisely target the spacecraft to its MOI aim point. Both of these burns will utilize the 4.4-N attitude control thrusters.

Contingency maneuver windows will exist at L+ 30 days for TCM1 and at MOI- 10 days for TCM4 if the primary opportunities are missed due to unforeseen circumstances.

4.2.1 TCM Implementation

Both TCM1 and TCM2 will utilize the main engine (596-N thrust) for primary thrust output with the smaller engines (4.4-N thrust) for attitude control. The small magnitude of the other two maneuvers, TCM3 and TCM4, will allow the attitude control thrusters to perform the burns.

Before both TCM1 and TCM2, the spacecraft will turn from its normal cruise attitude and point the -Z axis (engine exhaust thrust axis) opposite the direction of the desired Delta-V. In addition, the X and Y axes of the spacecraft in the maneuver attitude will point in a direction to optimize the Sun angle to the solar arrays. If this angle is too large, the spacecraft will need to operate on batteries during the maneuver.

Because TCMs are statistical maneuvers designed to correct errors in the trajectory, the spacecraft operations team will not know the exact direction and magnitude of the burn until several days before the scheduled maneuver date. Therefore, the turn may take up to 10 minutes if the magnitude of the turn angle measures as large as 180°. After reaching the maneuver attitude, the spacecraft will then hold a fixed inertial attitude while autonomous AACS checks occur. The spacecraft will then execute the burn autonomously with the capability to abort in the event of a malfunction.

During all TCMs, the spacecraft will control the burn duration by using accelerometer measurements as inputs. At the end of the burn, the spacecraft will turn back to its original cruise attitude. For all maneuvers, at least one of the recorders will continuously record engineering data for subsequent playback. In the case of TCM1, the spacecraft will be close enough to the Earth to also return data at the 2 ksps engineering rate in realtime, given a favorable maneuver direction in terms of sunlight on the solar array and low gain antenna (LGA) orientation with respect to the Earth. During the other three maneuvers, the link margin on the LGA will not support the 2-ksps engineering rate. Consequently, the transmitter will be shut off to conserve battery power. However, in the event that in-flight performance data shows an adequate power situation, the transmitter may be left on in "carrier only" mode during TCMs 2, 3, and 4.

Knowledge of the location of the Sun will factor heavily into the design of the TCMs because of a significant payload constraint on maneuver attitudes, required by the MOC and the MOLA instruments. These instruments do not have covers. Consequently, a flight rule exists that states that spacecraft's +Z axis (science instrument panel) cannot be pointed closer than 30° from the Sun. Due to the fact that the Delta-V direction for a TCM can be in any inertial direction, a performance penalty may exist for pointing the thrust in a non-optimal direction to satisfy the Sun pointing constraint.

4.2.2 Propulsion System Operations for TCMs

At launch the pressure in the entire propulsion system will sit at a blanket value of 100 psig. The mono-propellant lines will be launched wet down to the engine valves, allowing the attitude control thrusters to operate in blow-down mode to support spacecraft separation from the Delta and despin.

The main engine valve and latch valves below the tanks will remain closed at launch. Prior to TCM1, a dry-firing of the main engine for several minutes will vent the lines (between the latch and main engine valves) to space and bleed them dry. After venting, the ground team will command these latch valves to open. This action will allow the fuel and oxidizer to fill the lines up to the main engine valve at approximately the same blanket pressure used to operate the mono-propellant system in blow-down mode after launch.

Pressurization of the propulsion system will occur prior to TCM1. In order to pressurize the system, the high pressure latch valve located between the helium pressurant tank and the regulators will open, followed by the opening of one of the normally closed pyro valves located immediately below the pressurant tank. Tank pressurization will begin when the pressure downstream of the regulators reaches 175 psig and ruptures the two burst discs located upstream of the fuel and oxidizer tanks. Then, the regulator will come on line when the pressure in the tanks reaches the nominal value of 250 psig.

Due to concerns about regulator leakage with the tanks nearly full, one of the normally open pyro valves in the pyro ladder immediately downstream from the helium tank will be closed after TCM1, effectively isolating the high pressure helium from the rest of the system. Consequently, the remaining TCMs must occur in "blow-down" mode using the residual pressure level remaining in the tanks.

4.3 Inner Cruise Activities

Inner cruise will be devoted to characterizing the operation of the spacecraft, performing the first TCM, checking out the spacecraft, and calibrating the science instruments. Adequate link margins and continuous DSN coverage during this time period will support the return of data rates as high as the 40,000-sps, S&E-2 realtime rate for payload operations and check-out.

4.3.1 Major Spacecraft Activities and TCMs

Activities during the first half-month of flight will focus only on spacecraft check-out and preparations for executing the first trajectory correction maneuver (TCM) at 15 days after launch. No "special" check-out procedures exist other than planned intensive monitoring of each subsystem with contingency recording of telemetry on one of the recorders. In addition, telemetry recorded during launch will be replayed during this time.

Propulsion system activities will begin at L+ 7 days with the "dry-firing" of the main engine. The "dry-firing" will vent the lines (downstream of the tanks) to space and bleed them dry. Pressurization of the tanks with the high pressure helium will follow on flight day L+ 12 in preparation for the first TCM at L+ 15 days. This schedule implementation will provide margin before the pressurization and TCM in the event of an anomaly during the first few days of flight.

Upon successful completion and verification of TCM1, one of the pyro valves immediately downstream of the helium tank will be closed to effectively isolate both the oxidizer and fuel tanks from the high pressure assembly upstream. Re-pressurization of the tanks will not occur until just prior to Mars orbit insertion (MOI).

4.3.2 Instrument Calibrations

Instrument calibrations will begin on flight day L+ 16 with the activation and checkout of the payload data subsystem (PDS). This device is the computer control unit that collects data from the instruments and formats the information into data packets for storage on the SSRs or playback to Earth. The checkout sequence will include a flight software upload into the PDS RAM. Following the software upload, the different data-rate modes controlled by the PDS will be tested by broadcasting data to Earth at all of the S&E1 and S&E2 data rates from 4 ksps to 85 ksps. Previous experience has shown that the software upload and data-rate test will consume most of the prime-shift working day.

On day L+ 17, the flight team will activate all of the science instruments and will provide a brief time period for the science investigation teams to send flight-software uploads to their instruments. Later in the day, the spacecraft will perform an hour-long slew to allow the TES to image the Earth and the Moon. These images will allow the TES team to calibrate their instrument.

Later on in the day, the radio science team will perform a checkout of the ultra-stable oscillator (USO). This test will involve turning off the spacecraft's telemetry and broadcasting to Earth in carrier-only mode for two hours. The USO test will allow the radio science team to determine the frequency stability characteristics of the device. Two more USO tests will follow in mid-December 1996.

In addition, the Mars Relay (MR) will begin broadcasting its UHF relay beacon on day L+ 17. The MR test will consist of two parts. The primary portion of the test will involve use of the 46m antenna at Stanford to broadcast test data to the MR for storage. The spacecraft will store this data on its SSRs for subsequent playback to Earth. The secondary portion of the test will involve amateur radio enthusiasts around the world. These HAM operators will use their own radios to listen to MR's signal beacon and report their results back to JPL.

The MOC will begin its calibration activities on flight day L+ 18. During the morning, the spacecraft will perform a one-hour slew to allow the camera to image the Earth and Moon for calibration purposes. Then, on flight day L+ 19, the spacecraft will perform another one-hour slew for the MOC to image several stars. These two days worth of images will allow the MOC team to make minor focus adjustments by altering heater settings on the camera.

4.3.3 Important Modification to Baseline Schedule

For the first seven days of the launch period (November 6th through November 12th, inclusive), TCM1 and isolation will take place on flight days L+ 14 and L+ 15, respectively. Executing the maneuver one day earlier than in the baseline plan will allow the instrument calibration activities (L+ 16 to L+ 19) to finish by the Thanksgiving holiday.

4.4 Outer Cruise Activities

About two months after launch and sometime in January 1997, the angle between the Earth and Sun as seen from the spacecraft (values of 60° or less) will allow for pointing the undeployed high-gain antenna (HGA) directly at Earth and still provide for adequate illumination of the solar arrays. This geometry will allow the spacecraft to transition from using the LGA to the HGA.

From this time until Mars orbit insertion, excluding trajectory correction maneuvers, the normal spacecraft configuration will be outer-cruise array normal spin (ANS) in order to take advantage of the decreasing angles between the Earth and Sun. When configured in this mode, the spacecraft solar panels will be swept forward 30° above the Y axis toward the +X direction, the +X-axis of the spacecraft will point directly at the Earth, and the spacecraft will roll about the +X axis at the rate of one revolution every 100 minutes.

4.4.1 Normal Spacecraft Cruise Operations

Primary activities during outer cruise will involve routine monitoring of the spacecraft, collection of navigation data, and execution of the remaining three trajectory correction maneuvers. Because the spacecraft HGA will point directly at Earth during outer cruise, substantial capability will exist for returning data from the science payload. However, only a limited number of science data collection activities and calibrations are currently planned due to the limited staffing level of the flight team.

Normal operations of the spacecraft during cruise will consist of fairly uncomplicated tasks. Except for TCMs and limited science activities, minimal commanding will be sufficient to maintain the engineering sub-systems. The primary ground activity will involve analyzing trends in the performance of various sub-systems based on engineering telemetry and radio-metric data returned during the daily DSN tracking pass.

Although the spacecraft can operate for many days without contact, a command-loss timer will initiate fault protection responses if a command fails to arrive within a set amount of time. This command-loss time period is variable and can be changed. The only routine uplinks required to maintain the spacecraft will consist of "no-op" command-loss timer reset commands, and a bi-weekly star catalog and planetary ephemeris update. Data from this ephemeris will assist in the pointing of the spacecraft's +X axis in the array-normal-spin attitude mode, while the star catalog will be used to update and maintain the spacecraft's inertial reference.

4.4.2 Major Spacecraft Activities and TCMs

Until preparations for Mars orbit insertion (MOI) begin toward the end of outer cruise, no major spacecraft activities will occur other than the execution of TCM2, TCM3, and TCM4. Due to the isolation of the high-pressure helium tanks after TCM1, the three remaining maneuvers will occur in "blow-down" mode using the residual pressure level remaining in the fuel and oxidizer tanks. Seven days prior to MOI, the propulsion system will be repressurized to 250 psig by firing one of the normally-closed pyro valves immediately upstream of the regulator system. Seven days will allow margin in the event of an anomaly associated with the repressurization of the propulsion system.

4.4.3 Major Payload Activities

Instrument checkout activities will continue in early-January 1997 with the MOLA test. During this one-hour activity, the spacecraft will slew its +Z axis toward the Earth to allow MOLA to fire laser pulses at the Earth. The MOLA team will utilize ground stations equipped with laser-detection devices to verify proper operation of the laser altimeter.

One day after the laser test, the MOC will resume its focus testing. These tests will occur on four separate occasions spaced one or two days apart. Each test will consist of a single one-hour spacecraft slew to allow the camera to image several stars. The images of the stars will allow the MOC team to make minor focus adjustments by altering heater settings on the camera.

After performing the four focus-check tests, the MOC will enter a 60-day bakeout period. During this time, a special set of heaters in the camera will continuously "bake" the instrument at low heat levels to solidify the desired focus adjustments. Upon completion of the bakeout event, the camera will perform another focus-check test. These post-bakeout focus checks will be performed in a manner similar to the pre-bakeout focus test.

In addition to performing focus checks, the MOC will acquire approach images of Mars at various distances prior to arrival. The first of four imaging opportunities will take place 120 days prior to Mars orbit insertion (MOI). The remaining three will occur at MOI- 90, 60, and 30 days.

The radio science team will also conduct tests of the USO in outer cruise. The tests will be similar to the one conducted on flight day L+ 17 and will occur approximately once every two to four weeks. Exact dates will be announced during cruise.

5. Orbit Insertion Phase

The orbit insertion phase of the mission will begin with the Mars orbit insertion (MOI) burn in mid-September 1997. This burn will slow the spacecraft by approximately 980 m/s and allow Mars to capture it into a highly elliptical orbit with a 48-hour period. Unfortunately, the MGS spacecraft will not carry enough propellant to propulsively reach the required low-altitude, Sun-synchronous mapping orbit due to the relatively low interplanetary injected mass capability of the low-cost Delta booster.

Instead, the spacecraft will rely on aerobraking, an innovative mission-enabling technique, to trim the initial, highly elliptical, capture orbit down to mapping orbit altitudes. During aerobraking, the spacecraft will pass through the upper fringes of the Martian atmosphere on every periapsis pass. Air resistance from the atmosphere during the "drag pass" will cause the spacecraft to lose a small amount of momentum and will cause the altitude on the next apoapsis pass to slightly decrease.

Aerobraking will end in mid-January 1998. Over the course of the next two months, the spacecraft will use a combination of propulsive maneuvers and orbital perturbations from the Martian gravity field to reach the desired mapping orbit. The orbit insertion phase will end with the beginning of mapping in mid-March 1998. Figure 5-1 shows a timeline of the events that occur during orbit insertion.

5.1 Mars Orbit Insertion

MOI will slow the spacecraft and allow Mars to capture it into an elliptical orbit. Before the burn, the spacecraft's velocity relative to Mars will measure approximately 5,650 m/s. Near the closest approach point of the inbound hyperbolic trajectory, the 659-N main engine will fire for between 20 to 25 minutes to provide a Delta-V of about 980 m/s. Burn ignition will occur approximately about 10 minutes before Mars closest approach. Figure 5-2 shows a trajectory plot of the maneuver.

During the burn, the spacecraft will utilize a "pitch-over" maneuvering strategy to slew the spacecraft's attitude at a constant rate in an attempt to keep the thrust nearly tangent to the trajectory arc. Although this constant rate pitch will not allow the thrust vector to exactly follow an optimal steering profile, the strategy will provide a more optimal solution than a constant burn vector pointed in a fixed, inertial direction.

After MOI-burn cut-off 10 minutes after periapsis, the spacecraft will orbit Mars on a highly elliptical orbit with a period of 48 hours and periapsis altitude of about 300 km (periapsis radius of 3,700 km). Due to specification uncertainties for maneuver execution, the capture orbit period may vary from as low as 40.3 hours to as high as 59.7 hours. In addition, the periapsis altitude may vary from 229 km to 399 km. Both the period and altitude figures represent 3σ dispersions due to specification uncertainties. However, the true capability errors will be smaller and will cause these uncertainty figures to be reduced.

Figure 5-1: *Orbit Insertion Timeline (open of launch period scenario)*

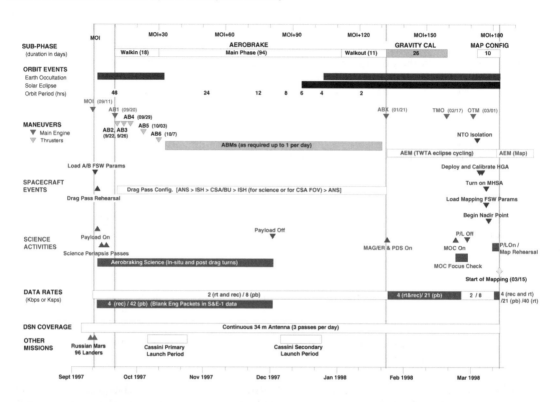

Figure 5-2: *Trajectory Plot of MOI*

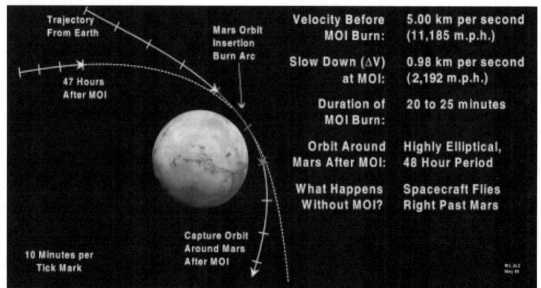

5.1.1 Maneuver Execution Time

Depending on the exact launch date, the spacecraft will arrive at Mars between 11 September 1997 and 22 September 1997. The incoming hyperbolic trajectory will be targeted for a Mars closest approach at roughly 1:00 a.m. UTC to take advantage of overlapping DSN coverage from Goldstone and Canberra. The exact time of Mars closest approach will vary depending on the arrival date, but will always occur approximately half-way into the overlap period between the two DSN stations. This timing strategy will provide about one hour of overlapping tracking coverage both before and after the MCI burn.

Goldstone will serve as the primary station for pre-MOI burn coverage, while Canberra will take over for post-burn tracking. This decision was based on allowing the station complex with the higher elevation angle view of Mars to be prime for uplink. However, both the 34m HEF and 70m antennas at both complexes will listen to the spacecraft's signal during the overlap period. If the 34m HEF at the prime complex fails, then the 70m will be able to automatically take over in listen-only mode. In the event that commanding capability or an uplink carrier is required, then prime tracking responsibility will switch to the 34m HEF at the other complex.

Table 5-1 lists the arrival date and time of Mars closest approach for each launch date. Under normal conditions, the MOI burn will be centered on the closest approach and will last for between 20 to 25 minutes. In the table, the closest approach and Goldstone set (DSN overlap end) occurs on the same calendar day as the arrival day. The Canberra rise time (DSN overlap begin) occurs on the calendar day before the arrival day. All times listed in the table are in ephemeris time (ET).

Table 5-1: *MOI Times*

Launch Date	Arrival Date	Canberra Rise	Closest Approach	Goldstone Set
6-Nov-96	11-Sep-97	23:43:50	01:27:53	03:40:32
7-Nov-96	12-Sep-97	23:42:02	01:26:09	03:38:32
8-Nov-96	12-Sep-97	23:42:02	01:26:09	03:38:32
9-Nov-96	13-Sep-97	23:40:15	01:24:28	03:36:33
10-Nov-96	13-Sep-97	23:40:15	01:24:28	03:36:33
11-Nov-96	14-Sep-97	23:38:29	01:22:47	03:34:35
12-Nov-96	14-Sep-97	23:38:29	01:22:47	03:34:35
13-Nov-96	14-Sep-97	23:38:29	01:22:47	03:34:35
14-Nov-96	15-Sep-97	23:36:44	01:21:07	03:32:38
15-Nov-96	15-Sep-97	23:36:44	01:21:07	03:32:38
16-Nov-96	16-Sep-97	23:35:00	01:19:29	03:30:42
17-Nov-96	16-Sep-97	23:35:00	01:19:29	03:30:42
18-Nov-96	17-Sep-97	23:33:17	01:17:52	03:28:48
19-Nov-96	18-Sep-97	23:31:35	01:16:14	03:26:54
20-Nov-96	18-Sep-97	23:31:35	01:16:14	03:26:54
21-Nov-96	19-Sep-97	23:29:54	01:14:40	03:25:02
22-Nov-96	20-Sep-97	23:28:15	01:13:05	03:23:10
23-Nov-96	20-Sep-97	23:28:15	01:13:05	03:23:10
24-Nov-96	21-Sep-97	23:26:36	01:11:32	03:21:20
25-Nov-96	22-Sep-97	23:24:59	01:10:01	03:19:30

MOI Time Changes for Russian Lander Support

During the four orbits immediately following MCI, the spacecraft may be required to relay data back to Earth from Russian landers on the surface by using the Mars Relay antenna. Unfortunately, the MOI times chosen for the dual-station overlap strategy will result in a ground track pattern incompatible with overflights of Russian landing sites after MOI. If relay operations are required, then the MOI time will be changed at TCM2 to produce a post-burn ground track that overflies the landing sites within the first four orbits after the burn.

Initial analysis shows that MOI will need to be moved between two and seven hours earlier to facilitate the relay of Russian data. Calculations of the exact MCI offset time will be performed after both the MGS spacecraft and the Russian landers have launched during the November 1996 opportunity. In general, MGS launches early in the launch period will result in a greater offset as compared to launches later in the launch period.

Moving MOI earlier will result in two major impacts to the mission. First, dual-station overlap for pre-burn and post-burn tracking redundancy will be lost because the Madrid to Goldstone overlap will last for less than 30 minutes in September 1997. Second, changing the time of MOI at TCM2 will cost between 3 to 6 m/s. The Delta-V penalty will have substantial ramifications to the ability to alter the MOI time because TCM2 will be limited to 6 m/s or less. This limitation results from the fact that TCM2 will be performed in "blow-down" mode using the residual pressure level remaining in the tanks following isolation of the high-pressure helium lines after TCM1.

Altering the MCI time at TCM1 will not be possible due to the fact that the Russian launch period opens on 16 November 1996. Due to a high desire to maintain the dual-station overlap during MOI strategy, the project will not commit to altering the Mars arrival time until it has been verified that the Russians have launched. Although November 16th will occur before TCM1, sufficient time will not exist to incorporate the navigational targeting changes into the flight sequence before the execution of the maneuver.

5.1.2 Spacecraft Activities During MOI

The final phase of the MOI burn sequence will begin with the start of 2-kbps engineering telemetry recording on two of the recorders. Prior to the start of the burn, the spacecraft will power on the catbed and main engine heaters, begin the turn to the pre-determined burn attitude using reaction wheel control, and then gimbal the solar arrays to their maneuver positions by rotating the outer gimbals -90° from their cruise configuration. At this time, the TWTA beam will be turned off both to conserve power and because the spacecraft will be occulted by Mars as viewed by Earth throughout much of the burn.

At the pre-specified burn start time, the flight software maneuver task will fire the 659-N main engine. During the burn, the spacecraft will execute a "pitch-over" steering strategy to maximize the Delta -V efficiency of the maneuver. This strategy will be implemented by using the attitude control thrusters to slew the spacecraft at a fixed rate during the burn in an attempt to keep the thrust tangent to the trajectory arc. In addition, the thrusters will provide attitude control during the burn.

The MOI maneuver will terminate after the spacecraft's accelerometers have sensed the proper amount of velocity change. Termination will normally occur between 20 to 25 minutes after ignition. However, the spacecraft will automatically cut-off the engine if the burn duration exceeds the maximum allowable time. This maximum time value is a changeable parameter that will be pre-determined by the flight team prior to the loading of the MCI sequence.

Immediately after the completion of the burn, the spacecraft will begin the slew back to ANS Earth-point configuration. Then, downlink telemetry will be turned on and 2-kbps, realtime engineering data will be broadcast to Earth. The realtime telemetry will allow the flight team to assess the post-burn health status of spacecraft and will provide tracking data for a navigation solution to the capture orbit.

The spacecraft has several levels of fault protection that can impact the success of the MCI maneuver. For example, redundancy management software will be enabled for all critical hardware components. Upon detection of a fault, the spacecraft will automatically switch to the redundant hardware. The spacecraft will also use special system-level fault-protection modes, designated as safe mode and contingency mode. If entered, these modes will abort the MCI sequence. Consequently, safe and contingency modes will be disabled at appropriate times prior to the start of the maneuver. These times will be determined by estimating the recovery times needed to reestablish the spacecraft to the required configuration for MOI.

5.2 Capture Orbit Activities

After successful completion of the MOI burn, the spacecraft will orbit Mars once every 48 hours with a periapsis and apoapsis altitude of approximately 300 km and 56,675 km, respectively. For the next nine days, the spacecraft will remain in this capture orbit as the flight team prepares for the beginning of aerobraking. Activities during this time period will include aerobraking rehearsal orbits and the collection of a limited amount of science data. Table 5-2 provides an orbit by orbit summary of the activities that will take place during the post-MOI orbits.

Table 5-2: *Summary of Post-MOI Activities*

Spacecraft Position	Time	Activity	Configuration
First Orbit			
Periapsis #0	MOI	Mars Orbit Insertion	Maneuver
Apoapsis #1	MOI+ 1 day	Post-MOI Spacecraft Health Assessment	Array Normal Spin
Rest of Orbit	MOI to MOI+2	Post-MOI Spacecraft Health Assessment	Array Normal Spin
Second Orbit			
Periapsis #1	MOI+ 2 days	Routine Spacecraft Health Monitoring	Array Normal Spin
Apoapsis #2	MOI+ 3 days	Routine Spacecraft Health Monitoring	Array Normal Spin
Rest of Orbit	MOI+ 2 to MOI+ 4	Routine Monitoring, Prepare for Drag Rehearsal	Array Normal Spin
Third Orbit			
Periapsis #2	MOI+ 4 days	Aerobrake Drag Rehearsal	Drag
Apoapsis #3	MOI+ 5 days	Routine Spacecraft Health Monitoring	Array Normal Spin
Rest of Orbit	MOI+ 4 to MOI+ 6	Activate Science Instruments, PDS	Array Normal Spin
Fourth Orbit			
Periapsis #3	MOI+ 6 days	Science Data Collection	Nadir
Apoapsis #4	MOI+ 7 days	Routine Spacecraft Health Monitoring	Array Normal Spin
Rest of Orbit	MOI+ 6 to MOI+ 8	Routine Monitoring, Playback of Science Data	Array Normal Spin
Fifth Orbit			
Periapsis #4	MOI+ 8 days	Science Data Collection	Nadir
Apoapsis #5	MOI+ 9 days	AB1 Burn (lower periapsis to start aerobraking)	Maneuver
Rest of Orbit	MOI+ 8 to MOI+ 10	Routine Monitoring, Playback of Science Data	Array Normal Spin
Sixth Orbit			
Periapsis #5	MOI+ 10 days	First aerobrake drag pass (150 km altitude)	Drag
Apoapsis #6	MOI+ 11 days	AB2 Burn (lower periapsis some more)	Maneuver
Rest of Orbit	MOI+ 10 to MOI+ 11	Routine Monitoring	Array Normal Spin

5.2.1 Spacecraft Checkout

The navigation team will use radiometric data to track the spacecraft and determine the exact solution to the capture orbit during the first revolution around Mars. In addition, the flight team will monitor the engineering telemetry to determine the health and status of the spacecraft and its subsystems. This performance characterization will be necessary largely because of the spacecraft's entry into the new thermal environment around Mars.

During the second revolution around Mars, the flight team will prepare to conduct an aerobrake drag-pass rehearsal. The command sequences and scripts for aerobraking, as well as various flight software parameter updates, will be generated and uploaded prior to the second periapsis pass after MOI. Just before the passage through that periapsis, the spacecraft will configure itself into the proper attitude and orientation for flight through the upper Martian atmosphere. This rehearsal will test all of the ground operational procedures and spacecraft activities that will take place during aerobraking without subjecting the spacecraft to an actual drag pass.

5.2.2 Science Activities

Upon completion of the aerobrake drag-pass rehearsal at the start of the third orbit after MOI, the payload data subsystem (PDS) will be powered on and then checked out for the first time at Mars. Check out activities will involve a flight software upload into the PDS RAM. Previous experience has shown that this upload will consume most of the prime-shift working day after the drag-pass rehearsal. Because the PDS controls the collection of data from the science payload, the instruments will not be powered on until after checkout of the PDS.

Six days after MOI, at the beginning of the fourth orbit, the spacecraft will slew to nadir point the science instruments at the surface of Mars for about 15 minutes centered at periapsis. During the 15 minutes, the MOC, TES, and MOLA will be able to record a limited amount of science. In addition, any use of the MR to collect data from Russian landers on the surface will also take place during this time. After the data collection period has expired, the spacecraft will return to array-normal-spin configuration and play back the recorded science at the highest available data rate. Science data collection will also be performed in this manner eight days after MOI at the beginning of the fifth orbit.

In order to minimize the effort required to execute the sequence of events required for periapsis data collection, the spacecraft will utilize a command script similar to the one that will be used during the aerobrake drag passes. Normally, the aerobraking command script temporarily slews the spacecraft from array normal spin (ANS) to drag pass configuration for a predetermined amount of time near periapsis. The only difference is that the spacecraft will fly with a 90° offset in attitude during data collection. This offset will allow the science instruments on the +Z panel to point downward at Mars during periapsis instead of at the normal aerobraking attitude of +Z backward along the velocity vector.

From the time of instrument activation, the MAG, ER, and radio science (gravity field studies) will collect data on a continuous basis because they do not require nadir pointing. Upon completion of the two science-observation periapsis passes, the instruments will remain active for a significant portion of aerobraking in order to acquire unique science data and to support aerobraking operations.

5.3 Aerobraking

After completing the two "science orbits," the spacecraft will spend the next four months aerobraking. This part of the mission will consist of three major phases called walk-in, main phase, and walkout. During walk-in, the periapsis of the orbit will be gradually lowered into the Martian atmosphere at an altitude of about 112 km. Then, the spacecraft will enter main phase and remain there for the majority of aerobraking. In main phase, air resistance from the atmosphere will slow the spacecraft on every periapsis drag pass and will gradually cause the apoapsis to shrink. After the apoapsis altitude has decayed to less than 2,000 km, the spacecraft will use the walk-out phase to gradually raise the periapsis of its orbit out of the Martian atmosphere while continuing to finish aerobraking.

5.3.1 The General Scenario

One of the major constraints driving the aerobraking timeline is the requirement to reach a 2:00 p.m. descending node orientation with respect to the fictitious mean Sun. The spacecraft's approach direction from its interplanetary trajectory will result in an initial node located at about 5:45 p.m. During most of aerobraking, the node time will move backward by about a half minute (0.524°) per day due to Mars' motion around the Sun.

Aerobraking must reduce the orbit size to mapping orbit altitudes in the exactly same amount of time that the node takes to move backward from 5:45 p.m. to 2:00 p.m. Because the nodal motion is to first order independent of orbit size for most of aerobraking (56,675 km to about 10,000 km apoapsis), a "too-rapid" apoapsis decay profile will result in arriving at the mapping orbit with a node time later than 2:00 p.m. Conversely, a "too-slow" profile will result in a mapping orbit earlier than 2:00 p.m. Figure 5-3 illustrates the orbit node geometry constraints affecting aerobraking.

5.3.2 Walk-in Phase

Aerobraking will begin nine days after MOI with a maneuver at the fifth apoapsis. This burn, called AB1, will be the first and largest of four to six mono-propellant maneuvers designed to lower periapsis into the upper Martian atmosphere in gradual steps. In addition, the burn will also correct for small inclination errors caused by MOI. After AB1, the spacecraft will take about one day to coast down to the new periapsis at an altitude of 150 km. The post-AB1 periapsis will represent both the start of the sixth orbit after MOI, and the first drag pass of aerobraking.

AB2 will occur at the sixth apoapsis and exactly one revolution after AB1. Afterward, the flight team will perform AB3, AB4, AB5, and AB6 at apoapsis on every other revolution. Therefore, AB6 will occur approximately 27 days after MOI at the 14th apoapsis. The goal is to use these six "walk-in" maneuvers to

Figure 5-3: Orbit Insertion Geometry

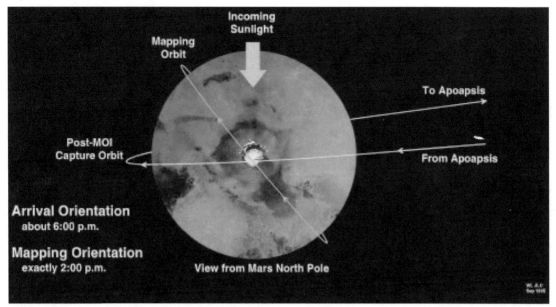

gradually lower periapsis to the target point where air resistance will cause the spacecraft to experience a heating rate of 0.38 W/cm². Current models predict that this heating rate will occur at an altitude of about 112 km and will amount to roughly 48% of the maximum tolerable rate of 0.79 W/cm².

Although AB1 will lower the periapsis to a pre-planned altitude, the navigation team will choose the size of the AB2 through AB6 burns in realtime during walk-in operations. The reason for the realtime selection is due to the large uncertainty in the atmospheric density model of Mars. In particular, the navigation team will need to estimate the scale height of the Martian atmosphere. This parameter quantifies the amount of altitude decrease needed to increase the atmospheric density by a factor of 2.718 (one exponential factor). Smaller scale heights indicate that the atmospheric density increases more rapidly as altitude decreases than for an atmosphere with a larger scale height number.

Unfortunately, post-drag-pass tracking data will yield the density of the atmosphere, but not the scale height. In order to determine the safety of a proposed walk-in maneuver, the navigation team will calculate a test parameter called the critical scale height. This parameter represents the scale height that would need to exist at Mars for the spacecraft to encounter the critical atmospheric density on the next drag pass after the proposed walk-in maneuver. The critical atmospheric density defined as 143 kg/km³, will cause the spacecraft to experience the maximum allowable drag-pass heating rate of 0.79 W/cm².

The navigation team will compare the critical scale height for the proposed walk-in maneuver to the scale height predicted by the Mars-Gram atmospheric model. If the critical scale height is less than the predicted actual scale height, then the proposed maneuver will probably not decrease the periapsis altitude into the "danger zone." The reason is that a smaller critical scale height will indicate that the increase in atmospheric density with altitude decrease will need to be greater than predicted to cause the spacecraft to encounter the critical atmospheric density at the periapsis immediately after the maneuver.

During walk-in operations, the navigation team will select a burn magnitude from the list of 0.05, 0.10, 0.20, 0.40, 0.60, and 0.80 m/s for each one of the maneuvers from AB2 through AB6. The maneuver sizes for these five burns will be selected both to provide a gradual lowering of the periapsis altitude toward the target value at about 112 km, and to satisfy the critical-scale-height safety test. Because of the uncertainty in the knowledge of the Martian atmosphere, the target altitude that corresponds to the desired heating rate of 0.38 W/cm² will probably be different than predicted. Consequently, the navigation team may decide to change, delete, or add walk-in maneuvers as necessary.

Figure 5-4: *Phases of Aerobraking*

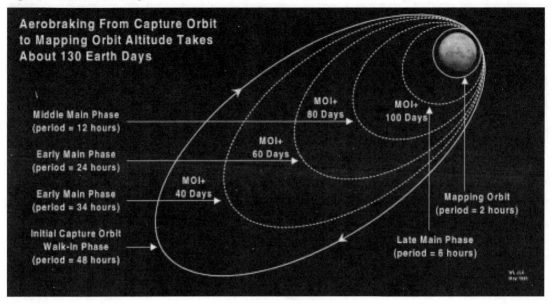

Figure 5-4: *Phases of Aerobraking*

5.3.3 Aerobrake Main Phase

After the completion of walk-in, the spacecraft will spend about 94 days in the main phase of aerobraking. During this time, repeated passes through the atmosphere at periapsis will cause the apoapsis altitude to shrink in size from about 56,675 km down to 2,000 km. As often as once per day, small propulsive maneuvers (ABMs) executed at apoapsis will maintain the periapsis altitude within a well-defined corridor low enough to produce enough drag to reduce the orbit within the time constraints to reach the 2:00 p.m. node, yet high enough to avoid spacecraft heating and maximum dynamic pressure limits. Due to the oblateness of Mars and the fact that periapsis will be migrating northward toward the pole during main phase, the altitude of periapsis will tend to rise. Consequently, most of the ABMs will be in the down direction to lower the periapsis altitude into the control corridor.

As compared to the walk-in and walk-out phases of aerobraking, spacecraft heating rates during the drag pass will reach a maximum during main phase. The spacecraft design can tolerate a top free-stream aerodynamic heating rate of 0.79 W/cm² at periapsis without violating the qualification limits of the solar arrays. Given the amount of time that the node takes to move from 5:45 p.m. to 2:00 p.m., the spacecraft will be able to aerobrake at an apoapsis decay rate that corresponds to a maximum drag-pass heating rate of 0.38 W/cm². Because atmospheric density is related to the heating rate, this baseline plan will provide margin for an orbit-to-orbit atmospheric variability of up to 90% (70% of true variability, 20% for navigation uncertainty) without violating the spacecraft's heating limits.

5.3.4 Walk-Out Phase

The two weeks of aerobraking following the main phase will represent an extremely critical period with respect to maintaining a viable orbit as the spacecraft lowers its apoapsis toward the target finish altitude of 450 km. During this time, the spacecraft will slowly "walk-out" of the atmosphere by gradually raising its periapsis altitude to 143 km. Daily ABMs will be performed to maintain a guaranteed worst-case, two-day orbit lifetime. In other words, in the absence of ABMs due to unforeseen events that inhibit the ability of flight controllers to send commands, the spacecraft will always be at least two days from crashing into the surface.

Walk-out phase will also represent an extremely critical time with respect to the power situation on the spacecraft. During most of the aerobraking main phase, the spacecraft will never encounter regions in the orbit where Mars eclipses the Sun from the solar arrays. Consequently, the only battery discharge times will occur during the drag pass when the solar arrays will be oriented in an aerodynamic favorable orientation rather than in a power-collection configuration. In walk-out, the spacecraft will experience both long drag passes and eclipse zones on every orbit. The maintenance of a favorable energy balance situation on the spacecraft will be further complicated by the fact that the short two-hour orbits in walk-out will provide little time to recharge the batteries between successive drag passes.

5.3.5 Aerobrake Termination

Aerobraking will end with a termination burn (ABX) performed sometime during mid-January 1998. This burn will raise the orbit periapsis out of the atmosphere to an attitude of approximately 450 km. At this time, the spacecraft will be circling in a 400 x 450 km orbit with a period slightly under two hours. In addition, the descending node location will have regressed from its original MOI position at 5:45 p.m. with respect to the fictitious mean Sun to nearly 2:00 p.m.

5.3.6 Spacecraft Drag-Pass Activities

Figure 5-5 shows a sample event profile plot for a typical aerobraking orbit. The events in the orbit will occur relative to the predicted time of periapsis passage and will be executed from an onboard, stored command sequence. Timing pads added to both sides of the periapsis passage will allow for navigation uncertainties in the periapsis crossing time. As the orbit period decreases, the drag-pass durations will increase. Consequently, the timing between events in the sequence will need to be updated on a periodic basis.

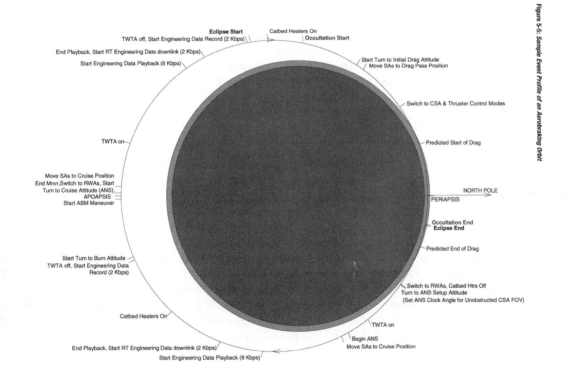

Figure 5-5: Sample Event Profile of an Aerobraking Orbit

Throughout the aerobrake drag-pass sequence, high-rate engineering telemetry at 2 kbps will be recorded on one of the SSRs. This sequence will start with a 20-minute warm-up of the AACS thruster's catbed heaters. Then, the transmitter will turn off to conserve power. Approximately 7.5 minutes prior to the predicted start of the drag pass (including margin for navigational uncertainty), the spacecraft will turn to a "tail-first" attitude under reaction wheel control. The desired entry attitude will point the -Z axis along the velocity vector to keep the science instruments and solar cells pointed away from the incoming air flow.

Once the turn completes, the solar arrays will then gimbal to their drag-pass positions of 30° above the Y-axis in the direction of +Z. The gimbaling to this configuration will be accomplished by rotating the inner (elevation) gimbals by 90° from their ANS positions. The 30° sweep-back angle will provide for maximum aerodynamic stability with minimal drag reduction by moving the center-of-pressure aft of the center-of-gravity. This configuration will be aerodynamically stable because the aerodynamic torque will push the attitude back toward the aerodynamic null, a point where the velocity will align with the -Z axis. Additionally, this orientation will also place the solar arrays against their gimbal hard-stop positions to prevent back driving the gimbal motors.

At the predicted start of the pass (including margin for navigational uncertainty), attitude control authority will switch from reaction wheel to the AACS thrusters, and the wheels will be placed in TACH-HOLD mode in order to maintain their current spin speeds. The reason for this switch is that aerodynamic torques on the spacecraft during the drag pass will be larger than the reaction wheels can counteract. In order to minimize the number of thruster firings and the amount of AACS propellant consumed during aerobraking, attitude-control deadband regions will be opened up to ±15°. This scheme will allow the spacecraft's attitude to oscillate within 15° of the aerodynamic null without consuming any propellant. At the predicted time of periapsis, the spacecraft will perform a controlled momentum desaturation by setting the TACH-HOLD wheel speeds to zero.

During the drag pass, the spacecraft will attempt to keep the -Z axis along the velocity vector and the +X axis nadir pointed by using the CSA in an aerobraking attitude-control mode. This mode will utilize the onboard ephemeris to provide position knowledge about Mars. Unfortunately, buffeting of the spacecraft by atmospheric turbulence will dramatically increase the chances of star misidentification by the CSA. In order to prevent this potential misidentification, star processing by the CSA will be disabled during the drag pass, and the gyros in the IMU will propagate the spacecraft's inertial-attitude reference.

At the predicted end of the drag pass, the attitude-control deadbands will be tightened back to normal, and the reaction wheels will assume attitude-control authority from the AACS thrusters. Then, the spacecraft will slew back to an ANS orientation and point the +X axis at the Earth. For very small orbit periods of less than three hours in duration, there exists a concern that Mars may block part of the CSA's field of view upon return to ANS. This blockage may further extend the amount of time that the spacecraft's inertial reference remains in a state not updated by star scans from the CSA.

One option to mitigate the Mars-blockage situation involves using ISH mode to rotate the spacecraft by 180° around its Y axis before returning to ANS. The net effect of the rotation will cause the +Z axis of the spacecraft to flip from pointing backward along the velocity vector to pointing forward. Because the CSA sits attached to one corner of the nadir panel on the +Z end, the "flip" will cause the CSA to point in an opposite direction. When the spacecraft returns to ANS from this flipped orientation, Mars will not block the CSA field of view.

The 180° Y-axis slew to flip the +Z side of the spacecraft and reposition the CSA will result in an added bonus. As the +Z panel moves from pointing backward along the velocity to pointing forward, the panel will slew across the Martian surface because the velocity vector will lie approximately tangent to the surface. Consequently, the science instruments will be able to take data during this 10-minute slew because they sit on the +Z panel of the spacecraft. During the slew, the view geometry will vary from a horizon view at the beginning, to a direct nadir view in the middle, and to the opposite horizon view at the end.

After returning to an Earth-pointed ANS orientation following the completion of the post-drag-pass slew, the onboard sequence will activate the transmitter and the spacecraft will begin a period of realtime broadcast of engineering telemetry at the 2-kbps rate. This broadcast will allow the flight team to assess the current status of the spacecraft and its systems. A playback of the engineering telemetry recorded on the SSRs during the drag pass will follow the realtime transmission period.

5.3.7 Corridor Control Maneuvers

As described earlier, the periapsis altitude during aerobraking must lie within a well-defined corridor low enough to produce enough drag to reduce the orbit within the time constraints to reach the 2:00 p.m. node, yet high enough to avoid spacecraft heating and maximum dynamic pressure limits. As often as once per day, the spacecraft will execute maneuvers at apoapsis to maintain the periapsis altitude within this corridor. These maneuvers are called ABMs.

The ABM command sequence will be pre-loaded in the spacecraft's memory and initiated by realtime command as dictated by the flight team. As part of the ABM command process, the flight team will upload key parameters to the onboard command script. Some of these parameters include the burn time (almost always at apoapsis), burn direction, and burn magnitude.

In order to expedite the ABM command process, the flight team will select the maneuver parameters from a predefined menu of choices. This menu will contain only two burn directions and three burn magnitudes. The burn direction will be either UP for a periapsis raise maneuver, or DOWN for a periapsis lowering maneuver. The list of burn magnitudes consists of HALF-SIZE, FULL-SIZE, and DOUBLE-SIZE.

Typically, the quantitative values that correspond to the parameter choices in the ABM maneuver menu will be valid for ABMs on several consecutive orbits. The flight team will update these values periodically throughout aerobraking. Updates will occur with increasing frequency toward the end of aerobraking due to the increasing pace of variation in the orbit geometry as main phase ends and walk-out begins.

5.3.8 Science Activities

The science instruments will collect a limited amount of data during the early and middle portions of aerobraking. Data collection will occur during the 10-minute time period after the aerobrake drag pass as the spacecraft slews its +Z panel across Mars from one horizon to the opposite horizon. This post-periapsis slew will represent both a unique opportunity for the scientists to gather high-resolution data due to the low drag-pass altitude (about 110 km as opposed to 378 km during mapping), and for the flight team to receive key information to support aerobraking operations. Due to spacecraft power constraints, the instruments will be powered off when the orbit period has shrunk to less than six hours.

Specifically, the flight team will use the science data generated during aerobraking to help in the understanding of the density and the orbit-to-orbit variability of the Martian atmosphere. For example, images generated by the MOC may provide insight into the detection of local and global dust storms. An early warning into the formation of these storms will be crucial because the storms tend to heat the atmosphere and increase its density. If the density of the atmosphere suddenly increases, then the flight team may need to immediately raise the altitude of periapsis to compensate.

The TES will also perform atmospheric scans during the post-drag slew. Additionally, because TES will sit on the spacecraft in a position pointing backward along the velocity vector during the drag-pass, the instrument will be able to directly scan the regions of the atmosphere that the spacecraft has just recently flown through. These scans will allow the TES team to supply quantitative information to the navigation team in the form of temperature profiles of the lower atmosphere with a one kilometer resolution. An understanding of the temperature profile of the lower atmosphere will yield clues into the density and variability of the upper atmosphere.

In addition to the MOC and TES, the MOLA and MAG will also take measurements during the post-drag-pass slew. Although this data will be useful from a scientific perspective, it will not support aerobraking operations because no insight can be gained into the density of the Martian atmosphere from these two instruments.

5.4 Transfer to Mapping

The transfer-to-mapping phase will be the final period of the orbit insertion phase. Three critical events will occur in this time period. The first is the acquisition of the mapping orbit. Second, a gravity calibration will be performed to update the gravity field model. Finally, the spacecraft will be deployed into its mapping configuration and the instruments will activate in preparation for the start of mapping.

5.4.1 Mapping Orbit Acquisition

The mapping orbit for MGS is a low-altitude, near-circular, near-polar orbit which is Sun-synchronous with the dayside equatorial crossing at 2:00 p.m. mean solar time, all of which have nearly been achieved upon completion of the aerobraking phase. In order to meet altitude variation requirements in mapping for the instruments and the horizon sensor (MHSA), the mapping orbit must also be a frozen orbit with a stationary periapsis location around the South Pole. Upon completion of ABX, the periapsis latitude is 6° S for the open of the launch period and is 42° N for the close of the launch period. Propulsively moving periapsis down to the South Pole is beyond the capability of the propulsion system. The effect of the gravity field, however, is to move periapsis towards the South Pole at approximately 3.25 °/ day. Thus periapsis reaches the South Pole in 26 days for the open of launch period case and 37 days for the close of launch period case. Once periapsis is in position, the Transfer to Mapping Orbit (TMO) maneuver is executed with the main engine to acquire the mapping orbit.

5.4.2 Gravity Calibration

In order to improve prediction and reconstruction of the spacecraft ephemeris early in the mapping phase, Navigation desires a gravity calibration period in order to update the gravity field model. Ideally the gravity calibration would be performed over a 7 sol or 7.2 day (1 sol equals a Martian day) period after acquisition of the final mapping orbit. The mapping orbit provides a ground track pattern that nearly repeats in 88 orbits or seven Martian days, providing uniform coverage of the planet with a maximum ground track spacing at the equator of about 240 km. This provides good resolution for the gravity field solutions. However, because of the extended orbit insertion timeline up to this point, in the interest of expediting the start of mapping, the gravity calibration and subsequent model update is instead performed during the extended orbit drift period between ABX and TMO.

During the gravity calibration period, the DSN will collect a continuous Doppler data set, except during occultations. The spacecraft will remain in the cruise configuration during the gravity calibration period, rotating slowly about the Earth-pointed +X axis. This configuration will minimize HGA disturbances that could degrade the Doppler measurements.

Twelve days after completion of TMO, the first Orbit Trim Maneuver (OTM1) is performed to freeze the mapping orbit based on the updated gravity model. As a result of the expected 3σ Delta-V magnitude for this maneuver, OTM1 is executed with the main engine. Due to contamination of the deployed HGA during a main engine burn, the HGA cannot be deployed to its mapping position until completion of OTM1. In other words, once the HGA has been deployed the main engine is no longer utilized for maneuvers.

5.4.3 Mapping Configuration

Upon completion of OTM1, a period of ten days is reserved for deployment of the spacecraft into the mapping configuration, powering on the instruments and to perform a mapping rehearsal. These events must be completed before declaring the start of the mapping phase.

On the first day of the mapping configuration period, the oxidizer side of the propulsion system is isolated by closing the normally open pyro valve. The next day the HGA will be deployed to its final mapping position. The following day, an HGA calibration is performed to determine the exact position of the HGA boom and to update, if necessary, the HGA gimbal zero-reference point offset parameters in the flight software.

On the fourth day of the mapping configuration period, the Mars Horizon Sensor Assembly (MHSA) is powered on for 48 hours prior to use in order to characterize its performance before initiating mapping control. Flight software parameter uploads required to configure the spacecraft for mapping operations are also uplinked to the spacecraft on this day. Finally on day five, the spacecraft is commanded to begin mapping nadir pointing. The spacecraft is monitored throughout the next day to characterize its operations in the mapping configuration.

On day seven, the instruments are powered on and their memories uploaded as required. The PDS is then commanded to the LRC mode (4 ksps data rate for realtime and/or recording) and continual recording initiated on the recorders. Over the next two days a mapping rehearsal is performed to verify the mapping data return strategy. Specifically the use of autonomous eclipse detection algorithm (AEM) to sequence the data return is verified. A brief description of AEM follows in Section 5.3.4 with regards to powering off the transmitter during eclipse periods during the post aerobraking period before the start of mapping. Detailed description of AEM and its use for sequencing the mapping data return is provided in Section 6. Upon completion of the mapping rehearsal, the spacecraft will be declared ready for mapping and the mapping phase initiated.

5.4.4 Spacecraft Activities

In the mapping orbit, the power subsystem cannot support the transmitter high power beam on during the approximately 40 minute solar eclipse period each orbit. During aerobraking, the transmitter was cycled on and off as part of the drag pass sequence of events. Upon completion of aerobraking, however, Autonomous Eclipse Management (AEM) is enabled to manage the cycling of the TWTA beam high power during the solar eclipse period of each orbit for the remainder of the mission. AEM is an autonomous eclipse detection algorithm which detects when the spacecraft enters and exits eclipse and initiates ground specified command scripts, in this case to power the transmitter off and on, respectively. Use of AEM during the transfer to mapping sub-phase allows characterization of this capability for use during mapping operations when AEM is additionally utilized to sequence the science data return events.

After reaching the mapping orbit and completing the gravity calibration period, the high-gain antenna (HGA) will be fully deployed and verified. The HGA deploy sequence begins with a 20 minute warm up of the thruster catbed heaters. Next communications are switched from the HGA to the primary LGA transmit antenna mounted on the HGA, to ensure downlink throughout most of the deployment and in the event the boom does not deploy properly. There is insufficient link margin to transmit the high rate engineering telemetry over the LGA, so the EDF is commanded to emergency mode to provide 10 bps downlink. A couple of minutes prior to the deployment, the spacecraft is commanded to the "deploy/ despin" attitude control mode and actuator control is switched to the thrusters. In this mode, the four reaction wheels are held in tach hold at or above 200 rpm to protect them from possible shock damage when the retention and release devices are fired. The spacecraft controls to a desired attitude throughout the deployment. The HGA is deployed by simultaneously actuating three retention and release devices. The boom/antenna assembly rotates roughly 150° and latches into place within 10 minutes. Actuator control is switched back to reaction wheels and the normal cruise "array-normal-spin" mode is reacquired. The HGA is then rotated around to align the boresight back to the Earth. This is accomplished by rotating the outboard or azimuth gimbal roughly 180° followed by rotating the inboard or elevation gimbal about 30°. Based on the signal strength an approximate measure of any boom displacement can be made and a determination to re-enable communications over the HGA. An HGA calibration is planned in order to determine the exact position of the boom. Upon determination of the actual boom position, the gimbals' zero reference points are updated as required.

After verification of the HGA deployment, the IMU is commanded to the "low rate" mode to meet required mapping pointing accuracy. Additionally, the MHSA is turned on 48 hours prior to initiation of mapping nadir pointing control, in order to verify its health and characterize its operation. Various flight software parameters updates for mapping operations are subsequently uploaded to the spacecraft.

At this point the spacecraft is now ready to acquire the mapping nadir pointing attitude. In order to do this, the spacecraft is first commanded to the "inertial slew/hold" attitude control mode to slew the spacecraft such that the +Z axis is pointed at Mars. Using this mode provides autonomous payload Sun avoidance protection during Mars acquisition, which is a capability not available in the mapping attitude control modes. Attitude control is then switched to "CSA/Mapping" mode to point and maintain the spacecraft +Z axis along the velocity vector. Once the MHSA has acquired "Mars Lock" in which all four quadrants are viewing Mars, attitude control is autonomously switched to "Primary" mode. In primary mode, roll and pitch control are maintained using the MHSA, while yaw is controlled using the IMU as a gyrocompass. Additionally, using the planetary ephemeris, autonomous HGA Earth tracking and solar array Sun tracking are enabled. After a period of on-orbit characterization, the spacecraft is declared ready for mapping.

5.4.5 Science Activities

The only science activities planned during the transfer-to-mapping phase are MAG data collection throughout most of the phase and the final MOC focus check, scheduled for 31 January 1998. The MAG/ ER and the PDS are powered back on after ABX and left on until the start of the spacecraft mapping deployment. The MOC focus check is the last science activity and represents the low temperature regime in establishing the focus heater control authority.

6. Mapping Phase

Mapping phase represents the period of concentrated return of science data from the mapping orbit. This phase will start on 15 March 1998 and last until 31 January 2000, a time period of one Martian year (687 Earth days). These dates will remain fixed and are independent of the actual day of lift-off within the launch period. During this phase, the spacecraft will keep its science instruments (+Z panel of the spacecraft) nadir pointed to enable data recording on a continuous basis. On a daily basis, the spacecraft will transmit its recorded data back to Earth during a single 10-hour Deep Space Network (DSN) tracking pass. An articulating high-gain antenna (HGA) on the spacecraft will allow data recording to proceed while downlink to Earth is in progress. A timeline of mapping activities is shown in Figure 6-1.

6.1 Mapping Orbit Design

Designing the proper mapping orbit to fulfill the science objectives, and to comply with Mars planetary protection requirements required taking many constraints into account. Section 3 of the MGS Investigation Description and Science Requirements Document (ID-SRD, JPL D-12487 or MGS 542-300) contains detailed information regarding the trajectory related science requirements. In general, the ID-SRD requirements are satisfied by a low-altitude, near-circular, near-polar, Sun-synchronous orbit with a "short" repeat cycle.

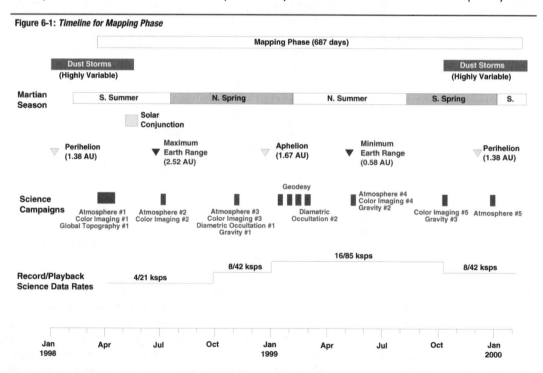

Figure 6-1: *Timeline for Mapping Phase*

The baseline mapping orbit chosen for the mission will utilize a descending node orientation of 2:00 p.m., index altitude of 378 kilometers (semi-major axis of 3774.998 km, near-circular eccentricity of 0.00953, equatorial inclination of 93.011°, and argument of periapsis of -90°. This combination will result in a "frozen orbit" with a nodal period of approximately 117.64 minutes. In other words, the frozen orbit condition will cause the argument of periapsis to always remain close to -90°, a location near the Martian South Pole. These orbital elements were derived using the Mars 50c 50 x 50 gravity field model. Prior to the start of mapping, this gravity field model will be refined using radiometric data collected during gravity calibration activities. Subsequently, the mapping orbit will be adjusted to be consistent with the new field model.

Figure 6-2: *View of the Mapping Orbit*

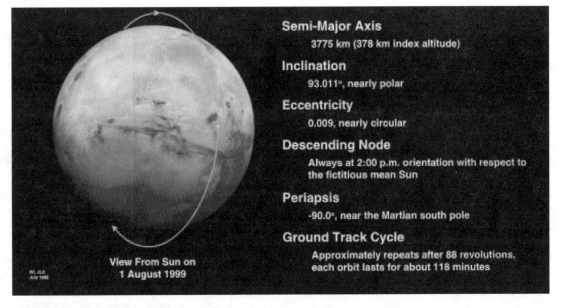

Semi-Major Axis
3775 km (378 km index altitude)

Inclination
93.011°, nearly polar

Eccentricity
0.009, nearly circular

Descending Node
Always at 2:00 p.m. orientation with respect to the fictitious mean Sun

Periapsis
-90.0°, near the Martian south pole

Ground Track Cycle
Approximately repeats after 88 revolutions, each orbit lasts for about 118 minutes

View From Sun on 1 August 1999

6.1.1 Basis for Mapping Orbit Selection

In general, the Mars Orbiter Camera (MOC) team desired an orbit with a node time late in the afternoon for longer shadows, while the Thermal Emission Spectrometer (TES) preferred an orbit closer to 1:00 p.m. to maximize the signal-to-noise ratio in their data. Also, the Magnetometer (MAG) team asked for an orbit fixed in local time to minimize the impact of large scale ionospheric currents on their observations. The 2:00 p.m. Sun synchronous, descending node orbit orientation represents a compromise between MOC and TES that was also acceptable to the MAG.

Constraints on mean altitude and inclination were primarily driven by the MOC and the Mars Orbiter Laser Altimeter (MOLA). The MOC requirements called for an average mapping orbit altitude less than 540 km to guarantee image resolutions better than two meters per pixel, while the MOLA needed an orbit inclination within 3° of polar with altitude variations less than 105 km. Use of a 378-km index altitude for mapping operations satisfies the MOC low-altitude requirement. Index altitude is a measure of the mean semi-major axis, and reflects the difference between the semi-major axis length and the equatorial radius of Mars (3397.2 km). This parameter is related to, but not the same as the average height of the spacecraft above the surface of Mars. The 378-km number was chosen in large part for its desirable properties with respect to ground track repeat cycles and spacing.

Setting the send-major axis to a specific value constrained the valid choices for eccentricity and inclination in terms of the spacecraft maintaining a Sun-synchronous orientation. Although many possible solutions exist, the spacecraft will fly in a 93.011° orbit with an eccentricity of 0.009. These parameters satisfy the MOLA constraint for a near polar, near circular orbit with minimal altitude variations.

One of the biggest advantages for the specific choice of semi-major axis, inclination, and eccentricity is that the mapping orbit can subsequently be "frozen" by setting the argument of periapsis to -90°, almost directly over the Martian south pole. This orbital configuration will keep the mean values for eccentricity and argument of periapsis fixed. Essentially, the secular motion of periapsis due to J2 is canceled by the long period variation due to the J3 component of the gravity field.

For the MGS frozen orbit, the range of areocentric altitudes over a single orbit will always remain between

345 km (south pole) to 417 km (north pole), satisfying the MOLA constraint for less than 105 km altitude variation. Without the frozen orbit, the variations in eccentricity and periapsis location will cause the altitude range to exceed the horizon sensor's (MHSA) operational limits of 335-km minimum to 455-km maximum. Such an excursion would also violate the MOLA's 105-km altitude variation constraint.

A frozen orbit with a +90° argument of periapsis (over the Martian north pole) is also mathematically possible. However, this option was not chosen because it would force the eccentricity to take on a higher value than 0.009. Consequently, the orbit would appear less circular in shape and the altitude variations would correspondingly increase.

6.1.2 Ground Track Repeat Cycle

A 378-km index mapping altitude yields an 88 revolution near-repeat cycle lasting about seven sols (Martian day). In other words, the pattern of ground tracks under the spacecraft will nearly repeat every 88 orbits. Technically, the mapping orbit was designed for a Q value of 6917/550. This parameter denotes the number of revolutions completed by the spacecraft in one Martian day. Therefore, if the mapping orbit is perfectly maintained, the ground track pattern will exactly repeat after a "super cycle" of 6917 orbits in exactly 550 sols. However, maintaining this pattern will not be possible due to navigational uncertainties in controlling and predicting the orbit. Atmospheric drag represents the most difficult orbit perturbation to measure.

In practice, the Q value of the mapping orbit will be roughly 88/7. Since 88/7 represents only an approximate fractional reduction of 6917/550 (12.57143 vs. 12.57636), the ground track pattern will not repeat exactly on every cycle. Instead, after every 88-orbit repeat cycle, the ground track pattern for that cycle will appear offset slightly to the east of the pattern laid down by the previous cycle. This offset amounts to 58.6 km when measured at the equator. For example, the equatorial location of ground track number 89 will lie 58.6 km to the east of track number 1. This 58.6-km differential is called the "orbitwalk."

Figure 6-3: *Mapping Orbit Ground Track Pattern for Seven-Sol Repeat Cycle*

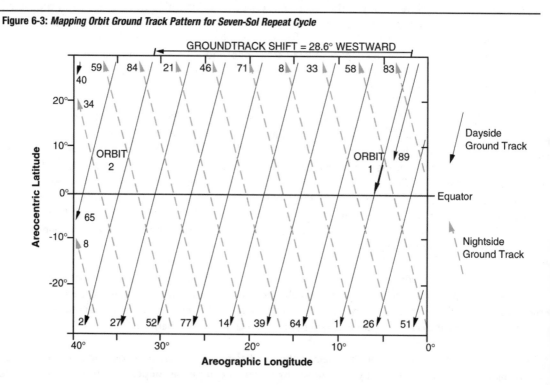

The 58.6-km orbit-walk at the equator does not represent the spacing between ground tracks on two successive orbits. During each mapping orbit lasting 117.65 minutes, Mars turns under the orbit from west to east at a rate of approximately 0.24° per minute. This rate of planetary rotation will cause the spacecraft to fly over locations about 28.62° to the west of the locations flown over on the previous orbit. At the

equator, 28.62° translates to a ground track spacing of 1,697 km between two successive orbits. Ground tracks laid down by later orbits within the 88-orbit repeat cycle will gradually fill the 1,697-km gap between two consecutive orbits. Figure 6-3 illustrates the concepts discussed in this paragraph.

6.1.3 Sun and Earth Obscuration Geometry

The percentage of time during the mapping orbit when Mars obscures the spacecraft's view of either the Earth or Sun will directly impact spacecraft sequencing and the data gathering and return strategy. Obscuration durations depend directly on the beta angle to the target in question. Target betas measure the angle between the plane of the orbit and a vector to the target, either the Sun or Earth. In other words, the beta angle provides a measure of the angular distance between the target and mapping orbit plane. Specifically, the beta angle is 90° minus the angle between the vector to the target and the angular momentum vector of the orbit.

If an orbit appears "face-on" to the target (view from the target perpendicular to the plane of the orbit), the beta angle will measure 90° and no obscuration of the target occurs as seen from the spacecraft. In other words, the spacecraft will spend no time "behind" Mars relative to the target. Beta values for no obscuration can fall lower than 90° with a high enough orbital altitude. If an orbit appears "edge-on" to the target (view from the target parallel to the plane of the orbit), the beta angle will measure 0° and the target will be obscured for a maximum amount of time. As a general rule of thumb, longer obscurations will result when the target is closer in position to the mapping orbit plane than when farther away. Figure 6-4 illustrates the range of angular separations between the Earth, Sun and the MGS mapping orbit.

Because the mapping orbit lies fixed at a 2:00 p.m. orientation relative to the fictitious mean Sun at Mars, a 90° beta angle means that the target lies at either 8:00 a.m. (90° west of the orbit) or 8:00 p.m. (90° east of the orbit). On the other hand, a 0° beta angle indicates that the target's position in the sky also lies at 2:00 p.m. relative to the mean Sun at Mars. In general, this type of analogy requires the use of caution because 15° of beta angle does not always translate into one hour of mean solar time differential. The reason is that while the fictitious mean Sun always lies in the Martian equatorial plane, the target may not lie there.

6.1.4 Earth Occultations

During the mapping phase, the position of the Earth will change drastically with respect to the mapping orbit plane (see Figure 6-4). Earth occultation durations will last for roughly 40 minutes out of every 117.64 minute orbit at the start of mapping in March 1998. At this time, the Earth's position will be at about 12:30 p.m. relative to the fictitious mean Sun at Mars, roughly 15° to the west of the mapping orbit plane.

Figure 6-5 plots the Earth occultation duration during mapping. As time progresses forward from the start of mapping, the Earth will gradually move eastward toward the mapping orbit plane, and occultation durations on every orbit will increase as the beta angle drops to 0° on 29 October 1998. On that day, the orbit plane will appear "edge-on" to the Earth, and occultation durations will reach a maximum of 41.3 minutes out of every revolution.

About three and one half months later on 19 February 1999, the Earth's position as seen from Mars will pass through the orbit plane yet another time, but this time on its way back west from the east. Again, occultation durations will reach a maximum due to the "edge-on" orbital geometry. Then, the occultations will slowly decrease in duration as the Earth continues to move westward, away from the mapping orbit plane.

Four months later on 19 June 1999, the Earth will reach a position nearly 70° to the west of the mapping orbit, roughly 9:05 a.m. relative to the fictitious mean Sun. This date is significant because despite not yet reaching a "face-on" orientation to Earth, the orbit lies at sufficiently high altitude that the spacecraft will no longer pass behind Mars relative to Earth. In other words, Earth occultations will no longer occur during any part of the mapping orbit, and a clear line of sight will always exist between the Earth and spacecraft.

The Earth's position as seen from Mars will continue to move westward until 22 September 1999 when it reaches an extreme of 8:50 a.m. relative to the fictitious mean Sun. Then, the Earth will begin to move back toward the east. Occultations will begin to occur again on 26 November 1999 and will increase in duration for the remainder of the mapping phase of the mission.

Figure 6-4: *Earth and Sun Positions Relative to Mapping Orbit*

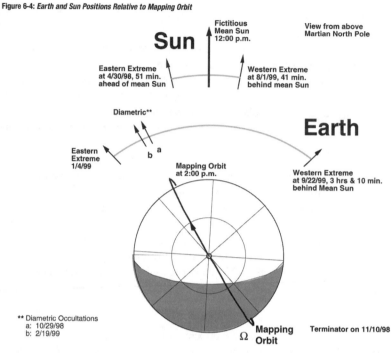

6.1.5 Solar Eclipses

Despite using a Sun-synchronous mapping orbit, the position between the spacecraft's orbit plane and the Sun will oscillate as a function of time. Consequently, both the solar beta angle and the amount of time during an orbit that Mars eclipses the Sun from the spacecraft will also vary (see Figure 6-5). The reason for this variation is that the 2:00 p.m. Sun-synchronous orientation is defined with respect to the fictitious mean Sun, and Mars' obliquity and its orbital eccentricity around the Sun cause the position of the true Sun to oscillate about the mean Sun. This position will vary depending on the Martian season and the location of Mars in its orbit around the Sun. During the mapping phase, the true Sun's position will vary from a maximum of roughly 51 minutes ahead (east of) to a minimum of about 41 minutes behind (west of) the fictitious mean Sun. Because the mapping orbit plane lies two hours ahead of the fictitious mean Sun, the true Sun will always be to the west during mapping despite the positional variations.

The longest amount of time during one orbit that Mars eclipses the Sun occurs when the true Sun reaches its most eastern position (closest to the mapping orbit) in late April, early May 1998. At that time, eclipses will last for 40.7 minutes out of every 117.64 minute revolution. On the other hand, the shortest eclipse durations occur when the true Sun arrives at its most western position (farthest away from the mapping orbit) in late July, early August 1999. Then, eclipses will only last for about 35.5 minutes every orbit.

6.1.6 Mapping Orbit Numerical Parameters

Table 6-1 lists the predicted orbital elements of the mapping orbit, both mean values and osculating at periapsis. All values in the table are referenced to the Mars centered, Mars mean equator and IAU vector of epoch coordinate system, reference epoch 15 January 1998 at 1:00 ephemeris time (ET) for both the orbit and IAU vector. These elements were derived using the Mars 50c 50 x 50 gravity field model. Although the elements listed in the table are extremely accurate with respect to the current knowledge of Mars' gravity field, the exact values are also highly dependent on the initial conditions. Consequently, the orbit will be adjusted prior to the start of mapping using radiometric navigation data collected during the post-aerobraking gravity calibration period.

Figure 6-5: *Earth Occultation and Solar Eclipse Durations During Mapping*

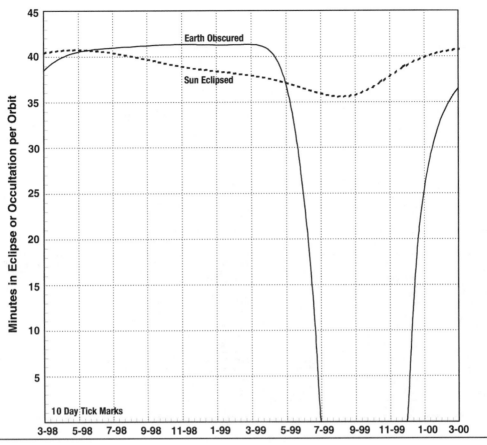

Table 6-1: *Mapping Orbital Elements (Pre-Launch Prediction)*

Orbital Element	Mean Value	Osculating at Periapsis
Semi-Major Axis	3774.998 km	3765.939690
Eccentricity	0.00953	0.006338
Inclination	93.011°	93.014163°
RA of the Ascending Node	varies (see text)	varies (see text)
Argument of Periapsis	270°	270°

During mapping, the orbital elements will be controlled to keep the spacecraft in a Sun-synchronous orientation with a descending node crossing within three minutes of 2:00 p.m. ($\pm$ 12° from the normal position). As a result, the right ascension of the ascending node (Ω) will advance with time at a rate that matches the mean motion of Mars around the Sun. The expression below provides a way to calculate Ω at any arbitrary time during the mission.

$$\Omega = -98.382323^\circ + 0.524040(t - t_o)^\circ$$

In this expression, t is number of Earth days, t_o is 10 September 1997 at 0:00 ET, and Ω is in degrees referenced to the Mars mean equator, IAU vector coordinate system. Keep in mind that the 2:00 p.m. orientation is with respect to the position of the fictitious mean Sun, not the location of the true Sun. With respect to the true Sun, the descending node time of the orbit will vary from about 1:08 p.m. to 2:41 p.m. depending on the Martian season and Mars' exact location in its orbit around the Sun.

Based on the pre-launch mapping orbital element predictions, a trajectory was integrated forward for two years to determine the important events from an Earth and Sun view geometry and Martian season perspective. Table 6-2 lists these events.

Table 6-2: *Important Geometrical Events During Mapping*

Date	Event	Comments
06-Feb-98	Winter Solstice	Start of southern summer
30-Apr-98	Solar Beta Angle Maximum	Beta angle at -15.909^0, solar eclipse maximum of 40.72 minutes
04-May-98	Begin No Commanding Period	Start solar conjunction period, Sun-Earth-Mars angle < 2^0
12-May-98	Conjunction	Sun-Earth-Mars angle reaches minimum of 0.04^0
21-May-98	End No Commanding Period	End solar conjunction period, Sun-Earth-Mars angle > 2^0
22-Jun-98	Mars to Earth Range Maximum	Distance at 2.518 AU
15-Jul-98	Vernal Equinox	Start of northern spring
08-Jul-98	Northern Declination Maximum	Mars declination at 24.04^0 as seen from Earth
29-Oct-98	Diametric Occultation (Edge-On-Orbit)	Earth beta angle at 0.0^0, Earth occultation maximum of 41.34 minutes
17-Dec-98	Mars Aphelion	Solar distance at 1.666 AU
04-Jan-99	Earth Beta Angle Maximum	Beta angle at 4.891^0
29-Jan-99	Summer Solstice	Start of northern summer
19-Feb-99	Diametric Occultation (Edge-On-Orbit)	Earth beta angle at 0.0^0, Earth occultation maximum of 41.34 minutes
24-Apr-99	Opposition	Sun-Earth-Mars angle at 178.62^0
02-May-99	Mars to Earth Range Minimum	Distance at 0.578 AU
19-Jun-99	Begin No Earth Occultations Period	Earth beta angle at -64.4^0
01-Aug-99	Autumn Equinox	Start of southern spring
22-Sep-99	Earth Beta Angle Minimum	Earth beta angle at -77.656^0
15-Oct-99	Southern Declination Maximum	Mars declination at -25.21^0 as seen from Earth
25-Nov-99	Perihelion	Solar distance at 1.382 AU
26-Nov-99	End No Earth Occultations Period	Earth beta angle at -63.2^0
25-Dec-99	Winter Solstice	Start of southern summer

6.2 Spacecraft Configuration

While in mapping, the two solar arrays and the HGA (deployed on a two meter boom) will be gimbaled in two degrees of freedom to track the Sun and Earth while the spacecraft moves around the 117.64 minute orbit. In the primary attitude control mode, data from the horizon sensors (MHSA) and the IMU will collectively be used to point the science instruments on the +Z spacecraft panel in the nadir direction, and to point the +X axis in the direction of orbital motion. A backup control mode will also be available to point the spacecraft using data from the star sensor (CSA) and an ephemeris prediction loaded from the ground. Such a scheme will probably require updating the ephemeris on a weekly basis in order to maintain the 10-mrad pointing requirement.

6.2.1 Orbit-Trim Maneuvers

A set of three reaction wheels will point the spacecraft under normal circumstances. However, due to atmospheric drag, gravity gradient, and solar pressure torques, the momentum in the wheels will build up and must be periodically unloaded by autonomous thruster firings of the attitude control jets. Most of the unloading burns will occur near periapsis (around the south pole). For typical orbit-trim maneuvers (OTM) with a velocity increment in the +X or -X direction, the spacecraft will turn under reaction wheel control to point the desired attitude control thruster in the proper burn direction. A typical OTM will last for several seconds.

6.2.2 Effect of HGA Gimbal Limit on Radio Science

A set of two gimbals, called the inner and outer, will allow the spacecraft to point its HGA directly at the Earth. These gimbals work on the theory that the vector to the Earth can be decomposed into an "out of the mapping orbit plane" component and an "in the mapping orbit plane" component. One of the gimbals, called the outer gimbal, will compensate for the out-of-plane component by rotating the antenna through an angle equal to the angular difference between the vector to the Earth and the plane of the mapping orbit. This angle is related to the Earth beta angle, changes slowly over the course of the mission, and can be considered constant for any given day during the mapping phase.

In contrast to the out-of-plane angle, the in-plane component can vary from -180° to +180° over the course of a single orbit. Due to size constraints on the inner gimbals, they will only be able to rotate through a range of -155° to +155°. This limitation does not pose a problem for the return of recorded data from the mapping orbit because the extreme range of in-plane angles will always occur as the spacecraft passes over the "far side" of Mars with respect to Earth.

However, the inner gimbal limit will affect the return of radio science occultation measurements. This experiment will gather data about the Martian atmosphere by observing the spacecraft's radio signal passing through the atmosphere as the spacecraft enters and exits the region of Earth occultation. During part of the mission's mapping phase, the in-plane angle will exceed +155° at the time of Earth occultation entry or exit.

Earth occultations take place on 528 out of the 687 days of the mapping mission with the gap occurring between 19 June 1999 and 25 November 1999, inclusive. The inner gimbal limit will reduce the total days of occultation experiments from 528 to 495, an amount equal to 6% of the total possible radio science days. All of these lost days will occur in June and November 1999 near the days when the orbit geometry causes Earth occultations to vanish and reappear, respectively. Radio science can be recovered on these days by moving the spacecraft off of its normal nadir point attitude. However, such a solution will impact the other instruments' ability to gather data for several minutes during the occultation ingress and egress time periods.

6.3 Data Collection Strategy

Two primary modes exist for collecting science data. Most of the time, data gathering will involve continuously recording the data on the solid state recorders and then playing it back to Earth during the daily Deep Space Network (DSN) tracking pass. The other mode will return realtime, high-rate telemetry at the 40 or 80 ksps rate during additional tracking passes scheduled every third day. In addition, collection of radiometric (Doppler and range) data will occur whenever the spacecraft downlinks telemetry to Earth.

As described in Section 2, the S&E-1 and S&E-2 data streams used for science data collection will be Reed-Solomon encoded, resulting in 250 encoded symbol bits for every 218 raw data bits collected by the Payload Data Subsystem (PDS) from the science instruments. Output streams will leave the PDS already Reed-Solomon encoded. Further detail regarding the bit allocations within the data stream can be found in Appendix A.

6.3.1 Date Collection Strategy for Recorded Data (S&E1)

The basic strategy for collecting science during the mapping phase will involve recording both science and engineering data on to one solid state recorder (SSR) on a continuous basis, and then playing back 24 hours worth of data during a 10-hour DSN track pass normally scheduled once per Earth day. Scientific data collection will continue during the playback period to avoid missing portions of the planet that the spacecraft travels over during the playback. This mode is called S&E1.

Playback Timing

During the standard, daily 10-hour DSN tracking pass for playback of recorded data (S&E1 mode), the spacecraft will circle Mars every 117.64 minutes and complete about five orbits. For most of the mapping phase, the spacecraft will enter zones of Earth occultation as it passes behind Mars relative to the direct line of sight to Earth (see Figure 6-6). The time that the spacecraft spends on the "back side of Mars" in the Earth occultation zone will typically eliminate 40 minutes out of every orbit from use for playback of the recorded science and engineering data.

Figure 6-6: *Typical Profile for Playback Orbit*

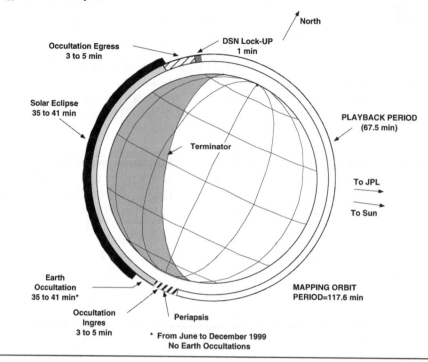

Some additional time when the spacecraft exits occultation will be used for the atmospheric occultation experiment. This process works by transmitting to the Earth through the Martian atmosphere for a range of atmospheric altitudes from zero to 200 kilometers. Because telemetry modulation will be turned off for this experiment to provide maximum signal strength, the duration of this experiment will subtract from the available playback time. Normally, the exit experiment will typically require about five minutes of transmission per orbit. Occultation exit will typically occur near the Martian north pole.

After the spacecraft exits the zone of Earth occultation and completes the radio science experiment, the DSN will require a short amount of time to gain a "signal lock" for modulated telemetry. This time will amount to a maximum of five minutes on the first observed orbit of a tracking pass, but drop to one minute on each subsequent orbit. Then, the spacecraft will be free to downlink its recorded science and engineering data to Earth as it orbits from north to south around the "front" side of Mars.

Another radio occultation experiment will occur on every playback orbit as the spacecraft approaches the Martian South Pole, but before it passes behind Mars relative to the Earth and enters the occultation zone. Like the occultation exit experiment previously described, the entry experiment will consume about five minutes of available time for transmission of recorded data.

All of this information leads to the fact that the total available playback time on each orbit of a DSN tracking pass can be determined by taking the orbit period, and subtracting the time the spacecraft spends in Earth occultation, the time allocated for the radio-science experiment, DSN lockup time, solid-state recorder (SSR) management activities, and navigational uncertainty. With these subtractions, only 67.5 minutes per playback orbit out of the total 117.64 will be available for data return on a worst-case orbit. Therefore, over a period of five playback orbits that elapse during a standard 10-hour DSN tracking pass, a theoretical total of 337.5 minutes (five 67.5-minute segments) will be available for downlink (worst case design constraint).

However, several other design constraints will reduce the true amount of available downlink time per tracking pass from the 337.5-minute theoretical total. First, for the purpose of design simplicity and operational cost savings, the playback command script will be programmed to downlink recorded data only in discrete 67.5-minute segments. Second, DSN tracking stations will require five minutes to gain a "signal lock" on modulated telemetry the first time the station attempts to listen to telemetry on any given tracking pass. Therefore, in order to avoid potential loss of recorded data due to this initial five minute lock-up period, the playback of recorded data will not begin until the second 67.5-minute segment of the DSN tracking pass. Satisfying this constraint will result in only four guaranteed, whole, discrete 67.5 minute segments (270 minutes total) available for playback per 10-hour tracking pass.

The net result is that the playback rate must exceed the data record rate by a 5.33 to 1.00 ratio (24 hours of data in 270 minutes). MGS will utilize playback rates 21.3, 42.7, and 85.3 ksps that correspond to record rates of 4.0, 8.0, and 16.0 ksps, respectively. These rate pairs were selected to match the 5.33 to 1.00 ratio, and to cover the range of expected telecommunications capability as the Earth-to-Mars distance varies over the mapping phase. At maximum Earth-to-Mars range, the link will only support the 21.3 ksps playback rate. At other times, the operations team will select the highest downlink rate supportable by the link margin.

If the DSN tracking pass begins when the spacecraft is in view of the Earth, then the pass will consist of one partial segment at the beginning, four whole segments in the middle, and one partial segment at the end. However, if the tracking pass begins during Earth occultation, then the pass will consist of five whole, unbroken segments. Either way, the beginning of the first segment (either whole or partial depending on the exact timing) will serve as the "lock-up" period, and playback will not begin until the second segment.

In order not to "waste" the first segment of a tracking pass, the spacecraft will downlink realtime S&EI data during that time. This data will be the same as that being recorded for the next day's playback. The first five minutes of the realtime data will serve as a "throw-away filler" for the initial DSN signal "lock-up" period. Any additional realtime data returned during the first downlink segment of the pass will provide engineering telemetry for the ground monitoring teams to assess the current spacecraft "health" and status. This data is important because the engineering data downlinked during the playback session will represent the spacecraft status up to one day prior.

From 19 June 1999 to 25 November 1999, the MGS mapping orbit will appear "face on" relative to Earth, and no periods of Earth occultation will occur. However, such a geometry will not substantially increase the amount of transmit time available per orbit because of solar eclipse periods. During times when the spacecraft passes through Mars' shadow, the batteries will provide power to run the electronics. Under the current design, running the transmitter throughout the solar eclipse period will drain the batteries to an unacceptably low depth of discharge level. Therefore, the data playback will still need to be broken up into the multiple 67.5-minute segment scheme previously discussed.

Autonomous Eclipse Management (AEM)

During a DSN tracking pass, the MGS spacecraft will use a capability called "autonomous eclipse management" (AEM) to trigger the execution of a command script that will perform the radio science and playback of recorded data, AEM takes the form of an algorithm that resides in the flight software power task. This algorithm will utilize power subsystem telemetry outputs (short circuit current, and solar array and battery currents) in a majority vote scheme to detect transitions into and out of eclipse to an accuracy within 12 seconds. The major advantage of using AEM to assist in autonomously returning recorded data will be manifested in significant savings in operational planning time and cost due to a reduction in the number of data playback command scripts needed over the mission's mapping phase.

One of the key aspects of using AEM to trigger a playback script involves determining when to begin the playback because downlink cannot occur during periods in the orbit when Mars occults the Earth. When the spacecraft flies through the zone of Earth occultation behind Mars, it will also fly through the zone of solar eclipse (see Figure 6-6 and Figure 6-7). The reason that these zones overlap is that the vectors from Mars to the Sun and Earth point in approximately the same direction. However, the durations of Earth occultation and solar eclipse, as well as the relative entry and exit times will vary throughout the mission's mapping phase. As shown in Figure 6-7, the occultation ingress and egress times may occur before or after the eclipse ingress and egress times, respectively. This fact is significant because choosing eclipse exit as the script trigger

Figure 6-7: *AEM Timing Relationship Between Earth Occultation and Solar Eclipse*

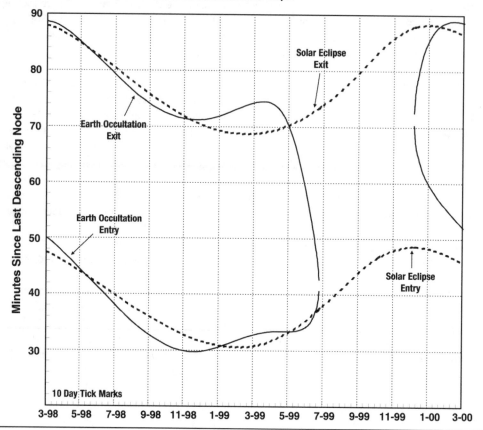

may result in a late playback start because occultation exit will have already occurred during many days of the mapping phase. Therefore, eclipse entry must serve as the triggering mechanism for the playback script.

Due to the timing variation between eclipse entry and occultation exit, the first command that the spacecraft will execute from the solar eclipse ingress triggered script will be to wait until two minutes before Earth occultation exit. This "delay" time will be implemented as a variable parameter loaded from the ground. Figure 6-8 shows the amount of time that elapses between eclipse ingress and occultation egress as a function of time during the mapping phase. Under normal circumstances, the delay parameter will be updated about once every six weeks.

When the "delay" expires, the spacecraft will turn on its transmitter to perform the radio science atmospheric occultation experiment. "Transmitter on" will occur about two minutes before the spacecraft exits Earth occultation to allow enough time for the downlink signal to stabilize prior to the start of radio science. The two minutes allocated for transmitter warm-up will be subject to change after integration and test of the telecommunications subsystem and during flight as actual performance data becomes available.

After transmitter warm-up, the command script will configure the telecommunications subsystem for radio science by turning off telemetry modulation and switching to one-way, non-coherent mode using the Ultra-Stable Oscillator (USO) as the downlink frequency source. This configuration will maximize the signal strength and stability as it passes through the Martian atmosphere on the way to Earth. The radio science experiment will last from Earth occultation exit to when the spacecraft has moved to a point in the orbit where the radio signal no longer passes through Mars' atmosphere on its way to Earth (defined at an altitude of 200 km above the Martian surface). This time will be determined from navigation predicts and will vary throughout the mapping phase.

Playback of recorded S&EI data from the appropriate SSR will begin immediately after the end of radio

Figure 6-8: *Elapsed Time Between Eclipse Ingress and Occultation Egress*

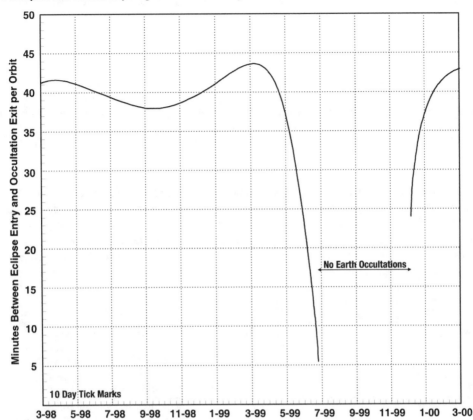

science and will last for 67.5 minutes. During this time, downlink will be configured for telemetry modulation in two-way coherent mode. After 67.5 minutes, the command script will reconfigure the downlink for another radio science occultation experiment as the spacecraft begins to pass behind Mars relative to the Earth. Simultaneously, the spacecraft will wait for the detection of solar eclipse entry to trigger the playback command script for the next orbit of the DSN tracking pass.

6.3.2 Data Collection Strategy for Realtime Data (S&E2)

The MCS mission will also utilize data rates of 40 and 80 ksps to permit the realtime return of high-bandwidth data that would otherwise be constrained to lower record rates (4, 8, or 16 ksps). Approximately every third day during the mapping phase, an additional 10-hour tracking pass (as compared to the daily tracking passes) will allow realtime data return. This mode is called S&E2, and only the Mars Orbiter Camera (MOC) and Thermal Emissions Spectrometer (TES) will utilize this realtime mode.

During periods of S&E2 operations, the solid state recorders will continue to record data in the S&E1 mode for daily playback (described in the previous section) to maintain continuous observations for Magnetometer (MAG) and Mars Orbiter Laser Altimeter (MOLA) data. Another reason for simultaneous S&E1 recording is that realtime data transmission cannot occur in the Earth occultation zone, primarily on the night side of the planet. However, the bit allocations for the different instruments within the S&E1 data stream will be slightly different when transmitting S&E2 (on the "day side" of the planet) than during normal S&E1 record operations. When the spacecraft flies over the night side of Mars, the data collection mode will temporarily switch back to normal S&E1 for the duration of flight through the Earth occultation zone. Detailed information regarding the bit allocations for MOC and MOLA within the S&E2 data stream and the modified bit allocations within the S&E1 data stream appear in Appendix A.

Although a 10-hour view period will be requested from the DSN for the every third day realtime S&E2 track, an eight-hour period will suffice when either the link margin will not support a 10-hour DSN pass, or the physical viewing geometry (horizon to horizon) will not support a 10-hour pass. The reason is that the return of realtime data will not require four discrete 67.5 minute transmit segments as in the case of S&E1 playback. Using three discrete segments, plus two partial segments, will work as long as the total realtime transmit time totals 270 minutes.

Similar to the strategy used to control the return of S&E1 data, S&E2 return will also utilize the spacecraft's autonomous eclipse management (AEM) capability. The scripts for both types of data modes will resemble each other. The main difference will be that instead of commanding playback from the SSRs after radio science at occultation egress, the spacecraft will downlink realtime telemetry.

6.3.3 Science Campaign Schedule

Science campaigns will consist of 10 individual instances of intensive periods of continuous, realtime S&E2 data return to observe Mars during periods when unique surface and atmospheric conditions occur due to seasonal changes, during periods of special orbit geometry, and for spatial or global coverage issues (see Table 6-3). Nine of the 10 campaigns will last for 88 revolutions over a period of roughly seven sols to provide global coverage for a complete orbit repeat cycle. One of the campaigns will last for four repeat cycles over a time period of 29 days.

Table 6-3: *Science Campaign Schedule*

Mission Dates	Code	L_s	Disciplines	Observation Targets and Comments
15-Mar-98 to 13-Apr-98	A	300^0	Atmosphere #1 Color Imaging #1 Global Topography #1	**Northern Winter / Southern Summer** CO_2 clouds in the northern polar regions, start of dust and winter storms, high-resolution topography map for MOLA.
29-Jun-98 to 6-Jul-98	B	354^0	Atmosphere #2 Color Imaging #2	**Northern Winter / Southern Summer** Residual southern polar cap, close of dust storm period.
26-Oct-98 to 2-Nov-98	C	47^0	Atmosphere #3 Color Imaging #3 Diametric Occultation #1 Gravity #1	**Northern Spring / Southern Fall** Mid-spring water sources and CO_2 clouds in the southern polar regions. Earth is in the plane of the mapping orbit for gravity campaign and diametric occultation.
5-Jan-99 to 12-Jan-99	D1	80^0	Geodesy Look 1, Pass 1	**Start of Southern Winter** Atmosphere in both hemispheres is colorist at this time
20-Jan-99 to 27-Jan-99	D2	87^0	Geodesy Look 2, Pass 1	for geodesy campaigns. Earth is in the plane of the mapping orbit for the diametric occultation. Each geodesy
3-Feb-99 to 10-Feb-99	D3	94^0	Geodesy Look 1, Pass 2	"week" must be separated by multiples of 88 orbits.
18-Feb-99 to 25-Feb-99	D4	101^0	Geodesy Look 2, Pass 2 Diametric Occultation #2	
3-May-99 to 10-May-99	E	137^0	Atmosphere #4 Color Imaging #4 Gravity #2	**Northern Summer / Southern Winter** Atmosphere profiles near the equator and water transport across the equatorial zone.
4-Oct-99 to 11-Oct-99	F	213^0	Color Imaging #5 Gravity #3	**Start of Northern Fall** Earth out of mapping orbit plane angle is at a maximum.
13-Dec-99 to 20-Dec-99	G	263^0	Atmosphere #5	**Start of Southern Summer** CO_2 clouds in northern polar regions and start of dust and winter storms.

As shown in Table 6-3, the 29-day campaign and each of the week-long campaigns will be dedicated to one or more disciplines in the fields of atmospheric studies, diametric occultations, global color imaging, geodesy, and Martian gravity field measurements in the form of radiometric data from radio science. In general, the science campaign schedule calls for as many of the "sub-campaigns" corresponding to the various individual disciplines to occur at the same time in order to minimize the amount of continuous tracking requested from the Deep Space Network (DSN).

The 29-day science campaign will start at the beginning of the mapping on 15 March 1998, and will primarily accommodate the production of the high-resolution MOLA topographic map of Mars. This production will provide an early, high-quality product from the MOLA investigation team and will ensure useful scientific results in case of laser degradation on the instrument. Although MOLA does not use realtime S&E2 downlink, continuous DSN tracking for this campaign is necessary in order to provide post-navigation reconstruction of the orbit at a fidelity high enough to achieve the necessary vertical accuracy for the construction of the map. Without continuous tracking (10 hours vs. 24 hours per day), the errors introduced into the modeling process due to atmospheric drag and solar radiation may be increased by a factor of five. In addition, the month-long campaign will enable the return of a high volume of data from all of the instruments at the beginning of mapping. Such a scenario is highly desirable because at the start of mapping, the spacecraft will have already spent nearly seven months at Mars without significant return of science data due to orbit insertion, aerobraking, and transition to mapping.

During each day of a campaign, the current plan involves transmitting S&E2 realtime for two-thirds of the day, and then downlinking the previous 24 hours of recorded S&E1 data in the remaining one-third of the day. In a manner similar to the strategy used for collecting realtime data during the "extra" pass every third day as described previously, the solid state recorders (SSRs) will continue to record data in the standard S&E1 mode for daily playback while the spacecraft simultaneously transmits the realtime, S&E2 data stream back to Earth. Because this daily playback will require eight hours to complete, realtime data return during science campaigns will be limited to eight out of the 12 Martian orbits that the spacecraft completes in a typical day.

6.3.4 Data Rates and Mapping Sequences

Daily playback of data recorded on the solid state recorders will utilize downlink transmission rates of 21.3, 42.7, and 85.3 ksps (corresponding to record rates of 4.0, 8.0, and 16.0 ksps, respectively). These rate pairs were selected to match the 5.333 to 1.0 playback-to-record ratio constraint, and to cover the range of expected telecommunications capability as the Earth-to-Mars distance varies over the mapping phase. Specifically, the spacecraft will always utilize the fastest playback rate supportable for an entire DSN station pass.

The choice of whether to utilize either the 40-ksps or 80-ksps transmission rate for realtime data return will also depend on choosing the fastest rate supportable for an entire DSN station pass. For example, on days when the link margin can only support the 42.7-ksps playback rate, the 40 ksps rate will be used for realtime operations. The 80-ksps rate will be reserved for realtime data return on days when the link margin and DSN view geometry of Mars can support the 85.3-ksps playback rate.

Between the dates of 15 March 1998 and 25 October 1998, the link margin will only support the 21.3-ksps data rate because Mars will be at maximum range relative to the Earth. Although this performance limitation will not impact the playback of recorded data, it will preclude using the 40-ksps rate for the return of realtime data under normal, two-way coherent communication conditions. During this time period, the DSN will drop the uplink carrier during realtime data return periods. This scheme will boost the signal gain to a level that can support the 40-ksps rate by removing the antenna transmitter noise from the link equation. Unfortunately, lack of an uplink will eliminate the possibility of spacecraft and instrument commanding during realtime data return periods. For these seven months, commanding will only be possible during the daily 10-hour DSN pass used for the playback of recorded data during normal mapping operations, or during the one-third of the day allocated to playback during science campaigns.

Table 6-4 shows the dates of usage for the different data rates over the course of the entire mapping phase. As shown in the table, the mapping phase of the mission will consist of 19 distinct time segments named M1 through M19. Each segment is called a "mapping sequence" and will last anywhere from 28 to 42 days. Prior to the start of the sequence, all of the basic spacecraft commands for that time period will be uplinked in a single batch. Consequently, the transmission rates used for playback and realtime data will not change during the course of a single sequence.

The boundary spacing of 28 to 42 days between sequences will minimize the number of command loads generated during mapping without sacrificing the prediction accuracy of the times that spacecraft events must occur relative to orbit events. However, commanding of science instruments may take place at any time during a sequence.

Table 6-4: *Data Rate Usage During Mapping*

Dates	ID Code	Range	Playback Mode (PB)	Realtime Mode (RT)	Elevation Mask
15-Mar-98 to 13-Apr-98	M1	2.43 AU	21.3 ksps	40 ksps	25^0 / 35^0 (8^0 for PB)
14-Apr-98 to 03-May-98	M2	2.47 AU	21.3 ksps	40 ksps (one way)	12^0 / 16^0
04-May-98 to 21-May-98	M3	2.50 AU	n/a (solar conjunction)	n/a (solar conjunction)	12^0 / 16^0
22-May-98 to 28-Jun-98	M4	2.52 AU	21.3 ksps	40 ksps (one way)	13^0 / 17^0
29-Jun-98 to 09-Aug-98	M5	2.52 AU	21.3 ksps	40 ksps (one way)	13^0 / 17^0
10-Aug-98 to 20-Sep-98	M6	2.46 AU	21.3 ksps	40 ksps (one way)	11^0 / 16^0
21-Sep-98 to 25-Oct-98	M7	2.30 AU	42.7 ksps	40 ksps	21^0 / 29^0
26-Oct-98 to 29-Nov-98	M8	2.08 AU	42.7 ksps	40 ksps	12^0 / 17^0
30-Nov-98 to 03-Jan-99	M9	1.80 AU	42.7 ksps	40 ksps	07^0 / 09^0
04-Jan-99 to 31-Jan-99	M10	1.46 AU	85.3 ksps	80 ksps	12^0 / 16^0
01-Feb-99 to 14-Mar-99	M11	1.18 AU	85.3 ksps	80 ksps	06^0 / 07^0
15-Mar-99 to 25-Apr-99	M12	0.79 AU	85.3 ksps	80 ksps	06^0 / 06^0
26-Apr-99 to 06-Jun-99	M13	0.68 AU	85.3 ksps	80 ksps	06^0 / 06^0
07-Jun-99 to 18-Jul-99	M14	0.92 AU	85.3 ksps	80 ksps	06^0 / 06^0
19-Jul-99 to 29-Aug-99	M15	1.17 AU	85.3 ksps	80 ksps	06^0 / 07^0
30-Aug-99 to 11-Oct-99	M16	1.41 AU	85.3 ksps	40 ksps	10^0 / 14^0 (6^0 for RT)
12-Oct-99 to 21-Nov-99	M17	1.63 AU	42.7 ksps	40 ksps	06^0 / 07^0
22-Nov-99 to 02-Jan-00	M18	1.85 AU	42.7 ksps	40 ksps	07^0 / 10^0
03-Jan-00 to 31-Jan-00	M19	2.01 AU	42.7 ksps	40 ksps	10^0 / 14^0

In Table 6-4, the column that contains elevation mask information contains two numbers that specify the minimum elevation angle of Mars, as seen from a 34m HEF antenna, that can support the data rates listed for the given sequence. The first number specifies the minimum elevation at Goldstone, while the second number specifies the same information for either Madrid or Canberra. In all cases, the minimum elevation angle required at Goldstone to support a given data rate is less than or equal to that at the other two DSN sites. The reason is that air around Goldstone contains less water vapor on average than either Madrid or Canberra. Therefore, received signal experiences less attenuation and produces a stronger link margin at Goldstone.

The elevation angle numbers presented in Table 6-4 for each mapping sequence represent values averaged across the entire sequence. Consequently, the actual values at any given instant in time will be slightly different. For the purposes of DSN scheduling and conflict resolution with other projects, the elevation masks will be computed on a week by week basis and supplied to the DSN at the appropriate times.

6.4 OTM Sequencing

Orbit trim maneuvers (OTMs) during the mapping phase will be performed for the purpose of controlling the spacing of the repeating ground tracks, and for correcting the orbit eccentricity and inclination. It is currently anticipated that OTMs will be scheduled approximately once every month. These maneuvers will normally be placed near periapsis, although they will occasionally occur near apoapsis. However, OTMs designed specifically for inclination control must be executed near the equator.

Four-hour windows will be provided in the mapping sequences for the OTMs. These windows will allow the maneuver time to be selected from one of two different orbits in the event that large gravity anomalies are present on one of the orbits. Spacecraft and payload activities which conflict with the maneuver will be kept out of the window.

6.5 Solar Conjunction

The mapping phase will be interrupted for a period centered about 12 May 1998 at 20:00 UTC. At that time,

Mars will pass through solar conjunction as seen from the Earth (minimum Sun-Earth-Mars angle of 0.04 degrees). Due to solar interference, both the uplink and downlink radio signals will be degraded for a period of several days on each side of the conjunction. Based on current analysis, it is assumed that the uplink signal will be unacceptably degraded for commanding when SEM angle is less than 2°.

This uplink degradation constraint will result in a 17-day command moratorium starting on 4 May 1998 and ending on 21 May. The spacecraft will operate autonomously during this time period and will maintaining all of its normal engineering functions without commanding over this 17-day period. Throughout most of solar conjunction, it will be highly unlikely that telemetry will be successfully received at Earth even when using the slowest S&E1 playback rate of 21.3 ksps. Consequently, solid-state recorder playbacks will not occur during this time period.

The current plan is to use a special AEM script that will provide realtime spacecraft telemetry and navigation Doppler data during a single DSN tracking pass each day. This strategy will allow for daily health assessment of the spacecraft and the payload, provide for radiometric monitoring of the orbit, and possibly allow for the return of limited science observations. Unfortunately, realtime performance at the lowest science collection rate of 4.0 ksps may not be achievable throughout all of the solar conjunction period. Consequently, a hybrid strategy may be used. For the first ten minutes of each orbit, engineering-only telemetry will be returned at 250 bps. For the rest of the orbit, 4.0 ksps S&E-1 will be returned in realtime. This strategy will permit the spacecraft to attempt to return science data and enhance the chance that some spacecraft engineering data will reach Earth.

Radio science atmospheric-occultation measurements will also be performed during solar conjunction. During the radio-science segments of each orbit, the transponder will be set in 1-way mode with telemetry and ranging modulation off. During the rest of the orbit, the transponder will be configured with telemetry and ranging on, and with 2-way coherent tracking enabled. As the conjunction geometry becomes more extreme, the uplink carrier may not be transmitted from the ground because removing the ground transmitter from the link and utilizing the "listen-only" mode will boost the effective gain of the downlink signal.

The data return profile described above will be repeated for each orbit during solar conjunction, despite the fact that tracking requirements call for only one 34m HEF pass per day. If anomalies or any other circumstances indicate that additional coverage is warranted, downlink from the spacecraft will be available without emergency commanding. When commanding becomes possible after conjunction, normal mapping data return will begin.

Appendix A - Data Rate Modes

This appendix describes the modes that the MGS spacecraft will use for the collection of science and engineering data. In the two tables in this appendix, "bps" stands for bits per second and "sps" stands for symbols per second. A symbol is essentially a Reed-Solomon encoded (250:218 ratio) data bit. Therefore, it takes approximately 1.147 bits of storage space to encode one bit of raw data.

The data rates corresponding to the science instruments represents an average rate because the Payload Data Subsystem (PDS) does not receive data continuously. Each one of the science instruments generates data in discrete packets collected by the PDS at regular intervals determined by the internal software. The packets from each instrument will include identification and timing information in addition to the raw science data. In addition to collecting instrument packets, the PDS will also gather spacecraft engineering packets and engineering packets from itself. After collecting the packets, the PDS will format them into transfer frames 10,000 symbols (8,720 bits) in length, including a header 80 bits in length.

Table A-1 lists the rates for the science instruments and engineering data streams corresponding to S&E1 operations. In this mode, the data can either be stored on the solid state recorders (SSRs) for later playback, or downlinked back to Earth in realtime. In S&E2 operations as shown in Table A-2, the PDS will simultaneously generate two data streams. The first data stream (S&E2) will be downlinked back to Earth in realtime. The second data stream (modified S&E1) will be stored onto the solid state recorders (SSRs) for later playback. In Table A-2, the numbers in parenthesis correspond to the modified S&E1 data rate. All numbers in Tables A-1 and A-2 are in bits per second (bps) unless otherwise specified.

Both tables in this appendix show data rates for GRS and PMIRR. These two scientific instruments were flown on the Mars Observer mission, but were not included as part of the MGS payload due to mass considerations. When MGS inherited its PDS from Mars Observer, the data rates for these two instruments were not reallocated.

Table A-1: *S&E1 Data Modes*

	LRC	MRC	HRC	MR	EOM
Description	Low Record	Medium Record	High Record	Mars Relay	Eng. Only
Total Symbol Rate (sps)	4000	8000	16000	4000	4000
Total Data Bit Rate	3488	6976	13952	3488	3488
MAG/ER Data Rate	324	648	1296	324	0
MOLA Data Rate	618	618	618	618	0
MOC Data Rate	700	2856	9120	2670	0
TES Data Rate	688	1664	1664	0	0
Unallocated (GRS)	665	665	665	0	0
Unallocated (PMIRR)	156	156	156	0	0
S/C Engineering	256	256	256	256	256
PDS Engineering	48	48	48	48	3072
T/F Header	32	64	128	32	32
Filler	1	1	1	2	128

Table A-2: *Combined S&E2 and S&E1 Data Modes*

	RTL		RTM	
Description	Real Med.	Low Rec.	Real Med.	Med. Rec.
Total Symbol Rate (sps)	40000	(4000)	40000	(8000)
Total Data Bit Rate	34880	(3488)	34880	(6976)
MAG/ER Data Rate	0	(324)	0	(648)
MOLA Data Rate	0	(618)	0	(618)
MOC Data Rate	29260	(1388)	29260	(4520)
TES Data Rate	4992	(0)	4992	(0)
Unallocated (GRS)	0	(665)	0	(665)
Unallocated (PMIRR)	0	(156)	0	(156)
S/C Engineering	256	(256)	256	(256)
PDS Engineering	48	(48)	48	(48)
T/F Header	320	(32)	320	(64)
Filler	4	(1)	4	(1)

Table A-2: *Combined S&E2 and S&E1 Data Modes (continued)*

	RTH1		TST	
Description	Real Med.	High Rec.	Real High	High Rec.
Total Symbol Rate (sps)	40000	(16000)	80000	(16000)
Total Data Bit Rate	34880	(13952)	69760	(13952)
MAG/ER Data Rate	0	(1296)	0	(1296)
MOLA Data Rate	0	(618)	0	(618)
MOC Data Rate	29260	(10782)	63808	(10782)
TES Data Rate	4992	(0)	4992	(0)
Unallocated (GRS)	0	(665)	0	(665)
Unallocated (PMIRR)	0	(156)	0	(156)
S/C Engineering	256	(256)	256	(256)
PDS Engineering	48	(48)	48	(48)
T/F Header	320	(128)	640	(32)
Filler	4	(3)	16	(3)

Appendix B – Payload Data Sheet

Table B-1: *Payload Data Sheet (1 of 2)*

	MAG/ER	MOC	MOLA
Instrument Name	Magnetometer and Electron Reflectometer	Mars Orbiter Camera	Mars Orbiter Laser Altimeter
Principal Investigator	M. H. Acuna	M. C. Malin	D. E. Smith
Instrument Manager	J. Scheifele	G. E. Danielson	B. L. Johnson
Build Source	Mars Observer Spare	Mars Observer Spare	Assemblies
Mass	5.3 kg	21 kg	25.9 kg
Power (Peak / Average)	4.63 / 4.63 W	29.75 / 10.50 W	34.94 / 30.94 W
Operating Voltage	28 ± 0.56 V	28 ± 6 V	28 ± 2 V
Operating Temp. Limit	-10° C to 50° C	-28° C to -2° C (NA) -62° C to -24° C (WA)	-20° C to 30° C
Non-Operating Temp. Limit	-40° C to 75° C	-30° C to 40° C	-30° C to 40° C
Field of View	4π steradian (MAG) $360^\circ \times 14^\circ$ (ER)	$140.2^\circ \times 3^\circ$ (WA) 0.44° (NA)	0.85 mrad
Resolution	200 km	250 meters/pixel (WA) 1.4 meters/pixel (NA)	160 meters horizontal 2 meters vertical
Approximate Size	20 x 17.6 x 15 cm	86 x 45 x 40 cm	71 x 58 x 62 cm

Table B-2: *Payload Data Sheet (2 of 2)*

	MR	TES	USO
Instrument Name	Mars Relay	Thermal Emission Spectrometer	Ultra-Stable Oscillator
Principal Investigator	J. Blamont	P.R. Christensen	G.L. Tyler (team leader)
Instrument Manager	A. Ribes	G. Mehall	C.L. Hamilton
Build Source	Print	Assemblies	Mars Observer Spare
Mass	8.0 kg	14.1 kg	1.3 kg
Power (Peak / Average)	13 / 13 W (includes MOC)	18.20 / 15.55 W	4.5 (warm-up) / 3.0 W
Operating Voltage	10 ± 0.50 V (needs MOC)	26 to 32 V	24 to 30 V
Operating Temp. Limit	-10° C to 40° C	-20° C to 30° C	-20° C to 30° C
Non-Operating Temp. Limit	-30° C to 40° C	-20° C to 50° C	-30° C to 40° C
Field of View	130°	$0.95^\circ \times 1.43^\circ$	n/a
Resolution	Limb to Limb	3 km per detector	n/a
Approximate Size	24 x 25 x 19 cm	36 x 24 x 40 cm	16 x 9 x 10 cm

Appendix C - Compliance With Project Requirements

This appendix details the compliance of the Mars Global Surveyor mission design with the Mission Requirements Document (MRD) and the Instrument Description and Science Requirements Document (IDSRD).

C.1 Compliance With the Mission Requirements Document

Table C-1 gives an assessment of compliance with the MRD, based on the September 1995 version of that document. As shown in the table, all requirements are met with the current mission plan.

C.2 Compliance With the Science Requirements Document

Table C-2 gives an assessment of compliance with the IDSRD, based on the February 1995 version of that document and shows that all requirements are met with the current mission plan. Many of the science desires, such as special observations during the cruise and orbit insertion phases, are being evaluated by the project. References are given in the table to specific sections in the Mission Plan that discuss each requirement.

Table C-1: *MRD Requirements Compliance*

MRD #	MRD Requirement	Achieved?	Reference and Comments
3.1.1	Launch Opportunity	Yes	Section 3.3 - Launch Strategy
3.1.2	Launch Vehicle	Yes	Section 3.1 - Launch Vehicle Description
3.1.3	Launch Period	Yes	Section 3.3 - Launch Strategy
3.1.4	NASA Planetary Protection	Pending	Planetary Protection Plan (542-402)
3.1.5	Cruise and Orbit Insertion Science	Yes	Section 4.3 - Inner Cruise Activities Section 4.4 - Outer Cruise Activities Section 5.3 - Transfer to Mapping
3.1.5.1	Payload Engineering Checkout	Yes	Section 4.3 - Inner Cruise Activities
3.1.5.2	MOC Post Bakeout Focus Check	Yes	Section 4.4 - Outer Cruise Activities
3.1.5.3	MAG Calibration	Yes	Section 4.4 - Outer Cruise Activities
3.1.5.4	MOC Pre-MOI Mars Approach Observations	Yes	Section 4.4 - Outer Cruise Activities
3.1.5.5	Science Observations During the Orbit Insertion Phase	Yes	Section 5.1 - Mars Capture Subphase Section 5.2 - Aerobrake Subphase MAG data may not be achievable during walkout
3.1.5.6	MOC Star Images Before Mapping Deployment	Yes	Section 5.3 - Transfer to Mapping Subphase
3.1.6	Mapping Orbit Design	Yes	Section 6.1 - Mapping Orbit Design Orbit has 7 sol repeat, index alt. of 378 km
3.1.7	Gravity Calibration Period	Yes	Section 5.3 - Transfer to Mapping Subphase
3.1.8	S/C and Instrument Checkout Period	Yes	Section 5.3 - Transfer to Mapping Subphase Checkout period of 10 days
3.1.9	Mapping Phase Commencement	Yes	Section 5.3 - Transfer to Mapping Subphase Section 6.0 - Mapping Phase
3.1.10	Mapping Phase	Yes	Section 6.0 - Mapping Phase
3.1.11	Relay Phase	Pending	Information to be updated as Mars-98 mission design reaches maturity
3.1.12	Science Data Return	Yes	Section 6.3 - Data Collection Strategy
3.1.13	Radio Science Data	Yes	Section 6.3 - Data Collection Strategy
3.1.14	Solar Conjunction Command Moratorium	Yes	Section 6.5 - Solar Conjunction 17-day command moratorium at conjunction
3.1.15	DSN Usage	Yes	Section 2.6 - DSN Utilization Consistent with DMR and MRR documents
3.1.16 - 1	First Maneuver Constraint	Yes	Section 4.2 - Trajectory Correction Maneuvers
3.1.16 - 2	Maneuver Interval Constraint	Yes	Section 4.0 - Cruise Timeline Section 4.2 - Trajectory Correction Maneuvers Section 5.0 - Orbit Insertion Timeline Section 5.1 - MOI Maneuver Section 5.2 - Aerobrake Subphase
3.1.16 -3	Sun-safe Maneuver Direction Constraint	Yes	Section 4.2 - TCM Schedule/Implementation Section 5.2 - Aerobrake Subphase
3.1.17	Aerobraking	Yes	Section 5.2 - Aerobrake Subphase
3.1.18	S/C Overheating During Capture	Yes	Section 5.2 - Aerobrake Subphase

Appendix D - Mass and Delta-V Budget Details

The tables and figures in this appendix show the technical details of the mission Delta-V budget corresponding to the open and the close of the launch period (6 November 1996 and 25 November 1996, respectively). In addition, this appendix lists the spacecraft mass and propellant breakdowns corresponding to the Delta-V budget.

The spacecraft's propellant load was developed from a single Delta-V budget that incorporated elements both independent and dependent of the actual launch date. Some of the independent elements include mapping orbit maintenance, relay orbit maintenance, and aerobraking walk-in maneuvers. Examples of launch date dependent elements include the MOI burn, TCMs 1 and 2, and aerobraking mainphase and walk-out maneuvers.

Because the same propellant load will be used for all launch dates, the pre-launch Delta-V reserves line item

Table C-2: *IDSRD Requirements Compliance*

IDSRD #	IDSRD Requirement	Achieved?	Reference and Comments
3.2.2.1 (MAG1&2))	1-Day drift orbit with calibration data, 28-day minimum	Yes	Section 5.0 - Orbit Insertion Phase MGS does not have a 1-day drift orbit due to utilization of aerobraking to achieve mapping orbit. MAG data will be returned throughout most of the orbit insertion phase.
3.2.2.2 (MAG3)	Mapping orbit 350 ± 50 km	Yes	Section 6.1 - Mapping Orbit Design Orbit index altitude is 378 km
3.2.2.3 (MAG4)	Successive ground track separation 1/3 to 1/20 of orbit altitude	Yes	Section 6.1 - Mapping Orbit Design Orbit walk is about 1/6 (58.6 km)
3.2.2.3 (MAG5)	Repeat period not commensurable with solar rotation period of 26.354 Earth days	No	Requirement cannot be met with current project cost and staffing assumptions
3.2.2.4 (MAG6)	Minimum 10 mapping cycles	No	Requirement cannot be met with current project cost and staffing assumptions
3.2.2.5 (MAG7)	Mapping orbit fixed in local time	Yes	Section 6.1 - Mapping Orbit Design Orbit is Sun-synchronous at 2:00 p.m. mean
3.2.3.1 (MOC1)	Altitude above true surface must be less than 540 km	Yes	Section 6.1 - Mapping Orbit Design Apoapsis areographic altitude is 426.6 km
3.2.5.1 (MOLA1&2)	Mapping orbit inclination within 3° of 90°, difference between orbit periapsis and apoapsis less than 105 km, mean height less than 380 km	Yes	Section 6.1 - Mapping Orbit Design See description of orbital elements
3.2.6.1 (RS1)	Mapping orbit periapsis between 250 and 400 km	Yes	Section 6.1 - Mapping Orbit Design Periapsis altitude bounds are 360 to 389 km
3.2.6.2 (RS2&3)	Gravity calibration orbit with 7-10 day repeat cycle and continuous DSN tracking over one full cycle	Pending	Section 5.3 - Transfer to Mapping Subphase Gravity calibration scheduled between ABX and TMO, but with not with 7-10 day repeat cycle
3.2.7.1 (TES1)	Maintain altitude between 250 km and 2000 km	Yes	Section 6.1 - Mapping Orbit Design Orbit e = 0.007, index altitude = 378 km
3.2.7.2 (TES2)	Mapping cycles offset more than 9 km in consistent direction	No	Requirement cannot be met with current project cost and staffing assumptions
3.2.8.1 (IDS1&2)	Near-polar, Sun-synchronous, low-altitude orbit with short mapping cycle	Yes	Section 6.1 - Mapping Orbit Design Repeat cycle of 7 sols
3.2.8.4 (IDS3)	Near-circular, 2:00 p.m. local solar time equator crossing Sun-synchronous orbit, 350 km < h < 450 km	Yes	Section 6.1 - Mapping Orbit Design Mapping orbit is Sun-synchronous with respect to the fictitious-mean Sun
3.4.2 (MAG9)	Continuous data acquisition for 4 weeks in drift orbit prior to mapping orbit insertion	Yes	Section 5.0 - Orbit Insertion Phase MGS does not have a 1-day drift orbit due to utilization of aerobraking to achieve mapping orbit. MAG data will be returned throughout most of the orbit insertion phase.
3.4.2 (MAG10)	ER data collection during cruise and drift orbit	No for Cruise Yes for Orb. Ins.	Section 5.0 - Orbit Insertion Phase MGS does not have a 1-day drift orbit due to utilization of aerobraking to achieve mapping orbit. ER data will be returned throughout most of the orbit insertion phase.
5.1.2 (MAG24)	ER on if MAG calibration rolls are performed	Yes	Section 4.4 - Outer Cruise Activities MAG rolls will be performed if needed
5.1.2.2 (MAG25)	Continuous operation of MAG during mapping phase	Yes	Section 6.3 - Data Return Strategy Continuous data collection except solar conjunction
5.1.6.1 (RS12)	RS data required during cruise phase to test equipment, procedures and software	Yes	Section 4.3 - Inner Cruise Activities
5.1.6.3 (RS13)	S/C will transmit X-band signals whenever it is visible from Earth using project allocated DSN passes subject to power constraints	Yes	Section 6.3 - Data Collection Strategy Radio Science planned during all mapping phase DSN passes

Table C-2: *IDSRD Requirements Compliance*

IDSRD #	IDSRD Requirement	Achieved?	Reference and Comments
5.1.6.4 (RS18)	In-flight RS tests prior to data acquisition to determine warm-up time required	Yes	Section 4.3 - Inner Cruise Activities Section 4.4 - Outer Cruise Activities RS ORTs will be scheduled by the project and DSN to meet this requirement
5.1.8.4.2 (IDS4)	Acquire atmospheric and polar data in a nearly continuous manner at all times during a full Martian year	Yes	Section 6.1 - Mapping Orbit Design Section 6.3 - Data Collection Strategy Continuous data collection from all science instruments except during solar conjunction
5.3.6 (RS20)	RS observations include 20 seconds of data while spacecraft is geometrically behind Mars, data from surface to 200-km altitude, 100 seconds of data when transmission path is above 200 km from the surface, and appropriate timing pads for insurance.	Yes	Section 6.3 - Data Collection Strategy Occultation ingress and egress experiments planned for all DSN passes during mapping except periods of no Earth occultations
5.6.1.1 (RS24)	Regular, continuous, and complete tracking of the mapping orbit, using at least one 34m HEF DSN pass per day	Yes	Section 2.6 - DSN Utilization Minimum requirement satisfied, but coverage may not be regular and continuous

Table D-1: *Spacecraft Mass Breakdown*

Spacecraft Component	Mass (kg)
Total Spacecraft Injected Mass	1062.1
Spacecraft Dry Mass (includes pressurant)	674.7
Science Payload	75.90
Spacecraft	597.40
Pressurant (helium)	1.40
Spacecraft Propellant Load	388.4
Usable N_2O_4	148.21 (86.03% tank fill level)
Usable N_2H_4	230.59 (95.50% tank fill level)
Unusable	7.58 (2% of total propellant fill)
Oxidizer Ballast	1.00
Load Error	0.99

Figure D-1: *ΔV Budget Used for Propellant Loading*

MARS GLOBAL SURVEYOR SPACECRAFT ΔV/PROPULSIVE PERFORMANCE
Mission ΔV Budget - Propellant Loading

MISSION ΔV BUDGET - ORR BASELINE (8/96)
SPACECRAFT INJECTED MASS (kg) 1062.10

MISSION PHASE	MANEUVER	MONOPROP TRANSLATIONAL Isp (s)	MONOPROP TRANSLATIONAL ΔV Reqd (m/s)	BIPROP TRANSLATIONAL Isp (s)	BIPROP TRANSLATIONAL ΔV Reqd (m/s)	MONOPROP ROTATIONAL Isp (s)	MONOPROP ROTATIONAL ΔV Reqd (m/s)	PROPELLANT BREAKDOWN M-HYDRZ (kg)	PROPELLANT BREAKDOWN B-HYDRZ (kg)	PROPELLANT BREAKDOWN NTO (kg)	POSTBURN S/C MASS (kg)	EST. BURN DURATION MONOPROP #T (s)	EST. BURN DURATION MONOPROP (s)	EST. BURN DURATION BIPROP (s)
CRUISE ΔV95	TCM-1/2	0.0	0.0	317.9	36.0	220.0	0.3	0.15	6.61	5.62	1049.72	-	-	57.7
	TCM-3/4	220.0	2.0	0.0	0.0	0.0	0.0	0.97	0.00	0.00	1048.75	4	117.9	-
	pre-Lnch ΔV Reserves	220.0	0.0	315.0	13.2	0.0	0.0	0.00	2.42	2.05	1044.28	-	-	21.0
CAPTURE	MOI	0.0	0.0	317.9	975.1	220.0	5.9	2.85	151.98	129.18	760.27	-	-	1339.0
	(Rp = 3700 km, Per = 48 Hr)													
Aerobraking Walk-in	AB-1	0.0	0.0	315.0	7.5	220.0	0.1	0.04	1.00	0.85	758.39	-	-	8.6
	AB-i (i=2,4)	220.0	2.5	0.0	0.0	0.0	0.0	0.88	0.00	0.00	757.51	4	106.5	-
Aerobraking Main Phase	ABM Translation	220.0	5.0	0.0	0.0	0.0	0.0	1.75	0.00	0.00	755.76	-	-	-
	ACS Rotation	0.0	0.0	0.0	0.0	190.0	5.0	2.03	0.00	0.00	753.73	-	-	-
Aerobraking Walk-out	ABM Translation	220.0	20.0	0.0	0.0	0.0	0.0	6.95	0.00	0.00	746.78	-	-	-
	ACS Rotation	0.0	0.0	0.0	0.0	190.0	30.0	11.93	0.00	0.00	734.85	-	-	-
Aerobraking Contingency	Aerobraking Pop-Up	220.0	2.5	315.0	14.5	0.0	0.0	0.85	1.86	1.58	730.56	-	-	15.9
	post-Lnch ΔV Reserves	220.0	0.0	315.0	0.0	0.0	0.0	0.00	0.00	0.00	730.56	-	-	-
Transition To Mapping	ABX	0.0	0.0	317.9	60.2	220.0	0.4	0.14	7.57	6.44	716.41	-	-	66.3
	TMO	0.0	0.0	317.9	16.1	220.0	0.1	0.03	2.00	1.70	712.68	-	-	17.5
	OTM-1 (Frzn)	0.0	0.0	315.0	7.5	0.0	0.0	0.00	0.93	0.79	710.95	-	-	8.2
MAPPING Drag Dens 95%	OTM (Drag/GTE)	220.0	3.9	0.0	0.0	0.0	0.0	1.28	0.00	0.00	709.66	-	-	0.0
	ACS Rotation	0.0	0.0	0.0	0.0	190.0	40.0	15.07	0.00	0.00	694.59	-	-	-
ADDITIONAL ΔV REQ'D	QUARANTINE ORBIT (PQ) - 2-BURN	220.0	22.8	0.0	0.0	0.0	0.0	7.30	0.00	0.00	687.29	4	885.0	-
RELAY Drag Dens 95%	OTM (Drag)	220.0	1.0	0.0	0.0	0.0	0.0	0.32	0.00	0.00	686.97	-	-	-
	ACS Rotation	0.0	0.0	0.0	0.0	190.0	10.0	3.68	0.00	0.00	683.29	-	-	-
SUB-TOTALS			59.7		1130.1		91.8	56.22	174.37	148.21				
TOTAL MISSION ΔV		1281.6		S/C Dry Mass - Actual (kg)	673.28			S/C Dry Mass Capability (kg)	674.73	ΔDry Mass (kg)	1.45		

Spacecraft Injected Mass Breakdown (kg)		Mission ΔV Breakdown		Prop Tanks	Load Error	Max Usable	Desired Usable	Total Load Plus Error [3]	Full Tanks (100% Fill)	Fill Level (Percent)	PROPELLANT (kg)	
S/C Dry Mass	673.28	Translational ΔV	1189.80								Usable	378.81
Dry Mass Margin	0.45	Rotational ΔV	91.80	N2O4	0.37	164.72	148.21	152.55	177.32	86.03%	Unusable 2%	7.58
Total Propellant	388.37	Bipropellant ΔV	1130.10	(2x) N2H4	0.62	229.29	230.59	235.82	246.94	95.50%	Load Error	0.99
Total Inj Mass	1062.10	Monopropellant ΔV	151.50			Ge (m/s²)	9.80665	Engine	BIPROP	558.7	Oxidizer Ballast	1.00
Uncertainties	+0.86 / -1.21	Desired Dry Mass Margin (kg)	0.45	BIPROP MIX RATIO	0.85	Thrust (N)	MPROP	4.45	Total Prop	388.37			

[1]5% Ullage [3]Usable+Unusable+Ld_Err+Excess [5]Per Dominick Oxidizer Ballast (1=Y/2=N) 1

MSN DV - FMP BaseLoad

will be different for each launch date. The reason for this difference is that the propellant load will remain constant, but the Delta-V requirements for the launch date dependent elements will change. Specifically, the bi-propellant Delta-V requirements become more demanding toward the end of the launch period.

Appendix E - Aerobraking Design Data

This appendix provides detailed data regarding the reference trajectory design for aerobraking. The design presented in the mission plan assumes a launch on 6 November 1996, critical scale height approach to determine the size of the walk-in maneuvers (see Section 5), and a two-day orbit lifetime during aerobraking walk-out.

Figure D-2: ΔV *Budget, 6 November 1996 Launch*

MARS GLOBAL SURVEYOR SPACECRAFT MANEUVER EXECUTION
Mission ΔV Budget - ORR Baseline (8/96)
Launch Date - 11/6/96

SPACECRAFT INJECTED MASS (kg) 1062.10
Spacecraft Dry Mass (kg) 673.28
Dry Mass Margin (kg) 0.45
Total Propellant (kg) 388.37

Spacecraft Propellant Load

Engine	Thrust (N)	Propellant Tanks	Load Error	Max Usable*	Desired Usable	Usable Plus Error	Unusable 2%	Oxidizer Ballast	Tank Totals	Tank Redlines
BIPROP	658.7	N2O4 (kg)	0.37	164.72	148.21	148.58	2.96	1.00	152.54	4.33
MPROP	4.45	(2x) N2H4 (kg)	0.62	229.29	230.59	231.21	4.61	-	235.82	5.23

Ge (m/s^2): 9.80665
BIPROP MIX RATIO: 0.85

*5% Ullage

MISSION PHASE	MANEUVER	MONOPROP TRANSLATIONAL Isp (s)	MONOPROP TRANSLATIONAL ΔV Reqd (m/s)	BIPROP TRANSLATIONAL Isp (s)	BIPROP TRANSLATIONAL ΔV Reqd (m/s)	MONOPROP ROTATIONAL Isp (s)	MONOPROP ROTATIONAL ΔV Reqd (m/s)	PROPELLANT USAGE M-HYDRZ (kg)	PROPELLANT USAGE B-HYDRZ (kg)	PROPELLANT USAGE NTO (kg)	PROPELLANT REMAINING HYDRZ (kg)	PROPELLANT REMAINING NTO (kg)	POSTBURN S/C MASS (kg)	EST. BURN DURATION #T	EST. BURN DURATION (s)
CRUISE ΔV95	TCM-1/2	0.0	0.0	317.0	36.0	220.0	0.3	0.15	6.61	5.62	229.06	146.93	1049.72	n/a	57.7
	TCM-3/4	220.0	2.0	0.0	0.0	0.0	0.0	0.97	0.00	0.00	228.09	146.93	1048.75	4	117.9
	pre-Lnch ΔV Reserves	220.0	0.0	315.0	20.1	0.0	0.0	0.00	3.68	3.13	224.41	143.80	1041.95	n/a	27.0
CAPTURE	MOI (Rp = 3700 km, Per = 48 Hr)	0.0	0.0	317.0	968.2	220.0	5.9	2.85	150.72	128.12	70.85	15.69	760.26	n/a	1329.5
Aerobraking Walk-in	AB-1	0.0	0.0	315.0	7.5	220.0	0.1	0.04	1.00	0.85	69.81	14.84	758.38	n/a	8.6
	AB-i (i=2,4)	220.0	2.5	0.0	0.0	0.0	0.0	0.88	0.00	0.00	68.94	14.84	757.50	4	106.5
Aerobraking Main Phase	ABM Translation	220.0	5.0	0.0	0.0	0.0	0.0	1.75	0.00	0.00	67.18	14.84	755.75	-	-
	ACS Rotation	0.0	0.0	0.0	0.0	190.0	5.0	2.03	0.00	0.00	65.16	14.84	753.72	-	-
Aerobraking Walk-out	ABM Translation	220.0	20.0	0.0	0.0	0.0	0.0	6.95	0.00	0.00	58.20	14.84	746.77	-	-
	ACS Rotation	0.0	0.0	0.0	0.0	190.0	30.0	11.93	0.00	0.00	46.27	14.84	734.84	-	-
Aerobraking Contingency	Aerobraking Pop-Up	220.0	2.5	315.0	14.5	0.0	0.0	0.85	1.86	1.58	43.56	13.26	730.55	n/a	15.9
	post-Lnch ΔV Reserves	220.0	0.0	315.0	0.0	0.0	0.0	0.00	0.00	0.00	43.56	13.26	730.55	4	0.0
Transition To Mapping	ABX	220.0	0.0	317.0	60.2	220.0	0.4	0.14	7.57	6.44	35.85	6.82	716.40	n/a	66.3
	TMO	0.0	0.0	317.0	16.1	220.0	0.1	0.03	2.00	1.70	33.82	5.12	712.67	n/a	17.5
	OTM-1 (Frzn)	0.0	0.0	315.0	7.5	0.0	0.0	0.00	0.93	0.79	32.89	4.33	710.94	n/a	10.0
MAPPING Drag Dens 95%	OTM (Drag/GTE)	220.0	3.9	0.0	0.0	0.0	0.0	1.28	0.00	0.00	31.60	4.33	709.66	-	-
	ACS Rotation	0.0	0.0	0.0	0.0	190.0	40.0	15.07	0.00	0.00	16.53	4.33	694.59	-	-
ADDITIONAL ΔV REQ'D	QUARANTINE ORBIT (PQ) - 2-BURN	220.0	22.8	0.0	0.0	0.0	0.0	7.30	0.00	0.00	9.23	4.33	687.28	4	885.0
RELAY Drag Dens 95%	OTM (Drag)	220.0	1.0	0.0	0.0	0.0	0.0	0.32	0.00	0.00	8.91	4.33	686.97	-	-
	ACS Rotation	0.0	0.0	0.0	0.0	190.0	10.0	3.68	0.00	0.00	5.23	4.33	683.29	-	-
	SUB-TOTALS		59.7		1130.1		91.8	56.21	174.38	148.22	S/C Dry Mass (kg)		673.73		
	MISSION TOTAL ΔV		1281.6												

Figure D-3: ΔV *Budget, 25 November 1996 Launch*

MARS GLOBAL SURVEYOR SPACECRAFT MANEUVER EXECUTION
Mission ΔV Budget - ORR Baseline (8/96)
Launch Date - 11/25/96

SPACECRAFT INJECTED MASS (kg) 1062.10
Spacecraft Dry Mass (kg) 673.28
Dry Mass Margin (kg) 0.45
Total Propellant (kg) 388.37

Spacecraft Propellant Load

Engine	Thrust (N)	Propellant Tanks	Load Error	Max Usable*	Desired Usable	Usable Plus Error	Unusable 2%	Oxidizer Ballast	Tank Totals	Tank Redlines
BIPROP	658.7	N2O4 (kg)	0.37	164.72	148.21	148.58	2.96	1.00	152.54	4.33
MPROP	4.45	(2x) N2H4 (kg)	0.62	229.29	230.59	231.21	4.61	-	235.82	5.23

Ge (m/s^2): 9.80665
BIPROP MIX RATIO: 0.85

*5% Ullage

MISSION PHASE	MANEUVER	MONOPROP TRANSLATIONAL Isp (s)	MONOPROP TRANSLATIONAL ΔV Reqd (m/s)	BIPROP TRANSLATIONAL Isp (s)	BIPROP TRANSLATIONAL ΔV Reqd (m/s)	MONOPROP ROTATIONAL Isp (s)	MONOPROP ROTATIONAL ΔV Reqd (m/s)	PROPELLANT USAGE M-HYDRZ (kg)	PROPELLANT USAGE B-HYDRZ (kg)	PROPELLANT USAGE NTO (kg)	PROPELLANT REMAINING HYDRZ (kg)	PROPELLANT REMAINING NTO (kg)	POSTBURN S/C MASS (kg)	EST. BURN DURATION #T	EST. BURN DURATION (s)
CRUISE ΔV95	TCM-1/2	0.0	0.0	317.0	49.0	220.0	0.3	0.15	8.98	7.63	226.70	144.91	1045.34	n/a	78.4
	TCM-3/4	220.0	2.0	0.0	0.0	0.0	0.0	0.97	0.00	0.00	225.73	144.91	1044.37	4	117.4
	pre-Lnch ΔV Reserves	220.0	0.0	315.0	7.0	0.0	0.0	0.00	1.28	1.09	224.45	143.83	1042.01	n/a	9.4
CAPTURE	MOI (Rp = 3700 km, Per = 48 Hr)	0.0	0.0	317.0	975.1	220.0	5.9	2.85	151.65	128.90	69.96	14.93	758.61	n/a	1334.5
Aerobraking Walk-in	AB-1	0.0	0.0	315.0	7.5	220.0	0.1	0.04	0.99	0.85	68.93	14.08	756.74	n/a	8.6
	AB-i (i=2,4)	220.0	2.5	0.0	0.0	0.0	0.0	0.88	0.00	0.00	68.05	14.08	755.86	4	106.2
Aerobraking Main Phase	ABM Translation	220.0	5.0	0.0	0.0	0.0	0.0	1.75	0.00	0.00	66.30	14.08	754.11	-	-
	ACS Rotation	0.0	0.0	0.0	0.0	190.0	5.0	2.02	0.00	0.00	64.28	14.08	752.09	-	-
Aerobraking Walk-out	ABM Translation	220.0	11.0	0.0	0.0	0.0	0.0	3.82	0.00	0.00	60.46	14.08	748.27	-	-
	ACS Rotation	0.0	0.0	0.0	0.0	190.0	26.5	10.57	0.00	0.00	49.89	14.08	737.70	-	-
Aerobraking Contingency	Aerobraking Pop-Up	220.0	15.8	315.0	0.0	0.0	0.0	5.38	0.00	0.00	44.51	14.08	732.32	n/a	0.0
	post-Lnch ΔV Reserves	220.0	0.0	315.0	0.0	0.0	0.0	0.00	0.00	0.00	44.51	14.08	732.32	4	0.0
Transition To Mapping	ABX	220.0	0.0	317.0	65.8	220.0	0.4	0.14	8.29	7.05	36.08	7.03	716.84	n/a	72.6
	TMO	0.0	0.0	317.0	17.9	220.0	0.1	0.03	2.22	1.89	33.82	5.14	712.69	n/a	19.4
	OTM-1 (Frzn)	0.0	0.0	315.0	7.5	0.0	0.0	0.00	0.93	0.79	32.89	4.35	710.97	n/a	10.0
MAPPING Drag Dens 95%	OTM (Drag/GTE)	220.0	3.9	0.0	0.0	0.0	0.0	1.28	0.00	0.00	31.60	4.35	709.68	-	-
	ACS Rotation	0.0	0.0	0.0	0.0	190.0	40.0	15.07	0.00	0.00	16.53	4.35	694.61	-	-
ADDITIONAL ΔV REQ'D	QUARANTINE ORBIT (PQ) - 2-BURN	220.0	22.8	0.0	0.0	0.0	0.0	7.30	0.00	0.00	9.23	4.35	687.31	4	885.0
RELAY Drag Dens 95%	OTM (Drag)	220.0	1.0	0.0	0.0	0.0	0.0	0.32	0.00	0.00	8.91	4.35	686.99	-	-
	ACS Rotation	0.0	0.0	0.0	0.0	190.0	10.0	3.68	0.00	0.00	5.23	4.35	683.31	-	-
	SUB-TOTALS		64.0		1129.8		88.3	56.24	174.35	148.20	S/C Dry Mass (kg)		673.73		
	MISSION TOTAL ΔV		1282.1												

MSN DV - FMP BaseLoad

Table E-1: *Trajectory Profile Summary for Aerobraking*

Event	Days From MOI	Date	Orbit Number	Periapsis / Apoapsis Alt. (km)
MOI	MOI+ 0	11-Sep-97	Orbit 1	300 / 56,675
Walk-in Maneuver AB1	MOI+ 9 days	20-Sep-97	Orbit 5	150 / 56,980
Walk-in Maneuver AB2	MOI+ 11 days	22-Sep-97	Orbit 6	133 / 57,000
Walk-in Maneuver AB3	MOI+ 15 days	26-Sep-97	Orbit 8	124 / 56,640
Walk-in Maneuver AB4	MOI+ 19 days	29-Sep-97	Orbit 10	118 / 56,040
Walk-in Maneuver AB5	MOI+ 23 days	3-Oct-97	Orbit 12	113 / 55,510
Walk-in Maneuver AB6	MOI+ 27 days	7-Oct-97	Orbit 14	112 / 53,600
Start of Main Phase	MOI+ 28 days	8-Oct-97	Orbit 15	112 / 52,650
Orbit Period at 40 Hours	MOI+ 32 days	13-Oct-97	Orbit 18	112 / 49,460
Orbit Period at 30 Hours	MOI+ 46 days	27-Oct-97	Orbit 28	113 / 39,820
Orbit Period at 24 Hours	MOI+ 57 days	7-Nov-97	Orbit 37	113 / 33,170
Orbit Period at 12 Hours	MOI+ 83 days	3-Dec-97	Orbit 74	114 / 18,610
Orbit Period at 6 Hours	MOI+ 102 days	22-Dec-97	Orbit 129	112 / 8,980
Orbit Period at 3 Hours	MOI+ 115 days	4-Jan-98	Orbit 204	104 / 3,190
Begin Walk-Out	MOI+ 121 days	10-Jan-98	Orbit 256	108 / 1,390
Aerobrake Termination Burn	MOI+ 132 days	21-Jan-98	Orbit 385	426 / 450

Figure E-1: *Apoapsis Altitude vs. Time (top), Local Mean Solar Hour vs. Time (bottom)*

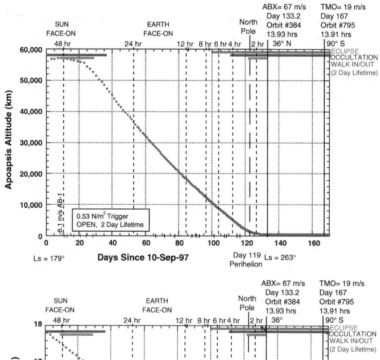

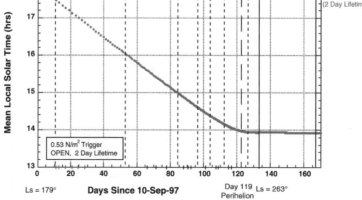

Figure E-2: *Periapsis Altitude vs. Time (top), IAU Argument of Periapsis vs. Time (bottom)*

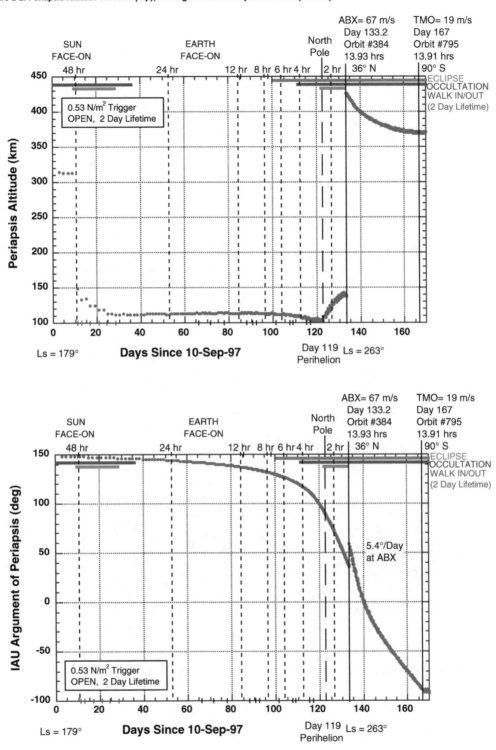

Figure E-3: *Heating Rate at Periapsis (top), Entry / Exit Times (bottom)*

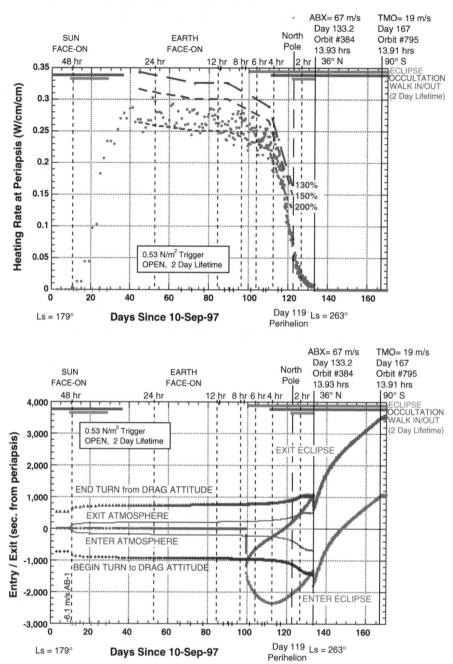

Donald Savage Headquarters, Washington, DC (Phone: 202/358-1727)
Mary Hardin Jet Propulsion Laboratory, Pasadena, CA (Phone: 818/354-0344)
Dr. Ken Edgett Malin Space Science Systems, San Diego, CA (Phone: 858/552-2650 x500)

RELEASE: 00-82 May 22, 2000

PUBLIC INVITED TO BROWSE 20,000 NEW ADDITIONS TO MARS PHOTO GALLERY

More than 20,000 new images of the planet Mars taken by NASA's Mars Global Surveyor spacecraft are now available in a web-based photo album — the single largest one-time release of images for any planet in the history of solar system exploration.

The 'picture postcard' scenes in the new images reveal the Red Planet, often said to be the most Earth-like planet, as an alien, bizarre and puzzling world.

"These are exciting times for Mars scientists and this release of images is in my opinion something unprecedented in the Mars science business," said Dr. Ken Edgett, staff scientist at Malin Space Science Systems, San Diego, CA. "People everywhere with Internet access will be able to take their own personal journey of exploration and discover Mars via these pictures. They can experience them the same way that Mars Global Surveyor scientists do — one at a time, no captions or explanations, just 'Here it is. What does it show me?'"

The archive of images now covers a period that spans one Mars year (687 Earth days), beginning in September 1997 with pictures taken during the aerobraking phase and extending through August 1999 when Global Surveyor was well into its mapping mission. Many of the pictures have such high resolution that objects on the surface the size of a school bus can be seen.

According to the Mars Orbiter Camera imaging team, placing these images within NASA's Planetary Data System for archiving is an important step in the Mars Global Surveyor mission that permits the public to examine the original data and make discoveries "for themselves."

"Putting these data into perspective is very difficult. We have focused on 'themes.' Layers on the Martian surface are the biggest 'theme' or 'finding' of the imaging investigation so far. To a geologist, layers record history and they are the most geologically important, profound thing we have seen," said Dr. Michael Malin, principal investigator for the camera system at Malin Space Science Systems. "We see layers in the walls of canyons, craters, and troughs. We see layers in both the north and south polar regions. We see them in the craters on top of volcanoes, we see them in pits at the bottoms of impact craters, we see them virtually everywhere that some process has exposed the subsurface so that we can see it from above."

"Seeing Mars up close through the narrow angle camera has been a humbling experience. We often find surfaces for which there are no obvious analogs on Earth, like certain ridges that look like dunes. Our terrestrial geologic experience seems, at times, to fail us," Edgett said. "Perhaps it is because water is the dominant force of erosion on Earth, even in the driest desert regions. But on Mars that force of change may have been something else, like wind. The ridges seen in places like the Valles Marineris floors are strange. They aren't dunes because they occur too close together, their crests are too sharp, their slopes too symmetrical. They often appear to be a specific layer of material that has undergone erosion — we just wish we knew what processes are involved that cause this kind of erosion."

The camera system uses a "push-broom" technique that systematically builds up pictures of the surface directly below one line at a time as the spacecraft orbits Mars. The wide-angle lens provides a complete low-resolution global map of the planet every day showing surface features and clouds at a resolution of about 4.6 miles (7.5 kilometers). The narrow-angle telescope takes close-up pictures of small areas with a resolution of about 5 feet (1.5 meters). Because of the extremely high data volume of the high-resolution images, controllers cannot use this mode continuously. Instead, they painstakingly plan which areas they want to target.

Mars Global Surveyor was launched on November 7, 1996 and arrived at Mars on September 12, 1997. The spacecraft has made more than 5,000 orbits of and has been systematically mapping the Red Planet since March 1999.

Mars Global Surveyor is managed by the Jet Propulsion Laboratory for NASA's Office of Space Science, Washington, DC. The camera system was built and is operated by Malin Space Science Systems, San Diego, CA. JPL's industrial partner is Lockheed Martin Astronautics, Denver, CO, which developed and operates the spacecraft. JPL is a division of the California Institute of Technology, Pasadena, CA. - end -

NOTE TO EDITORS: The archive of images can be found at: http://www.msss.com/moc_gallery/ A subset of the images can be seen at: http://photojournal.jpl.nasa.gov/new and http://mars.jpl.nasa.gov/mgs

Mars Climate Orbiter Arrival Press Kit

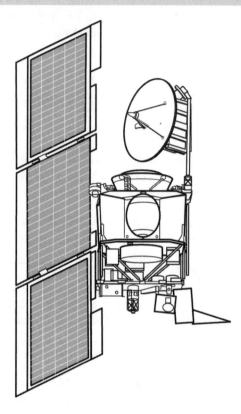

NATIONAL AERONAUTICS AND SPACE ADMINISTRATION

RELEASE: September 1999

Contacts

Douglas Isbell
Policy/Program Management
Headquarters,
Washington, DC
202/358-1753

Franklin O'Donnell
Mars Climate Orbiter Mission
Jet Propulsion Laboratory,
Pasadena, CA
818/354-5011

Mary Hardin
Mars Climate Orbiter Mission
Jet Propulsion Laboratory,
Pasadena, CA
818/354-0344

NASA'S MARS CLIMATE ORBITER: FIRST MARTIAN WEATHER SATELLITE

Mars Climate Orbiter, the first of two NASA spacecraft to reach Mars this year, is set to go into orbit around the red planet to become the first interplanetary weather satellite and a communications relay for the next lander mission to explore Mars.

The orbiter will fire its main engine at 1:50 a.m. Pacific Daylight Time on Thursday, September 23, 1999, to slow itself down so that it can be captured in orbit around the planet. Signals confirming the event will be received on Earth about 11 minutes later at 2:01 a.m.

"The curtain goes up on this year's Mars missions with the orbit insertion of Mars Climate Orbiter," said Dr. Sam Thurman, flight operations manager for the orbiter at NASAL Jet Propulsion Laboratory, Pasadena, CA. "If all goes well, the happily-ever-after part of the play will be the successful mission of the Mars Polar Lander that begins in December followed by the mapping mission of the orbiter that is set to begin next March."

Once captured in orbit around Mars, the orbiter will begin a period of aerobraking. During each of its long, elliptical loops around Mars, the orbiter will pass through the upper layers of the atmosphere each time it makes its closest approach to the planet. Friction from the atmosphere on the spacecraft and its wing-like solar array will cause the spacecraft to lose some of its momentum during each close approach. As the spacecraft slows with each pass, the maximum altitude of the orbit will decrease and the orbit will become more circular.

Mars Climate Orbiter's first assignment after it completes aerobraking will be to serve as the communications relay for its sibling spacecraft, Mars Polar Lander; set to land near the south pole on December 3. After the Lander's surface mission ends in February 2000, the orbiter's science mission begins with routine monitoring of the atmosphere, surface and polar caps for a complete Martian year (687 Earth days), the equivalent of almost two Earth years.

"We're interested in what happens during all the seasons of a Mars year. Weather is what happens from day-to-day and the long-term effect of all of that is climate," said Dr. Richard Zurek, project scientist for the orbiter at JPL. "Mars Climate Orbiter will do what weather satellites do — it will take pictures of clouds, it will look for storms and it will try to understand the atmospheric winds by measuring temperature and pressure and by watching how the atmospheric distributions of dust and water vapor change with time."

Today the Martian atmosphere is so thin and cold that it does not rain; liquid water placed on the surface would quickly freeze into ice or evaporate into the atmosphere. The temporary polar frosts which advance and retreat with the seasons are made mostly of condensed carbon dioxide, the major constituent of the Martian atmosphere. But the planet also hosts both water-ice clouds and dust storms, the latter ranging in scale from local to global. If typical amounts of atmospheric dust and water were concentrated today in the polar regions, they might deposit a fine layer on the ground year after year Consequently, the top meter (or yard) of the polar layered terrain could be a well-preserved tree-ring-like record showing tens of thousands of years of Martian geology and climatology.

The orbiter carries two science instruments: the Pressure Modulator Infrared Radiometer, a copy of the atmospheric sounder on the Mars Observer spacecraft lost in 1993; and the Mars Color Imager, a new, light-weight imager combining wide- and medium-angle cameras. The radiometer will measure temperatures, dust, water vapor and clouds by using a mirror to scan the atmosphere from the Martian surface up to 80 kilometers (50 miles) above the planet's limb. The radiometer was provided by JPL, supported by Oxford University and Russia's Space Research Institute; its principal investigator is Dr. Daniel McCleese of JPL.

Meanwhile, the imager will gather horizon-to-horizon images at up to kilometer-scale (half-mile) resolutions, which will then be combined to produce daily global weather images. The camera will also image surface features and produce a map showing objects the size of a football field with 40-meter (130-foot) resolution in several colors, providing global views that will help create a season-to-season portrait of the planet. The camera was provided by Malin Space Science Systems, San Diego; CA; its principal investigator is Dr. Michael

Malin, who also provided the camera for the currently orbiting Mars Global Surveyor

Mars Climate Orbiter is managed by the Jet Propulsion Laboratory for NASA's Office of Space Science, Washington, DC. Lockheed Martin Astronautics, Denver, CO, is the agency's industrial partner for development and operation of the Orbiter. JPL is a division of the California Institute of Technology.

[End of General Release]

Media Services Information

NASA Television Transmission

NASA Television is broadcast on the satellite GE-2, transponder 9C, C band, 85 degrees west longitude, frequency 3880.0 MHz, vertical polarization, audio monaural at 6.8 MHz.

Status Reports

Status reports on Mars Climate Orbiter mission activities will be issued by the Jet Propulsion Laboratory's Media Relations Office. They may be accessed online as noted below. Audio status reports are available by calling (800) 391-6654 or (818) 354-4210.

Briefing

A summary of Mars Climate Orbiter arrival events will be presented in a news briefing broadcast on NASA Television originating from the Jet Propulsion Laboratory at 8 a.m. PDT September 23, 1999.

Internet Information

Extensive information on Mars Climate Orbiter, including an electronic copy of this press kit, press releases, fact sheets, status reports and images, is available from the Jet Propulsion Laboratory's World Wide Web home page at http://www.jpl.nasa.gov/. The Mars Climate Orbiter project maintains a home page at http://mars.jpl.nasa.gov/msp98/orbiter/.

Quick Facts

Spacecraft

Dimensions: Main bus 6.9 feet (2.1 meters) tall, 5.4 feet (1.6 meters) wide and 6.4 feet (2 meters) deep; wingspan of solar array 18 feet (5.5 meters) tip to tip — Weight: 1,387 lbs (629 kg) total, consisting of 745-lb (338-kg) spacecraft plus 642 lbs (291 kg) fuel — Science instruments: Pressure Modulator Infrared Radiometer, Mars Color Imager — Power: Solar array providing up to 1,000 watts just after launch, 500 watts at Mars

Launch/Cruise

Launch: December 11, 1998, at 1:45:51 p.m. Eastern Standard Time (18:45:51 UTC) from Launch Complex 17A at Cape Canaveral Air Station, Florida Launch vehicle: Delta 11 Model 7425 Cruise: 9½ months Earth-Mars distance at launch: 158.6 million miles (255.2 million km) Total distance traveled Earth to Mars: 416 million miles (669 million km)

Mars Arrival

Arrival: September 23, 1999
Engine firing begins: 2:01 a.m. Pacific Daylight Time (PDT) (Earth-received time)
Spacecraft passes behind Mars: 2:06 a.m. PDT (Earth-received time)

Engine firing ends: 2:17 a.m. PDT (Earth-received time)
Spacecraft reappears from behind Mars: 2:27 a.m. PDT (Earth-received time)
Speed before engine firing (relative to Mars): 12,300 mph (5.5 km/sec)
Speed after engine firing (relative to Mars): 9,840 mph (4.4 km/sec)
Change in speed from engine firing: 3,065 mph (1.37 km/sec)
Earth-Mars distance at arrival: 121.9 million miles (196.2 million km)
One-way speed-of-light time from Mars to Earth at arrival: 10 minutes, 49 seconds
Mars seasons at arrival: Fall in northern hemisphere, spring in southern hemisphere

Aerobraking and Mapping Mission

Aerobraking: September 25 to November 9, 1999
Time to orbit Mars once in initial orbit: 12 to 17 hours
Time to orbit Mars once in final orbit: 2 hours
Final orbit: Circular, Sun-synchronous, altitude 262 miles (421 km)
Relay for Mars Polar Lander: December 1999-February 2000
Primary science mapping mission: March 2000-January 2002
Mars seasons when mapping begins: Winter in northern hemisphere, summer in south

Mars '98 Project

Cost: $193.1M spacecraft development, $42.8M mission operations; total $235.9 million for both Mars Climate Orbiter and the Mars Polar Lander spacecraft (not including launch vehicles or lander's Deep Space 2 microprobes)

Mars at a Glance

General
*One of five planets known to ancients; Mars was Roman god of war, agriculture and the state
*Reddish color; occasionally the 3rd brightest object in night sky after the Moon and Venus

Physical Characteristics
*Average diameter 6,780 kilometers (4,217 miles); about half the size of Earth, but twice the size of Earth's Moon
*Same land area as Earth
*Mass 1/10th of Earth's; gravity only 38 percent as strong as Earth's
*Density 3.9 times greater than water (compared to Earth's 5.5 times greater than water)
*No planet-wide magnetic field detected; only localized ancient remnant fields in various regions

Orbit
*Fourth planet from the Sun, the next beyond Earth
*About 1.5 times farther from the Sun than Earth is
*Orbit elliptical; distance from Sun varies from a minimum of 206.7 million kilometers (128.4 million miles) to a maximum of 249.2 million kilometers (154.8 million miles); average distance from Sun, 227.7 million kilometers (141.5 million miles)
*Revolves around Sun once every 687 Earth days
*Rotation period (length of day) 24 hours, 37 min, 23 sec (1.026 Earth days)
* Poles tilted 25 degrees, creating seasons similar to Earth's

Environment
* Atmosphere composed chiefly of carbon dioxide (95.3%), nitrogen (2.71%) and argon (1.6%)
* Surface atmospheric pressure less than 1/100th that of Earth's average
* Surface winds up to 80 miles per hour (40 meters per second)
* Local, regional and global dust storms; also whirlwinds called dust devils
* Surface temperature averages -53 C (-64 F); varies from -128 C (-199 F) during polar night to 27 C (80 F) at equator during midday at closest point in orbit to Sun

Features

* Highest point is Olympus Mons, a huge shield volcano about 16 miles (26 kilometers) high and 600 kilometers (370 miles) across; has about the same area as Arizona
* Canyon system of Valles Marineris is largest and deepest known in solar system; extends more than 4,000 kilometers (2,500 miles) and has 5 to 10 kilometers (3 to 6 miles) relief from floors to tops of surrounding plateaus
* "Canals" observed by Giovanni Schiaparelli and Percival Lowell about 100 years ago were a visual illusion in which dark areas appeared connected by lines. The Mariner 9 and Viking missions of the 1970s, however, established that Mars has channels possibly cut by ancient rivers

Moons

* Two irregularly shaped moons, each only a few kilometers wide
* Larger moon named Phobos ("fear"); smaller is Deimos ("terror"), named for attributes personified in Greek mythology as sons of the god of war

Historical Mars Missions

Mission, Country, Launch Date, Purpose, Results

[Unnamed], USSR, 10/10/60, Mars flyby, did not reach Earth orbit
[Unnamed], USSR, 10/14/60, Mars flyby, did not reach Earth orbit
[Unnamed], USSR, 10/24/62, Mars flyby, achieved Earth orbit only
Mars 1, USSR, 11/1/62, Mars flyby, radio failed at 65.9 million miles (106 million km)
[Unnamed], USSR, 11/4/62, Mars flyby, achieved Earth orbit only
Mariner 3, U.S., 11/5/64, Mars flyby, shroud failed to jettison
Mariner 4, U.S. 11/28/64, first successful Mars flyby 7/14/65, returned 21 photos
Zond 2, USSR, 11/30/64, Mars flyby, passed Mars but radio failed, returned no planetary data
Mariner 6, U.S., 2/24/69, Mars flyby 7/31/69, returned 75 photos
Mariner 7, U.S., 3/27/69, Mars flyby 8/5/69, returned 126 photos
Mariner 8, U.S., 5/8/71 Mars orbiter, failed during launch
Kosmos 419, USSR, 5/10/71, Mars lander, achieved Earth orbit only
Mars 2, USSR, 5/19/71, Mars orbiter/lander arrived 11/27/71, no useful data, lander destroyed
Mars 3, USSR, 5/28/71, Mars orbiter/lander, arrived 12/3/71, some data and few photos
Mariner 9, U.S., 5/30/71, Mars orbiter, in orbit 11/13/71 to 10/27/72, returned 7,329 photos
Mars 4, USSR, 7/21/73, failed Mars orbiter, flew past Mars 2/10/74
Mars 5, USSR, 7/25/73, Mars orbiter, arrived 2/12/74, lasted a few days
Mars 6, USSR, 8/5/73, Mars orbiter/lander, arrived 3/12/74, little data return
Mars 7, USSR, 8/9/73, Mars orbiter/lander, arrived 3/9/74, little data return
Viking 1, U.S., 8/20/75, Mars orbiter/lander, orbit 6/19/76-1980, lander 7/20/76-1982
Viking 2, U.S., 9/9/75, Mars orbiter/lander, orbit 8/7/76-1987, lander 9/3/76-1980; combined, the
      Viking orbiters and landers returned 50,000+ photos
Phobos 1, USSR, 7/7/88, Mars/Phobos orbiter/lander, lost 8/89 en route to Mars
Phobos 2, USSR, 7/12/88, Mars/Phobos orbiter/lander, lost 3/89 near Phobos
Mars Observer, U.S., 9/25/92, lost just before Mars arrival 8/21/93
Mars Global Surveyor, U.S., 11/7/96, Mars orbiter, arrived 9/12/97, currently conducting prime
      mission of science mapping
Mars 96, Russia, 11/16/96, orbiter and landers, launch vehicle failed
Mars Pathfinder, U.S., 12/4/96, Mars lander and rover, landed 7/4/97, last transmission 9/27/97
Nozomi (Planet-B), Japan, 7/4/98, Mars orbiter, currently in orbit around the Sun; Mars arrival
      delayed to 12/03 due to propulsion problem
Mars Climate Orbiter, U.S., 12/11/98; due to enter orbit 9/23/99
Mars Polar Lander/Deep Space 2, U.S., 1/3/99, lander/descent probes; lander to set down near
      south pole 12/3/99; Deep Space 2 microprobes to smash into surface the same day

Mars, Water and Life

The planet Mars landed in the middle of immense public attention on July 4, 1997, when Mars Pathfinder touched down on a windswept, rock-laden ancient flood plain. Two months later, Mars Global Surveyor went into orbit, sending back pictures of towering volcanoes and gaping chasms at resolutions never before seen.

In December 1998 and January 1999, another orbiter and lander were sent to Mars. And every 26 months over the next decade, when the alignment of Earth and Mars are suitable for launches, still more robotic spacecraft will join them at the red planet.

These spacecraft carry varied payloads, ranging from cameras and other sensors to rovers and robotic arms. Some of them have their roots in different NASA programs of science or technology development. But they all have the goal of improving our understanding of Mars, primarily by delving into its geology, climate and history.

With the announcement in 1996 by a team of scientists that a meteorite believed to have come from Mars contained what might be the residue of ancient microbes, public interest became regalvanized by the possibility of past or present life there. The key to understanding whether life could have evolved on Mars, many scientists believe, is understanding the history of water on the planet.

Mars today is too cold, with an atmosphere that is too thin, to support liquid water on its surface. Yet scientists who studied images from the Viking orbiters kept encountering features that appeared to be formed by flowing water — among them deep channels and canyons, and even features that appeared to be ancient lake shorelines. Added to this are more recent observations by Mars Pathfinder and Mars Global Surveyor that suggest widespread flowing water in the planet's past. Some scientists identified features that they believe appear to be carved by torrents of water with the force of 10,000 Mississippi Rivers.

There is no general agreement, however, on what form water took on the early Mars. Two competing views are currently popular in the science community. According to one theory, Mars was once much warmer and wetter, with a thicker atmosphere; it may well have boasted lakes or oceans, rivers and rain. According to the other theory, Mars was always cold, but water trapped as underground ice was periodically released when heating caused ice to melt and gush forth onto the surface.

In either case, the question of what happened to the water remains a mystery. Most scientists do not feel that Mars' climate change was necessarily caused by a cataclysmic event such as an asteroid impact that, perhaps, disturbed the planet's polar orientation or orbit. Many believe that the demise of flowing water on the surface could have resulted from gradual climate change over many millennia as the planet lost its atmosphere.

Under either the warmer-and-wetter or the always-cold scenario, Mars must have had a thicker atmosphere in order to support water that flowed on the surface, even only occasionally If the planet's atmosphere became thinner, liquid water would rapidly evaporate. Over time, carbon dioxide gas reacts with elements in rocks and becomes locked up as a kind of compound called a carbonate. What is left of Mars' atmosphere today is overwhelmingly carbon dioxide.

On Earth, shifting tectonic plates are continually plowing carbonates and other minerals under the surface; heated by magmas, carbon dioxide is released and spews forth in volcanic eruptions, replenishing the carbon dioxide in the atmosphere. Although Mars has no known active volcanoes and there are no signs of fresh lava flows, it had abundant volcanic activity in its past. However, Mars appears to have no tectonic plates, so a critical link in the process that leads to carbon dioxide replenishment in Earth's atmosphere is missing. In short, Mars' atmosphere could have been thinned out over many eons by entrapment of carbon dioxide in rocks across its surface.

That scenario, however, is just a theory. Regardless of the history and fate of the atmosphere, scientists also do not understand what happened to Mars' water. Some undoubtedly must have been lost to space. Water ice has been detected in the permanent cap at Mars' north pole, and may exist in the cap at the south pole. But much water is probably trapped under the surface — either as ice or, if near a heat source, possibly in liquid form well below the surface.

Current and Future Missions

Mars Climate Orbiter and Mars Polar Lander are designed to help scientists better understand the climate history of Mars, not to look for life. They do not, for example, contain any biology experiments similar to the chemistry lab on the Viking landers. However, their focus on Mars' climate and the role of water will have an impact on the life question. Water is also important as a resource for eventual human expeditions to the red planet.

In addition, Mars Climate Orbiter and the currently orbiting Mars Global Surveyor will aid the search for likely sites for future Mars robotic landers. Scientists are interested in three types of Martian environments which are potentially most favorable to the emergence and persistence of life. They are:

* Ancient groundwater environments. Early in the planet's history, liquid water appears to have been widespread beneath the surface. During the final stages of planetary formation, intense energy was dissipated by meteor impacts. This, along with active volcanoes, could have created warm groundwater circulation systems favorable for the origin of life.

* Ancient surface water environments. Also during early Martian history, water was apparently released from subsurface aquifers, flowed across the surface and pooled in low-lying regions. Evidence of the early climate of Mars and of ancient life, if any may be preserved in sedimentary rocks in these environments.

* Modern groundwater environments. Life may have formed at any time, including recently, in habitats where subsurface water or ice is geothermally heated to create warm groundwater circulation systems. In addition, life may have survived from an early epoch in places beneath the surface where liquid water is present today.

Where to Next?

In 1998 NASA conducted a review of its multi-year Mars program, calling in outside experts to help evaluate and refine the architecture for the future direction of the effort. Some of the details are subject to change as plans evolve, but the following are the basic robotic missions currently planned to Mars:

* December 3, 1999: Mars Polar Lander sets down near the planet's south pole, equipped with a robotic arm to dig into the soil in search of water ice. Other science instruments will take pictures and study the weather and atmosphere. Two microprobes created under the New Millennium Program's Deep Space 2 project will detach from the lander shortly before arrival and crash into the surface of Mars to test new technologies for probe/penetrators.

* 2001: NASA will launch an orbiter and lander broadly similar to the 1998 Mars spacecraft. The orbiter, equipped with three science instruments, will be the first planetary spacecraft to be launched from the west coast of the United States. The lander will touch down near Mars' equator, carrying a spare Mars Pathfinder rover, a robotic arm and several other science instruments, including three that will return data in support of eventual human exploration.

* 2003: NASA will begin the series of sample-return missions with launch of a lander and rover that will search for soil and rock samples and return them to a mini-ascent vehicle that will loft the samples to Mars orbit. The European Space Agency (ESA), in collaboration with the Italian space agency (ASI), will provide an orbiter. ASI may also provide a drill and other robotic elements for landers beginning in this year NASA's Human Exploration and Development of Space (HEDS) enterprise, as well as international partners, may also begin providing science and technology experiments for the large lander. As early as this year, NASA and the French space agency, CNES, may also launch a "micromission" spacecraft designed to deliver payloads such

as a robotic airplane or a small telecommunications orbiter.

* 2005: NASA and CNES will launch an orbiter, lander and rover on a French-provided Ariane 5 rocket. The lander and rover will search for soil and rock samples, and return them to a mini ascent vehicle that will loft the samples to Mars orbit. CNES will provide the orbiter spacecraft that will retrieve the 2003 and 2005 samples from Mars orbit and return them to Earth in a vehicle provided by NASA. The samples will reach Earth in 2008. The CNES spacecraft will also carry four miniature landers called "netlanders." In addition, NASA and CNES may collaborate on two "micromissions."

* 2007-2009: As currently envisioned, NASA's strategy will be to continue to collect samples and place them in Mars orbit for later retrieval. In 2007, NASA will launch another lander, rover and ascent vehicle. NASA and CNES may collaborate on two more "micromissions," and HEDS may provide more experiments. In 2009, NASA will launch a lander and rover that will search for soil and rock samples and return them to an ascent vehicle. Current plans call for another orbiter provided by CNES that will retrieve the 2007 and 2009 samples from Mars orbit and return them to Earth in a vehicle provided by NASA.

* 2011-2013: Although plans for these years are uncertain, a repeat of the 2007-2009 strategy is currently envisioned.

These Martian environments can be investigated in several ways. We can get a glimpse of underground environments by using rovers to explore young craters and what appear to be the remains of water-eroded channels, and by drilling from lander spacecraft. Sensors on orbiters will search for the most likely reservoirs of water in these regions.

To investigate these scientific themes, NASA's Mars program will carry out the following implementation strategy for the initial phases of Mars exploration:

* The Mars orbiters launched in 1996, 1998 and 2001 will provide sufficient information to guide an early sample return from an ancient groundwater environment.

* Ancient surface-water environments will be explored in greater depth. When a sample-return mission is sent to Mars, it is extremely important to be able to identify minerals formed by water.

* Ancient and modern sites exhibiting evidence of hydrothermal activity will be studied, followed eventually by efforts to drill as deeply as possible below the surface.

Samples will be collected using rovers capable of extensive searches and of collecting and storing samples of rock and soil. Sophisticated sensors onboard rovers will help insure that diverse rock types are collected. Drills capable of reaching several meters (or yards) below the surface will also be used to analyze subsurface material. It is likely that it will be some time before space technologies will permit drilling to depths of a kilometer (half-mile) and more to access subsurface water.

Samples of the Martian atmosphere will also be brought back to Earth. The possible origin and evolution of life on Mars must be linked to the evolution of its atmosphere.

In 2003, NASA and its international partners will see the first launch of a mission to collect samples and place them in Mars orbit to await their transport back to Earth. A mission in 2005 will include two spacecraft — a lander like the 2003 mission to collect surface samples, and a French-built orbiter to return both the 2003 and 2005 samples to Earth. A series of at least three sample return missions similar to this are expected to be carried out over the following decade.

Even if it turns out that Mars never harbored life, study of the planet can help in understanding life on our own. Much of the evidence for the origin of life on Earth has been erased by movement of the planet's crust and by weathering. Fortunately, large areas of Mars' surface date back to the very earliest period of planetary evolution — about 4 billion years ago, overlapping the period on Earth when pre-biotic chemical evolution first gave rise to life. Thus, even if life never developed on Mars, studies of the planet may yield crucial information about the prebiotic chemistry that led to life on Earth.

Mission Overview

Mars Climate Orbiter carries two science instruments, one a copy of an instrument carried by the Mars Observer spacecraft lost in 1993. Between them, Mars Climate Orbiter and the currently orbiting Mars Global Surveyor carry all but one of the entire suite of science instruments from Mars Observer. In addition, both orbiters will provide radio relay support for Mars Polar Lander.

Launch and Cruise

Mars Climate Orbiter was launched aboard a Delta II Model 7425 rocket from Cape Canaveral Air Station's Launch Complex 17 A on December 11, 1998, at 1:45:51 p.m. Eastern Standard Time (18:45:51 UTC).

The spacecraft will have been traveling for 286 days, or about 9½ months, when it reaches Mars on September 23, 1999. The spacecraft's flight path is called a Type 2 trajectory because it took the orbiter more than 180 degrees around the Sun; this results in a slower speed at Mars arrival. By comparison, the Mars Pathfinder lander; which followed a Type 1 trajectory that took it less than 180 degrees around the Sun, spent only seven months in flight to Mars. During the first leg of its journey, Mars Climate Orbiter flew slightly inward toward the Sun before spiraling out beyond Earth's orbit to Mars.

Arrival Events

Time	Event
9/15/99	Final thruster firing to adjust flight path to Mars
9/20/99	Flight team transmits sequence of commands that will control the spacecraft during the arrival burn. Team begins around-the-clock monitoring.
9/23/99	1:41 a.m.: Solar array stowed
	1:50 a.m.: Spacecraft turns to orientation for main engine burn
	1:56 a.m.: Pyrotechnic devices fired to open valves to begin pressurizing fuel and oxidizer tanks
	2:01 a.m.: Main engine begins firing
	2:06 a.m.: Spacecraft passes behind Mars, out of view of Earth
	2:17 a.m.: Main engine concludes firing
	2:19 a.m.: Spacecraft turns to orientation to allow Earth contact when it emerges from behind Mars
	2:27 a.m.: Spacecraft emerges from behind Mars, contact reestablished
	2:30 a.m.: Solar array unstowed

All times are in Pacific Daylight Time in "Earth-received time," when the signal confirming an event would be received on Earth. Actual events on the spacecraft occur about 11 minutes earlier.

During the first phase of cruise, the spacecraft maintained contact with Earth using its low-gain or medium-gain antenna while keeping its solar arrays pointed at the Sun. For the first seven days, the spacecraft was tracked 24 hours per day. During the second through fourth weeks after launch, the spacecraft was tracked a minimum of 12 hours per day using the 34-meter-diameter (112-foot) Deep Space Network antennas. During quiescent times of the cruise phase, the orbiter was tracked a minimum of four hours each day. The tracking rate increased again to 12 hours per day starting 45 days before the spacecraft is captured in orbit around Mars.

Twelve days into flight, the radiator door on one of the science instruments, the Pressure Modulator Infrared Radiometer, was moved to the vented position to acclimate the instrument's passive radiative cooler to the environment of space.

During interplanetary cruise, the orbiter fired its thrusters a total of four times to adjust its flight path. The first trajectory correction maneuver was carried out 15 days after launch and corrected launch injection errors and adjusted the Mars arrival aimpoint.

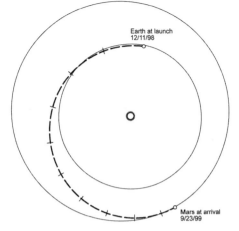

Orbiter's interplanetary trajectory

The remainder of the trajectory correction maneuvers were used to shape the trajectory and direct the spacecraft to the proper aimpoint for the orbit insertion burn. The second course maneuver was conducted 45 days later, on January 25, 1999; the third occurred 60 days before the spacecraft is captured in orbit around Mars, on July 25, 1999; and the fourth took place on September 15, 1999, eight days prior to Mars orbit insertion.

In addition to the course corrections, the science payload was powered on, tested and calibrated during cruise. The Pressure Modulator Infrared Radiometer and Mars Color Imager were calibrated during a week-long checkout 80 days after launch. During this checkout activity the camera was commanded to turn and scan across a specific star cluster as part of a star calibration exercise.

As the spacecraft neared Mars, 16 days before orbit insertion, the camera took a picture of the planet, available at :

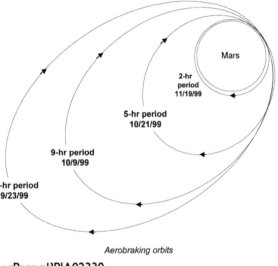

Aerobraking orbits

http://photojournaljpl.nasa.gov/cgi-bin/PIAGenCatalogPage.pl?PIA02330

Mars Orbit Insertion

As the spacecraft nears its closest point to the planet, coming in over the northern hemisphere, the spacecraft will fire its 640-newton main engine for 16 to 17 minutes to brake into an elliptical capture orbit. After the burn, the orbiter will loop around Mars once every 12 to 17 hours.

About 22 minutes after completion of the burn, the spacecraft will turn to point its high-gain antenna at Earth. During the first two days after orbit insertion, the spacecraft will communicate continuously with the Deep Space Network's 70-meter diameter (230-foot) and 34-meter diameter (112-foot) antennas.

Based on the details of the initial capture orbit, the spacecraft will fire its thrusters during its next closest pass by Mars to lower the orbit and reduce the orbit period by two to four hours. The flight team will work to achieve as low an orbit as possible to reduce the amount of time required for aerobraking.

Aerobraking

In aerobraking, a spacecraft is slowed down by frictional drag as it flies through the upper part of a planet's atmosphere. The technique was first tested by NASA's Magellan spacecraft at Venus at the end of its prime mission in 1994 and it was used with great success by the currently orbiting Mars Global Surveyor spacecraft.

Mars Climate Orbiter will be starting from an orbit much lower than Mars Global Surveyor — with an orbit period only one-third as long — so it will require much less aerobraking. The spacecraft will reach its final, low altitude, science-gathering orbit two weeks before its sibling spacecraft — Mars Polar Lander — touches down within about 900 kilometers (590 miles) of the southern polar cap.

During each of its long, elliptical loops around Mars, the orbiter will pass through the upper layers of the atmosphere each time it makes its closest approach to the planet. Friction from the atmosphere on the spacecraft and its wing-like solar array will cause the spacecraft to lose some of its momentum during each close approach, known as an "aeropass." As the spacecraft slows down during each close approach, the orbit will gradually lower and circularize.

Before the beginning of each aeropass, the orbiter's solar wing will be braced against the body of the spacecraft for mechanical stability. The spacecraft there will not receive any solar energy until after the aeropass is over and the solar array can again be turned toward the Sun.

Mars Climate Orbiter may begin aerobraking within a day or two after it enters Martian orbit. The first thruster firing to adjust the orbit for aerobraking may be performed as early as the first time the spacecraft swings away from the planet after entering orbit. During the next several days, the flight team will fire the thrusters each time the spacecraft reaches the point in the orbit most distant from the planet. This will bring the point of its closest approach to Mars down within the planet's upper atmosphere.

The main aerobraking phase will begin once the point of the spacecraft's closest approach to the planet — known as the orbit's periapsis — has been lowered to within about 100 kilometers (60 miles) above the Martian surface. As the spacecraft's orbit is reduced and circularized over approximately 200 aeropasses in 44 days, the periapsis will move northward from 34 degrees north latitude to 89 degrees north latitude, which will take it almost directly over the north pole. Small thruster firings when the spacecraft is most distant from the planet will keep the aeropass altitude at the desired level to limit heating and dynamic pressure on the orbiter

Although the orbiter's transmitter can be on continuously during aerobraking, contact with Earth will not be possible during the aeropasses or when the spacecraft passes behind Mars as seen from Earth. As a result, the spacecraft will be out of contact with Earth for 45-to-60 minutes per orbit. The Mars Climate Orbiter will be monitored and tracked during each orbit, but sometimes share Deep Space Network antenna time with Mars Global Surveyor Concurrent observations from Global Surveyor will provide early warning of dust storms and other atmospheric changes occurring over the planet.

The most challenging part of aerobraking will occur during what the flight team calls the "end game": the last few days of aerobraking, when the period of the spacecraft's orbit is the shortest. At that time, the aeropasses will be occurring most frequently and lasting longer than previous aeropasses.

The final science orbit will be a "late afternoon" orbit that takes the spacecraft over the Martian equator at approximately 4:30 p.m. local mean solar time on the day side of the planet, and at 4:30 a.m. on the night side of the planet.

Lander Approach Navigation

Mars Climate Orbiter's first assignment after it completes aerobraking will be to serve as the communications relay for its sibling spacecraft, Mars Polar Lander Ground controllers will implement a program of near-simultaneous tracking of Mars Climate Orbiter and Mars Polar Lander to support precision approach navigation for the lander as it prepares for its entry, descent and landing on Mars on December 3, 1999.

The 34-meter (112-foot-diameter) antennas of the Deep Space Network will track the orbiter just before or after tracking the lander. This will allow engineers to reduce effects of errors from locations of ground stations, modeling of solar plasma and other sources. First demonstrated in late 1997 with the Mars Pathfinder and Mars Global Surveyor spacecraft, the technique is expected to increase significantly the spacecraft team's ability to control the entry angle of the lander and thus reduce the size of the landing zone footprint. Mars Global Surveyor may also be used as a navigation aid for the approaching lander.

Mapping Orbit

The orbiter will fire its thrusters in two maneuvers at the end of aerobraking. The first and largest of these maneuvers will raise the spacecraft's periapsis — the point in its orbit where it passes closest to the planet — out of the atmosphere. The second maneuver will place the spacecraft into an orbit designed to fine-tune the timing of the orbiter's passage over the Mars Polar Lander's landing site. Shortly before the lander arrives, additional small maneuvers will be executed to place the orbiter into its final mapping orbit, a nearly circular, Sun-synchronous polar orbit with an average altitude of approximately 421 kilometers (262 miles).

Once the spacecraft has reached the mapping orbit, the high gain antenna will be deployed. The antenna, which was stowed during the nine-and-a-half-month trip to Mars and during aerobraking may require testing to assess its performance in the deployed position.

Lander Support

For the first three months after insertion into the mapping orbit, the orbiter's main task will be to act as a radio relay for Mars Polar Lander. During this time the orbiter will relay commands to the lander, as well as data from the lander to Earth. The orbiter may spend limited time collecting science data and downlinking the information during daily, 10-hour communications sessions. The lander support is scheduled to last through the end of February 2000.

Mapping

From March 2000 through January 2002, the orbiter will carry out its primary science mission making systematic observations of the atmosphere and surface of Mars using its two science instruments, the Pressure Modulator Infrared Radiometer and Mars Color Imager. The length of the science mission was chosen to span one Martian year, or 687 Earth days, so that scientists can observe seasonal variations in the Martian weather. Data will be sent to Earth during the daily, 10-hour communications sessions.

Relay for Future Missions

Once its mapping mission is complete, the orbiter will be available for up to two years as a communications relay for future Mars landers. During this phase, the orbiter will fire its thrusters to increase its altitude. This maneuver fulfills planetary protection regulations by increasing the length of time the spacecraft remains in orbit before eventually entering the atmosphere, breaking up and crashing into the planet's surface.

Spacecraft

Mars Climate Orbiter's main structure, or bus, is 2.1 meters (6.9 feet) tall, 1.6 meters (5.4 feet) wide and 2 meters (6.4 feet) deep. At launch it will weigh 629 kilograms (1,387 pounds), consisting of the 338-kilogram (745-pound) dry spacecraft plus 291 kilograms (642 pounds) of fuel.

The framework of the spacecraft bus is made of a combined graphite composite/aluminum honeycomb structure similar to that used in the construction of commercial aircraft.

Most systems on the spacecraft are fully redundant; for example, there are two onboard computers, two radio transmitter/receivers and so on, to compensate in case any device fails. The main exceptions are the electrical battery and the main engine used to brake the spacecraft into orbit upon arrival at Mars.

Onboard Computer

The spacecraft's computer system is greatly simplified compared with computers on planetary spacecraft a few years ago. By the late 1980s, spacecraft boasted the equivalent of onboard local area networks, with a main computer communicating with other computers used to run various subsystems and science instruments. By contrast, the Mars '98 orbiter and lander have a single onboard computer that runs all spacecraft activities (the science instruments contain microprocessor chips but not complete computer systems).

The spacecraft's computer uses a RAD6000 processor, a radiation-hardened version of the PowerPC chip used on some models of Macintosh computers. It can be switched between clock speeds of 5 MHz, 10 MHz or 20 MHz The computer includes 128 megabytes of random-access memory (RAM); unlike many other spacecraft, the orbiter does not have an onboard tape recorder or solid-state data recorder, but instead stores data in its RAM for transmission to Earth. The computer also has 18 megabytes of "flash" memory that can store data even when the computer is powered off. Eight megabytes of the flash memory will be used to store triplicate copies of high-priority data; when the computer accesses the data later, it will check all three copies of each byte to make sure that information has not become corrupted.

Attitude Control

The attitude control system manages the spacecraft's orientation, or "attitude." Like most planetary spacecraft, the orbiter is three-axis stabilized, meaning that its orientation is held fixed in relation to space, as opposed to spacecraft that stabilize themselves by spinning. The orbiter determines its orientation at any given time using a star camera, two Sun sensors and one of two inertial measurement units, each of which consists of three ring-laser gyroscopes and three accelerometers. The orientation is changed by firing thrusters or by using three reaction wheels, devices similar to flywheels.

During most of the interplanetary cruise, the spacecraft will be in "all stellar" mode, relying only on its star camera and Sun sensors for attitude determination without using the inertial measurement unit. To save wear and tear on their gyros, the inertial measurement units will be primarily used in Mars orbit.

Telecommunications

The spacecraft will communicate with Earth in the microwave X band, using a transponder (transmitter and receiver) based on a design used on the Cassini mission, along with a 15-watt radio frequency power amplifier. It uses a 1.3-meter-diameter (4.3-foot) dish-shaped high-gain antenna to transmit and receive, a medium-gain antenna with transmit-only capability and a receive-only low-gain antenna. A receiver and 15-watt transmitter in the UHF radio band will support two-way communications with Mars Polar Lander and future Mars landers.

Power

The spacecraft obtains its power from a solar array consisting of gallium arsenide solar cells mounted on three panels that form a single wing spanning 5.5 meters (18 feet) from tip to tip. Shortly after launch, the solar array will provide up to 1,000 watts of power; in Mars orbit it will provide up to about 500 watts. Power is stored in a 12-cell, 16-amp-hour nickel-hydrogen battery. In addition to providing power, the solar array acts as the spacecraft's "brakes" during aerobraking. Wing flaps have been added to the array to increase the amount of surface area and improve aerobraking performance.

Thermal Control

The thermal control system uses electrical heaters, thermal radiators and louvers to control the temperature inside the spacecraft. Multi-layer insulation, Kapton blankets and protective coatings are used to shield electronics from the harsh environment of space.

Propulsion

The propulsion system is similar to Mars Global Surveyor's, featuring both sets of small thrusters for maneuvers as well as a main engine that will fire to place the spacecraft in orbit around Mars. The main engine uses hydrazine propellant with nitrogen tetroxide as an oxidizer to produce 640 newtons (144 pounds) of thrust. The thrusters, which use hydrazine as a monopropellant, are divided into two sets. Four larger thrusters, each of which puts out 22 newtons (5 pounds) of thrust, will be used for trajectory correction maneuvers or turning the spacecraft. Four smaller thrusters producing 0.9 newton (0.2 pound) of thrust each will be used exclusively for attitude control.

Science Objectives

One of the chief scientific issues that Mars Climate Orbiter will study is the question of water distribution on Mars. On a planet with temperatures that rarely rise above freezing (0 C or 32 F) and plummet to lows of about -88 C (-126 F), water ice and carbon dioxide ice remain year-round in its permanent polar caps.

Over time, some water has surely been lost to space; some has been added by the infall of comets and meteorites. Water is also likely to be stored in the ground, chemically and physically bound to soil particles and as ice. Models of subsurface temperatures indicate that ground ice should be near the surface in the polar regions.

Instruments onboard the orbiter and lander will analyze the composition of surface materials, characterize daily and seasonal weather patterns and frost deposits,

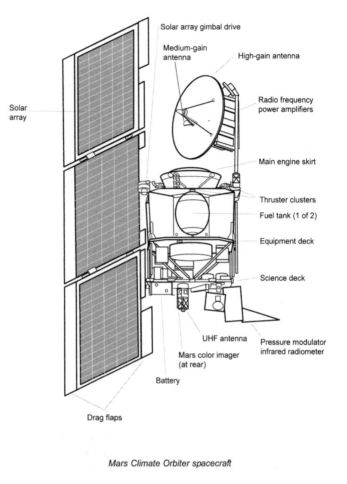

Mars Climate Orbiter spacecraft

and monitor surface and atmospheric interactions to better understand the planet as a global system.

Other major goals of the mission are:

* Study variations in atmospheric dust and volatile materials, such as carbon dioxide and water, in both their vapor and frozen forms. Mars Climate Orbiter will track these variations over a full Martian year (687 Earth days).

* Identify surface reservoirs of volatile material and dust, and observe their seasonal variations. The orbiter's imager and sounder will be able to characterize surface compositional boundaries and changes that might occur with time or seasons.

* Explore climate processes that stir up or quell regional and global dust storms, as well as atmospheric processes that transport volatiles such as water ice clouds and dust around the planet.

* Search for evidence of Mars' ancient climate, which some scientists believe was temperate and more

Earth-like with a thicker atmosphere and abundant flowing water. Layered terrain in the polar regions suggests more recent, possibly cyclic, climate change. Studies of Mars' early climate compared with Earth's may explain whether internal or external factors (such as changes in Mars' orbit) are primary drivers of climate change.

The orbiter carries two science instruments, a radiometer and camera.

Pressure Modulator Infrared Radiometer

This instrument is a sounder that will scan Mars' thin atmosphere, measuring temperatures, dust, water vapor and condensate clouds. It can scan the planet's atmosphere at the horizon or straight down underneath the spacecraft.

The instrument detects radiation in a total of nine channels. One of them detects visible light, while the other eight detect various spectral bands of infrared radiation at wavelengths between 6 and 50 microns. These data will allow scientists to construct vertical profiles of the atmosphere from near the surface to as high as 80 kilometers (50 miles) above the surface. Bands of water vapor and carbon dioxide will be detected, for the first time, at a vertical resolution of 5 kilometers (3 miles). A radiative cooler will keep the detectors located on the instrument's focal plane assembly at temperatures of about -193 C (-315 F).

The instrument weighs 42 kilograms (93 pounds) and uses 41 watts of power. The main instrument box is 23 by 30 by 74 centimeters (9 by 12 by 29 inches); a smaller cooler attached to the instrument is 58 by 65 by 30 centimeters (23 by 26 by 12 inches).

The joint principal investigators for the radiometer are Dr. Daniel McCleese of JPL and Dr. Vassili Moroz of the Space Research Institute (IKI), Moscow, Russia.

Mars Color Imager

The imager combines a wide-angle and a medium-angle camera. Both are designed to snap a series of overlapping frames as the spacecraft sweeps over the planet. Each camera has a charge-coupled device (CCD) detector overlaid with spectral or color filter strips. The camera shutters are electronically controlled at intervals timed so that the spacecraft motion overlaps the filter strips in order to produce a series of color images covering the same surface area.

The wide-angle camera detects light in seven spectral bands — five in the visible spectrum from 425 to 750 nanometers, and two in the ultraviolet spectrum from 250 to 330 nanometers. The camera is capable of taking pictures with a resolution of up to about 1 kilometer (six tenths of a mile) in Mars orbit when the data rate of the communications system allows sending large numbers of images. At other times, the camera will average adjacent pixels together to result in pictures with a resolution as low as 7.2 kilometers (4.5 miles) per pixel. The camera can also image the limb of the planet to detail the structure of clouds and hazes at a resolution of about 4 kilometers (2.5 miles) per pixel.

The medium-angle camera detects light in eight spectral bands from violet to near infrared (425 to 930 nanometers wavelength). Pointed down at the planet, the camera will take pictures with a resolution of about 40 meters (130 feet) per pixel across a six-degree field-of-view covering 40 kilometers (25 miles).

Once the spacecraft is in its mapping orbit, the Mars Color Imager will provide daily global images of the Martian atmosphere and surface with the wide-angle camera, and monitor surface changes with the medium-angle camera when high downlink data rates are possible. Each of the imager's two cameras is 6 by 6 by 12 centimeters (2.35 by 2.35 by 4.7 inches); together they weigh 2 kilograms (4.4 pounds). The imager uses 4 watts of power.

Principal investigator for the Mars Color Imager is Dr. Michael Malin of Malin Space Science Systems Inc., San Diego, CA.

Program/Project Management

Mars Climate Orbiter is managed by the Jet Propulsion Laboratory, Pasadena, CA, for NASA's Office of Space Science, Washington, DC. At NASA Headquarters, Dr. Edward Weiler is associate administrator for space science. Dr. Carl Pilcher is science director for solar system exploration. Ken Ledbetter is director of the Mission and Payload Development Division, and Steven Brody is program executive for Mars Climate Orbiter Joseph Boyce is program scientist for Mars Climate Orbiter.

At the Jet Propulsion Laboratory, Mars Climate Orbiter has been operated after launch by JPL's Mars Surveyor Operations Project; Richard Cook is project manager and Dr. Sam Thurman is flight operations manager. Dr. Richard Zurek is project scientist. Dr. John McNamee was project manager for development of Mars Climate Orbiter At Lockheed Martin Astronautics, Denver, CO, Dr. Edward A. Euler is the company's program director for Mars Climate Orbiter.

9-20-99 HQ

Mars Climate Orbiter Mission Status

MEDIA RELATIONS OFFICE
JET PROPULSION LABORATORY
CALIFORNIA INSTITUTE OF TECHNOLOGY
NATIONAL AERONAUTICS AND SPACE ADMINISTRATION
PASADENA, CALIFORNIA 91109.
TELEPHONE (818) 354-5011 http://www.jpl.nasa.gov

FOR IMMEDIATE RELEASE September 23, 1999

NASA'S MARS CLIMATE ORBITER BELIEVED TO BE LOST

NASA's Mars Climate Orbiter is believed to be lost due to a suspected navigation error.

Early this morning at about 2 a.m. Pacific Daylight Time the orbiter fired its main engine to go into orbit around the planet. All the information coming from the spacecraft leading up to that point looked normal. The engine burn began as planned five minutes before the spacecraft passed behind the planet as seen from Earth. Flight controllers did not detect a signal when the spacecraft was expected to come out from behind the planet.

"We had planned to approach the planet at an altitude of about 150 kilometers (93 miles). We thought we were doing that, but upon review of the last six to eight hours of data leading up to arrival, we saw indications that the actual approach altitude had been much lower. It appears that the actual altitude was about 60 kilometers (37 miles). We are still trying to figure out why that happened," said Richard Cook, project manager for the Mars Surveyor Operations Project at NASA's Jet Propulsion Laboratory. "We believe that the minimum survivable altitude for the spacecraft would have been 85 kilometers (53 miles)."

"If in fact we have lost the spacecraft it is very serious, but it is not devastating to the Mars Surveyor Program as a whole. The program is flexible enough to allow us to recover the science return of Mars Climate Orbiter on a future mission. This is not necessarily science lost; it is science delayed," said Dr. Carl Pilcher, science director for Solar System Exploration at NASA Headquarters, Washington, D.C. "We have a robust program to explore Mars that involves launching on average one mission per year for at least a decade. It began with the launch of Mars Pathfinder and Mars Global Surveyor in 1996, continued with Mars Climate Orbiter and Mars Polar Lander and will be followed by more missions in 2001, 2003 and 2005. In fact, Mars Polar Lander will arrive in just over two months and its mission is completely independent of the Mars Climate Orbiter. The science return of that mission won't be affected."

Flight controllers at NASA's Jet Propulsion Laboratory in Pasadena, CA and Lockheed Martin Astronautics in Denver, CO will continue their efforts to locate the spacecraft through the Deep Space Network during the next several hours. A special investigation team has been formed by JPL to further assess the situation.

Mars Climate Orbiter is one of a series of missions in a long-term program of Mars exploration known as the Mars Surveyor Program that is managed by the Jet Propulsion Laboratory for NASA's Office of Space Science, Washington, DC. JPL is a division of the California Institute of Technology, Pasadena, CA.

MEDIA RELATIONS OFFICE
JET PROPULSION LABORATORY
CALIFORNIA INSTITUTE OF TECHNOLOGY
NATIONAL AERONAUTICS AND SPACE ADMINISTRATION
PASADENA, CALIFORNIA 91109.
TELEPHONE (818) 354-5011 http://www.jpl.nasa.gov

Mars Climate Orbiter Mission Status

September 24, 1999

Flight controllers for NASA's Mars Climate Orbiter are planning to abandon the search for the spacecraft at 3 p.m. Pacific Daylight Time today. The team has been using the 70-meter-diameter (230-foot) antennas of the Deep Space Network in an attempt to regain contact with the spacecraft.

Engineers now estimate that the altitude of the spacecraft's closest approach to Mars as it was firing its engine to enter orbit around the planet was 57 kilometers (35 miles). The original target altitude had been about 140 kilometers (about 90 miles). The spacecraft team estimates that the minimum survivable altitude for the spacecraft was between 85 and 100 kilometers (about 53 to 62 miles).

The project is moving swiftly to determine the causes of this error, assisted by an internal review team. Expert independent review teams are being formed by JPL and NASA.

Mars Climate Orbiter is one of a series of missions in a long-term program of Mars exploration known as the Mars Surveyor Program that is managed by the Jet Propulsion Laboratory for NASA's Office of Space Science, Washington, DC. JPL is a division of the California Institute of Technology, Pasadena, CA.

Douglas Isbell Headquarters, Washington, DC Sept. 30, 1999 (Phone: 202/358-1753)

Mary Hardin Jet Propulsion Laboratory, Pasadena, CA (Phone: 818/354-5011)

Joan Underwood Lockheed Martin Astronautics, Denver, CO (Phone: 303/971-7398)

RELEASE 99-113

MARS CLIMATE ORBITER TEAM FINDS LIKELY CAUSE OF LOSS

A failure to recognize and correct an error in a transfer of information between the Mars Climate Orbiter spacecraft team in Colorado and the mission navigation team in California led to the loss of the spacecraft last week, preliminary findings by NASA's Jet Propulsion Laboratory internal peer review indicate.

"People sometimes make errors," said Dr. Edward Weiler, NASA's Associate Administrator for Space Science.

"The problem here was not the error, it was the failure of NASA's systems engineering, and the checks and balances in our processes to detect the error. That's why we lost the spacecraft."

The peer review preliminary findings indicate that one team used English units (e.g., inches, feet and pounds) while the other used metric units for a key spacecraft operation. This information was critical to the maneuvers required to place the spacecraft in the proper Mars orbit.

"Our inability to recognize and correct this simple error has had major implications," said Dr. Edward Stone, director of the Jet Propulsion Laboratory. "We have underway a thorough investigation to understand this issue."

Two separate review committees have already been formed to investigate the loss of Mars Climate Orbiter: an internal JPL peer group and a special review board of JPL and outside experts. An independent NASA failure review board will be formed shortly.

"Our clear short-term goal is to maximize the likelihood of a successful landing of the Mars Polar Lander on December 3," said Weiler. "The lessons from these reviews will be applied across the board in the future."

Mars Climate Orbiter was one of a series of missions in a long-term program of Mars exploration managed by the Jet Propulsion Laboratory for NASA's Office of Space Science, Washington, DC. JPL's industrial partner is Lockheed Martin Astronautics, Denver, CO. JPL is a division of the California Institute of Technology, Pasadena, CA.

Douglas Isbell/Don Savage
Headquarters, Washington, DC
Nov. 10, 1999
(Phone: 202/358-1547)
Embargoed until 2 p.m. EST

RELEASE: 99-134

MARS CLIMATE ORBITER FAILURE BOARD RELEASES REPORT, NUMEROUS NASA ACTIONS UNDERWAY IN RESPONSE

Wide-ranging managerial and technical actions are underway at NASA's Jet Propulsion Laboratory, Pasadena, CA, in response to the loss of the Mars Climate Orbiter and the initial findings of the mission failure investigation board, whose first report was released today.

Focused on the upcoming landing of NASA's Mars Polar Lander, these actions include: a newly assigned senior management leader, freshly reviewed and augmented work plans, detailed fault tree analyses for pending mission events, daily telecons to evaluate technical progress and plan work yet to be done, increased availability of the Deep Space Network for communications with the spacecraft, and independent peer review of all operational and contingency procedures.

The board recognizes that mistakes occur on spacecraft projects, the report said. However, sufficient processes are usually in place on projects to catch these mistakes before they become critical to mission success. Unfortunately for MCO, the root cause was not caught by the processes in place in the MCO project.

"We have mobilized the very best talent at the Jet Propulsion Laboratory (JPL) to respond thoroughly to the specific recommendations in the board's report and the other areas of concern highlighted by the board," said Dr. Edward Stone, director of JPL. "Special attention is being directed at navigation and propulsion issues, and a fully independent 'red team' will review and approve the closure of all subsequent actions. We are committed to doing whatever it takes to maximize the prospects for a successful landing on Mars on Dec. 3."

The failure board's first report identifies eight contributing factors that led directly or indirectly to the loss of the spacecraft. These contributing causes include inadequate consideration of the entire mission and its post-launch operation as a total system, inconsistent communications and training within the project, and lack of complete end-to-end verification of navigation software and related computer models.

"The 'root cause' of the loss of the spacecraft was the failed translation of English units into metric units in a segment of ground-based, navigation-related mission software, as NASA has previously announced," said Arthur Stephenson, chairman of the Mars Climate Orbiter Mission Failure Investigation Board. "The failure review board has identified other significant factors that allowed this error to be born, and then let it linger and propagate to the point where it resulted in a major error in our understanding of the spacecraft's path as it approached Mars.

"Based on these findings, we have communicated a range of recommendations and associated observations to the team planning the landing of the Polar Lander, and the team has given these recommendations some serious attention," said Stephenson, director of NASA's Marshall Space Flight Center, Huntsville, AL.

The board's report cites the following contributing factors:

Errors went undetected within ground-based computer models of how small thruster firings on the spacecraft were predicted and then carried out on the spacecraft during its interplanetary trip to Mars.

The operational navigation team was not fully informed on the details of the way that Mars Climate Orbiter was pointed in space, as compared to the earlier Mars Global Surveyor mission.

A final, optional engine firing to raise the spacecraft's path relative to Mars before its arrival was considered but not performed for several interdependent reasons.

The systems engineering function within the project that is supposed to track and double-check all interconnected aspects of the mission was not robust enough, exacerbated by the first-time handover of a Mars-bound spacecraft from a group that constructed it and launched it to a new, multi-mission operations team.

Some communications channels among project engineering groups were too informal.

The small mission navigation team was oversubscribed and its work did not receive peer review by independent experts.

Personnel were not trained sufficiently in areas such as the relationship between the operation of the mission and its detailed navigational characteristics, or the process of filing formal anomaly reports.

The process to verify and validate certain engineering requirements and technical interfaces between some project groups, and between the project and its prime mission contractor, was inadequate.

The failure board will now proceed with its work on a second report due by Feb. 1, 2000, which will address broader lessons learned and recommendations to improve NASA processes to reduce the probability of similar incidents in the future.

Mars Climate Orbiter and its sister mission, the Mars Polar Lander, are part of a series of missions in a long-term program of Mars exploration managed by the Jet Propulsion Laboratory for NASA's Office of Space Science, Washington, DC. JPL's industrial partner is Lockheed Martin Astronautics, Denver, CO. JPL is a division of the California Institute of Technology, Pasadena, CA.

Mars Polar Lander/ Deep Space 2 Press Kit

NATIONAL AERONAUTICS AND SPACE ADMINISTRATION

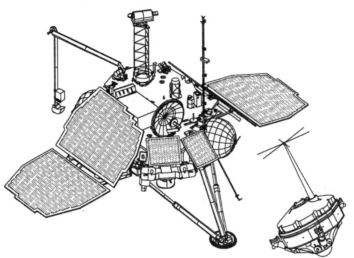

RELEASE: December 1999
Contacts

Douglas Isbell
Policy/Program Management
Headquarters, Washington, DC
202/358-1753

Franklin O'Donnell
Mars Polar Lander and Deep Space 2 Missions
Jet Propulsion Laboratory,
Pasadena, CA
818/354-5011

Mary Hardin
Mars Polar Lander and Deep Space 2 Missions
Jet Propulsion Laboratory,
Pasadena, CA
818/354-0344

Harlan Lebo/Stuart Wolpert
Mars Volatiles and Climate Science Package
University of California
Los Angeles, CA
310/206-0510

Susan Lendroth
Mars Microphone
The Planetary Society
Pasadena, CA
626/793-5100, x214

MARS POLAR LANDER, DEEP SPACE 2 SET FOR ARRIVAL

NASA returns to the surface of Mars on December 3 with a spacecraft that will land on the frigid, windswept steppe near the edge of Mars' south polar cap. Piggybacking on the lander are two small probes that will smash into the Martian surface to test new technologies.

The lander mission is the second installment in NASA's long-term program of robotic exploration of Mars, which was initiated with the 1996 launches of the currently orbiting Mars Global Surveyor and the Mars Pathfinder lander and rover, and included the recently lost Mars Climate Orbiter.

Mars Polar Lander will advance our understanding of Mars' current water resources by digging into the enigmatic layered terrain near one of its poles for the first time. Instruments on the lander will analyze surface materials, frost, weather patterns and interactions between the surface and atmosphere to better understand how the climate of Mars has changed over time.

Polar Lander carries a pair of basketball-sized microprobes that will be released as the lander approaches Mars and dive toward the planet's surface, penetrating up to about 3 feet (1 meter) underground to test 10 new technologies, including a science instrument to search for traces of water ice. The microprobe project, called Deep Space 2, is part of NASA's New Millennium Program.

A key scientific objective of the two missions is to determine how the climate of Mars has changed over time and where water, in particular, resides on Mars today. Water once flowed on Mars, but where did it go? Clues may be found in the geologic record provided by the polar layered terrain, whose alternating bands of color seem to contain different mixtures of dust and ice. Like growth rings of trees, these layered geological bands may help reveal the secret past of climate change on Mars and help determine whether it was driven by a catastrophic change, episodic variations or merely a gradual evolution in the planet's environment.

Today the Martian atmosphere is so thin and cold that it does not rain; liquid water does not last on the surface, but quickly freezes into ice or evaporates into the atmosphere. The temporary polar frosts which advance and retreat with the seasons are made mostly of condensed carbon dioxide, the major constituent of the Martian atmosphere. But the planet also hosts both water-ice clouds and dust storms, the latter ranging in scale from local to global. If typical amounts of atmospheric dust and water were concentrated today in the polar regions, they might deposit a fine layer every year so that the top yard (or meter) of the polar layered terrains could be a well-preserved record showing 100,000 years of Martian geology and climatology.

The lander and microprobes will arrive December 3, 1999. They are aimed toward a target sector within the edge of the layered terrain near Mars' south pole. The exact landing site coordinates were selected in August 1999, based on images and altimeter data from the currently orbiting Mars Global Surveyor.

Like Mars Pathfinder, Polar Lander will dive directly into the Martian atmosphere, using an aeroshell and parachute scaled down from Pathfinder's design to slow its initial descent. The smaller Polar Lander will not use airbags, but instead will rely on onboard guidance and retro-rockets to land softly on the layered terrain near the south polar cap a few weeks after the seasonal carbon dioxide frosts have disappeared. After the heat shield is jettisoned, a camera will take a series of pictures of the landing site as the spacecraft descends. These are recorded onboard and transmitted to Earth after landing.

As the lander approaches Mars about 10 minutes before touchdown, the two Deep Space 2 microprobes are released. Once released, the projectiles will collect atmospheric data before they crash at about 400 miles per hour (200 meters per second) and bury themselves beneath the Martian surface. The microprobes will test the ability of very small spacecraft to deploy future instruments for soil sampling, meteorology and seismic monitoring. A key instrument will draw a tiny soil sample into a chamber, heat it and use a miniature laser to look for signs of vaporized water ice.

About 35 miles (60 kilometers) away from the microprobe impact sites, Mars Polar Lander will dig into the top of the terrain using a 6½ foot-long (2-meter) robotic arm. A camera mounted on the robotic arm will image the walls of the trench, viewing the texture of the surface material and looking for fine-scale layering. The robotic arm will also deliver soil samples to a thermal and evolved gas analyzer, an instrument that will heat the samples to detect water and carbon dioxide. An onboard weather station will take daily readings of wind temperature and pressure, and seek traces of water vapor. A stereo imager perched atop a 5-foot (1.5 meter) mast will photograph the landscape surrounding the spacecraft. All of these instruments are part of an integrated science payload called the Mars Volatiles and Climate Surveyor.

Also onboard the lander is a light detection and ranging (lidar) experiment provided by Russia's Space Research Institute. The instrument will detect and determine the altitude of atmospheric dust hazes and ice clouds above the lander. Inside the instrument is a small microphone, furnished by the Planetary Society, Pasadena, CA, which will record the sounds of wind gusts, blowing dust and mechanical operations onboard the spacecraft itself.

The lander is expected to operate on the surface for 60 to 90 Martian days through the planet's southern summer (a Martian day is 24 hours, 37 minutes). The mission will continue until the spacecraft can no longer protect itself from the cold and dark of lengthening nights and the return of the Martian seasonal polar frosts.

Mars Polar Lander and Deep Space 2 are managed by the Jet Propulsion Laboratory for NASA's Office of Space Science, Washington, DC. Lockheed Martin Astronautics Inc., Denver, CO, is the agency's industrial partner for development and operation of the orbiter and lander spacecraft. JPL designed and built the Deep Space 2 microprobes. JPL is a division of the California Institute of Technology, Pasadena, CA.

[End of General Release]

Media Services Information

NASA Television Transmission

NASA Television is broadcast on the satellite GE-2, transponder 9C, C band, 85 degrees west longitude, frequency 3880.0 MHz, vertical polarization, audio monaural at 6.8 MHz. The schedule for Mars arrival television transmissions will be available from the Jet Propulsion Laboratory, Pasadena, CA; Johnson Space Center, Houston, TX; Kennedy Space Center FL, and NASA Headquarters, Washington, DC.

Status Reports

Status reports on mission activities will be issued by the Jet Propulsion Laboratory's Media Relations Office. They may be accessed online as noted below. Audio status reports are available by calling (800) 391-6654.

Landing Media Credentialing

The Jet Propulsion Laboratory will operate a newsroom with facilities for journalists and broadcast crews from November 29 to December 10, 1999. Requests to cover the Mars Polar Lander and Deep Space 2 arrival must be faxed in advance to the JPL newsroom at 818 354-4537. Requests must be on the letterhead of the news organization and must specify the editor making the assignment to cover the launch. Reporters may also make arrangements to cover subsequent science operations at the University of California at Los Angeles.

Briefings

An extensive schedule of news and background briefings will be held at JPL during the landing period, with later briefings originating jointly from JPL and UCLA. A schedule of briefings is available on the Internet at JPL's Mars News site (below).

(Editor's note: The following five pages are very similar to the equivalent pages in the Pathfinder Press Kit. There are, however, subtle differences.)

Internet Information

Extensive information on Mars Polar Lander and Deep Space 2, including an electronic copy of this press kit, press releases, fact sheets, status reports, briefing schedule and images, is available from the Jet Propulsion Laboratory's Mars News web site at http://www.jpl.nasa.gov/marsnews . The Mars program also maintains a home page at http://marslander.jpl.nasa.gov/.

Quick Facts

Mars Polar Lander

Dimensions: 3½ feet (1.06 meters) tall by 12 feet (3.6 meters) wide
Weight: 1,270 lbs (576 kg) total, consisting of 639-lb (290-kg) lander plus 141 lbs (64 kg) propellant,
        181-lb (82-kg) cruise stage, 309-lb (140-kg) aeroshell and heat shield

Science instruments: Mars Volatiles and Climate Surveyor (integrated package with surface imager, robotic arm with camera, meteorology package, and thermal and evolved gas analyzer); Mars Descent Imager; Lidar (including Mars microphone)

> Power: Solar panels providing 200 watts on Mars' surface
> Launch date: January 3, 1999
> Earth-Mars distance at launch: 136.6 million miles (219.9 million km)
> Earth-Mars distance at arrival: 157.2 million miles (253 million km)
> One-way speed of light time Mars to Earth on landing day: 14 minutes, 4 seconds
> Total distance traveled Earth to Mars: 470 million miles (757 million km)

Mars landing: December 3, 1999, at 12:01 p.m. Pacific Standard Time (actual event time in outer space; signals confirming flight events would be received on Earth about 14 minutes later)

Landing site: 76 degrees south latitude, 195 degrees west longitude, about 500 miles (800 km) from
        Mars' south pole
Estimated temperature at landing site: -73 F (-58 C)
Primary mission period: December 3, 1999 - March 1, 2000

Mars '98 Project

Mars Polar Lander cost: $110M spacecraft development, $10M mission operations; total $120 million
        (not including launch vehicle or Deep Space 2 microprobes)
Mars Climate Orbiter cost: $80M spacecraft development, $5M mission operations; total $85 million
        (not including launch vehicle)

Deep Space 2

Dimensions: aeroshell 11 inches (275 mm) high, 14 inches (350 mm) diameter; enclosing a forebody (penetrator) 4.2 inches (105.6 mm) long, 1.5 inches (39 mm) diameter; and an aftbody (ground station) 4.1 in (105.3 mm) high (plus 5-in (127-mm) antenna), 5.3 inches (136 mm) diameter

Weight: forebody 1.5 lbs (670 grams), aftbody 3.8 lbs (1,737 grams), aeroshell 2.6 lbs (1,165 grams); total 7.9 lbs (3,572 grams)

Power: Two lithium-thionyl chloride batteries providing 600 milliamp-hours each

Science instruments: sample collection/water detection experiment, soil thermal experiment, atmospheric descent accelerometer, impact accelerometer

Technologies: Total of 10 new technologies being flight-tested

Impact: December 3, 1999, at approximately 12:01 p.m. Pacific Standard Time (actual event time in outer space)

Estimated distance of probe impacts from Polar Lander: About 35 miles (60 km) northwest (1 degree north, 1 degree west of the lander)

Estimated distance of probe impacts from each other: Roughly 1 mile (2 km)
Duration of mission: 1 to 2 days
Cost: pre-launch development $28M, data analysis $1.6M; total $29.6 million

Mars at a Glance

General
* One of five planets known to ancients; Mars was Roman god of war, agriculture and the state
* Reddish color; occasionally the 3rd brightest object in night sky after the Moon and Venus

Physical Characteristics
* Average diameter 4,217 miles (6,780 kilometers); about half the size of Earth, but twice the size of Earth's Moon
* Same land area as Earth
* Mass 1/10th of Earth's; gravity only 38 percent as strong as Earth's
* Density 3.9 times greater than water (compared to Earth's 5.5 times greater than water)
* No planet-wide magnetic field detected; only localized ancient remnant fields in various regions

Orbit
* Fourth planet from the Sun, the next beyond Earth
* About 1.5 times farther from the Sun than Earth is
* Orbit elliptical; distance from Sun varies from a minimum of 128.4 million miles (206.7 million kilometers) to a maximum of 154.8 million miles (249.2 million kilometers); average distance from Sun, 141.5 million miles (227.7 million kilometers)
* Revolves around Sun once every 687 Earth days
* Rotation period (length of day) 24 hours, 37 min, 23 sec (1.026 Earth days)
* Poles tilted 25 degrees, creating seasons similar to Earth's

Environment
* Atmosphere composed chiefly of carbon dioxide (95.3%), nitrogen (2.7%) and argon (1.6%)
* Surface atmospheric pressure less than 1/100th that of Earth's average
* Surface winds up to 80 miles per hour (40 meters per second)
* Local, regional and global dust storms; also whirlwinds called dust devils
* Surface temperature averages -64 F (-53 C); varies from -199 F (-128 C) during polar night to 80 F (27 C) at equator during midday at closest point in orbit to Sun

Features
* Highest point is Olympus Mons, a huge shield volcano about 16 miles (26 kilometers) high and 370 miles (600 kilometers) across; has about the same area as Arizona
* Canyon system of Valles Marineris is largest and deepest known in solar system; extends more than 2,500 miles (4,000 kilometers) and has 3 to 6 miles (5 to 10 kilometers) relief from floors to tops of surrounding plateaus
* "Canals" observed by Giovanni Schiaparelli and Percival Lowell about 100 years ago were a visual illusion in which dark areas appeared connected by lines. The Mariner 9 and Viking missions of the 1970s, however, established that Mars has channels possibly cut by ancient rivers

Moons
* Two irregularly shaped moons, each only a few kilometers wide
* Larger moon named Phobos ("fear"); smaller is Deimos ("terror"), named for attributes personified in Greek mythology as sons of the god of war

Historical Mars Missions

Mission, Country, Launch Date, Purpose, Results

[Unnamed], USSR, 10/10/60, Mars flyby, did not reach Earth orbit
[Unnamed], USSR, 10/14/60, Mars flyby, did not reach Earth orbit
[Unnamed], USSR, 10/24/62, Mars flyby, achieved Earth orbit only
Mars 1, USSR, 11/1/62, Mars flyby, radio failed at 65.9 million miles (106 million km)
[Unnamed], USSR, 11/4/62, Mars flyby, achieved Earth orbit only
Mariner 3, U.S., 11/5/64, Mars flyby, shroud failed to jettison
Mariner 4, U.S. 11/28/64, first successful Mars flyby 7/14/65, returned 21 photos
Zond 2, USSR, 11/30/64, Mars flyby, passed Mars but radio failed, returned no planetary data
Mariner 6, U.S., 2/24/69, Mars flyby 7/31/69, returned 75 photos
Mariner 7, U.S., 3/27/69, Mars flyby 8/5/69, returned 126 photos
Mariner 8, U.S., 5/8/71, Mars orbiter, failed during launch
Kosmos 419, USSR, 5/10/71, Mars lander, achieved Earth orbit only
Mars 2, USSR, 5/19/71, Mars orbiter/lander arrived 11/27/71, no useful data, lander destroyed
Mars 3, USSR, 5/28/71, Mars orbiter/lander, arrived 12/3/71, some data and few photos
Mariner 9, U.S., 5/30/71, Mars orbiter, in orbit 11/13/71 to 10/27/72, returned 7,329 photos
Mars 4, USSR, 7/21/73, failed Mars orbiter, flew past Mars 2/10/74
Mars 5, USSR, 7/25/73, Mars orbiter, arrived 2/12/74, lasted a few days
Mars 6, USSR, 8/5/73, Mars orbiter/lander, arrived 3/12/74, little data return
Mars 7, USSR, 8/9/73, Mars orbiter/lander, arrived 3/9/74, little data return
Viking 1, U.S., 8/20/75, Mars orbiter/lander, orbit 6/19/76-1980, lander 7/20/76-1982
Viking 2, U.S., 9/9/75, Mars orbiter/lander, orbit 8/7/76-1987, lander 9/3/76-1980; combined, the Viking orbiters and landers returned 50,000+ photos
Phobos 1, USSR, 7/7/88, Mars/Phobos orbiter/lander, lost 8/89 en route to Mars
Phobos 2, USSR, 7/12/88, Mars/Phobos orbiter/lander, lost 3/89 near Phobos
Mars Observer, U.S., 9/25/92, lost just before Mars arrival 8/21/93
Mars Global Surveyor, U.S., 11/7/96, Mars orbiter, arrived 9/12/97, currently conducting prime mission of science mapping
Mars 96, Russia, 11/16/96, orbiter and landers, launch vehicle failed
Mars Pathfinder, U.S., 12/4/96, Mars lander and rover, landed 7/4/97, last transmission 9/27/97
Nozomi (Planet-B), Japan, 7/4/98, Mars orbiter, currently in orbit around the Sun; Mars arrival delayed to 12/03 due to propulsion problem
Mars Climate Orbiter, U.S., 12/11/98; lost on arrival at Mars 9/23/99
Mars Polar Lander/Deep Space 2, U.S., 1/3/99, lander/descent probes; lander to set down near south pole 12/3/99; Deep Space 2 microprobes to smash into surface the same day

Mars, Water and Life

The planet Mars landed in the middle of immense public attention on July 4, 1997, when Mars Pathfinder touched down on a windswept, rock-laden ancient flood plain. Two months later, Mars Global Surveyor went into orbit, sending back pictures of towering volcanoes and gaping chasms at resolutions never before seen.

In December 1999, another lander will arrive at Mars. And every 26 months over the next decade, when the alignment of Earth and Mars are suitable for launches, still more robotic spacecraft will join them at the red planet.

These spacecraft carry varied payloads, ranging from cameras and other sensors to rovers and robotic arms. Some of them have their roots in different NASA programs of science or technology development. But they all have the goal of understanding Mars better, primarily by delving into its geology, climate and history.

With the announcement in 1996 by a team of scientists that a meteorite believed to have come from Mars contained what might be the residue of ancient microbes, public interest became regalvanized by the possibility of past or present life there. The key to understanding whether life could have evolved on Mars, many scientists believe, is understanding the history of water on the planet.

Mars and Life

Mars perhaps first caught public fancy in the late 1870s, when Italian astronomer Giovanni Schiaparelli reported using a telescope to observe "canali," or channels, on Mars. A possible mistranslation of this word as "canals" may have fired the imagination of Percival Lowell, an American businessman with an interest in astronomy. Lowell founded an observatory in Arizona, where his observations of the red planet convinced him that the canals were dug by intelligent beings - a view which he energetically promoted for many years.

By the turn of the century, popular songs told of sending messages between Earth and Mars by way of huge signal mirrors. On the dark side, H.G. Wells' 1898 novel "The War of the Worlds" portrayed an invasion of Earth by technologically superior Martians desperate for water. In the early 1900s novelist Edgar Rice Burroughs, known for the "Tarzan" series, also entertained young readers with tales of adventures among the exotic inhabitants of Mars, which he called Barsoom.

Fact began to turn against such imaginings when the first robotic spacecraft were sent to Mars in the 1960s. Pictures from the first flyby and orbiter missions showed a desolate world, pockmarked with craters like Earth's Moon. The first wave of Mars exploration culminated in the Viking mission, which sent two orbiters and two landers to the planet in 1975. The landers included experiments that conducted chemical tests in search of life. Most scientists interpreted the results of these tests as negative, deflating hopes of a world where life is widespread.

The science community had many other reasons for being interested in Mars apart from searching for life; the next mission on the drawing boards, Mars Observer; concentrated on a study of the planet's geology and climate. Over the next 20 years, however new developments in studies on Earth came to change the way that scientists thought about life and Mars.

One was the 1996 announcement by a team from Stanford University, NASA's Johnson Space Center and Quebec's McGill University that a meteorite believed to have originated on Mars contained what might be the fossils of ancient microbes. This rock and other so-called Mars meteorites discovered on several continents on Earth are believed to have been blasted away from the red planet by asteroid or meteor impacts. They are thought to come from Mars because gases trapped in some of the rocks match the composition of Mars' atmosphere. Not all scientists agreed with the conclusions of the team announcing the discovery of fossils, but it reopened the issue of life on Mars.

Other developments that shaped scientists' thinking included new research on how and where life thrives on Earth. The fundamental requirements for life as we know it are liquid water, organic compounds and an energy source for synthesizing complex organic molecules. Beyond these basics, we do not yet understand the environmental and chemical evolution that leads to the origin of life. But in recent years it has become increasingly clear that life can thrive in settings much different from the long held notion of a tropical soup rich in organic nutrients.

In the 1980s and 1990s, biologists found that microbial life has an amazing flexibility for surviving in extreme environments - niches that by turn are extraordinarily hot, or cold, or dry, or under immense pressures - that would be completely inhospitable to humans or complex animals. Some scientists even concluded that life may have begun on Earth in heat vents far under the ocean's surface.

This in turn had its effect on how scientists thought about Mars. Life might not be so widespread that it would be found at the foot of a lander spacecraft, but it may have thrived billions of years ago in an underground thermal spring. Or it might still exist in some form in niches below the frigid, dry, windswept surface wherever there might be liquid water.

NASA scientists also began to rethink how to look for signs of past or current life on Mars. In this new view, the markers of life may well be so subtle that the range of test equipment required to detect it would be far too complicated to package onto a spacecraft. It made more sense to collect samples of Martian rock, soil and air to bring back to Earth, where they could be subjected to much more extensive laboratory testing with state-of-the-art equipment.

Mars and Water

Mars today is too cold, with an atmosphere that is too thin, to support liquid water on its surface. Yet scientists who studied images from the Viking orbiters kept encountering features that appeared to be formed by flowing water - among them deep channels and canyons, and even features that appeared to be ancient lake shorelines. Added to this were more recent observations by Mars Pathfinder and Mars Global Surveyor which suggested widespread flowing water in the planet's past. Some scientists identified features which they believe appear to be carved by torrents of water with the force of 10,000 Mississippi Rivers.

There is no general agreement, however, on what form water took on the early Mars. Two competing views are currently popular in the science community. According to one theory, Mars was once much warmer and wetter, with a thicker atmosphere; it may well have boasted lakes or oceans, rivers and rain. According to the other theory, Mars was always cold, but water trapped as underground ice was periodically released when heating caused ice to melt and gush forth onto the surface.

In either case, the question of what happened to the water remains a mystery. Most scientists do not feel that Mars' climate change was necessarily caused by a cataclysmic event such as an asteroid impact that, perhaps, disturbed the planet's polar orientation or orbit. Many believe that the demise of flowing water on the surface could have resulted from gradual climate change over many millennia as the planet lost its atmosphere.

Under either the warmer-and-wetter or the always-cold scenario, Mars must have had a thicker atmosphere in order to support water that flowed on the surface even only occasionally If the planet's atmosphere became thinner, liquid water would rapidly evaporate. Over time, carbon dioxide gas reacts with elements in rocks and becomes locked up as a kind of compound called a carbonate. What's left of Mars' atmosphere today is overwhelmingly carbon dioxide.

On Earth, shifting tectonic plates are continually plowing carbonates and other minerals under the surface; heated by magmas, carbon dioxide is released and spews forth in volcanic eruptions, replenishing the carbon dioxide in the atmosphere. Although Mars has no known active volcanoes and there are no signs of fresh lava flows, it had abundant volcanic activity in its past. However, Mars appears to have no tectonic plates, so a critical link in the process that leads to carbon dioxide replenishment in Earth's atmosphere is missing. In short, Mars' atmosphere could have been thinned out over many eons by entrapment of carbon dioxide in rocks across its surface.

That scenario, however, is just a theory. Regardless of the history and fate of the atmosphere, scientists also do not understand what happened to Mars' water. Some undoubtedly must have been lost to space. Water ice has been detected in the permanent cap at Mars' north pole, and may exist in the cap at the south pole. But much water is probably trapped under the surface - either as ice or, if near a heat source, possibly in liquid form well below the surface.

Polar Lander and Future Missions

Mars Polar Lander is designed to help scientists better understand the climate history of Mars, not to look for life. It does not, for example, contain any biology experiments similar to the chemistry lab on the Viking landers. However, its focus on Mars' climate and the role of water will have an impact on the life question. Water is also important as a resource for eventual human expeditions to the red planet.

In addition, the currently orbiting Mars Global Surveyor will aid the search for likely sites for future Mars robotic landers. Scientists are interested in three types of Martian environments which are potentially most favorable to the emergence and persistence of life. They are:

* **Ancient groundwater environments.** Early in the planet's history, liquid water appears to have been widespread beneath the surface. During the final stages of planetary formation, intense energy was dissipated by meteor impacts. This, along with active volcanoes, could have created warm groundwater circulation systems favorable for the origin of life.

* **Ancient surface water environments.** Also during early Martian history, water was apparently released from subsurface aquifers, flowed across the surface and pooled in low-lying regions. Evidence of the early climate of Mars and of ancient life, if any may be preserved in sedimentary rocks in these environments.

* **Modern groundwater environments.** Life may have formed at any time, including recently, in habitats where subsurface water or ice is geothermally heated to create warm groundwater circulation systems. In addition, life may have survived from an early epoch in places beneath the surface where liquid water is present today.

These Martian environments can be investigated in several ways. We can get a glimpse of underground environments by using rovers to explore young craters and what appear to be the remains of water-eroded channels, and by drilling from lander spacecraft. Sensors on orbiters will search for the most likely reservoirs of water in these regions.

To investigate these scientific themes, NASA's Mars program will carry out the following implementation strategy for the initial phases of Mars exploration:

*The Mars orbiters in 1996 and 2001 will provide sufficient information to guide an early sample return from an ancient groundwater environment.

* Ancient surface-water environments will be explored in greater depth. When a sample-return mission is sent to Mars, it is extremely important to be able to identify minerals formed by water.

* Ancient and modern sites exhibiting evidence of hydrothermal activity will be studied, followed eventually by efforts to drill as deeply as possible below the surface.

Samples will be collected using rovers capable of extensive searches and of collecting and storing samples of rock and soil. Sophisticated sensors onboard rovers will help insure that diverse rock types are collected. Drills capable of reaching several yards (or meters) below the surface will also be used to analyze subsurface material. It is likely that it will be some time before space technologies will be able to drill to depths of a half-mile (kilometer) and more to access subsurface water.

Samples of the Martian atmosphere will also be brought back to Earth. The possible origin and evolution of life on Mars must be linked to the evolution of its atmosphere.

In 2003 NASA and its international partners will see the first launch of a mission to collect samples and place them in Mars orbit to await their transport back to Earth. A mission in 2005 will include two spacecraft — a lander like the 2003 mission to collect surface samples, and a French-built orbiter to return both the 2003 and 2005 samples to Earth. A series of at least three sample return missions similar to this are expected to be carried out over the following decade.

Even if it turns out that Mars never harbored life, study of the planet can help in understanding life on our own. Much of the evidence for the origin of life on Earth has been erased by movement of the planet's crust and by weathering. Fortunately, large areas of Mars' surface date back to the very earliest period of planetary evolution — about 4 billion years ago, overlapping the period on Earth when pre-biotic chemical evolution first gave rise to life. Thus, even if life never developed on Mars, studies of the planet may yield crucial information about the prebiotic chemistry that led to life on Earth.

NASA Programs

Although they are targeted at the same planet, some Mars missions have their roots in different NASA programs. The following are the programs responsible for U.S. Mars missions in the present and recent past:

* Mars Surveyor Program. In 1994, NASA created a program to send spacecraft to the planet during each launch opportunity every 26 months over the next decade. The first spacecraft under the program, Mars Global Surveyor, was launched in 1996 and is currently in orbit at the planet. The Mars Surveyor Program missions in 1998 were Mars Polar Lander and the recently lost Mars Climate Orbiter, collectively known as the Mars Surveyor '98 project. All of NASA's Mars missions now planned for the first decade of the next century also fall under this programmatic umbrella. In order to save costs, a single industrial partner, Lockheed Martin Astronautics, was chosen to build and operate all of the Mars Surveyor spacecraft over several years. In addition, an ongoing project office called the Mars Surveyor Operations Project was created at JPL, consolidating management of mission operations across the multi-year program.

* Discovery Program. Created in 1992, NASA's Discovery Program competitively selects proposals for low-cost solar system exploration missions with highly focused science goals. Mars Pathfinder was the second mission approved and launched under the Discovery Program. Originally conceived as an engineering demonstration of a way to deliver a spacecraft to the surface of Mars with a novel approach using airbags to land, the mission evolved to include a science payload focused primarily on geology

* New Millennium Program. Technology, rather than science, is at the center of NASA's New Millennium Program, created in 1994. The goal of the program is to identify and flight-test new technologies that will enable science missions of the early 21st century. Teams are formed with partners from government, private industry, academia and the nonprofit sector to develop promising technologies in spacecraft autonomy, telecommunications, microelectronics, science instruments and mechanical systems. Deep Space 2, the project that is sending two microprobes to piggyback on Mars Polar Lander, is the second mission under New Millennium. In addition to "Deep Space" missions to the solar system, a series of "Earth Orbiter' missions is also planned to test new technologies for Earth-observing spacecraft. These missions may also collect science data, but technology is always at the forefront.

Mars Pathfinder Science Highlights

Launched December 4, 1996, Mars Pathfinder landed July 4, 1997, in Ares Vallis, an ancient flood plain in Mars' northern hemisphere. The spacecraft deployed a small robotic rover named Sojourner to study rocks at the landing site. Key science findings included:

* Chemical analyses returned by Mars Pathfinder indicate that some rocks at the landing site appear to be high in silica, suggesting differentiated parent materials. These rocks are distinct from the meteorites found on Earth that are thought to be of Martian origin.

* The identification of rounded pebbles and cobbles on the ground, and sockets and pebbles in some rocks, suggests conglomerates that formed in running water, during a warmer past in which liquid water was stable.

* Some rocks at the landing site appear grooved and fluted, suggesting abrasion by sand-sized particles. Dune-shaped deposits were also found in a trough behind the area of the landing site known as the Rock Garden, indicating the presence of sand.

* The soil chemistry of the landing site appears to be similar to that of the Viking I and 2 landing sites, suggesting that the soil may be a globally deposited unit.

* Radio tracking of Mars Pathfinder indicates that the radius of the planet's central metallic core is greater than 800 miles (1,300 kilometers) but less than roughly 1,250 miles (2,000 kilometers).

* Airborne dust is magnetic with each particle about I micron in size. Interpretations suggest the magnetic mineral is maghemite, a very magnetic form of iron oxide, which may have been freeze-dried on the particles as a stain or cement. The iron may have been leached out of materials in the planet's crust by an active water cycle.

* Whirlwinds called dust devils were frequently measured by temperature, wind and pressure sensors, suggesting that these gusts are a mechanism for mixing dust into the atmosphere.

* Imaging revealed early morning water ice clouds in the lower atmosphere, which evaporate as the atmosphere warms.

* Abrupt temperature fluctuations were recorded in the morning, suggesting that the atmosphere is warmed by the planet's surface, with heat convected upwards in small eddies.

* The weather was similar to weather encountered by Viking I; there were rapid pressure and temperature variations, down-slope winds at night and light winds in general. Temperatures at the surface were about 18 F (10 C) warmer than those measured by Viking I.

* The atmosphere was a pale pink color due to fine dust mixed in the lower atmosphere, as was seen by Viking. Particle size and shape estimates and the amount of water vapor in the atmosphere are also similar to Viking observations.

Mars Global Surveyor Science Highlights

Launched November 7, 1996, Mars Global Surveyor entered orbit around Mars on September 12, 1997, and achieved its final orbit by skimming through the planet's upper atmosphere in a technique called aerobraking. The prime science mapping mission began in spring 1999, although the spacecraft had collected much science data during the aero braking period. Key findings include:

* The planet's magnetic field is not globally generated in the planet's core, but is localized in particular areas of the crust. Multiple magnetic anomalies were detected at various points on the planet's surface, indicating that magma solidified in an ancient magnetic field as it came up through the crust and cooled very early in Mars' evolution.

* Mars' very localized magnetic field also creates a new paradigm for the way in which it interacts with the solar wind, one that is not found with other planets. While Earth, Jupiter and other planets have large magnetospheres, and planets like Venus have strong ionospheres, Mars' small, localized magnetic fields are likely to produce a much more complicated interaction process as these fields move with the planet's rotation.

* New temperature data and close-up images of the Martian moon Phobos show its surface is composed of powdery material at least 3 feet (1 meter) thick, caused by millions of years of meteoroid impacts. Measurements of the day and night sides of Phobos show extreme temperature variations on the sunlit and dark sides of the moon. Highs were measured at 25 F (-4 C) and lows registered at -170 F (-112 C).

* New images suggest that some areas previously thought to be shorelines of ancient lakes or oceans do not show landforms expected at shorelines. Some areas of the northern hemisphere are the flattest terrain yet observed in the solar system, with elevations that vary only a few feet over many miles, suggesting formation by a sedimentary process.

* An area near the Martian equator was found to have an accumulation of the mineral hematite, consisting of tiny crystallized grains of iron oxide that on Earth typically originate in standing bodies of water. The material has been previously detected on Mars in more dispersed concentrations, and is widely thought to be an important component of the materials that give Mars its red color.

* The spacecraft's altimeter has given scientists their first global three-dimensional map of the Martian surface. The instrument's profiles of the northern ice cap show an often striking surface topology of canyons and spiral troughs in the polar cap, made of water ice and carbon dioxide ice, that can reach depths as great as 3,600 feet (1 kilometer) below the surface. Many of the larger and deeper troughs display a staircase structure, which may ultimately be correlated with seasonal layering of ice and dust that was observed by NASA's Viking mission orbiters in the late 1970s. At 86.3 degrees north, the highest latitude yet sampled, the cap achieves an elevation of 6,600 to 7,900 feet (1.25 to 1.5 miles, or 2 to 2.5 kilometers) over the surrounding terrain.

* Images reveal much more layering than previously known in the terrain of the south polar region, where Mars Polar Lander will land on December 3, 1999. Pictures show swirling bands of eroded, layered rock, reminiscent of the edges of Alaskan ice sheets, and an array of light and dark mottled patterns blanketing the frigid floor of the south pole. They reveal that Mars Polar Lander's target landing zone at about 76 degrees south latitude is more rugged and geologically diverse than scientists had previously thought.

Where to Next?

In 1998, NASA conducted a review of its multi-year Mars program, calling in outside experts to help evaluate and refine the architecture for the direction of the effort. The review resulted in a new approach to achieve a series of sample-return missions beginning early in the next decade. Details are subject to change as plans evolve, but the following are the basic missions foreseen:

* 2001: NASA will launch a lander and orbiter broadly similar to Mars Polar Lander and Mars Climate Orbiter. The orbiter, equipped with three science instruments, will be the first planetary spacecraft to be launched from the west coast of the United States. The lander will touch down near Mars' equator, carrying a spare Mars Pathfinder rover, a robotic arm and several other science instruments, including three that will return data in support of eventual human exploration.

* 2003: NASA will begin the series of sample-return missions with launch of a lander and rover that will search for soil and rock samples and return them to a mini-ascent vehicle that will loft the samples to Mars orbit. The European Space Agency (ESA), in collaboration with the Italian space agency (ASI), will provide an orbiter. ASI may also provide a drill and other robotic elements for landers beginning in this year. NASA's Human Exploration and Development of Space (HEDS) enterprise may also begin providing science and technology experiments for the large lander. As early as this year, NASA and the French space agency, CNES, may also launch a "micromission" spacecraft designed to deliver payloads such as communications/navigation relays, penetrators, small landers or exploration balloons.

* 2005: NASA and CNES will launch an orbiter, lander and rover on a French-provided Ariane 5 rocket. The lander and rover will search for soil and rock samples, and return them to a mini-ascent vehicle that will loft the samples to Mars orbit. CNES will provide the orbiter spacecraft that will retrieve the 2003 and 2005 samples from Mars orbit and return them to Earth in a vehicle provided by NASA. The samples will reach Earth in 2008. The lander may also carry an ASI-provided drill and additional payloads, perhaps contributed by ASI, CNES, HEDS or other partners. The CNES spacecraft will also carry four miniature landers called "netlanders." In addition, NASA and CNES may collaborate on two "micromissions."

* 2007-2009: As currently envisioned, NASA's strategy will be to continue to collect samples from different types of Martian terrain and place them in Mars orbit for later retrieval. In 2007, NASA would launch another lander, rover and ascent vehicle. NASA and CNES may collaborate on two more "micromissions," and HEDS may provide more experiments. In 2009, NASA would again launch a lander, rover and ascent vehicle. Current plans call for another orbiter provided by CNES that will retrieve the 2007 and 2009 samples from Mars

orbit and return them to Earth in a vehicle provided by NASA.

* 2011-2013: The plans for these years are being formulated. They may include "robotic outposts" interacting robotic units undertaking complex science investigations.

Mars Polar Lander

Mars Polar Lander will settle onto the surface of the red planet, much as Mars Pathfinder did in 1997. But instead of inflating airbags to bounce on the surface as it lands, Mars Polar Lander will use retro-rockets to slow its descent, like the Viking landers of the 1970s. Instead of a rover, Mars Polar Lander is equipped with a robotic arm that will dig into the soil near the planet's south pole in search of subsurface water and fine-scale layering that may physically record past changes in climate.

The lander will also conduct experiments on soil samples acquired by the robotic arm and dumped into small ovens, where the samples will be heated to drive off water and carbon dioxide. Surface temperatures, winds, pressure and the amount of dust in the atmosphere will be measured on a daily basis, while a small microphone records the sounds of wind gusts or mechanical operations onboard the spacecraft.

Mission Overview

Launch. Mars Polar Lander was launched January 3, 1999, at 3:21 p.m. Eastern Standard Time on a Delta II rocket from Space Launch Complex 17B at Cape Canaveral Air Station, FL. The Delta II was a model 7425 with two liquid-fuel stages augmented by four strap-on solid-fuel boosters, and a third-stage Thiokol Star 48B solid-fuel booster.

At the time of launch, the lander was encased within an aeroshell attached to a round platform called the cruise stage. Because the lander's solar panels are folded up within the aeroshell, a second set of solar panels is located on the cruise stage to power the spacecraft during its interplanetary cruise. Shortly after launch, these hinged solar panels unfolded and the spacecraft fired its thrusters to orient the solar panels toward the Sun. Fifty-eight minutes after launch, the 112-foot-diameter (34-meter) antenna at the Deep Space Network complex in Canberra, Australia, acquired Polar Lander's signal.

Interplanetary cruise. By the time it reaches Mars on December 3, 1999, Polar Lander will have spent 11 months in cruise. The spacecraft's flight path is called a Type 2 trajectory because it has taken the lander more than 180 degrees around the Sun, enabling it to target a landing zone near Mars' south pole. By comparison, Mars Pathfinder followed a Type 1 trajectory which took it less than 180 degrees around the Sun, reaching Mars in only seven months. During the first leg of its trip, Mars Polar Lander flew slightly inward toward the Sun before spiraling out beyond Earth's orbit to Mars. Toward the end of cruise, it has been flying slightly out past the orbit of Mars before returning inward to intersect the planet's orbit.

Throughout cruise, the spacecraft has been communicating with Earth using its X-band transmitter and the medium-gain horn antenna on the cruise stage. During the first 30 days after launch, the spacecraft was tracked from 10 to 12 hours per day. During quiet phases of the flight, when spacecraft operations are at a minimum, one four-hour tracking session per day has been conducted.

Forty-five days before Mars arrival, tracking will be increased. At least three four-hour sessions per day are required for high-precision navigation, with continuous tracking when possible. Starting 30 days before arrival, nearly continuous tracking sessions have switched between Mars Polar Lander and the currently orbiting Mars Global Surveyor in order to fine tune the lander's final approach to Mars.

During interplanetary cruise, Polar Lander was scheduled to fire its thrusters in up to six maneuvers to adjust its flight path. The first of these trajectory correction maneuvers was carried out January 21, 1999. This maneuver, which lasted 3 minutes, removed a small bias in the lander's trajectory that was introduced at launch to send the third stage of the Delta II rocket, which was trailing behind the spacecraft, past Mars rather than directly toward the planet. The maneuver also corrected minor injection errors caused by the spacecraft's liftoff. In the second trajectory maneuver, carried out March 15, 1999, the spacecraft's thrusters fired for about 10 seconds.

The next maneuver, which took place September 1, 1999, was called a site adjustment maneuver. Designed to fine-tune the spacecraft's landing site based on recent images of the Mars south pole area from the currently orbiting Mars Global Surveyor; this required the thrusters to fire for 30 seconds.

The next maneuver was carried out October 30, 1999, when the thrusters fired for 12 seconds. Another trajectory maneuver is scheduled at about 10 a.m. Pacific Standard Time on Tuesday, November 30, 1999. An optional sixth maneuver is tentatively scheduled to take place if required at about 5:30 a.m. landing day — Friday, December 3, 1999.

(For bookkeeping purposes, the Polar Lander project refers to these thruster firings as follows: January 21, trajectory correction maneuver #1; March 15, trajectory correction maneuver #2; September 1, site adjustment maneuver; October 30, trajectory correction maneuver #3; November 30, trajectory correction maneuver #4; and December 3, (optional) trajectory correction maneuver #5.)

Science instruments were tested and calibrated during two week-long periods during cruise. Five of the instruments were checked out the week of April 5-9, 1999. The second period of tests took place September 8-9, 1999.

Science instrument checkout data and spacecraft engineering data gathered during the cruise was transmitted to Earth via the Deep Space Network's 230-foot-diameter (70-meter) antennas. Use of these large dish antennas allowed ground controllers to receive data at higher data rates than possible with the smaller 112-foot (34-meter) antennas.

The meteorological package's pressure transducer was powered on for a few minutes each month during cruise for calibration. The surface stereo imager twice took images of dark space inside the lander's aeroshell during cruise to calibrate its charge-coupled device (CCD) detectors.

Pre-entry events. Preparations for the lander's entry into the Martian atmosphere will begin 14 hours in advance, when the final tracking coverage of the cruise period begins. This will be the final opportunity for ground controllers to gather navigation data before entry. About 18 hours before entry, software which normally puts the spacecraft in safe mode in reaction to unexpected events will be disabled for the remainder of the spacecraft's flight and descent to the surface.

An opportunity to transmit commands for thruster firings to fine-tune the flight path occurs between nine and seven hours before entry. If a final trajectory correction maneuver is required, computer commands for that thruster firing could be sent to the spacecraft during this window. The maneuver would be executed at 6½ hours before entry.

Starting about five hours before entry, heaters on the lander's thrusters will be turned on. A one-hour tracking session will begin 75 minutes before entry. This session will be used to monitor spacecraft health and status and perform tracking after the final thruster firing. A pyro valve will fire at 40 minutes before entry to pressurize the descent engines. Fifteen minutes before entry, software controlling the Mars Descent Imager will be initialized.

About 10 minutes before entry, the spacecraft will be commanded to switch to inertial navigation — computing its position, course and speed from gyroscopes and accelerometers. Six minutes before entry, the spacecraft will begin pulsing its thrusters for 80 seconds to turn it to its entry orientation. Five minutes before entry and 10½ minutes before landing, the cruise stage will separate from the aeroshell-encased lander. Cut off from the cruise stage's solar panels, the lander will rely on its internal battery until it can unfold its own solar panels on the planet's surface. The Deep Space 2 microprobes, piggybacking on the lander's cruise stage, will be jettisoned about 18 seconds later. The lander will then be commanded to assume the correct orientation for atmospheric entry.

Entry, descent and landing. Traveling at about 15,400 miles per hour (6.9 kilometers per second), the spacecraft will enter the upper fringe of Mars' atmosphere some 33 to 37 seconds later. Onboard accelerometers, sensitive enough to detect "G" forces as little as 3/100ths of Earth's gravity, will sense when friction from the atmosphere causes the lander to slow slightly. At this point, the lander will begin using its thrusters to keep the entry capsule aligned with its direction of travel.

The spacecraft's descent from the time it hits the upper atmosphere until it lands takes about five minutes and 30 seconds to accomplish. As it descends, the spacecraft will experience G forces up to 12 times Earth's gravity, while the temperature of its heat shield rises to 3000 F (1650 C).

About two minutes before landing, the lander's parachute will be fired from a mortar (or small cannon) when the spacecraft is moving at about 960 miles per hour (430 meters per second) some 4.5 miles (7.3 kilometers) above the Martian surface. Ten seconds after the parachute opens, the Mars Descent Imager will be powered on and the spacecraft's heat shield will be jettisoned. The first descent image will be taken 0.3 seconds before heat-shield separation. The imager will take a total of about 30 pictures during the spacecraft's descent to the surface.

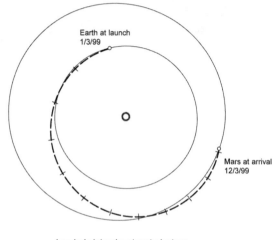

Earth at launch
1/3/99

Mars at arrival
12/3/99

Lander's interplanetary trajectory

About 70 to 100 seconds before landing, the lander legs will be deployed; 1.5 seconds after that, the landing radar will be activated. The radar will be able to gauge the spacecraft's altitude about 44 seconds after it is turned on, at an altitude of about 1.5 miles (2.5 kilometers) above the surface.

Shortly after radar ground acquisition, when the spacecraft is traveling at about 170 miles per hour (75 meters per second) some 4,600 feet (1.4 kilometers) above the surface, the thrusters that the spacecraft has used for maneuvers throughout its cruise will be turned of and the backshell will separate from the lander. The descent engines will be turned on one-half second later, turning the lander so its flight path gradually becomes vertical.

The pulse-modulated descent engines will maintain the spacecraft's orientation as it descends. The engines will fire to roll the lander to its proper orientation so that it lands with the solar panels in the best orientation to generate power as the Sun moves across the sky. The radar will be turned off at an altitude of about 130 feet (40 meters) above the surface, and the spacecraft continues using its gyros and accelerometers for inertial guidance as it lands.

Once the spacecraft reaches either an altitude of 40 feet (12 meters) or a velocity of 5.4 miles per hour (2.4 meters per second), the lander will drop straight down at a constant speed. The descent engines will be turned off when touchdown is detected by sensors in the footpads. The engines will have been on for a total of about 40 seconds during final descent to the surface.

Post-landing. The lander is expected to touch down at 4:20 a.m. local time at the Mars landing site (12:01 p.m. Pacific Standard Time). (Because radio signals take 14 minutes to travel from Mars to Earth, during the landing the mission team will be watching events in "Earth-received time," with landing noted at 12:15 p.m.. To avoid confusion, all subsequent times of mission events discussed below are stated in "Earth-received time," when a signal would be received on Earth. Actual events will have taken place on the spacecraft about 14 minutes earlier in each case.)

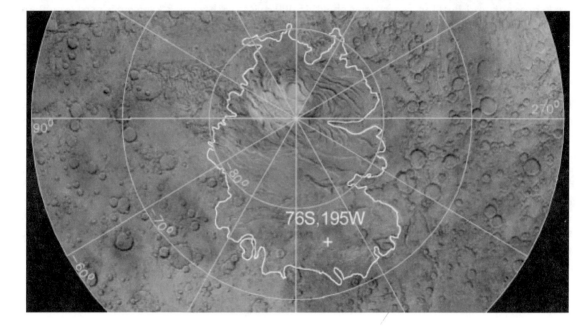

76S,195W
+

Lander's target landing site

Shortly after landing, software which puts the spacecraft in safe mode in reaction to unexpected events will be reenabled. The descent imager will be turned off 60 seconds after landing.

After waiting five minutes to allow for dust kicked up by the landing to settle, the lander's solar arrays will be unfolded. Eight minutes after landing, while the medium-gain antenna is being turned to point at Earth, the spacecraft's gyros will be used like compasses to determine which way is north. The spacecraft's inertial measurement units will then be powered off.

After gyrocompassing is completed, the medium-gain antenna will turn to point to Earth. This antenna slew may take up to 16 minutes to complete. A vertical scan will then be taken by the surface stereo imager before its boom is deployed. Both the meteorological and imager masts will then be raised.

First signal. The first opportunity to hear from the lander will take place when it begins transmitting directly to Earth using its dish-shaped medium-gain antenna about 23 minutes after landing; this signal is expected to be received at 12:39 p.m. PST. This transmission session will end 45 minutes later, or 1:24 p.m. PST, and would include engineering data on the lander's entry, descent and landing, as well as a possible low-resolution black-and-white picture from the undeployed camera on the lander's deck. At 1:46 p.m. PST, the lander shuts down and "sleeps" for four hours and 40 minutes while its solar panels recharge its onboard battery. No wakeups occur during this time.

Assuming that all is normal with the spacecraft, it will power up again at 6:26 p.m. PST and turn on its receiver. At this time, mission controllers expect that they would send the lander commands such as what data rate to use for later radio transmissions. The receiver will continue listening for commands from Earth until 7:41 p.m. PST.

At 8:09 p.m., the lander will begin transmitting to Earth until 10:45 p.m.. After that session concludes, the lander will run through a sequence that takes about half an hour as it prepares to shut down for the night. At 11:24 p.m. PST the lander will power down.

Nighttime activities. The lander's onboard computer and meteorological package will be activated several times during the Martian night. Because of the extreme latitude during the Martian southern summer, the

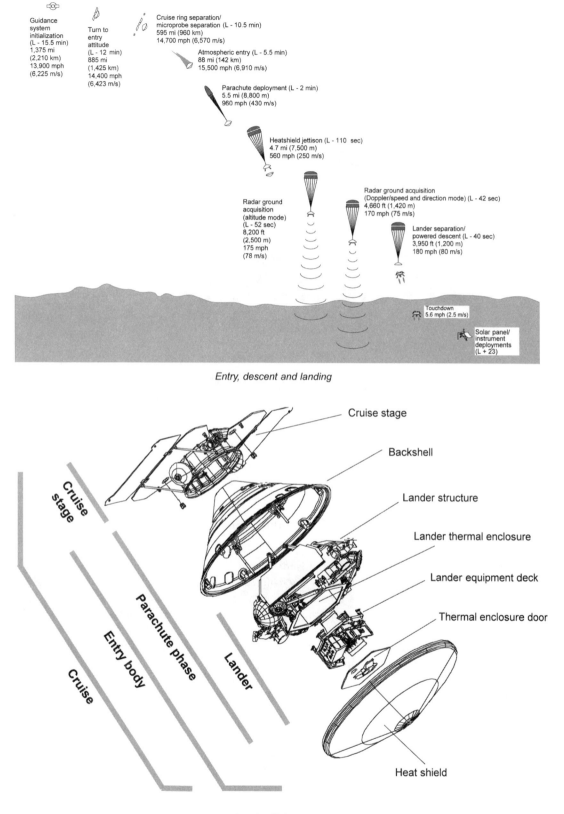

Guidance
system
initialization
(L - 15.5 min)
1,375 mi
(2,210 km)
13,900 mph
(6,225 m/s)

Turn to
entry
attitude
(L - 12 min)
885 mi
(1,425 km)
14,400 mph
(6,423 m/s)

Cruise ring separation/
microprobe separation (L - 10.5 min)
595 mi (960 km)
14,700 mph (6,570 m/s)

Atmospheric entry (L - 5.5 min)
88 mi (142 km)
15,500 mph (6,910 m/s)

Parachute deployment (L - 2 min)
5.5 mi (8,800 m)
960 mph (430 m/s)

Heatshield jettison (L - 110 sec)
4.7 mi (7,500 m)
560 mph (250 m/s)

Radar ground acquisition
(Doppler/speed and direction mode) (L - 42 sec)
4,660 ft (1,420 m)
170 mph (75 m/s)

Radar ground
acquisition
(altitude mode)
(L - 52 sec)
8,200 ft
(2,500 m)
175 mph
(78 m/s)

Lander separation/
powered descent (L - 40 sec)
3,950 ft (1,200 m)
180 mph (80 m/s)

Touchdown
5.6 mph (2.5 m/s)

Solar panel/
instrument
deployments
(L + 23)

Entry, descent and landing

Cruise stage

Backshell

Lander structure

Lander thermal enclosure

Lander equipment deck

Thermal enclosure door

Heat shield

Cruise
stage

Parachute phase

Entry body

Cruise

Lander

Lander flight system

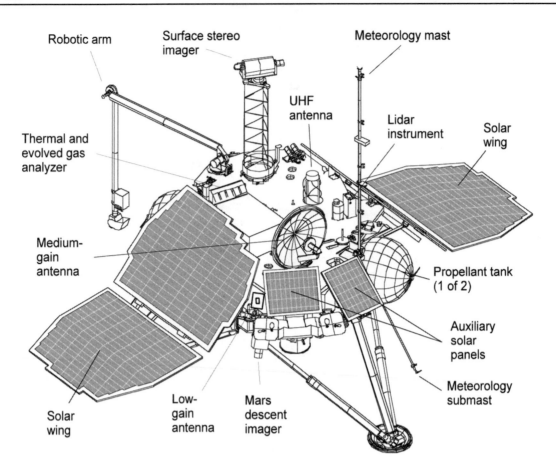

Mars Polar Lander spacecraft

Sun actually will not set at any time during the lander's prime mission. The normal schedule calls for the meteorological package to be powered on for a few minutes for temperature and pressure observations, at about 9 p.m., 1 a.m. and 5 a.m. local Mars time. Each time the experiment is turned on, data will be sent via the UHF antenna to the orbiter for relay to Earth. The lidar instrument may also be activated during the night.

During the first Martian night following landing, these wakeups will take place Saturday, December 4, at 12:16-12:41 a.m., 4:22-4:48 a.m., 8:29-8:55 a.m. and 12:36-1:01 p.m.

Sol 1. For mission planning purposes, engineers refer to Martian days, or "sols," each of which is 24 hours, 37 minutes in length. Landing day is referred to as "sol 0," while the next Martian day, which corresponds to evening California time on Saturday, December 4, is called "sol 1."

The schedule of activities from this point forward depends greatly on the condition of the spacecraft. If everything is operating very well, mission controllers will carry out a nominal mission calling for lander wakeup at 7:06 p.m. PST Saturday, December 4. The lander would transmit to Earth for nearly five hours from 7:32 p.m. to 12:19 a.m. PST. If all is going well, the lander could send any remaining data on entry, descent and landing from the descent imager and other science instruments. If the mission team has to deal with any spacecraft anomalies, however, some of these activities may be delayed.

Initial science operations that would be conducted during the first day or two of the landed mission would normally include the first motion test of the robotic arm, a health check of the thermal and evolved gases analyzer; observations by the lidar instrument, and additional pictures by the surface stereo imager and the camera on the robotic arm. Periodic meteorological temperature and pressure observations will continue to be taken, and an additional set of water vapor and carbon dioxide readings will be gathered from laser sensors on the meteorological mast.

Within the first few days of the mission, the team hopes to assemble an initial 360-degree color panorama of the landing site from the camera on the lander's deck. In addition, picture frames from the descent imager would be released and later assembled into a "movie."

Payload activities. When the flight team concludes that the lander is ready to begin science operations, the spacecraft's robotic arm will be turned on and instructed to acquire a sample of soil on the surface. If the mission is going extremely well, this sample would be delivered to the lander's small test ovens no earlier than sol 4 and 5 (late evening on Tuesday and Wednesday, December 7-8); this could be scheduled later for a variety of reasons based on the spacecraft's state. While the arm is working, the thermal and evolved gas analyzer will be powered on for warm-up and calibration. The soil sample will be imaged by the camera mounted on the robotic arm and delivered to one of the analyzer's "ovens." This experiment will use an LED indicator to confirm automatically that a soil sample has been delivered.

After the science team receives verification that the sample has been delivered, a low temperature cook sequence which heats the sample to 80 F (27 C) will begin on the next Martian day after the sample is delivered. If the sample has not been successfully delivered, another attempt will be made. The robotic arm will return to the trench to measure soil temperature by inserting a probe into the ground. At the conclusion of the soil cook sequence, the thermal and evolved gas analyzer will be powered off for the night and the soil sample will remain in the oven to cool overnight. On the next day the experiment will run a high-temperature cook sequence, heating the soil sample to 1,880 F (1,027 C).

The meteorological package will take weather measurements throughout each day. Lidar observations are also planned. Throughout the 90-day primary mission, science experiments will collect data on atmospheric conditions and weather patterns, observe changes in the landscape and search for evidence of subsurface or surface water.

Telecommunications modes. Originally most data from Polar Lander was to be sent first to Mars Climate Orbiter for relay to Earth, but that spacecraft was lost upon arrival at Mars on September 23, 1999. Fortunately, the lander has two other modes for communication with Earth. The lander is able to communicate directly with Earth — either sending data or receiving commands — using its X-band radio and medium-gain antenna. The lander can also send data (but not receive commands) via the currently orbiting Mars Global Surveyor. The lander can communicate directly with Earth via X-band at up to 12,600 bits per second over the Deep Space Network's 230-foot (70-meter) antennas, or at 2,100 bits per second over 112-foot (34-meter) antennas. Using Mars Global Surveyor as a relay, the lander can send data at 128,000 bits per second.

For the first few days of the mission, the lander will exclusively use its X-band radio to transmit directly to Earth. If the mission is proceeding normally, the relay through Global Surveyor would first be tested on about sol 4 (the evening of Tuesday, December 7). If both the direct-to-Earth and Global Surveyor relay paths are working well, the mission team expects to use both of them for different types of data. Information that the mission team wants to receive quickly on critical operations such as robot arm digs would be sent directly to Earth, whereas data that is gathered over a longer period of time for products such as photo panoramas might be sent via Global Surveyor.

Spacecraft engineering data will be delivered to the spacecraft team at Lockheed Martin Astronautics in Denver, CO, and at JPL in Pasadena, CA. Science data will be delivered to the experiment principal investigators at their home institutions. Experimenters will be able to obtain data on a daily basis and send commands for the next day's activities.

Extended mission. If the lander science payload continues to operate well, the primary mission of three months may be extended. Lander activities in an extended mission would include continued use of the robotic arm's camera and surface stereo imager; ongoing temperature, pressure, dust and atmospheric opacity measurements; and continued monitoring of spacecraft performance in the harsh environment of Mars' southern polar region, as summer wanes and seasonal frosts begin to return.

Contingencies. If the spacecraft is not behaving normally immediately upon landing, there are several contingency scenarios from which the mission team could recover and still carry out a full mission.

If no signal is received within the first few minutes when Polar Lander is scheduled to transmit after touchdown at 12:39 p.m. PST Friday, December 3, the team will first check the ground system to make sure that the problem is not on Earth. They would then listen again during a one-hour period beginning at approximately 2:20 p.m. the same day. Polar Lander would transmit at this time if it entered a "safe mode" immediately upon touchdown. Like all other planetary spacecraft, the lander has onboard fault-protection software that puts it into a standby state, or "safe mode," in response to certain types of abnormal events. In safe mode, the lander is placed into a protective state and awaits commands from Earth.

If no signal is received during this time period, the next time that mission controllers will attempt to listen for a signal is a 2½ hour period beginning at approximately 8:09 p.m. Friday, December 3. Mission controllers will have sent commands to the lander instructing it to transmit at this time while it moves its dish antenna sequentially in order to find Earth.

if the lander is still silent, mission controllers will attempt to listen for a signal is a one hour period beginning at approximately 9 p.m. Saturday, December 4. Polar Lander would transmit at this time if it entered a safe mode after touchdown but before its first scheduled transmission session.

The next time that mission controllers will attempt to listen for a signal is at approximately 10 p.m. Sunday, December 5. At this time, Polar Lander would automatically switch to its UHF transmitter and attempt to send data via Mars Global Surveyor if it has not been instructed otherwise in commands from Earth.

If the lander receives no commands from Earth for six days, its "command loss timer" would time out and it would automatically begin switching out hardware subsystems and attempting to radio Earth. Mission managers believe it is more likely, however, that they would be actively sending commands to the spacecraft to execute such hardware swaps if contact is not established within the first day or two after landing.

Spacecraft

The lander stands 3½ feet (1.06 meters) tall from the ground to the top of the science deck and measures 12 feet (3.6 meters) wide. The dry spacecraft weighs 1,129 pounds (512 kilograms); loaded with propellant, the weight is 1,270 pounds (576 kilograms). The spacecraft is constructed of a composite material with a honeycomb aluminum core and graphite-epoxy facesheets bonded to each side of the bus.

The landing legs are made of aluminum and are equipped with compression springs to deploy the legs from the stowed position. Tapered, crushable aluminum honeycomb inserts in each leg provide the shock absorption necessary for landing. The lander's central enclosure houses the onboard computer, power distribution, the 12-cell nickel-hydrogen battery, a unit that controls battery charging, and radio equipment.

A separate component deck outside of the central electronics enclosure contains gyroscopes, electronics to fire pyrotechnic devices used to deploy instruments, and radar equipment that will be used only during entry, descent and landing at Mars arrival.

The lander's solar arrays are inverted and shaped like gull wings, extending about 12 feet (3.6 meters) when fully deployed. They are expected to provide up to 200 watts of electrical power on the Martian surface.

Most systems on the lander are redundant; it contains two computers, two radios, and so on, so that if one fails the other can take over. The main equipment that is not redundant are the landing radar, battery and science instruments.

Cruise stage. During the flight from Earth to Mars, the lander has been attached to a circular cruise stage and propulsive lander/entry assembly. After the 11-month flight to Mars, the cruise stage will be jettisoned just before atmospheric entry, providing a clean aerodynamic shape for the spacecraft's plunge toward the surface.

Onboard the cruise stage are two solar arrays used to generate power during cruise. Telecommunications during cruise will be routed through an X-band medium-gain horn antenna and a low-gain antenna. The cruise stage also contains a radio frequency power amplifier

Entry, descent and landing system. When it dives into the atmosphere, the lander is encased in an aeroshell featuring a 7.9-foot-diameter (2.4-meter) heat shield. The heat shield shares the same nose radius and cone angle of the shield on Mars Pathfinder; but the Mars Polar Lander heat shield is smaller in diameter.

The 27-foot-diarneter (8.4-meter) parachute made of polyester fabric will be deployed by a mortar, or small cannon, to ensure that it separates properly and inflates instantaneously

Propulsion. The lander is equipped with four clusters of thrusters, each of which contains one large thruster producing 5 pounds (22 newtons) of force and one small thruster providing 1 pound (4.4 newtons) of force. The lander's final descent will be controlled by 12 descent engines, or retro-rockets, each delivering 60 pounds (266 newtons) of force, arranged in three groups on the underside of the lander; they will be pulsed on and of as the lander descends. Two spherical propellant tanks underneath the lander's solar arrays carry 141 pounds (64 kilograms) of hydrazine for all of the spacecraft's engines and thrusters.

Power. The lander obtains its power from a total of six solar panels. The four larger panels are arranged as a pair of wings on either side of the lander; and are deployed after landing. Two smaller panels fixed to the side of the lander were added to increase the total power output after the main solar panels were made as large as possible while still fitting in the launch vehicle's fairing or nose cone. During the southern Martian summer at the time of arrival, the sun never sets below the horizon at the landing site. A rechargeable 16-amp-hour nickel-hydrogen battery will keep the central electronics enclosure relatively warm (above -22 F(-30 C)) during -110 F (-80 C) night-time temperatures near the Martian pole. The lifetime of the battery will probably be the main factor determining how long the lander operates. As nights grow colder in late Martian summer, the battery will eventually be unable to provide enough power to keep the spacecraft enclosure warm at night. The lander would then freeze, ending the mission.

Telecommunications. The lander contains two radio systems, one in the UHF (ultrahigh-frequency) band close to the upper channels on a conventional television set, and the other in the microwave X-band. Each system includes both a transmitter and receiver as a combined unit called a transponder. The UHF system, which is used only to communicate with Mars Global Surveyor when that orbiter is acting as a relay between the lander and Earth, has a single dedicated antenna.

The X-band system communicates directly with Earth through one of two antennas mounted on the lander deck — a dish-shaped medium-gain antenna that must be pointed at Earth, or a non-directional low-gain antenna. During interplanetary cruise, the X-band system communicates with Earth using a horn-shaped medium-gain antenna and radio frequency power amplifier mounted on the cruise stage.

Science Objectives

Mars Polar Lander will touch down in a unique region of Mars near the border of the southern polar cap at a latitude of about 76 degrees south. The lander is the only spacecraft planned by any space agency to study an area of Mars this far south or north.

Polar Lander is designed to study the volatiles (materials that change form readily, such as water) and climate history of Mars. To that end, its scientific goals are to:

* Land on the layered terrain in Mars' south polar region.

* Search for evidence related to ancient climates and more recent periodic climate change.

* Give a picture of the current climate and seasonal change at high latitudes and, in particular, the exchange of water vapor between the atmosphere and ground.

* Search for near-surface ground ice in the polar regions, and analyze the soil for physically and chemically bound carbon dioxide and water.

* Study surface morphology (forms and structures), geology, topography and weather of the landing site.

Mars has polar caps at both its north and south poles. Both caps include a permanent or residual cap visible year-round, and a temporary or seasonal cap that appears in winter and disappears in summer. In the north, the permanent cap is water ice, while in the south the permanent cap is mostly carbon dioxide ice with perhaps some water The north's permanent cap is 10 times larger than the south's; it remains a mystery as yet why the caps differ so. The south's seasonal cap is larger than the north's, which is caused by the fact that the southern winter takes place when Mars is farthest in its orbit from the Sun.

Both poles show signs of an unusual layered terrain, whose alternating bands of color may contain different mixtures of dust and ice. Like growth rings of trees, these layered geological bands may help unravel the mystery of past climate change on Mars. They may also help determine whether geologically recent climate change may be driven by changes in the tilt of the planet's rotation axis, or by changes in the shape of Mars' orbit around the Sun. This may lend insight on how the planet evolved from a wetter ancient climate — whether this was driven by a catastrophic change, or merely a gradual evolution in the planet's environment. One of the lander's primary science objectives is to conduct a visual survey of this largely unknown dome of ice and dust, and provide a technical portrait of the mineralogical makeup of the layered terrain.

Landing site. In planning the lander mission, scientists desired to place the spacecraft onto the layered terrain near the south pole but land on bare soil, not the seasonal carbon dioxide frost. The longitude chosen is the area where the south pole's layered terrain extends the farthest north. At this longitude, seasonal frost also dissipates earliest in the southern spring.

Polar Lander was launched toward the center of a swath of terrain measuring about 4,000 square kilometers (1,500 square miles) centered at 75 degrees south latitude, 210 degrees west longitude. Scientists expected to make the final landing site selection once Mars Global Surveyor could return important data on the appearance and the altitude of the south polar layered terrain. Because of its prolonged aerobraking phase, Global Surveyor was unable to start its systematic mapping of Mars until March 1999, some two months after the launch of Polar Lander. Except for a handful of images taken in January 1998, the next opportunity for Global Surveyor to observe the landing target sector in sunlight was in June 1999, when the south polar region again emerged into sunlight.

Based on data from Global Surveyor's altimeter and camera, scientists selected a primary and secondary landing site for Polar Lander. A thruster firing in September 1999 targeted the lander to 76 degrees south latitude, 195 degrees west longitude. Imaging has continued as the seasonal frosts have retreated across the landing site, revealing a strangely textured terrain with relatively low relief, but with complicated patterns of bright and dark ground. With a thruster firing carried out on October 30, 1999, Polar Lander remains targeted to the primary site.

Science payload. Mars Polar Lander carries three science investigations: the Mars Volatiles and Climate Surveyor (MVCS); the Mars Descent Imager; and a Light Detection and Ranging (Lidar) instrument.

MVACS. The Mars Volatiles and Climate Surveyor is an integrated instrument package designed to carry out a variety of studies of the surface environment, weather and geology of the south pole region. Dr. David Paige of UCLA is the principal investigator. MVCS includes the following four component systems:

* **Surface stereo imager.** The imager, mounted on top of a 5-foot (1.5-meter) mast that will pop up from the lander's deck, is identical to the imager on the Mars Pathfinder lander. It will capture panoramas of the landing site, and provide imaging support for other payload elements such as the robotic arm and the thermal and evolved gas analyzer. The imager will also take pictures of magnets attached to the lander's deck to identify any magnetic material that collects there. It can perform imaging of the Sun to study aerosols or water vapor in the atmosphere. The imager has a spectral range from violet to near-infrared (400 to 1,100-nanometer wavelength). Two sets of lenses are mounted slightly more widely apart than a normal pair of human eyes in order to capture stereo images; the two sets of optics share a single charge-coupled device (CCD) detector. Two filter wheels allow the imager to take pictures in various spectral ranges. Drs. Peter Smith of the University of Arizona, Tucson, and H. Uwe Keller of the Max Planck Institut fuer Aeronomie, Germany, are co-investigators for the imager. The magnetic properties experiments are supplied by the University of Copenhagen, Denmark.

* **Robotic arm.** This jointed 6½ foot (2-meter) arm attached to the lander's deck has an articulated member on its end with a digging scoop, camera and temperature probe. The scoop will dig trenches and deliver soil samples to the thermal and evolved gas analyzer on the lander's deck. Pictures taken by the camera will show fine-scale layering of surface and subsurface materials, if any, as well as fine-scale texture of soil samples and sides of the trenches dug by the scoop. The temperature probe will measure the ambient temperature and thermal conductivity of the soil.

* **Meteorology package.** The lander's weather station includes a 4-foot (1.2-meter) mast with a wind speed and direction sensor, temperature sensors and tunable diode lasers that detect water vapor and isotopes of water and carbon dioxide. A 3-foot (0.9-meter) submast with a wind speed sensor and two temperature sensors points downward from the lander's deck. The submast is designed to study atmospheric effects in the zone just 4 to 6 inches (10 to 15 centimeters) above the surface to determine the threshold wind speed required for dust storms to start. Drs. David Crisp and Randy May of JPL and Ari-Matti Harri of the Finnish Meteorological Institute are co-investigators for the weather package.

* **Thermal and evolved gas analyzer.** This instrument heats soil samples and analyzes them to determine concentrations of volatiles such as water or carbon dioxide, whether present as ice or in volatile-bearing minerals. The robotic arm deposits soil samples in a receptacle, which is then mated with a cover to form a small oven; heater wires like the coils in a toaster heat the sample gradually across a range of temperatures up to 1,880 F (1,027 C). The heater is able to achieve this heat with the limited power of the solar-powered spacecraft because the sample chamber holds only 1/300th of an ounce (0.1 gram) of material. As it is heated, gases are released (or "evolved") from the sample. A tunable diode laser, meanwhile, emits beams of light which passes through the gases to a detector. Any carbon dioxide or water vapor in the gas absorbs some of the laser light, which is measured by the detector. Once used, the ovens cannot be used again; the instrument can perform a total of eight soil analyses during the lander mission. Dr. William Boynton of the University of Arizona, Tucson, is co-investigator for the instrument.

Mars Descent Imager. This imager will take approximately 30 pictures as the lander descends toward Mars' surface, beginning just before heat-shield ejection at an altitude of about 5 miles (8 kilometers) and continuing until landing. The first pictures will show areas of Mars 5.6 miles (9 kilometers) square with a resolution of 25 feet (7.5 meters) per pixel, while the final pictures will show an area 30 feet (9 meters) square with a resolution of 1/3 inch (9 mm) per pixel. The imager has a single camera head with an electronically shuttered charge-coupled device (CCD) which will capture black-and-white images 1,000 by 1,000 pixels. Dr Michael Malin of Malin Space Science Systems Inc., San Diego, CA, is principal investigator.

Lidar. This instrument is a distant cousin to radar, emitting pulses of energy and then detecting their echo as they bounce off material in the atmosphere. But instead of the radio pulses used in radar, the lidar instrument sends out pulses of light from a laser. The transmitter uses a gallium aluminum arsenide laser diode to emit 2,500 pulses of near-infrared light each second upward from the lander's deck. A detector then times how long it takes the pulses to return, allowing scientists to locate and characterize ice and dust hazes in the lower part of Mars' atmosphere (below about 1 to 2 miles (2 to 3 kilometers)). The instrument is fixed, pointing straight up from the lander. The lidar instrument is provided by the Russian Academy of Science's Space Research Institute (IKI) under the sponsorship of the Russian Space Agency. The principal investigator is Dr. V S. Linkin of IKI. The lidar investigation is the first Russian experiment to be flown on a U.S. planetary spacecraft.

A unique feature of Mars Polar Lander is that it will be the first planetary spacecraft to carry a microphone to capture the sounds of another world. Despite Mars' extremely thin atmosphere, the microphone may pick up sounds of winds and mechanical events on the lander.

Contained within the electronics box for the lidar experiment, the microphone is enclosed in a package 2 by 2 by ½ inches (5 by 5 by 1 centimeters), weighs less than 1.8 ounces (50 grains) and uses less than 100 milliwatts of power. Designed to accommodate the limited rate at which lander data are returned to Earth, the microphone records and returns the loudest 10-second signal heard during a listening period. Later in the mission, longer sound records may be returned. The package takes advantage of many off-the-shelf technologies, such as a sound processor chip used in talking toys and educational computers that listen and respond to spoken words. The microphone itself is a type used in hearing aids.

A project of the Planetary Society, Pasadena, CA, the microphone was approved by NASA to be flown as part of the Russian lidar experiment payload. Planetary Society Executive Director Dr. Louis Friedman is responsible for the microphone, which was designed, constructed and tested at UC Berkeley's Space Sciences Laboratory under the direction of Dr Janet Luhmann.

Planetary Protection

The U.S. is a signatory to the United Nations' 1966 Treaty of Principles Governing the Activities of States in the Exploration and Use of Outer Space, Including the Moon and Other Celestial Bodies. Known as the "Outer Space Treaty," this document states in part that exploration of the Moon and other celestial bodies shall be conducted "so as to avoid their harmful contamination and also adverse changes in the environment of the Earth resulting from the introduction of extraterrestrial matter."

NASA policy establishes basic procedures to prevent contamination of planetary bodies. Different requirements apply to different missions, depending on which solar system object is targeted and the spacecraft or mission type (flyby, orbiter, lander, sample-return and so on). For some bodies such as the Sun, Moon and Mercury, there are no outbound contamination requirements. Current requirements for the outbound phase of missions to Mars, however, are particularly rigorous. Planning for planetary protection begins during pre-mission feasibility planning.

Planetary protection requirements called for the surfaces of the Mars Polar Lander spacecraft to contain a maximum of about 250 spores per square yard (300 spores per square meter) and 300,000 spores total. To meet this goal, the spacecraft was cleaned to the same level as the Viking landers before they were sterilized.

Technicians at Lockheed Martin Astronautics in Denver continually cleaned the spacecraft throughout development by rubbing down surfaces with ethyl alcohol. Large surface areas, such as the thermal blankets and parachute, had to be baked for about 50 hours at 230 F (110 C). The number of spores is determined by sampling the surfaces and conducting a special microbiological assay. In general, the procedure was the same as the sterilization methods used on the Mars Pathfinder lander. The spacecraft was checked constantly during processing at Lockheed Martin Astronautics and was given a final inspection just before it was encapsulated in its aeroshell. Results of that effort produced a spacecraft with an average spore burden density of no more than 250 spores per square yard (300 spores per square meter).

A final inspection and spore count was performed at NASA's Kennedy Space Center FL, before the spacecraft is integrated with the Delta II launch vehicle. That test assured that the lander contained fewer than 300,000 spores total.

Deep Space 2

The Deep Space 2 project is sending two identical microprobes along on the Mars Polar Lander spacecraft. Released shortly before the lander enters the planet's atmosphere, the probes will dive toward the surface and bury themselves up to about three feet (1 meter) underground.

As a project under NASA's New Millennium Program, the main purpose of Deep Space 2 is to flight-test new technologies to enable future science missions — demonstrating innovative approaches to entering a planet's atmosphere, surviving a crash-impact and penetrating below a planet's surface. As a secondary goal, the probes will search for water ice under Mars' surface.

Mission Overview

At the time of launch, the two Deep Space 2 probes were attached to the cruise stage on the Mars Polar Lander spacecraft. To simplify hardware and operations, there are no electrical interfaces between the probes and the lander's cruise stage. The probes are powered off during cruise, so there is no communication with them from installation on the launch pad until after impact on the Martian surface.

Five minutes before the lander enters Mars' upper atmosphere on December 3, 1999, the lander will jettison the cruise stage. The force of separation will initiate mechanical pyro devices, which in turn will separate the microprobes from the cruise stage about 18 seconds later. Each Deep Space 2 entry system consists of a basketball-size aeroshell containing a probe somewhat larger than a softball.

Upon release from the lander's cruise stage, the probes switch on power from their lithium batteries, and an onboard computer microcontroller powers up. The microcontroller performs a series of measurements of onboard subsystems to verify their health after the 11 month cruise to Mars.

About four minutes after power-up, the probes will enter Mars' atmosphere. A descent accelerometer is turned on and samples "G" forces 20 times a second until impact. Four minutes after entering the atmosphere, an impact accelerometer begins sampling "G" forces 25,000 times a second. When impact is detected, data from the event is stored in computer memory, and the impact accelerometer is turned off.

The two probe systems will hit the Martian surface about the same time as Mars Polar Lander's landing some 35 miles (60 kilometers) away. Upon impact, the acorn-shaped aeroshell will shatter, and the probe inside will separate into two parts. The bullet-shaped forebody will penetrate as far as 3 feet (1 meter) below the surface, depending on the hardness of the soil. The aftbody will remain on the surface to relay data back to Earth via the Mars Global Surveyor spacecraft, which has been orbiting Mars since September 1997. The forebody and aftbody communicate with each other via a flexible cable.

Unlike any spacecraft before, the Deep Space 2 probes smash into the planet at speeds of up to 400 miles per hour (200 meters per second). The probe's electrical and mechanical systems must withstand this crushing impact. This is achieved with a combination of advanced materials, mechanical designs and microelectronic packaging techniques developed based on extensive testing. After impact, the systems must withstand extreme temperatures. The forebody buried in the Martian soil must withstand temperatures as low as -184 F (-120 C), while the aftbody that remains on the surface is exposed to an environment as low as -112 F (-80 C).

Landed mission. Following impact, the probes collect data to flight-validate their microelectronic and micromechanical technologies. Minimum data to validate most of these technologies will be collected within the first 30 minutes after impact, while minimum data from the sample/water experiment will be collected within about 10 hours after impact. Data collection will continue until the probe batteries are depleted in about one to three days.

Each probe will transmit data to the orbiting Mars Global Surveyor using a radio in the UHF band (at frequencies near the upper channels of a conventional TV set) at a rate of 7,000 bits per second. The first such communications session is expected within eight hours after impact. Normally each probe will be in a low-power listening mode until it receives a signal from Mars Global Surveyor telling it to transmit data. The orbiter may either transmit the data to Earth immediately, or store the data temporarily and transmit them as soon as possible.

The first opportunity for a communication pass between Global Surveyor and the microprobes will take place at about 7:27 p.m. PST Friday, December 3. During a pass about 15 minutes, 16 seconds long, Global Surveyor will switch back and forth between communicating with each of the microprobes for about two minutes apiece, relaying the data to Earth immediately. It is possible that contact may not be made on the first pass due to the orientation of the microprobes in the Martian soil. If this is the case, contact may be established during any of a series of communications passes carried out every two hours for the first day or two after arrival.

The microprobe data are buffered onboard Global Surveyor using the memory of its camera. The data therefore are first sent to Malin Space Science Systems in San Diego, CA, which is responsible for the Global Surveyor camera. The Malin team may be able to determine about an hour after the communication pass if any data from the microprobes are present. The data will then be forwarded to JPL, where scientists may get their first look at the contents perhaps another hour later.

Science mission. Deep Space 2 has a secondary goal of collecting science data. Accelerometer data from the descent and impact will provide an estimate of the density of the atmosphere and hardness of the soil. After impact, the probes will measure the thermal conductivity and potential water content of the subsurface soil adjacent to the bullet-like probe forebody.

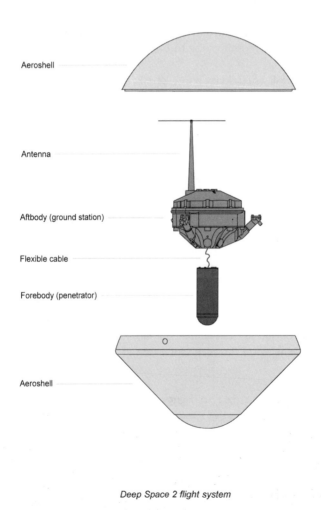

Aeroshell

Antenna

Aftbody (ground station)

Flexible cable

Forebody (penetrator)

Aeroshell

Deep Space 2 flight system

Following the first successful transmission to Global Surveyor, a micromotor will drive a small drill bit out the side of the probe's forebody. Bits of soil engaged by the drill bit will fall into a small heater cup, which is sealed by firing a pyro which closes a door. The soil is then heated, driving any water vapor into the analysis chamber. If water is present, it will be detected by measuring the difference in light intensity of a laser shining through the vapor. The tunable diode laser is set so that its light is at the point in the spectrum where water absorbs light.

Soil conductivity is determined by measuring the rate at which the forebody cools after plunging into the ground. Temperature readings are taken throughout the landed mission by two sensors mounted at opposite ends of the probe's forebody.

End of prime mission. The prime mission ends when the probes transmit to Mars Global Surveyor one set of data evaluating the project's engineering technologies. This transmission is expected within 10 hours after impact, but may take place up to 36 hours after impact. At the end of the prime mission, the probes will continue to collect and transmit data on soil thermal conductivity as the probes gradually cool, as well as soil temperature variations, until their batteries are depleted.

Technologies

Deep Space 2 is the second mission of NASA's New Millennium Program, whose goal is to greatly increase the efficiency and lower costs of space science missions through new technologies. Each New Millennium mission is designed to test specific technologies never before used in space missions.

Deep Space 2 will test technologies that could pave the way for future missions featuring multiple landers released from a single spacecraft, possibly distributed around an entire planet or other body. Such networks of probes offer a unique window on global processes such as weather or seismic activity of a planet.

To meet this goal, the mission was challenged to develop an entry and landing system that is very small, lightweight and capable of conducting experiments on both the surface and subsurface of a planet or similar body while surviving environmental extremes. Deep Space 2 will validate the following new technologies:

Entry system. Unlike other probes, the Deep Space 2 aeroshell is not required to be pointed or spin-stabilized when it enters Mars' atmosphere. Its design uses the same principle as a shuttlecock, or birdie, in badminton; most of the weight is placed well ahead of the aeroshell's center of pressure to insure that the heat shield passively aligns itself even if it is tumbling when it enters Mars' atmosphere. In addition, the entry system is "single-stage" from atmospheric entry until impact — there are no parachutes, retro rockets or airbags to slow the probes down. In fact, the aeroshell is not even jettisoned by the probe, but accompanies it to the surface of the planet where it shatters on impact. This very simple system greatly reduces the number of tests required to demonstrate that the design works, and thus greatly reduces the costs to the mission. The entry system was designed at JPL. Aerodynamic analysis was performed at NASA's Langley Research Center, Hampton, VA.

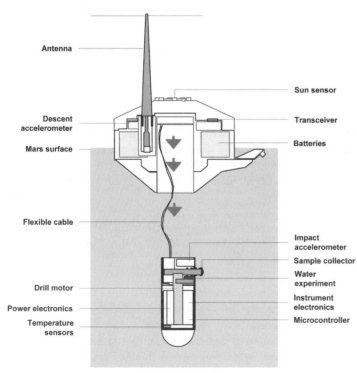

Deep Space 2 probe

The aeroshell's heat shield is made of an advanced thermal protection system known as SIRCA-SPLIT (silicon-impregnated, reusable ceramic ablator — secondary polymer layer-impregnated technique). This material is capable of maintaining the probe's internal temperature to within a few degrees of -40 F (-40 C) while the heat shield surface experiences temperatures of up to 3,500 F (2,000 C). This material was developed and tested by NASA's Ames Research Center, Moffet Field, CA. The silicon carbide aeroshell structure was developed by Poco Specialty Materials, Decatur, TX

Testing of the entry system design went through many phases. Early tests included dropping test articles made of clay pots, Styrofoam or Pyrex from airplanes two miles (3 kilometers) high. Silicon carbide, the same material used in sandpaper;

was selected as the material for the aeroshells, each of which weighs less than 2.6 pounds (1.2 kilogram). Final tests of a prototype aeroshell with a probe model were conducted using an airgun at Eglin Air Force Base, Fort Walton Beach, FL, where the probe system was shot into the ground at speeds up to 400 miles per hour (200 meters per second).

Penetrator system. Deep Space 2 is the first penetrator sent by NASA to another planet. Development of the penetrator system required an aggressive test program with a continuous design/develop/test/fix approach. The probe's bullet-like forebody is designed with a half circle nose to ensure penetration over a wide range of entry conditions. The aftbody features a wide frontal area to limit penetration and a "lawn dart" face which helps the aftbody anchor to Mars' surface. Tests were performed using an airgun in Socorro, New Mexico, in partnership with Sandia National Laboratories and the New Mexico Institute of Mining and Technology's Energetic Materials Research and Test Center.

High-G packaging techniques. Crashing into a planet at 400 miles per hour (200 meters per second) presents a unique challenge in the design of electrical and mechanical systems. Decelerations could reach levels up to 30,000 G's in the forebody and 60,000 G's in the aftbody (one "G" is equivalent to the force of gravity on Earth's surface). This is the same as requiring a desktop computer to operate after being hit by a truck at 400 miles per hour. In comparison, Mars Pathfinder experienced forces of about 17 G's during its landing.

There are two standard approaches for insuring high-G survival. One is to cushion the object, and the other is to provide a very rigid structure that allows the shock wave to pass through the object without deflecting it enough to break any of its components. For Deep Space 2, cushioning is impractical because of the extreme decelerations and the small size of the probes; engineers thus chose a rigid structure approach.

The mechanical design features a "prism" electronics assembly, a science "block" and selective use of materials to maximize structural rigidity. The electrical design features chip-on-board and three-dimensional high-density-interconnect packaging, encapsulated wire bonds and extensive use of flexible interconnects instead of wires. Assemblies are also typically bonded together to minimize potential loose parts and to distribute loads evenly

Micro-telecommunications system. Each probe features a miniaturized radio transmitter and receiver system weighing less than 1¾ ounces (50 grains), 9.9 square inches (64 square centimeters) in size, and consuming less than 500 milliwatts in receive mode and 2 watts in transmit mode. The system was developed at JPL.

Ultra-low-temperature lithium battery. The probe's batteries must be able to provide 600 milliamp-hours of power at temperatures as low as -112 F (-80 C). To meet those extreme needs, the Deep Space 2 project developed a new non-rechargeable lithium-thionyl chloride battery. The cells use a lithium tetrachlorogallate salt instead of the more conventional lithium aluminum chloride salt to improve low-temperature performance and reduce voltage delays.

Each probe uses two batteries composed of four "D-sized" cells weighing less than 1.4 ounce (40 grams) each. The batteries operate within a range of 6 to 14 volts and have a shelf life of three years. The batteries were developed by Yardney Technical Products, Pawcatuck, CT.

Power microelectronics. Power conditioning, regulation and switching for electronics in the bullet-shaped forebody are controlled by a power microelectronics unit making use of application-specific integrated circuits (ASICs) in which both digital and analog components are incorporated onto a single chip. The unit weighs less than 1/5th of an ounce (5 grains), has a volume of one-third cubic inch (5.6 cubic centimeters) and requires 5/100ths of 1 milliwatt to operate. The unit was developed by Boeing Missiles & Space, Kent, WA.

Advanced microcontroller. The Spartan computer system on the probes centers around an 80C51 microprocessor, a low-power chip used in products ranging from microwave ovens and videocassette recorders to cars and computer peripherals. The 8-bit system includes 128K of random access memory (RAM), 128K of permanent memory, and 32 digital-to-analog and analog-to-digital converters. The system is designed to use very low power (less than 6 milliwatts running at 10 megahertz, one-half milliwatt in sleep mode) with small volume (0. 13 cubic inch (2.2 cubic centimeters)) and mass (1/10th of an ounce (3.2 grams)). Electronic circuits are embedded in plastic to ensure survival during the 30,000-G impact event. The microcontroller was developed by a consortium led by the U.S. Air Force's Phillips Laboratory and including Mission Research Corp., the Boeing Co., NASA's Langley Research Center, Technology Associates, General Electric and the University of Tennessee.

Flexible interconnects. Normal wire cabling could easily break under the extreme "G" forces that the probes will endure, so a different approach was required. The flexible interconnect are strips made of alternating layers of copper traces and polyamide. The latter, a type of thin polymer or plastic film, is also used in thermal blankets on spacecraft, while the copper traces are similar to the thin copper paths on a computer or radio circuit board. Flexible interconnects are much lighter, more compact and more flexible than standard wire cables. On the Deep Space 2 probes, they are used between all electronic subsystems, and for the umbilical which connects the forebody (penetrator) to the aftbody (ground station). During flight, the umbilical is folded in a canister like a fire hose; at impact, the umbilical unfolds as the penetrator pulls away. The flexible interconnect system was developed by JPL in partnership with Lockheed Martin Astronautics, Denver, CO. The units were fabricated at Electrofilm Manufacturing Co., Valencia, CA, and Pioneer Circuits Inc., Santa Ana, CA.

Sample collection/water detection experiment. Each probe will obtain a sample of subsurface soil using a small, ruggedized drill run by an electric motor. When the motor is powered on, a latch is released and the drill shaft extends sideways from the forebody (penetrator), pulling less than 1/250th of an ounce (100 milligrams) of soil into a small cup which is then sealed. The sample is then heated, turning any water ice in the soil into water vapor A small tunable diode laser emits a beam of light through the vapor to a detector; if water vapor is present, it will absorb some of the light. The laser assembly is similar to tunable diode lasers flown on meteorology experiments on Mars Polar Lander, but is much smaller (about the size of a thumbnail) and thus has lower sensitivity. During operation, the water detection experiment requires a peak power of 1.5 watts. The sampling collection system is about 1 cubic inch (11 cubic centimeters) in size and weighs less than 1.6 ounce (5 0 grains). The instrument electronics is about one-third cubic inch (4.8 cubic centimeters) in size and weighs less than one-third ounce (10 grains). The tunable diode laser is about 1/50th of a cubic inch (0.3 cubic centimeter) in size and weighs less than 1/30th of an ounce (1 grain). The sample collection/water detection experiment was developed by JPL.

Soil thermal conductivity experiment The probes will use temperature sensors to measure how fast the forebody or penetrator cools down after impact, revealing how quickly heat dissipates in the soil. This approach requires far less energy than similar previous experiments on planetary missions, which have used onboard heaters to test the soil. On the Deep Space 2 probes, two platinum-resistor temperature sensors are mounted in the forebody.

Design, development and testing. Because of the many challenges associated with developing NASA's first planetary penetrator system, Deep Space 2 embarked on a rigorous design and test program. This started in spring 1995 to evaluate early design concepts before the project was formally approved in the fall of that year. Early tests included releasing prototype probes from airplanes and helicopters.

As test articles became more sophisticated and expensive, the need for a more controlled test environment became necessary. To accomplish this, the project teamed with Sandia National Laboratories and the New Mexico Institute of Mining and Technology's Energetic Materials Research and Test Center in Socorro, New Mexico, to use a Sandia airgun. This massive airgun has a 18-foot-long (5.5-meter), 6-inch-diameter (15-centimeter) barrel and rests on an 18-wheeler truck. After mounting the test article in the barrel, pressurized air is used to hurl the probe into the desert floor at speeds of up to 400 miles per hour (200 meters per second). More than 70 airgun tests over a period of two years were performed to validate the probe design under worst-case entry conditions. The last test of impact survivability was performed in September 1998.

A variety of tests were performed to validate the aeroshell design. Tests were performed at Eglin Air Force Base in Fort Walton Beach, FL, to verify that the aeroshell shatters on impact, leaving the probe to penetrate the surface. Eglin provided a large airgun 20 feet (6 meters) long and 15 inches (38 centimeters) in diameter and led the test operations. Tests of the aeroshell's aerodynamic properties were performed initially at Eglin's Wright Laboratory Ballistic Range and later at a transonic wind tunnel in Kalingrad, Russia, capable of simulating Martian atmospheric pressures. The aeroshell's heat shield material was tested at an arcjet facility at NASA's Ames Research Center, Moffet Field, CA.

Science Objectives

As a mission under NASA's New Millennium Program, the main focus of Deep Space 2 is testing new technologies on behalf of future science missions. In the process, however the probes collect data of interest not only to engineers developing technologies but also to scientists studying the environment of Mars. NASA thus organized a team of scientists to work with the data that the probe's instruments will deliver.

The objectives for Deep Space 2's science measurements dovetail with those of Mars Polar Lander, which is focused on understanding the climate of Mars. Deep Space 2 will attempt to:

* Determine if ice is present in the subsurface soil;

* Estimate the thermal conductivity of the soil at depth;

* Determine the atmospheric density throughout the probes' entire descent;

* Characterize the hardness of the soil and possibly the presence of any layering on a scale of many inches to a few feet (tens of centimeters).

The layered terrain around Mars' south pole is believed to consist of alternating layers of wind-deposited dust and water and/or carbon dioxide ice condensed out of the atmosphere. These deposits are thought to record the evidence of climate variations on Mars, much like the growth rings of a tree. Deep Space 2 will help give clues about where water ice is located today on Mars and how materials are deposited in the polar layered terrains. Since the two Deep Space 2 probes and Mars Polar Lander will touch down at different locations up to about 35 miles (60 kilometers) apart, data from each of them will tell scientists how much the polar terrain varies from one site to another.

Science activities on Deep Space 2 are organized as four investigations:

* **Sample collection/water detection experiment.** This experiment will obtain a tiny soil sample and heat it to detect any water that may be present. The presence or absence of water ice at a given depth will be compared to analysis of soils excavated by the robotic arm on Mars Polar Lander. One hypothesis is that much of the water that once flowed on Mars' surface is now frozen underground; this experiment will help to refine theories of the fate of Martian water. Science team members selected for this experiment are Dr. Bruce Murray, California Institute of Technology, and Dr. Aaron Zent, NASA Ames Research Center.

* **Soil thermal conductivity experiment.** Temperature sensors in the forebody or penetrator will show how quickly the probe's heat dissipates into the surrounding soil. This will provide information about Mars' polar layered deposits. A very low conductivity would indicate very fine-grain material, likely to have been wind-deposited. On the other hand, a very high conductivity would indicate large amounts of ice in the soil. Soil conductivity has a strong influence on the subsurface temperature, and thus the depth at which ice is predicted to be stable over many annual cycles. Dr. Paul Morgan, Northern Arizona University, and Dr. Marsha Presley, Arizona State University, were selected to analyze data from this experiment.

* **Atmospheric descent accelerometer.** The aftbody houses a descent accelerometer that will measure the drag on the probes as they descend through the Martian atmosphere. This single piece of information can allow scientists to develop profiles of many meteorological factors in Mars' atmosphere, including density,

temperature and pressure at various altitudes. Science team members for atmospheric science are Dr. David C. Catling and Dr Julio A. Magalhaes of NASA Ames Research Center.

* **Impact accelerometer.** The impact accelerometer will provide an estimate of the hardness of the soil, and possibly the presence of small-scale layers that can be compared with the materials encountered by the robotic arm on Mars Polar Lander. Scientists can interpret these terrain layers in terms of the geologic materials they are probably made of, such as ice layers, wind-blown dust and sediments. Data on the small-scale strata of Mars' polar layered terrains could yield important information on climate evolution. Dr Ralph D. Lorenz, University of Arizona, and Dr. Jeffrey E. Moersch, NASA Ames Research Center were selected for this experiment.

Program/Project Management

Mars Polar Lander and Deep Space 2 are managed by the Jet Propulsion Laboratory, Pasadena, CA, for NASA's Office of Space Science, Washington, DC. At NASA Headquarters, Dr. Edward Weiler is associate administrator for space science. Dr. Carl Pilcher is science director for solar system exploration. Ken Ledbetter is director of the Mission and Payload Development Division, and Dr. William Piotrowski is acting director of the Mars Exploration Program. Steven Brody is Mars Polar Lander program executive, and Joseph Boyce is Mars Polar Lander program scientist. Lia LaPiana is Deep Space 2 program executive, and Dr Michael Meyer is Deep Space 2 program scientist.

At the Jet Propulsion Laboratory, Dr. Charles Elachi is director of the Space and Earth Science Programs Directorate. For Mars Polar Lander, Dr. John McNamee was project manager for spacecraft development, Richard Cook is project manager for operations, Dr Sam Thurman is flight operations manager and Dr. Richard Zurek is project scientist. At Lockheed Martin Astronautics, Denver, CO, Dr. Edward A. Euler is the company's program director for Mars Polar Lander.

At JPL, Dr. Fuk Li is manager of the New Millennium Program, and Dr. David Crisp is New Millennium program scientist. For Deep Space 2, Sarah Gavit is project manager Kari Lewis is chief mission engineer and Dr. Suzanne Smrekar is project scientist.

11-30-99 HQ

Mars Polar Lander Mission Status

MEDIA RELATIONS OFFICE
JET PROPULSION LABORATORY
CALIFORNIA INSTITUTE OF TECHNOLOGY
NATIONAL AERONAUTICS AND SPACE ADMINISTRATION
PASADENA, CALIF. 91109.
TELEPHONE (818) 354-5011 http://www.jpl.nasa.gov

Mars Polar Lander Mission Status December 3, 1999 5 a.m. PST

Mars Polar Lander flight controllers opted to perform the last trajectory adjustment of the mission early this morning, sending commands to the spacecraft that will result in a short engine firing to target the spacecraft to the desired landing site near layered terrain in the red planet's south polar region.

Mars Polar Lander is scheduled to land on Mars shortly after noon Pacific time on Friday, December 3. The first signal is expected to be received at 12:39 p.m. The entry, descent, and landing sequence is the most complex and risky part of the mission.

The engine firing was scheduled for 5:39 a.m. Pacific time for 8 seconds, said flight operations manager Dr. Sam Thurman. "This maneuver will increase the entry flight path angle by 0.25 degrees, moving the flight path from our most recent estimate, 12 hours prior to entry, of minus 13 degrees back to the target value of minus 13.25 degrees," Thurman said. "We decided to perform the maneuver in order to ensure that the entry flight path achieved will be very close to the planned trajectory. It puts it just about right on top of the target point, which is in an area chosen because the terrain provides for a safe touchdown."

During descent, the spacecraft will enter the Martian atmosphere traveling at 6.9 kilometers per second (15,400 miles per hour). Onboard accelerometers will sense when friction from the atmosphere causes the lander to slow. From that time, it will be approximately 5 minutes and 30 seconds until touchdown on the surface, during which time the spacecraft will experience G forces up to 12 times Earth's gravity and the temperature of the heat shield's exterior will rise to 1,650 C (3,000 degrees F).

The Deep Space 2 microprobes, which are piggybacking on the lander, will be jettisoned to the planet about 5 minutes before the lander enters the Martian atmosphere. They will impact the Martian surface about 60 kilometers (about 30 miles) away from spot where Mars Polar Lander will set down.

Mars Polar Lander is part of a series of missions in a long-term program of Mars exploration managed by JPL for NASA's Office of Space Science, Washington, D.C. JPL's industrial partner is Lockheed Martin Astronautics, Denver. JPL is a division of the California Institute of Technology in Pasadena.

Mars Polar Lander Mission Status December 3, 1999 11 a.m. PST

NASA's Mars Polar Lander is performing flawlessly and poised to land on the layered terrain near the red planet's south polar region shortly after noon Pacific time today, the mission team reported.

The Mars Polar Lander navigation team reported on the success of this morning's trajectory adjustment, which took place at 5:39 a.m. PST. "It seems to be coming in pretty much right on the target line," said Michael Watkins, manager of JPL's navigation and mission design section.

Flight controllers opted to perform the final trajectory adjustment of the mission early this morning. "It was as smooth and clean a maneuver as we've done," said Project Manager Richard Cook. "We're a gnat's eyelash away from our target."

Wind speeds at NASA's Deep Space Network complex at Goldstone, Calif., were expected to remain in an acceptable range and not force stowage of the large antennas used to receive Polar Lander's signal after landing, Cook said. Winds in the range of about 32 kilometers (20 mph) were reported. The antennas would be stowed if there were sustained winds of about 80 kilometers per hour (50 mph) or gusts from about 73 to 88 kilometers per hour (45-55 mph).

The entry, descent and landing sequence is the most complex and risky part of the mission. During descent, the spacecraft will enter the Martian atmosphere traveling at 6.9 kilometers per second (15,400 miles per hour). Onboard accelerometers will sense when friction from the atmosphere causes the lander to slow. From that time, it will be approximately 5 minutes and 30 seconds until touchdown on the surface, during which time the spacecraft will experience G forces up to 12 times Earth's gravity and the temperature of the heat shield's exterior will rise to 1,650 C (3,000 degrees F).

Based on images from the camera on NASA's Mars Global Surveyor, the landing site, near the south polar layered terrain is expected to be devoid of rocks, generally flat and rolling, and fields of sand or dust dunes may be present, said Polar Lander Project Scientist Dr. Richard Zurek.

The Deep Space 2 microprobes, which are piggybacking on the lander, will be jettisoned to the planet about 5 minutes before the lander enters the Martian atmosphere. They will impact the Martian surface about 60 kilometers (about 35miles) northwest from spot where Mars Polar Lander will set down. The probes, called Scott and Amundsen after early Antarctic explorers, will hit the Mars surface about 1 kilometer (less than a mile) from each other.

The earliest signal from the spacecraft on Mars would be received at 12:39 p.m. PST, said Cook.

Mars Polar Lander is part of a series of missions in a long-term program of Mars exploration managed by JPL for NASA's Office of Space Science, Washington, D.C. JPL's industrial partner is Lockheed Martin Astronautics, Denver. JPL is a division of the California Institute of Technology in Pasadena.

Mars Polar Lander Mission Status December 3, 1999, 5 p.m.

Mission controllers for NASA's Mars Polar Lander mission are awaiting the next opportunity to communicate with the spacecraft, whose transmissions have not yet been received since it landed on Mars shortly after noon Pacific time today.

"I'm very confident the lander survived the descent," said Mars Polar Lander Project Manager Richard Cook at JPL. "Everything looked very good. I think we're a long way from getting concerned. It is not unexpected that we would not hear from it during the first opportunity." A variety of hardware problems from which the lander could recover may be responsible for the delay in initial telecommunications.

During the last telecommunications opportunity, which began at 2:04 p.m. PST, the spacecraft would have automatically moved its steerable antenna in a search pattern designed to find the Earth. The next communications window opens at 6:27 p.m. PST today when the team will again send commands to the lander instructing it to maneuver its medium gain antenna in another attempt to look for Earth. The lander would then carry out that procedure to transmit to Earth beginning at 8:08 p.m. until 10:40 p.m. tonight.

Even if no transmissions are heard today mission controllers have another opportunity to hear from the lander on Saturday. This is the time the spacecraft would be transmitting if it went into a safe mode shortly after landing. Engineers would also listen for it on Sunday evening, when the spacecraft would automatically switch to its UHF radio and transmit via Mars Global Surveyor. After that, they will send commands instructing the spacecraft to swap between various hardware subsystems in case one is damaged.

The Deep Space 2 microprobes, which impacted Mars about 60 kilometers (about 35 miles) from the lander, will transmit data through Mars Global Surveyor. The team will be listening tonight at about 7:30 p.m. when contact is expected with the microprobes.

The flight team's best flight path estimates are that lander most likely touched down at about 76.1 degrees south latitude, 195.3 degrees west longitude. The estimates for the Deep Space 2 microprobe impacts are 75.0 degrees south latitude, 163.5 degrees east longitude with the two probes being separated from each other by only a few kilometers.

Mars Polar Lander is part of a series of missions in a long-term program of Mars exploration managed by JPL for NASA's Office of Space Science, Washington, D.C. JPL's industrial partner is Lockheed Martin Astronautics, Denver. JPL is a division of the California Institute of Technology in Pasadena.

Mars Polar Lander Mission Status December 4, 1999 4:15 a.m.

Mission controllers for NASA's Mars Polar Lander and the accompanying Deep Space 2 microprobes will continue attempting to communicate with the lander and the probes throughout the weekend.

Controllers did not hear from the spacecraft in their first few attempts to communicate with the lander and the probes during the first 12 hours after the scheduled landing time.

The Deep Space 2 team will try to contact the probes approximately every two hours. The next opportunity for the Mars Polar Lander to contact Earth will be on Saturday evening, Dec. 4 at about 8:30 p.m. PST.

"We're remaining upbeat," said Mars Polar Lander Project Manager Richard Cook at JPL. "We have prepared for various scenarios, and we're trying all the options."

Cook pointed out that if the spacecraft entered a standby, or safe mode, about 20 minutes after landing at 12:15 p.m. PST on Fri., Dec. 3, it would not be able to receive any communications until it takes itself out of safe mode on Saturday evening.

So far, mission controllers have been attempting to communicate with the lander by using its medium gain antenna. If contact has not been established by Sunday morning., Dec. 5, they will try to communicate with the lander by using NASA's currently-orbiting Mars Global Surveyor spacecraft as a relay system. In that scenario, the lander would automatically switch to its UHF radio.

On Fri., Dec. 3, ground controllers attempted to "talk to" the lander at 6:27 p.m. PST. They tried to contact the Deep Space 2 microprobes at 7:30, 9:30, 11:30 p.m., 1:30 and 3:30 a.m. PST.

The Deep Space 2 microprobes, which impacted Mars about 60 kilometers (about 35 miles) north of the lander, will transmit data through Mars Global Surveyor.

"We know that the Mars Global Surveyor relay system is working," said Deep Space 2 Project Manager Sarah Gavit.

Latest estimates indicate that Mars Polar Lander touched down on a gentle slope of about two degrees within an "amphitheater" near the edge of a ridge, according to Dr. David Paige of UCLA, principal investigator for the Mars Volatiles and Climate Surveyor science package on the lander.

Mars Polar Lander is part of a series of missions in a long-term program of Mars exploration managed by JPL for NASA's Office of Space Science, Washington, D.C. JPL's industrial partner is Lockheed Martin Astronautics, Denver. JPL is a division of the California Institute of Technology in Pasadena.

Mars Polar Lander Mission Status December 4, 1999, 5:45 p.m. PST

Flight controllers for NASA's Mars Polar Lander have another opportunity to listen for a signal from the spacecraft beginning tonight at 8:30 p.m. PST. In a meeting late this afternoon they decided to listen for the lander during the first 30 minutes of the communications window, then they would transmit commands to the medium-gain antenna telling the spacecraft to search for Earth.

One scenario that would explain why engineers have not yet heard from the lander is that the spacecraft entered standby, or "safe mode," about 20 minutes after landing shortly after 12 noon PST Friday, Dec. 3. If the lander entered safe mode at that time, it would not be able to receive any communication until it "wakes up" this evening. It would be preprogrammed by onboard software to start looking for Earth starting at noon on Mars, or about 8:30 p.m. PST. The communication window lasts until 10:45 p.m. PST.

If contact has not been established by Sunday morning, Dec. 5, flight controllers will listen to see if the lander transmits via a UHF radio to the currently orbiting Mars Global Surveyor spacecraft. The lander would do that at about 10:50 a.m. PST Sunday if it did not receive commands from Earth telling it not to do so.

Engineers working on NASA's Deep Space 2 microprobes have additional opportunities to hear from the probes this evening at about 5:30 p.m. and 7:30 p.m. PST through the Global Surveyor relay. The probes would automatically begin to transmit at those times if their radio receivers were unable to pick up commands from Global Surveyor.

Mars Polar Lander is part of a series of missions in a long-term program of Mars exploration managed by JPL for NASA's Office of Space Science, Washington, D.C. JPL's industrial partner is Lockheed Martin Astronautics, Denver. JPL is a division of the California Institute of Technology in Pasadena.

Mars Polar Lander Mission Status December 4, 1999 11:15 p.m.

Mission controllers for NASA's Mars Polar Lander are proceeding with their checklist in a continuing attempt to communicate with the spacecraft.

On Sunday, Dec. 5 from 10:50 to 11:00 a.m. Pacific Standard Time, they will try to hear the lander's signal by using NASA's currently-orbiting Mars Global Surveyor spacecraft as a relay system for the lander's UHF radio. Until this point, engineers have tried to reach the lander via its medium gain antenna.

Controllers did not hear from the spacecraft during a communications opportunity on Saturday, Dec. 4 at 8:30 p.m. PST. They hoped to make contact during that window if, after landing, the spacecraft had successfully pointed its antenna toward Earth, then entered a safe, or standby mode.

"Now we can cross that scenario off the list," said Mars Polar Lander project manager Richard Cook of JPL. "We're ready to move on to the next possibility on Sunday morning, which we hope will work if the spacecraft is not in safe mode, but has its antenna pointed incorrectly. We're sprouting ideas as we go along about how to contact the lander."

If contact is not established during that attempt, additional attempts scheduled at this point will be made as follows:

Sunday, Dec. 5, from 10:10 to 11:10 p.m. using the lander's medium gain antenna scan if it is in safe mode but its antenna is not pointed correctly.

Tuesday, Dec. 7, at 12:20 a.m. PST using Mars Global Surveyor if Mars Polar Lander is in safe mode.

Analysis of the landing site reveals the spacecraft would have touched down within 10 kilometers (6 miles) of the target site on the Martian south pole, according to Dr. Sam Thurman at JPL, the lander's flight operations manager. He said they see no surface features that would obstruct the lander's view of Earth and therefore hamper its communications capabilities.

Engineers for the Deep Space 2 microprobes are continuing their attempts to communicate with the probes every two hours. The microprobes, designed to impact Mars about 60 kilometers (about 35 miles) north of the lander, will transmit data through Mars Global Surveyor.

"The probes may have arrived in an area of high slopes, rough terrain or sand dunes," said Deep Space 2 project manager Sarah Gavit.

Mission engineers believe the probes have entered a phase where they broadcast their data automatically for one minute out of every five. "It's also possible that the probes' batteries have not warmed sufficiently to power up the communications system. We're checking into all possibilities."

Mars Polar Lander is part of a series of missions in a long-term program of Mars exploration managed by JPL for NASA's Office of Space Science, Washington, D.C. JPL's industrial partner is Lockheed Martin Astronautics, Denver. JPL is a division of the California Institute of Technology in Pasadena.

Mars Polar Lander Mission Status December 5, 1999

Mission controllers for NASA's Mars Polar Lander have revised their strategy as they continue trying to make contact with the spacecraft.

"We're nearing the point where we've used up our final silver bullets," said the mission's project manager, Richard Cook of JPL, after Sunday night's unsuccessful attempt to communicate with the spacecraft.

Engineers will try to contact the lander again on Tuesday, Dec. 7 at 12:20 a.m. Pacific Standard Time, by directing Mars Polar Lander to use its UHF radio to communicate through a relay system onboard NASA's currently-orbiting Mars Global Surveyor. Most of the attempts to receive a signal from the lander over the past few days have used its medium gain antenna.

"Our probability of success will diminish significantly after this next attempt," Cook said, "but the team is still exploring all possibilities for establishing comunications with the lander."

Controllers are preparing a set of computer commands to have the lander conduct a full sky search for Earth within the next couple of days.

Mars Polar Lander is part of a series of missions in a long-term program of Mars exploration managed by JPL for NASA's Office of Space Science, Washington, D.C. JPL's industrial partner is Lockheed Martin Astronautics, Denver. JPL is a division of the California Institute of Technology in Pasadena.

Mars Polar Lander Mission Status December 5, 1999

Another telecommunications strategy to hear from NASA's Mars Polar Lander produced no results today, but the mission flight team is proceeding through its contingency checklist in continuing attempts to communicate with the spacecraft.

From about 11 to 11:30 a.m. PST today, the team listened for but detected no signals from the lander's UHF transmitter, which would have been relayed through the already-orbiting Mars Global Surveyor spacecraft. This was the first attempt at using the Global Surveyor; until then, engineers had tried to use the lander's medium gain antenna to transmit directly to Earth.

"We continue to work through our plan, which gives us confidence that we haven't exhausted all the possibilities," said Mars Polar Lander Project Manager Richard Cook. "But clearly, the team is getting more frustrated" as attempts to reach the lander yield no results.

No contact has yet been made in continuing efforts to communicate with the two Deep Space 2 Mars microprobes that also impacted Mars about 60 kilometers (abut 35 miles) north of the lander on Dec. 3, said Deep Space 2 Project Manager Sarah Gavit. Mission engineers believe the probes have entered a phase where they broadcast their data automatically for one minute out of every five.

Gavit said that data from last night's try at hearing signals from the probes that could have been recorded on NASA's Mars Global Surveyor orbiter would be reviewed later today. The team will also look for microprobe signals that could be relayed by Global Surveyor during another transmission today, she said. "If we haven't heard from them in the next 24 hours (by about noon Monday PST), we will have exhausted our opportunities to hear from them."

Tonight, Sunday, Dec. 5 from 10:10 to 11:10 p.m., an attempt will be made to listen for signals from the lander that would be sent through its medium gain antenna if the lander is in safe, or standby mode, and its antenna is not pointed correctly.

On Tuesday, Dec. 7, at 12:20 a.m. PST, attempts to hear signals from the lander's UHF transmitter will be made again using Mars Global Surveyor. This attempt would detect signals if Polar Lander is in safe mode. After Tuesday's post-midnight attempt, Cook said, "I think we will be at the point of diminishing returns in terms of getting in contact with the lander."

Mars Polar Lander is part of a series of missions in a long-term program of Mars exploration managed by JPL for NASA's Office of Space Science, Washington, D.C. JPL's industrial partner is Lockheed Martin Astronautics, Denver. JPL is a division of the California Institute of Technology in Pasadena.

Mars Polar Lander Mission Status December 7, 1999, 1:45 a.m.

Mission controllers for NASA's Mars Polar Lander acknowledge that they hold out very little hope of communicating with the spacecraft, but they vow to learn from the experience and continue exploring the Red Planet.

"The Mars Polar Lander flight team played its last ace," said the lander's project manager Richard Cook of JPL following an unsuccessful attempt early Tuesday morning to get the lander to talk to Earth via NASA's currently orbiting Mars Global Surveyor.

Cook said the team will continue trying to communicate with the lander for another two weeks or so, but that expectations for success are remote. Nonetheless, Cook praised the flight team for its heroic attempts to contact the spacecraft, even sleeping on the floors of their offices at times. "We're certainly disappointed, but we're extremely determined to recover from this and go on."

The next communication attempt will take place late Tuesday afternoon, when a 46-meter (about 150-foot) antenna at Stanford University, Stanford, Calif., will listen for a signal from the lander's UHF antenna. Engineers will command the spacecraft to use its medium-gain antenna on Wednesday to begin a scan of the entire sky. During the scan, the antenna is being asked to bend and stretch in every possible direction, in essence "craning its neck" in an effort to be heard by mission controllers on Earth.

Engineers are also considering a plan to command Mars Global Surveyor to fly over the landing site for Mars Polar Lander in coming weeks and take pictures of the area in hopes of spotting the spacecraft.

The Deep Space 2 microprobes that accompanied Mars Polar Lander have also been silent, and project manager Sarah Gavit said she couldn't envision any failure scenario in which the batteries could still hold a charge after four days on Mars.

"Just getting the probes to the launch pad was a measure of success," Gavit said, pointing out that as part of NASA's New Millennium program, the probes were designed to develop and test new technologies in preparation for future missions.

Review boards will be set up within JPL and at NASA to study the cause of the apparent loss and explore ways to prevent a recurrence.

"What we're trying to do is very, very difficult," Cook said. "We hope people, and children in particular, will see from this experience that the mark of a great person, or group of people, is the ability to persevere in the face of adversity."

Mars Polar Lander is part of a series of missions in a long-term program of Mars exploration managed by JPL for NASA's Office of Space Science, Washington, D.C. JPL's industrial partner is Lockheed Martin Astronautics, Denver. JPL is a division of the California Institute of Technology in Pasadena.

Mars Polar Lander Mission Status December 10, 1999

Flight controllers for Mars Polar Lander continued their attempts to communicate with the spacecraft yesterday and today so that they can be certain they have exhausted all possibilities before they conclude their search. While a recovery is still a possibility, the likelihood of hearing from the lander is considered remote at this point.

Yesterday morning at about 2:45 a.m. PST, the team sent commands to begin a lengthy "big sweep" during which the lander uses its steerable medium-gain antenna to scan across the sky. Presumably, it would eventually scan across the area where Earth is and its carrier wave signal would be heard by the Deep Space Network.

Other communication attempts took place today at 3:00 and 6:00 p.m. PST with the 46-meter (about 150-foot) antenna at Stanford University, Palo Alto, Calif., which listened for a signal from the lander's UHF antenna. An earlier attempt by Stanford that had been scheduled for Tuesday was postponed when the Stanford antenna experienced mechanical problems.

The "big sweep" will conclude tonight. Engineers will then begin a process of sending commands to the spacecraft to switch to back-up hardware and will then repeat some of the communications attempts they have already tried.

Mission planners are also working to implement a plan to use Mars Global Surveyor to take pictures of the landing site for Mars Polar Lander starting sometime next week in hopes of spotting the spacecraft or parachute.

Review boards will be set up within JPL and at NASA to study the cause of the apparent loss and explore ways to prevent a recurrence.

Mars Polar Lander is part of a series of missions in a long-term program of Mars exploration managed by JPL for NASA's Office of Space Science, Washington, D.C. JPL's industrial partner is Lockheed Martin Astronautics, Denver. JPL is a division of the California Institute of Technology, Pasadena, Calif.

Mars Polar Lander Mission Status December 22, 1999

Flight controllers for Mars Polar Lander are continuing to work through their fault-tree scenarios in their ongoing attempts to communicate with the spacecraft. Chances of recovering the lander remain remote.

Team members plan to continue looking for a signal from the lander through mid-January, and at that point they will be in a position of having exhausted all possible recovery modes.

Late last week, NASA's orbiting Mars Global Surveyor spacecraft began an imaging campaign to look for evidence of the lander, parachute or aeroshell. So far, nothing has been detected.

The Jet Propulsion Laboratory has appointed a special review board to evaluate the apparent loss of Mars Polar Lander and the Deep Space 2 microprobes. The board will attempt to determine the possible root causes for these losses and identify actions needed to assure success in future Mars landings.

The 12-member JPL board will be chaired by John Casani and is made up of members from JPL, Caltech, other NASA centers and industry. The findings of the board will be presented in a written report due by March 3, 2000. The board will offer its cooperation and assistance to related NASA efforts including the agency's Mars Program Independent Assessment Team.

Mars Polar Lander is part of a series of missions in a long-term program of Mars exploration managed by JPL for NASA's Office of Space Science, Washington, D.C. JPL's industrial partner is Lockheed Martin Astronautics, Denver. JPL is a division of the California Institute of Technology, Pasadena, Calif.

Mars Polar Lander Mission Status January 25, 2000

Mission managers have decided to send another set of commands to Mars to investigate the possibility that a signal detected by a radio dish at California's Stanford University came from Mars Polar Lander.

The commands were sent at 10 a.m. PST today. They will instruct the lander, if it is operating, to send a signal directly to Earth to the antenna at Stanford on Wednesday, January 26, at approximately 1 p.m. PST. The Stanford receiving station will listen again during the window on Wednesday to see if it picks up a signal that could originate from Mars. The results of this test will not be immediate and it will take the team several days to process the data.

Mission managers sent commands several times in December and January instructing Polar Lander to send a radio signal to the 45-meter (150-foot) antenna at Stanford. Although no signal was detected in real-time, the team in charge of the Stanford antenna says that after additional processing of the data they may have detected a signal that could have come from Mars during tests on December 18 and January 4. Because the signal was so weak, it took several weeks for the Stanford team to process their data and reach this conclusion.

"This week's test is a real long-shot, and I wouldn't want to get anyone too excited about it," said Richard Cook, Polar Lander project manager at NASA's Jet Propulsion Laboratory, Pasadena, CA. "The signal that the Stanford team detected is definitely artificial, but there are any one of a number of places it could have originated on or near Earth. Still, we need to conduct this test to rule out the possibility that the signal could be coming from Polar Lander."

If in fact the signal were from Polar Lander, two failures would have had to occur. First, the lander's X-band radio that it would use to transmit directly to Earth would have to be broken. Second, there would have to be a problem somewhere in the relay with Mars Global Surveyor that prevented the signal from being picked up and relayed by the orbiter. It is unlikely that a broken transmitter on the lander could be fixed, and unclear whether a problem with the relay could be resolved.

Although the Stanford data from the previous tests took several weeks to process, the team expects to have results within several days now that they know what they are looking for.

Even if the signal were coming from the lander, there is little hope that any science could be returned. However, it would give the team a few more clues in trying to eliminate possible failure modes.

Mars Polar Lander is managed by the Jet Propulsion Laboratory for NASA's Office of Space Science, Washington, DC. Lockheed Martin Astronautics Inc., Denver, CO, is the agency's industrial partner for development and operation of the spacecraft. JPL is a division of the California Institute of Technology, Pasadena, CA.

Mars Polar Lander Mission Status January 27, 2000

Radio scientists at California's Stanford University are continuing to process data from communications attempts made yesterday and today to determine if they have picked up a signal coming from Mars Polar Lander using their 45-meter (150-foot) antenna.

There were three 30-minute communications windows yesterday and three more listening windows today. It takes about 18 hours to process the data from each window. So far, Stanford scientists have looked at one of the three data sets taken yesterday and say they have not detected anything unusual. It will take several days to complete the processing and the researchers do not expect to have confirmation of a signal until some time next week.

"The signal we are looking for is very, very weak, about 1 watt of power — or like looking for a Christmas tree light on Mars," said Richard Cook, Polar Lander project manager at NASA's Jet Propulsion Laboratory, Pasadena, Calif. "Because of the weakness of the signal, we want to be absolutely sure we have something so we will check and double check these data before we will be willing to confirm there is a signal."

Mars Polar Lander is managed by the Jet Propulsion Laboratory for NASA's Office of Space Science, Washington, D.C. Lockheed Martin Astronautics Inc., Denver, Colo., is the agency's industrial partner for development and operation of the spacecraft. JPL is a division of the California Institute of Technology in Pasadena.

Mars Polar Lander Mission Status January 31, 2000

Mission managers for Mars Polar Lander report that radio scientists at Stanford University have not detected a signal from the spacecraft in data they collected last week. Stanford will continue to analyze the data and it is still possible that more detailed analysis might reveal a signal.

In the meantime, additional radio telescopes around the world have offered their assistance in helping to confirm if the signal picked up by Stanford is from Polar Lander. The project has accepted offers of help from an array of fourteen 25-meter (82-foot) antennas at Westerbork in The Netherlands as well as the 76-meter (about 250-foot) antenna at Jodrell Bank, near Manchester, England and an array located near Bologna, Italy.

"The international community has shown a real interest in being involved in our search. We appreciate their efforts and I think it shows that Mars is something that captivates everyone's imagination," said Richard Cook, project manager for Mars Polar Lander at NASA's Jet Propulsion Laboratory, Pasadena, CA.

New commands will be sent to the lander from NASA's Deep Space Network around the clock on Tuesday and Wednesday this week, Feb. 1 and 2. These commands will essentially tell the spacecraft, if it is functioning, to reset its clock and send a signal to Earth. On Friday, Feb. 4, windows will open for the antennas in the Netherlands, England and Italy to begin listening. The antenna at Stanford may also listen during these windows.

The one-way light time from Earth to Mars is currently about 16 minutes. Mars is presently about 300 million kilometers (181 million miles) from Earth.

Mars Polar Lander is managed by the Jet Propulsion Laboratory for NASA's Office of Space Science, Washington, D.C. Lockheed Martin Astronautics Inc., Denver, Colo., is the agency's industrial partner for development and operation of the spacecraft. JPL is a division of the California Institute of Technology, Pasadena, Calif.

Mars Polar Lander Mission Status February 4, 2000

Radio telescopes in The Netherlands, England and at Stanford University in California have begun listening for a possible signal from Mars Polar Lander today.

The array of fourteen 25-meter (82-foot) antennas at Westerbork in The Netherlands as well as the 76-meter (about 250-foot) antenna at Jodrell Bank, near Manchester, England have three 30-minute listening opportunities today. The 45-meter (150-foot) antenna at Stanford University is also able to listen during these windows. An array located near Bologna, Italy is not being used today.

Mission managers for Polar Lander say it will take each of the stations some time to review their data. "We want to make sure we have checked and double-checked these data before we can confirm whether or not there is a signal," said Richard Cook, project manager for Mars Polar Lander at NASA's Jet Propulsion Laboratory, Pasadena, Calif. "I don't think we'll know anything either way until sometime next week."

Mars Polar Lander is managed by the Jet Propulsion Laboratory for NASA's Office of Space Science, Washington, D.C. Lockheed Martin Astronautics Inc., Denver, Colo., is the agency's industrial partner for development and operation of the spacecraft. JPL is a division of the California Institute of Technology in Pasadena.

Mars Polar Lander Mission Status February 7, 2000

Radio telescopes in the Netherlands, the United Kingdom and at Stanford University in California are preparing for a second set of observations on Tuesday, February 8, to continue to listen for a possible signal from Mars Polar Lander.

"We have received tremendous support from the observatories at Westerbork, Jodrell Bank and Stanford. They have been working around the clock to help us and we are grateful for their efforts," said Richard Cook, project manager for Mars Polar Lander at NASA's Jet Propulsion Laboratory, Pasadena, Calif.

A second round of observations is required in order to eliminate remaining uncertainty about the operational status of the batteries on the lander. The operations Tuesday will consist of two 30-minute listening windows with a two-hour "cooling down" period in between. JPL will send a new set of commands to the spacecraft Monday night through the Deep Space Network. An additional antenna near Bologna, Italy, will also be used to listen on Tuesday.

Mission managers at JPL theorize that the lander may be in a different configuration than expected, and as a result the spacecraft might not have executed or received the commands that were sent last week.

Results from the listening windows on Friday, February 4, have not been conclusive. Both radio telescopes at Westerbork in the Netherlands and Jodrell Bank in the United Kingdom have operated optimally throughout the experiments. Observational data from both telescopes have been analyzed extensively, but nothing has been found in the data to suggest transmissions from Polar Lander. The two telescopes have a similar sensitivity for detecting signals from the lander, and thus far all signals they have detected are thought to be of terrestrial origin.

Analysis of the data from Stanford has also not yielded any conclusive results, and scientists there are still continuing to review that data.

Exhaustive analysis of the new data taken on Tuesday will take at least until the end of this week.

The Westerbork Synthesis Radio Telescope is operated by Astron, the Netherlands Foundation for Research in Astronomy, and is financed by the Netherlands Organization for Scientific Research.

The Lovell Telescope at the Jodrell Bank Observatory is operated by the University of Manchester's Department of Physics and Astronomy.

Mars Polar Lander is managed by the Jet Propulsion Laboratory for NASA's Office of Space Science, Washington, D.C. Lockheed Martin Astronautics Inc., Denver, Colo., is the agency's industrial partner for development and operation of the spacecraft. JPL is a division of the California Institute of Technology, Pasadena, Calif.

Mars Polar Lander Mission Status February 11, 2000

Initial analysis of data taken on Tuesday by radio telescopes in the Netherlands and Italy has shown no obvious signal from Mars Polar Lander, but exhaustive review of the data is continuing with a final report due next week.

Analysis of data taken at Stanford University in California is ongoing with no signal detected so far. A telescope at Jodrell Bank in the United Kingdom was not able to collect any data due to high winds at that facility.

"Our plan for the next week is to temporarily end active efforts to listen for a signal," said Richard Cook, project manager for Mars Polar Lander at NASA's Jet Propulsion Laboratory, Pasadena, Calif. "We are evaluating several scenarios for future listening attempts that could take place at the end of this month." Mission managers are also reviewing information about the Mars relay link between Mars Global Surveyor and the lander.

Mars Polar Lander is managed by the Jet Propulsion Laboratory for NASA's Office of Space Science, Washington, D.C. Lockheed Martin Astronautics Inc., Denver, Colo., is the agency's industrial partner for development and operation of the spacecraft. JPL is a division of the California Institute of Technology, Pasadena, Calif.

Mars Polar Lander Mission Status February 16, 2000

Radio scientists at NASA's Jet Propulsion Laboratory have made a detailed analysis of data taken by a radio telescope at Stanford University on Jan. 4 and believe the suspect signal is more likely of terrestrial origin and not from Mars Polar Lander.

Further analysis of data taken by radio telescopes in the Netherlands, Italy and at Stanford on Feb. 8 has not yielded any signal from Mars Polar Lander.

Extensive analysis of all data taken during the last few weeks is ongoing.

"We saw something in the Jan. 4 data that had all the earmarks of a signal and we felt we had to check it out. In parallel, we started to perform analysis to determine if the signal came from Mars," said Richard Cook, project manager for Mars Polar Lander at JPL. "Based on the latest results, it is unlikely that we will attempt to listen again."

Mars Polar Lander is managed by the Jet Propulsion Laboratory for NASA's Office of Space Science, Washington, D.C. Lockheed Martin Astronautics Inc., Denver, Colo., is the agency's industrial partner for development and operation of the spacecraft. JPL is a division of the California Institute of Technology, Pasadena, Calif.

PROPOSED FUTURE
MISSIONS

Donald Savage Headquarters, Washington, DC May 12, 2000 (Phone: 202/358-1547)

RELEASE: 00-81

NASA IDENTIFIES TWO OPTIONS FOR 2003 MARS MISSIONS; DECISION IN JULY

Mars Surveyor 2001 - Mars Mobile Lander - Mars Scientific Orbiter

In 2003, NASA may launch either a Mars scientific orbiter mission or a large scientific rover which will land using an airbag cocoon like that on the successful 1997 Mars Pathfinder mission. The two concepts were selected from dozens of options that had been under study. NASA will make a decision on the options, including whether or not to proceed to launch, in early July.

Two teams, one centered at NASA's Jet Propulsion Laboratory (JPL), Pasadena, CA, and the other at Lockheed Martin Astronautics, Denver, CO, will conduct separate, intensive two- month studies to further define the concepts. In the studies the teams also will evaluate risk, cost, and readiness for flight, allowing 36 months of development leading to a May 2003 launch date.

The reports will be submitted for review to Mars Program Director Scott Hubbard at NASA Headquarters,

Washington, DC. Dr. Ed Weiler, Associate Administrator for Space Science at NASA Headquarters, will make the final decision of which mission — if any — to launch in the 2003 opportunity. If selected, the cost of the 2003 mission will be about the same as the successful 1997 Mars Pathfinder mission (adjusted for inflation).

"Our budget will support only one of these two outstanding missions for the 2003 launch opportunity, and it will be a very tough decision to make," said Dr. Weiler. "Following this decision, later in the year we will have a more complete overall Mars exploration program to present to the American public which will represent the most exciting, most scientifically rich program of exploration we have ever undertaken of the planet Mars."

"These two mission concepts embody the requirements we have learned through the hard lessons of two recent Mars mission failures, and either one will extend the tremendous scientific successes we have had with the Mars Global Surveyor and Mars Pathfinder," said Hubbard.

The Mars Surveyor Orbiter is a multi-instrument spacecraft similar in size to the currently operating Mars Global Surveyor. It is designed to recapture all the lost science capability of the Mars Climate Orbiter mission as well as to seek new evidence of water-related materials. The orbiter's mission will be to study the martian atmosphere and trace the signs of ancient and modern water. Its instruments potentially will include a very high- resolution imaging system, a moderate-to-wide-angle multicolor camera, an atmospheric infrared sounder, a visible-to-near-infrared imaging spectrometer, an ultraviolet spectrometer, and possibly a magnetometer and laser altimeter. Telecommunications relay equipment that could be used to support Mars missions for 10 years also would be included.

The rover is a based on the Athena rover design, which already has been operated in field tests and previously was considered for the cancelled 2001 lander mission. The concept being proposed for the 2003 mission involves packaging the 286-pound (130-kilogram) rover in a system similar to the 1997 Mars Pathfinder structure, which would be cushioned on landing by airbags. Unlike the 1997 mission, however, the four-petal, self- righting enclosure would serve only as a means to deliver the rover to the surface and not function as a science or support station.

After landing, the Mars Mobile Lander would serve as a self-contained mission, communicating directly with Earth or with an orbiting spacecraft band as the rover traverses the martian terrain. The rover would be capable of travelling up to 100 yards (100 meters) a day, providing unprecedented measurements of the mineralogy and geochemistry of the martian surface, particularly of rocks, using a newly developed suite of instruments optimized to search for clues about ancient water on Mars. The mobile surface-laboratory will be able to gain access to a broad diversity of rocks and fine-scale materials for the first time on the surface of Mars, in its search for evidence of water-related materials. The rover's mission would last for at least 30 days on the surface.

"We are opening up a new frontier on the Red Planet, and we can't afford to overlook anything," Weiler added. "We have to make sure we plan it well, provide our people with the tools they need, and do whatever it takes to ensure the best possible chances for success."

MANNED MARS LANDING

PRESENTATION TO THE SPACE TASK GROUP
BY DR. WERNHER von BRAUN
AUGUST 4. 1969

(Color pictures associated with this document are at the end of this book. Picture numbers are in square brackets after each title.)

INTEGRATED PROGRAM

With the recent accomplishment of the manned lunar landing, the next frontier is manned exploration of the planets. Perhaps the most significant scientific question is the possibility of extraterrestrial life in our solar system. Manned planetary flight provides the opportunity to resolve this universal question thus capturing international interest and cooperation. The information presented here describes a method of landing man on the planet Mars in 1982. The scientific goals of the mission are described and the key decision dates are identified. The unmanned planetary missions described in Part 1 are critical to the final designs selected.

The 1981 manned Mars mission (1982 landing on Mars) is shown as an integral part of the total space program for the next two decades. The systems and experience resulting from the Apollo program and the missions proposed for the 1970's provide the technical and programmatic foundation for this undertaking. A 1982 manned Mars landing is a logical focus for the programs of the next decade. Although the undertaking of this mission will be a great national challenge, it represents no greater challenge than the commitment made in 1961 to land a man on the moon.

MANNED PLANETARY MISSION - 1981 MARS LANDING MISSION PROFILE [1]

Several different modes are possible for accomplishing Mars landing missions, each with its peculiar advantages and disadvantages. The typical Mars Landing Mission begins with the boost of the planetary vehicle elements into Earth orbit (1) utilizing the Saturn V and Space Shuttle vehicles. Following assembly of the complete planetary vehicle in Earth orbit (2), the Earth departure phase of the mission is initiated (3). The Mars vehicle then begins a 270-day journey to Mars. This is by no means an idle phase of the mission. In addition to observations of Mars, many other experiments and measurements will be made on both the Earth-to-Mars and Mars-to-Earth legs of the trip that are of prime scientific importance. The spacecraft represents a manned laboratory in space, free of the disturbing influences of the Earth. The fact that there will be two observation points, Earth and the spacecraft, permits several possible experiments regarding the temporal and spatial features of the interplanetary environment. In addition, the spacecraft can be used to supplement and extend numerous observations conducted from Earth orbital space stations, particularly in the field of astronomy. It is possible, for example, that as yet unidentified comets might be observed for the first time.

Upon arriving at Mars, the space ship is propulsively braked into orbit (4) in the same fashion that the Apollo moon ship is placed into lunar orbit. The ship remains in Mars orbit for about 80 days (5), during which time the Mars Surface Sample Return (MSSR) probes and the Mars Excursion Modules (MEM) are deployed and the surface exploration takes place. At the end of the Mars capture period, the spacecraft is boosted out of Mars orbit (6). The return leg of the trip lasts about 290 days, during which many experiments and observations are again conducted. A unique feature of the homeward trip is a close encounter with Venus (7), about 120 days after departing Mars. Probes will be deployed at Venus during this passage, in addition to the radar mapping measurements that will be made.

The approximate two-year journey ends with the return to Earth orbit (8), and following medical examinations the crew will be returned to Earth via a Space Shuttle.

EARTH ORBIT DEPARTURE [2]

The Mars Landing Mission can be accomplished with a single planetary vehicle assembled in Earth orbit. There are, however, advantages in deploying two ships on the mission because of the long duration. One obvious advantage is crew safety, each spacecraft being designed to accept the crew of the sister ship in the event of a major failure. This approach also allows more exploration equipment to be carried on the expedition and enhances the probability of achieving mission objectives.

In the current concept, each vehicle assembled in earth orbit consists of three nuclear propulsion modules (Nuclear Shuttles) side-by-side, with the planetary spacecraft docked to the center module. Each spacecraft is nominally capable of sustaining a crew of six people for two years, or a crew of twelve for an extended period, in case of emergency. The earth orbit departure maneuver is initiated with the firing of the outer two propulsion modules as illustrated.

EARTH ORBIT DEPARTURE MANEUVERS [3]

The two planetary vehicles are assembled in a circular Earth orbit. Each vehicle consists of a spacecraft, two nuclear shuttle vehicles for Earth departure propulsion, and one nuclear shuttle for the remaining propulsion requirements through the Mars mission.

Following assembly and checkout in Earth orbit, each of the planetary ships is accelerated by the two outer nuclear shuttles to trans-Mars injection velocity. The two outer nuclear shuttles are then shut down, separated from the planetary vehicle, oriented for retro-fire, and then retro-fired to place them on a highly elliptic path returning to the original assembly orbit altitude. After a coast of several days, the nuclear shuttles arrive at the original assembly orbit altitude and are retro-fired again to place them into a circular orbit. The nuclear shuttles rendezvous with a space station to be checked out and refueled for further utilization. The nuclear shuttles which return to Earth orbit will be available for transfer of fuel and supplies to geosynchronous orbit or to lunar orbit.

MISSION WEIGHT HISTORY [4]

Each completely assembled and fueled vehicle for the 1981 Manned Mars Landing Mission weighs approximately 1.6 million pounds just prior to the Earth orbit departure maneuver. As the vehicle reaches the necessary earth departure velocity and the outer two Nuclear Shuttles are separated, its weight has decreased to 675,000 pounds. Thus nearly 60 of the initial vehicle weight is required for the Earth departure maneuver.

During the Earth-to-Mars phase of the mission, the only significant weight losses are due to midcourse corrections and life support consumables which amount to about 25,000 pounds. Thus at Mars arrival, just prior to the propulsive braking into Mars orbit, the vehicle weighs 650,000 pounds. While in Mars orbit, all the surface sample return probes and the Mars Excursion Module are deployed. These and other losses reduce the vehicle weight to 380,000 pounds just prior to Mars departure.

Midcourse corrections and life support consumables are again lost during the Mars-to-Earth phase of the trip. Also lost are the probes deployed at Venus encounter. As the vehicle approaches Earth for the final propulsive maneuver, it weighs 190,000 pounds. The total weight returned to earth orbit, consisting of the empty Nuclear Shuttle and the Mission Module (minus expendables) is about 160,000 pounds. Of the initial weight of 1.6 million pounds in Earth orbit, approximately 1.2 million, or nearly 75%, are propellant.

EN ROUTE SPACECRAFT CONFIGURATION [5]

The forward compartment of the spacecraft is an unpressurized area housing the Mars Excursion Module (MEM), an airlock to provide for pressurized transfer to the MEM and for extra vehicular activities, and unmanned probes. Six Mars Surface Sample Return probes and two Venus probes are carried on each spacecraft. Immediately aft of the airlock is the Mission Module which provides the crew with a shirt sleeve environment, living quarters, space vehicle control capability, experiment laboratories, radiation shelter, etc. The functional areas are distributed on four decks. This compartment is occupied by the crew of six for the entire mission except during the Mars surface activity.

At the aft end of the mission module adjacent to the nuclear shuttle is a biological laboratory where the Mars surface samples are received and analyzed. The bio-lab is sterilized prior to Earth departure and remains sealed until initial remote analyses of the samples have been accomplished.

EN ROUTE SPACECRAFT CONFIGURATION ARTIFICIAL GRAVITY MODE [6]

The ability of man to withstand a zero gravity environment for periods of time exceeding a few weeks is still an unknown. The Saturn Workshop to be flown in the early seventies will determine man's capabilities in a zero gravity environment for a few months. It will remain, however, for the Space Station to demonstrate man's capabilities for the longer periods required for the Manned Mars Landing Mission.

The option to provide artificial gravity for the crew during the planetary trip must be kept open until conclusive results of man's abilities are established. If early missions indicate the need for artificial gravity the two spaceships can be docked end-to-end and rotated in the plane of the longitudinal axis during extended cost periods.

MARS SURFACE SAMPLE RETURN [7]

The currently funded Viking Project is aimed at placing soft landers on the surface of Mars in 1973 and perhaps in follow-on opportunities. These probes will provide important clues concerning the existence of life on Mars, but will not fully answer the questions as to the possible pathogenic nature of such life. Hence, on the first manned mission, it may be desirable to obtain surface samples prior to the actual landing of man and subsequent contamination of the planet.

Surface samples can be obtained with sterile unmanned probes deployed from the manned spacecraft. The probe would descend from the orbiting spacecraft, land on the Martian surface, automatically gather a sample and return it to the biological laboratory in the spacecraft for analysis. If the analysis revealed no significant biological hazards, man can then proceed to the surface and the samples could be returned to Earth for more detailed analysis, along with the more selective (but perhaps Earth-contaminated) samples obtained by man.

To provide a reasonably representative coverage of the various sections of the planet, six surface sample probes are carried in each spacecraft.

MARS SURFACE EXCURSION [8]

The three-man landing party from each ship is carried from the orbiting spacecraft to the surface in the Mars Excursion Module (MEM). Except for the effects of the Martian atmosphere, the landing and return to orbit sequence is analogous to the Apollo lunar landing operation utilizing the Lunar Module. In the case of Apollo 11, the Mars orbit spacecraft is analogous to Columbia and the MEM is analogous to Eagle.

Following final checkout, the MEM is separated from the spacecraft and de-orbited (1) by the retro-firing of a small rocket motor. The MEM is then aerodynamically decelerated (2) as it falls toward the surface. As the MEM approaches the surface, the protective shroud and a portion of the heat shield are jettisoned (3). Jettisoning of the shroud allows use of the ascent stage as an abort vehicle, if required before landing. The descent stage engine then provides terminal braking (4) and hovering just prior to touchdown (5). The MEM then-spends 30 to 60 days on the surface of Mars (6).

At the conclusion of surface operations, the ascent stage is fired (7) to initiate the return to Mars orbit and the waiting spacecraft. Propellant tanks are staged (discarded) during ascent to effect weight saving (8). After achieving proper orbit conditions, the MEM will rendezvous and dock with the spacecraft (9). Following docking, the crew will transfer to the Mission Module, and the MEM will be discarded.

MARS EXCURSION MODULE CONFIGURATION [9]

The Apollo-shaped Mars Excursion Module (MEM) is designed to carry three men to the surface of Mars and return the crew, scientific data, and samples to the spaceship. It provides living quarters and a laboratory during the 30-60 day stay on the Mars surface. The MEM consists of descent and ascent stages. The ascent stage houses the three-man crew during entry, descent, landing, and ascent. The ascent stage consists of the control center, ascent engine and propellant tanks. The descent stage contains the crew living quarters and laboratory for use while on Mars, the descent engine and propellant tanks, landing gear and an outer heat shield for the aerodynamic entry phase of the descent. A small one-man rover vehicle is provided in the descent stage for surface mobility. All descent stage equipment is left on the Martian surface. The capability is provided for one man to land a MEM and bring a stranded crew back to the ship orbiting Mars.

The diameter of the MEM at its base is 30 feet. At departure from the spacecraft, the MEM weighs about 95,000 pounds.

MARS SURFACE ACTIVITY [10]

Man's first step on Mars will be no less exciting than Neil Armstrong's first step on the moon. The Mars surface activity on the first mission will be similar in many ways to the Apollo 11 moon surface activity. Notable, however, is the much longer stay time (30-60 days per MEM), thus allowing more extensive observations, experimentation and execution of mission scientific objectives. The small rover vehicle allows trips to interesting surface features beyond the immediate landing area. Surface operations include experiments to be performed in the MEM laboratory as well as the external operations on Mars' surface.

During the planetary surface operations, the men in the orbiting spacecraft continue their experimentation observations, monitor the surface operations, and maintain the necessary spacecraft operations.

SCIENTIFIC OBJECTIVES - MANNED MARS LANDING MISSION

Geological and geophysical investigations of Mars are significant because Mars probably closely paralleled the earth in origin and the development. Basically the data required about Mars are: (1) its physical, mineralogical, and chemical composition, (2) distribution of surface material and the processes by which features and material were formed, altered, transported, and distorted, (3) the record of any life there, and (4) any major events preserved in Martian rocks. These investigations require the presence of a skilled observer functioning as an interpretative scientist..

Perhaps the single, most consuming scientific question of the space program is: "Does extraterrestrial life exist in our solar system?" Has life ever existed on Mars? Does it exist now? Are conditions such that some form of life could exist? Preliminary data indicate that some lower forms of life can survive in the Martian environment, and conceivably in isolated areas higher forms of life may exist. Man on Mars will be able to study not only the forms of life indigenous to Mars, but also the behavior of terrestrial life forms transplanted to the Martian environment. Drilling for or locating water will be an early objective on Mars, and its discovery would open many possibilities for utilization of Mars. For example, it might become possible to produce rocket fuel for the return trip on later missions.

SCIENTIFIC OBJECTIVES MANNED MARS LANDING MISSION

- MAKE GEOPHYSICAL OBSERVATIONS
- COLLECT SOIL AND ATMOSPHERIC SAMPLES
- STUDY LIFE FORMS
- STUDY BEHAVIOR OF TERRESTRIAL LIFE FORMS IN MARS ENVIRONMENT
- SEARCH FOR WATER AND USABLE NATURAL RESOURCES

ASCENT FROM MARS SURFACE TO MARS ORBIT [11]

Using the MEM descent stage as a launch platform, the ascent stage delivers the crew, scientific data and samples back to the orbiting spaceship. The return payload, consisting mainly of samples, data, and miscellaneous equipment weighs approximately 900 pounds.

MARS DEPARTURE [12]

At the completion of the 80-day period at Mars, the planetary spaceships will begin the return leg of the journey. The nuclear stage is ignited for this propulsive maneuver, boosting each spaceship, out of Mars orbit.

With the extensive Mars exploration activities behind them, the crew at this point can begin a more thorough analysis of the data and samples gathered at Mars, and prepare for the next major milestone of the trip a close encounter with the planet Venus.

RELEASE VENUS PROBES [13]

In this mission profile, a passage close to Venus during the return leg results in lower approach velocities upon Earth return, and thus lower weight requirements. In addition, this provides an opportunity for close-proximity observations and experiments at Venus.

Since clouds obscure the surface of Venus, radar mapping will be conducted to obtain information on surface features. Two probes are provided on each spacecraft, and will be deployed during the Venus passage.

LONG RANGE PLAN

EARTH RETURN [14]

The manned Mars landing mission concludes with the return to earth orbit, using the last of the propellant in the nuclear stage for the braking maneuver. An optional earth return mode would allow the crew to make direct aerodynamic entry (Apollo-style). Until a better assessment can be made of the back contamination hazard (the return by man of pathogens that might prove harmful to earth inhabitants), a more conservative approach has been planned, i.e. the return of the crew to earth orbit for a quarantine period. Another advantage of the orbit return mode is that the nuclear stage and mission module are available for possible reuse.

Once the spacecraft achieves the desired earth orbit, it will rendezvous with the waiting space base, where the crew will receive thorough medical examinations before returning to Earth via the Space Shuttle. The return samples could also be further examined prior to their return to Earth.

EARTH LAUNCH AND ORBITAL ASSEMBLY [15]

The Mars spacecraft and the Nuclear Shuttles will be launched into Earth orbit using two-stage Saturn V's. The Mars mission crews, expendables (food, water, etc.) and a portion of the fuel for the Nuclear Shuttles are carried to orbit with Space Shuttles.

Nuclear Shuttles for the maneuvers at Mars and for Earth-return (the center shuttle) will be launched from Earth surface, with approximately one-half of its propellant on board. The remaining fuel will be provided via Space Shuttle flights.

It is envisioned that Nuclear Shuttles already in orbit for other uses can be employed for the Earth-escape maneuver (the outer shuttles). Fuel for these shuttles will also be provided from Earth surface via Space Shuttle flights. The Mars spacecraft would be assembled and checked out on the ground prior to (unmanned) launch. Then after orbital assembly, checkout, and crew boarding, the vehicles are ready for departure.

LONG RANGE PLANNING SCHEDULE

The manned Mars landing mission that has been presented is but one mission in a total space program. It is desirable at this time to consider this mission in context of the total program. This is shown on the facing page. Missions depicted for the decade of the 1970's are taken from the Integrated Space Program presented recently to the Space Task Group. These missions have been extrapolated into the 1980's in order to give a continuous program which could be used in determining funding requirements for the 1970-80 time period. The rationale for the extrapolation was to project an aggressive manned program in all areas utilizing the systems developed in the 1970's. Annual funding requirements for this program are shown on the following chart.

LONG RANGE PLAN FUNDING

This funding shown in 1969 dollars reflects major emphasis in the mid 1970's on an expanding Earth orbital and lunar program. By 1980 the planetary program becomes prominent in the funding requirements while the lunar and Earth orbital funding remain essentially constant.

Major Earth orbital cost items include development of the common mission module, space shuttle, earth orbital experiments, and the operational costs associated with these items. The lunar costs include funds for development and operation of the nuclear shuttle, space tug, common mission module modifications for lunar surface, and accompanying lunar experiment costs.

The planetary costs include funds for development and operation of the Mars excursion module and planetary experiments plus the operation of the nuclear stage, mission module, launch vehicles and unmanned payloads. An institutional NASA base approximately equal to that which exists today is included in the funding. Additionally, funds for continuing technical efforts in the areas of aeronautics and space technology are shown.

LONG RANGE PLAN FUNDING

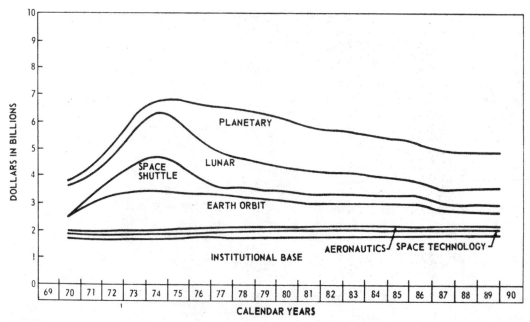

MSFC-69-PD-SA 197

SPACE AND THE GNP

The accompanying curves indicate the Gross National Product and the NASA budget plotted as a percentage of this GNP for a 30-year span from 1960 to 1990. The GNP line reflects actual data through mid-1969 and is increased by 4% per annum thereafter. This is a projected "real" growth rate without inflation.

The NASA percentage of the GNP reached a peak of 4/5 of 1% in 1964. It has dropped steadily since then to approximately 2/5 of 1% at present. The projected space program funding requirements would cause this value to increase to 3/5 of 1% in 1974 and decline slowly throughout the remaining years. By 1990, the NASA budget would only equal 1/5 of 1% of the GNP.

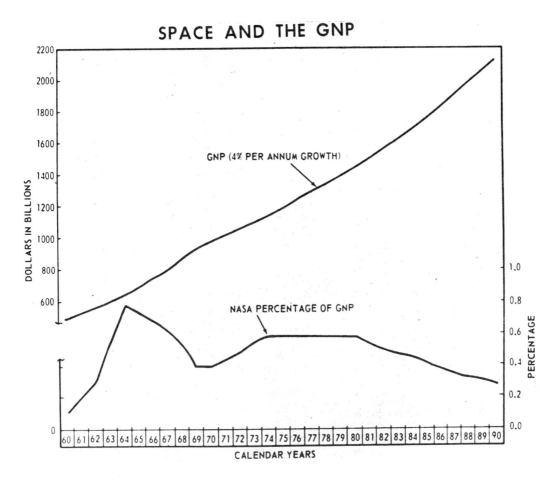

MAJOR SCHEDULE MILESTONES 1982 MANNED MARS LANDING

The development schedule for the 1982 Manned Mars Landing culminates with the December 1981 Earth departure. The five major hardware elements required are shown. The Nuclear Shuttle, Space Shuttle and the mission module are required to carry out the planned program for the 1970's. The Mars Excursion Module and the Mars Surface Sample Return Probe are needed exclusively for this landing mission.

"Go-ahead" dates refer to the start of a hardware development program and "1st flight" indicates the first test flight of the system. Sufficient test time has been included to allow an earth orbital practice mission because of the long mission duration and the infrequency of Mars mission opportunities.

The option to make the 1982 Mars Landing can be kept open until 1974 (Mars Excursion Module go-ahead) provided the Nuclear Shuttle, the Space Shuttle, and the Common Mission Module are designed to be responsive to the ultimate requirements of the Mars Landing Mission.

INTEGRATED PROGRAM LOGIC

In order to accomplish the Integrated Space Program for the funding shown, it was necessary to develop an integrated program logic. This logic required maximum use of Apollo systems, multi-mission application of common modules, and the introduction of reusable transportation systems. The application of the logic in formulating this Integrated Program made it economically feasible.

The long duration manned planetary missions will be made possible by the operational experience gained and the systems developed for the Integrated Program in the 1970's. The Earth orbital program will qualify both man and systems for long duration while the lunar program will provide techniques and experience for surface operations. The automated precursor missions provide Mars atmospheric and surface characteristics necessary to design and develop mission peculiar systems. In order to accomplish this high energy mission, it will be necessary to augment the chemical propulsion systems with a more efficient nuclear system.

MAJOR SCHEDULE MILESTONES
1982 MANNED MARS LANDING

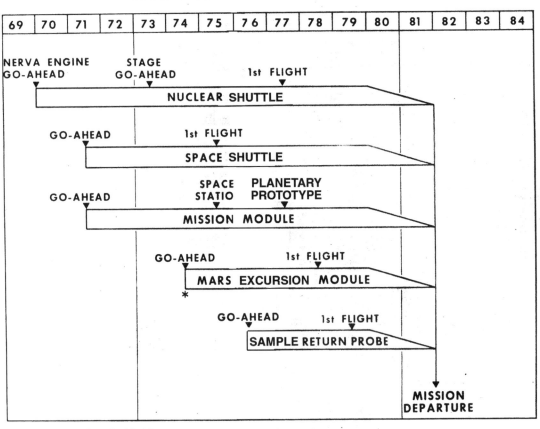

MSFC-69-PD-SA 181

With the systems developed and the experience gained from the early Mars mission, it will be possible to evolve to a temporary base by the end of the decade of the eighties if early explorations prove interesting.

INTEGRATED PROGRAM LOGIC

- MAXIMUM USE OF APOLLO SYSTEMS
- APPLICATION OF COMMON MISSION MODULES
- DEVELOPMENT OF REUSABLE TRANSPORTATION SYSTEMS
- EARTH ORBIT DEMONSTRATION OF MANNED LONG LIFE SYSTEMS
- LUNAR SURFACE ACTIVITY AS PREPARATION FOR MARS SURFACE OPERATIONS
- UTILIZATION OF AUTOMATED PRECURSORS
- ADD NUCLEAR PROPULSION TO CHEMICAL PROPULSION FOR DEEP SPACE OPERATIONS

EVOLUTION TO MARS BASE INTEGRATED PROGRAM 1970-1990

The application of the described logic results in the program shown on the next 2 charts. This chart depicts the integrated program for the 1970's already submitted to the Space Task Group. The next chart represents the logical extension of this program including the Manned Mars Landing Mission.

Maximum utilization of Apollo systems in the early 1970's is demonstrated by the use of two Saturn V Workshops in Earth orbit and additional Apollo type missions for further lunar exploration. The Saturn V vehicle is used to launch these missions. The Apollo systems are shown in blue.

Three new systems are required during the 1970's to provide the commonality and reusability necessary to increase the capabilities in the last half of the decade. The systems are a new mission module (green), a space shuttle (red) and the tug (yellow). Two mission modules are integrated to serve as a space station in Earth orbit until the end of the decade. This same basic mission module is used in lunar orbit and as a lunar surface base.

The space shuttle is the key element to future space operations. It is used to transport people, equipment, and supplies to Earth orbit in support of all subsequent missions. The space tug is required to support lunar missions and will also be used as a maneuvering unit at the Earth orbital space station. Reuse of these two systems will greatly reduce operational costs over that of the present Apollo-type throw-away systems.

In the late 1970's, the reusable nuclear shuttle (orange) is introduced as the primary space propulsion system. Its initial use is to support increased lunar activity.

The systems of the 1970's are the foundation for building major space facilities in the 1980's. The 1975 space station evolves into a Space Base that can support up to 100 people by the early 1980's. This facility allows extensive multi-disciplinary scientific activities as indicated. A geosynchronous station is practical in this time period with the availability of the nuclear shuttle. Similarly, these new systems permit increased lunar operations.

The logical culmination of the next decade is the 1981 Manned Mars Landing Mission. The systems and experience gained in the 1970's make this a feasible undertaking.

In addition to serving as a focus for the next decade, the 1981 Mars landing is the threshold for manned planetary exploration of the 1980's.

INTEGRATED PROGRAM

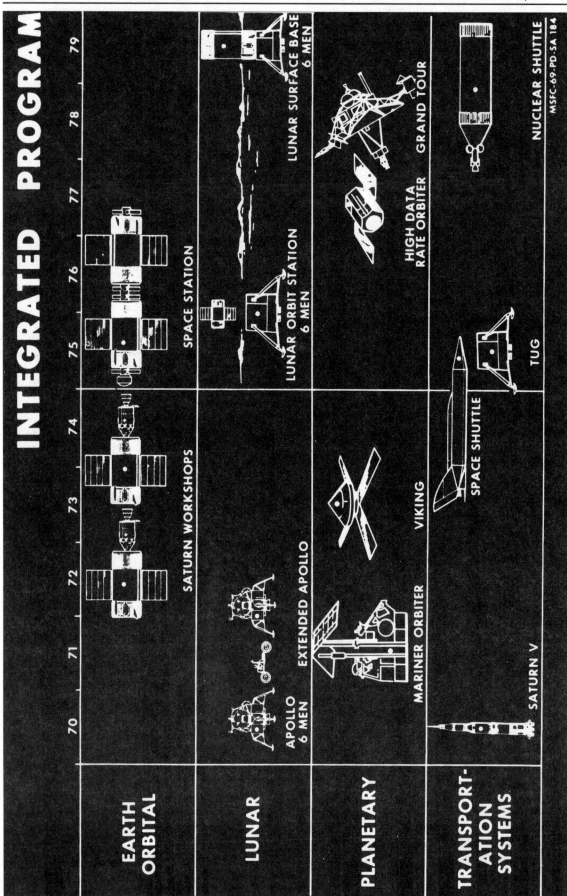

MSFC-69-PD-SA 184

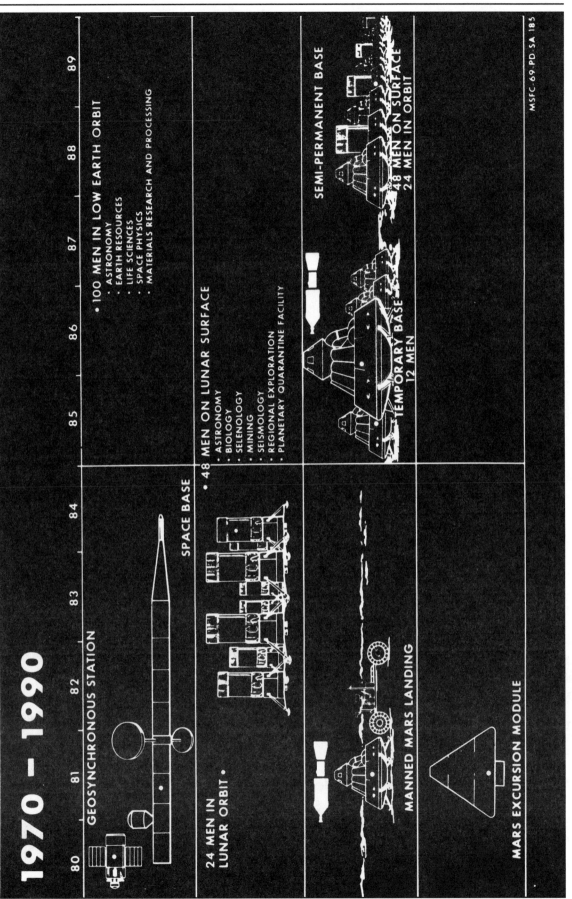

1970 – 1990

GEOSYNCHRONOUS STATION

SPACE BASE

- 100 MEN IN LOW EARTH ORBIT
  - ASTRONOMY
  - EARTH RESOURCES
  - LIFE SCIENCES
  - SPACE PHYSICS
  - MATERIALS RESEARCH AND PROCESSING

24 MEN IN LUNAR ORBIT •

- 48 MEN ON LUNAR SURFACE
  - ASTRONOMY
  - BIOLOGY
  - SELENOLOGY
  - MINING
  - SEISMOLOGY
  - REGIONAL EXPLORATION
  - PLANETARY QUARANTINE FACILITY

SEMI-PERMANENT BASE

TEMPORARY BASE
12 MEN

48 MEN ON SURFACE
24 MEN IN ORBIT

MANNED MARS LANDING

MARS EXCURSION MODULE

MSFC-69-PD-SA 185

The NASA Mission Reports

The NASA Mission Report books come with a bonus Windows CDROM featuring related movies and pictures. Forthcoming books in the NASA Mission Reports Series:-

Apollo 8 - 1-896522-50-5 10" x 7" 192 pg (8 color) $14.95 AVAILABLE NOW!
Apollo 9 - 1-896522-51-3 10" x 7" 240 pg (8 color) $14.95 AVAILABLE NOW!
Friendship 7 - 1-896522-60-2 10" x 7" 216 pg (8 color) $14.95 AVAILABLE NOW!
Apollo 10 - 1-896522-52-1 10" x 7" 184 pg (8 color) $14.95 AVAILABLE NOW!
Apollo 11 Volume 1 - 1-896522-53-X 10" x 7" 248 pg (8 color) $16.95 AVAILABLE NOW!
Apollo 11 Volume 2 - 1-896522-49-1 10" x 7" 168 pg $13.95 AVAILABLE NOW!
Apollo 12 - 1-896522-54-8 10" x 7" 248 pg (8 color) $16.95 AVAILABLE NOW!
Gemini 6 - 1-896522-61-0 10" x 7" 200 pg (8 color) $16.95 AVAILABLE NOW!
Apollo 13 1-896522-55-6 10" x 7" 256 (8 color) $16.95 AVAILABLE NOW!
Apollo 14 November 2000 1-896522-56-4 10" x 7" TBA (w. color) (approx $16.95)
Apollo 15 TBA 1-896522-57-2 10" x 7" TBA (w. color) (approx $16.95)

The series will continue on through Gemini, Mercury, Skylab, STS and unmanned missions.
All prices are in US dollars

Available from Apogee Books, Box 62034, Burlington, Ontario, L7R 4K2, Canada.
http://www.cgpublishing.com

Also available on-line at all good booksellers.

"A unique reference providing details available heretofore only to researchers with access to the archives." . . .
Library Journal - Feb '99

"(the series) will serve as an invaluable reference tool for aficionados of human spaceflight." . . .
Astronomy - Jul '99

"The package is guaranteed to put space enthusiasts into orbit." . .
Today's Librarian - Jul '99

". . budding Tsiolkovskys, Goddards, and von Brauns will devour each title." . . .
Booklist - Jul '99

"Highly recommended for space buffs who want detailed information on these flights." . . . **Choice - Nov '99**

"This series is highly recommended . . . A bargain at twice the price!" . . .
Ad Astra - National Space Society - Jan '00

". . . the most ambitious look at our space program to date." . . .
Playboy - Mar '00

The Watch

You've probably heard about Hollywood's latest version of Doomsday.

It involves either a very large Asteroid or Comet headed on a collision course for Planet Earth. The prospects for humanity are grim indeed, as our very civilization is threatened. Of course along comes either Bruce Willis or Robert Duvall to save the day and everything ends up with a happy ending.

The fact is that the scenarios of a large object possibly hitting the Earth is all too real! Unfortunately we don't have any secret spaceships to deal with this threat. More to the point, we have very little idea of where these objects actually are. Current projections based on empirical data suggest that there are in excess of 2,000 Near Earth Objects (NEO's) currently crossing Earth's orbit of a magnitude and size capable of ending life on Earth as we know it. We know where about 10% of them are, or about 200. We know that these 200 pose no threat to us at present. Of the other 1800, one could hit at any time and we would be defenseless against it.

We must find these objects as a matter of urgency as our very survival as a species depends on it.

The Watch has been formed as a project of The Space Frontier Foundation to raise funds as quickly as possible, to enable our most brilliant astronomers to access the necessary machinery and observatories as quickly as possible. This will enable a global search of the heavens, from the USA to Europe, India, Chile, Australia, Canada and wherever else we need observing stations. The rate of discovery has to increase 10 fold in order to catalog most of these killer objects within the next ten years. The difference that this information makes, just might give us a long-term shot at survival on this small planet. At some point in the near future we might be able to deflect these projectiles from impacting our homeworld, but we can't even think about this in the first place without knowing where they are.

You might agree that this is a possible scenario, but that the chances are that it won't happen in our lifetime, that these events only happen every 10 million years or so. This is just not the case! There are substantial impacts each and every year. Besides when mentioning "Chances" all this indicates is that we just "Don't Know!" Substantial impacts have occurred in this century in Siberia and the Amazon, we have no idea how many have occurred at ocean impacts, and we have had no idea (until late this century) how many tsunamis in the past have been caused by impacts.

This is not panic! It is being prudent! We must find these objects as quickly and efficiently as possible. If we don't, the downside is unimaginable.

 Remember as a species all of our eggs are in one basket called Earth.

The Watch as mentioned is a project of The Space Frontier Foundation, a not for profit organisation of private individuals dedicated to opening the Space Frontier to human settlement as rapidly as possible. Our goals include the protection of the Earth's fragile biosphere and creating a freer and more prosperous life for each generation by utilizing the unlimited energy and material resources of space. Our purpose is to unleash the power of free enterprise and lead a united humanity permanently into the Solar System.

The Watch organisation is comprised of an executive staff consisting of Richard Godwin as Executive Director and Rick Tumlinson President of the SFF overseeing the project. There is also The Watch Advisory Council which consists of some of the world's most renowned Planetary Scientists and Astronomers, who provide an active input into the workings of The Watch, and who will advise as to where raised funds can be most actively employed.

The Watch council consists of:

Dr. Richard Binzel, Professor at the Department of Earth, Atmospheric and Planetary Sciences at Massachusetts Institute of Technology (MIT) a leading authority on the Spectra and compositions of the near Earth and Belt asteroids.

Dr. Tom Gehrels: Professor of Planetary Sciences at the University of Arizona. Is the father and leader of the Spacewatch asteroid search programme, which pioneered the use of CCD imaging and real time computer analysis to increase discovery rates of NEO's

Dr. Eleanor Helin: renowned astronomer affiliated with Jet Propulsion Laboratory and for many years a pioneer in the discovery of NEO's.

Dr. John Lewis: University of Arizona, Professor of Planetary Sciences and Co. Director of the Space Engineering Research Centre, U of Arizona. Author of 150 research publications as well as the popular science books, "Rain of Iron & Ice" and "Mining the Sky."

Dr. Brian Marsden: Heads the Minor Planet Center in Massachusetts. A specialist in tracking and categorization of Asteroids and Comets. Member of IAU Nomenclature Committee for Minor planets.

CD-ROM

The accompanying CD-ROM is designed to use your World Wide Web browser to be viewed. It is programmed to not leave any footprint on your computer (i.e. no drivers, no installation, no updating your Windows registry etc.)

On inserting the disc in your CD-ROM drive you may be prompted to locate the file "Autorun.exe". This file is located in the root directory of your CD-ROM drive. (i.e. it is usually drive "D" but may be "E", "F" etc.) Most computers will find the Autorun program unassisted.

Autorun will open your default Web browser and you can then navigate the contents of the disc just like any web page.

Included on the disc are hundreds of images as well as hours of video. All of the video is in MPEG1 format and should automatically launch your default media player software (such as Windows Media Player).

STOP PRESS! NASA RELEASE: 00-91 June 13 2000 NEW ROCKET TECHNOLOGY COULD CUT MARS TRAVEL TIME

An agreement to collaborate on development of an advanced rocket technology that could cut in half the time required to reach Mars, opening the solar system to human exploration in the next decade, has been signed by NASA and MSE Technology Applications Inc. The technology could reduce astronauts' total exposure to space radiation and lessen time spent in weightlessness, perhaps minimizing bone and muscle mass loss and circulatory changes.

Called the Variable Specific Impulse Magnetoplasma Rocket (VASIMR), the technology has been under development at Johnson's Advanced Space Propulsion Laboratory. Called, "A precursor to fusion rockets, the VASIMR provides a power-rich, fast-propulsion architecture."

The VASIMR engine consists of three linked magnetic cells. The forward cell handles the main injection of propellant gas and its ionization. The central cell acts as an amplifier to further heat the plasma. The aft cell is a magnetic nozzle, which converts the energy of the fluid into directed flow.

Neutral gas, typically hydrogen, is injected at the forward cell and ionized. The resulting plasma is electromagnetically energized in the central cell by ion cyclotron resonance heating. In this process radio waves give their energy to the plasma, heating it in a manner similar to the way a microwave oven works.

After heating, the plasma is magnetically exhausted at the aft cell to provide modulated thrust. The aft cell is a magnetic nozzle, which converts the energy of the plasma into velocity of the jet exhaust, while protecting any nearby structure and ensuring efficient plasma detachment from the magnetic field.

On a mission to Mars, such a rocket would continuously accelerate through the first half of its voyage, then reverse its attitude and slow down during the second half. The flight could take slightly over three months. A conventional chemical mission would take seven to eight months and involve long periods of unpowered drift en route.

Errata: On the accompanying CDROM in the Mars Global Surveyor images section the last image is an animation which currently reads "A topographic map of Mars." it should read, "This movie is an animation of TES-derived global broadband (0.3 - ~3.0 microns) visible and near-infrared reflectance, also known as albedo."

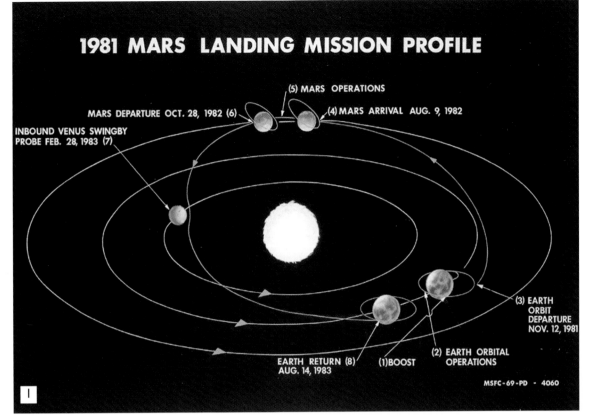

The following color pages are reference images used by Wernher von Braun during his presentation "Manned Mars Landing" to the Space Task Group in August 1969 (page 410).
Some of the pictures were updated in 1970. In those cases the original pictures from 1969 are inset.

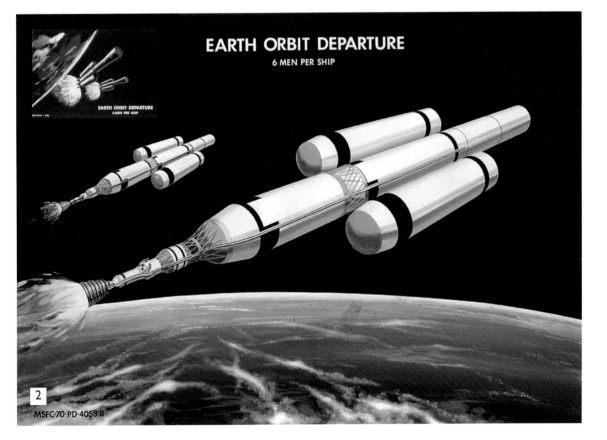

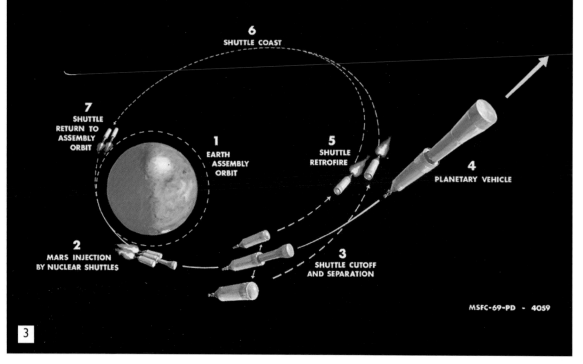

EARTH ORBIT DEPARTURE MANEUVERS

6 SHUTTLE COAST

7 SHUTTLE RETURN TO ASSEMBLY ORBIT

1 EARTH ASSEMBLY ORBIT

5 SHUTTLE RETROFIRE

4 PLANETARY VEHICLE

2 MARS INJECTION BY NUCLEAR SHUTTLES

3 SHUTTLE CUTOFF AND SEPARATION

MSFC-69-PD - 4059

3

Von Braun's Mars concept called for a 1.6 million pound Mars departure vehicle and Nuclear Propulsion. Many of these concepts were built upon the EMPIRE studies formulated as early as 1964. (See CD-ROM)

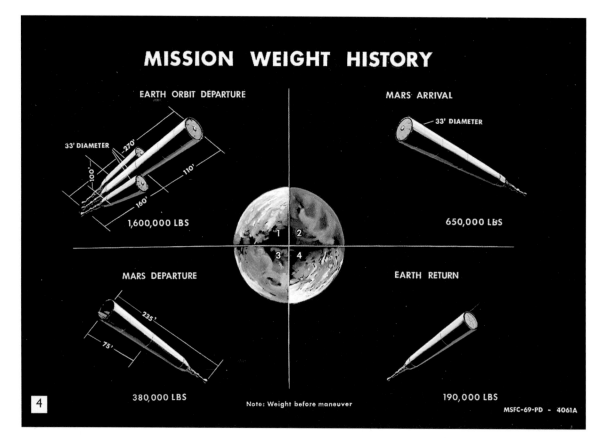

MISSION WEIGHT HISTORY

EARTH ORBIT DEPARTURE

33' DIAMETER · 270' · 110' · 100' · 160'

1,600,000 LBS

MARS ARRIVAL

33' DIAMETER

650,000 LBS

MARS DEPARTURE

235' · 75'

380,000 LBS

EARTH RETURN

190,000 LBS

Note: Weight before maneuver

4

MSFC-69-PD - 4061A

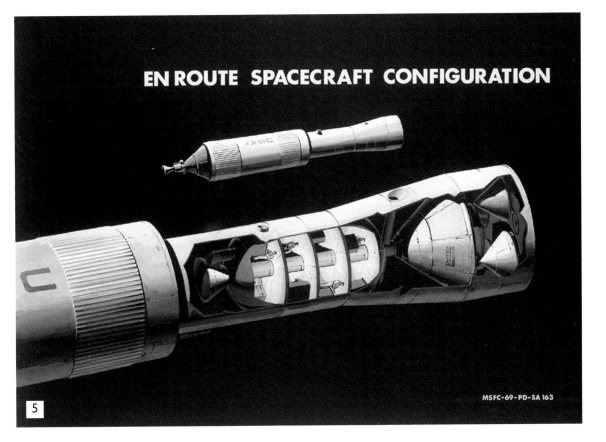

EN ROUTE SPACECRAFT CONFIGURATION

MSFC-69-PD-SA 163

5

These images clearly illustrate the large scale of von Braun's plans. Today's Mars planners are forced to think in smaller, cheaper terms but the essence of a long-duration mission to establish a permanent presence remains.

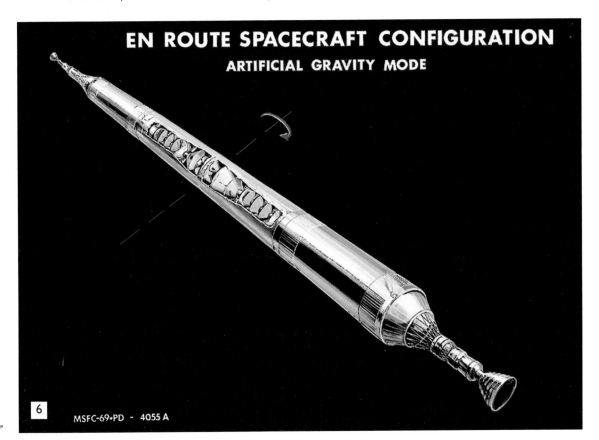

EN ROUTE SPACECRAFT CONFIGURATION
ARTIFICIAL GRAVITY MODE

6

MSFC-69-PD - 4055 A

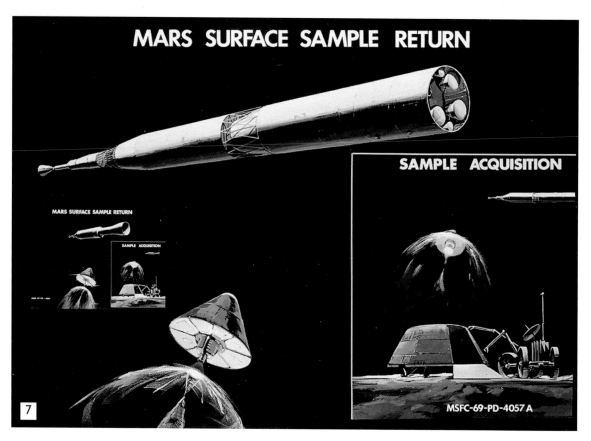

MARS SURFACE SAMPLE RETURN

SAMPLE ACQUISITION

MSFC-69-PD-4057 A

The design used above is slightly different to the original which von Braun discussed in 1969. The original picture had the same number (MSFC-69-PD-4057A) although the designs were altered slightly (inset left)

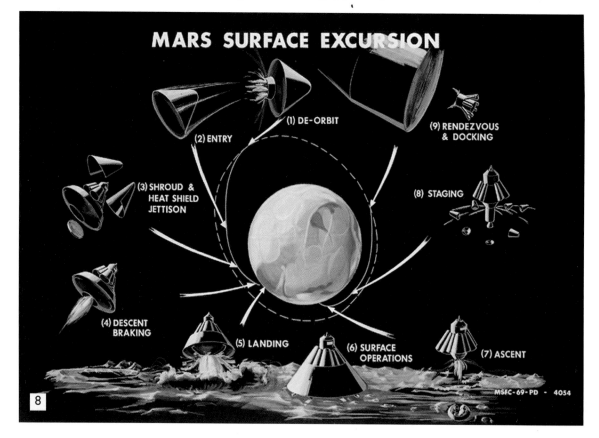

MARS SURFACE EXCURSION

(1) DE-ORBIT

(2) ENTRY

(9) RENDEZVOUS & DOCKING

(3) SHROUD & HEAT SHIELD JETTISON

(8) STAGING

(4) DESCENT BRAKING

(5) LANDING

(6) SURFACE OPERATIONS

(7) ASCENT

MSFC-69-PD - 4054

MARS EXCURSION MODULE CONFIGURATION

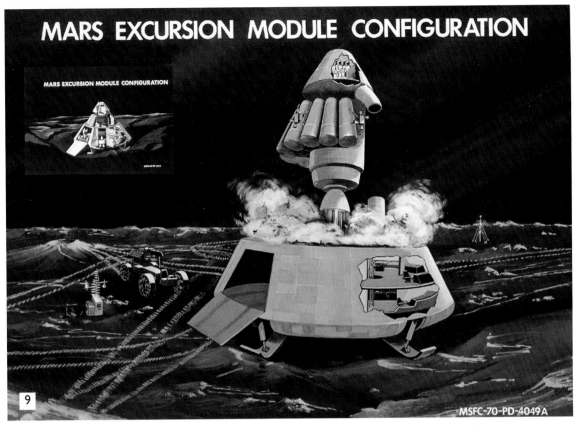

Note the change of the lander design from 1969 (inset) to 1970. The use of the lower half as a launch platform undoubtedly was a reflection of the success of a similar design used on the LM. The original pictures are inset above left and below above right.

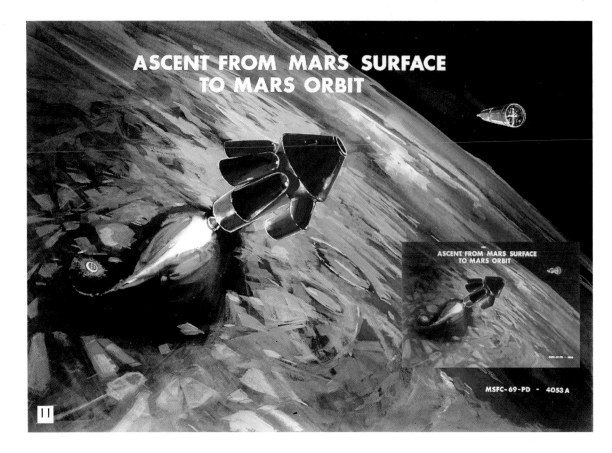

Von Braun's original presentation referred to a slightly different picture # MSFC-69-PD-SA-169 (inset right above)

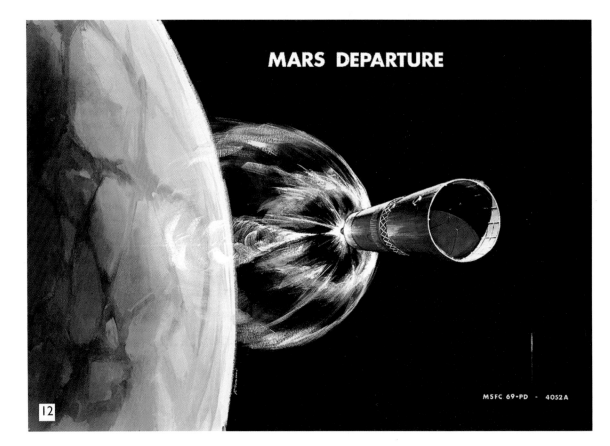

RELEASE OF VENUS PROBE

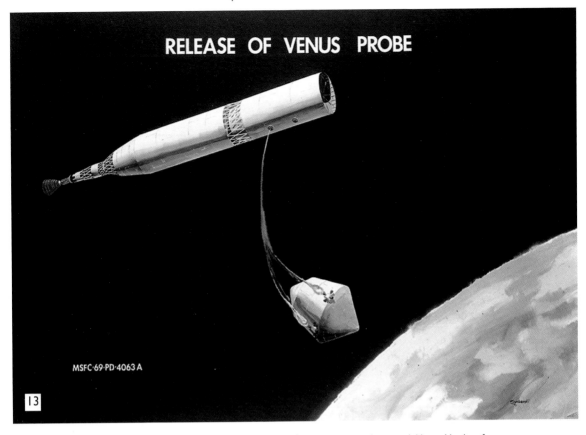

MSFC·69·PD·4063 A

13

Von Braun's Mars plan called for a return trajectory inwards towards Venus. He therefore postulated that an unmanned probe be launched down to the planet. (above) Von Braun's original presentation referred to a slightly different picture # MSFC-69-PD-SA-193

RETURN EARTH

MSFC-69-PD - 4064

14

EARTH LAUNCH AND ORBITAL ASSEMBLY

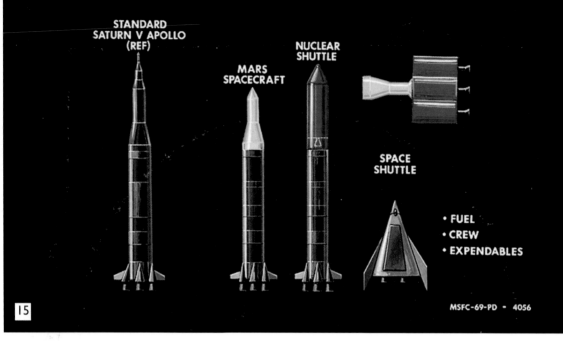

STANDARD
SATURN V APOLLO
(REF)

MARS
SPACECRAFT

NUCLEAR
SHUTTLE

SPACE
SHUTTLE

• FUEL
• CREW
• EXPENDABLES

MSFC-69-PD • 4056

It is interesting to note that von Braun's plans planned on using existing technology, such as the Saturn V, as well as using the upcoming Space Shuttle which was already in the planning stages. He also advocated the use of Nuclear propulsion. The exhaust velocities that chemical engines were capable of in 1969 were not thought to be high enough to mount a Mars mission. (above)

The final picture (below) was featured in the book "The Exploration Of Mars" by von Braun and Willy Ley. Published in 1956 the book featured the visionary artwork of Chesley Bonestell. In the painting below the lander is winged like an aircraft and carries an almost V2-style return vehicle which is tilted to a vertical position after landing. This idea can also be seen in the science-fiction film "The Conquest Of Space" for which Bonestell was the art designer. *Reproduced courtesy of Bonestell Space Art.*